Springer Series on Environmental Management

Series Editors

Bruce N. Anderson, VIC, Australia
Robert W. Howarth, NY, USA
Lawrence R. Walker, NV, USA

For further volumes:
http://www.springer.com/series/412

James S. Latimer · Mark A. Tedesco
R. Lawrence Swanson · Charles Yarish
Paul E. Stacey · Corey Garza
Editors

Long Island Sound

Prospects for the Urban Sea

Springer

Editors

James S. Latimer
Office of Research and Development
US Environmental Protection Agency
Narragansett, RI
USA

Mark A. Tedesco
Long Island Sound Office
US Environmental Protection Agency
Stamford, CT
USA

R. Lawrence Swanson
School of Marine
    and Atmospheric Sciences
Stony Brook University
Stony Brook, NY
USA

Charles Yarish
Department of Ecology
    and Evolutionary Biology
University of Connecticut
Stamford, CT
USA

Paul E. Stacey
Great Bay National
    Estuarine Research Reserve
New Hampshire Fish
    and Game Department
Durham, NH
USA

Corey Garza
Division of Science and Environmental
    Policy
California State University,
Monterey Bay
Seaside, CA
USA

ISSN 0172-6161
ISBN 978-1-4614-6125-8   ISBN 978-1-4614-6126-5   (eBook)
DOI 10.1007/978-1-4614-6126-5
Springer New York Heidelberg Dordrecht London

Library of Congress Control Number: 2013951776

Printed on acid-free paper

Springer is part of Springer Science+Business Media (www.springer.com)

Dedication

To Gerard M. Capriulo

While this book was being written, it was with deep regret that we learned that Gerard M. Capriulo, the Fletcher Jones Professor of Marine Biology, and chairman of the Biology Department at Saint Mary's College of California, died suddenly and unexpectedly on November 14, 2009. He was 56. Gerry, as he was known to his friends and colleagues, was an active scientist engaged in many activities involved in the biological oceanography of Long Island Sound and was one of the key chapter authors of the biology and ecology of the Sound.

Gerry was born in Brooklyn, New York, and was introduced to marine science as a student intern at the New York Aquarium. He graduated, magna cum laude, with his Bachelor of Sciences Degree in Biology from St. John's University where he developed a keen interest in microzooplankton, in particular, tintinnid ciliates. He received his Masters and Doctoral Degrees from Stony Brook University and accepted his initial academic appointment at Purchase College (SUNY) in 1983. It was at Purchase College where he developed an active teaching and research program dealing with Long Island Sound. He moved through the academic ranks at Purchase College, ultimately earning his professorship and was the chair of the environmental science program until his departure to Saint Mary's College in 1997.

Gerry was an activist and an enthusiast for his science. He was fervently devoted to his students and gave them unreserved attention. Gerry supervised more than 60 undergraduate research theses and was an active member of graduate student committees at the University of Connecticut. On the day of his passing, he had taken his students on a trip to the Pacific Ocean! While at Purchase College, Gerry was actively involved in many institutions and organizations around Long Island Sound including: Westchester County's Long Island Sound Advisory Committee; The Science and Technical Advisory Committee of the Long Island Sound Study; The Science Advisory Committee of the Connecticut Audubon Society; he served on the Board of Directors and the Executive Committee of Sound Waters; he was a member of the Board of Trustees of Save the Sound Inc. (formerly Long Island Sound Taskforce of the Oceanic Society); and he was a member of the Education, Program, and Science Committee of the Board of

Directors for the Maritime Aquarium at Norwalk. He was also a visiting scientist at The Bermuda Biological Station, a research scientist at the Department of Marine Sciences at The University of Connecticut and a member of the summer core faculty at the Shoals Marine Laboratory, Cornell University, and University of New Hampshire. In 1988, Gerry was a session chairman and keynote speaker for a NATO Advanced Study Institute held in England on protozoan ecology. After he relocated to California, he joined the Board of Directors of the Aquarium of the Pacific's Marine Conservation Research Institute.

Gerry's scholarship was as varied as his personal interests. He authored numerous scientific publications dealing with Long Island Sound. Gerry was the lead proponent that there was an alteration of the planktonic food web structure of Long Island Sound due to eutrophication via the paradigm of the microbial loop. In 2002, he was the lead author of a major treatise on the biological oceanography of Long Island Sound entitled "The Planktonic Food Web Structure of a Temperate Zone Estuary, and Its Alteration Due to Eutrophication" (Hydrobiologia). In 1990, Gerry published *The Ecology of Marine Protozoa* (Oxford University Press), and in 2003, he published a nonfiction book on the role of symbiosis in the origin of the universe, life, and the development of ecosystems entitled *The Golden Braid: The Symbiotic Nature of the Universe* (Global Outlook Publishing). At the time of his death, he was particularly interested in the role of symbiosis in the structuring of marine ecological interactions. He was focusing his research on the relationship between the aggregating sea anemone, *Anthopleura elegantissima (Brandt)* and its zooxanthellae. He was also writing a very popular newspaper column called *ECOfocus* for Clayton Pioneer newspapers (Clayton, California). Gerry was truly a very talented and gifted writer, who was able to communicate with his colleagues as well as the general public.

The loss of Gerry Capriulo is widely and deeply felt, most certainly by his family including his sister, parents-in-law, his daughters Lauren and Rebecca, and his wife Amelia. We are all so fortunate that through Gerry's insights we have a better understanding of the structure and functioning of Long Island Sound's estuarine and coastal ecosystems and food webs. Although we are all greatly saddened by the abrupt and untimely end of a very productive colleague, we should find a degree of solace in knowing that some of Gerry Capriulo's scientific thought and inspiration has passed to his students and colleagues. The editors wish to dedicate this volume to the memory of Prof. Gerard M. Capriulo. His spirit will be a guiding force for researchers and students interested in the Long Island Sound ecosystem for years to come.

Charles Yarish

Preface

Endeavoring to synthesize the estuarine science of Long Island Sound (LIS) from the perspective of public policy and management is not new. Published more than 35 years ago, *The Urban Sea: Long Island Sound* (Koppelman et al. 1976) and *Long Island Sound: An Atlas of Natural Resources* (CTDEP 1977) provided cross-disciplinary perspectives with an eye toward laying a foundation for still developing coastal management programs. Contemporaneous to these efforts, though not directly linked, was the development of comprehensive, interdisciplinary regional management plans, either centered around LIS (New England River Basin Commission 1975), or directed at portions of the watershed (e.g., Nassau-Suffolk Regional Planning Board 1978). Reflecting perhaps both the optimism and level of resources available at the time, those planning efforts were ambitious in scope and remarkable for the breadth of federal, state, local, and public involvement in their development. Even today they are instructive reading.

Yet the need for synthesis remains and has, if anything, grown stronger (Carpenter et al. 2009). Consider the ever increasing amount of specialized data and information about coastal and estuarine systems (Valiela and Martinetto 2005). Consider the complexity of new and enduring challenges such as climate change, coastal development and use conflicts, emerging contaminants, fisheries management, invasive species, and nutrient pollution. And also consider the increased demand by policy makers and the public to translate science, so that it is understood and applied more efficiently and effectively to address these challenges in a world of limited public resources.

But the challenges to synthesis for management are formidable. Resolving discrete environmental problems with directed management in isolation from other social and economic needs may no longer be feasible. The scale of human pressures on the ecosystem is shifting from local land modification and resource harvest to global pollutant deposition and climate change. Understandably, the consequent interaction of cause and effect has become blurred despite improving science, fueling debates about probable outcomes of action and the compromises of cost. The future variability of natural systems such as stream flow or storm intensity may no longer be predictable from historical conditions due to anthropogenic changes in the Earth's climate (Milly et al. 2008). If so, management models will need to account for those structural changes and integrate them more fully with human societal and economical needs. Failure to adapt management approaches accordingly could threaten or slow the progress achieved in the 40 years since enactment of the Clean Water Act.

These factors and trends are evident in the continuing management of LIS. The practical genesis of this book can be traced back to a 2007 workshop sponsored by the Science and Technical Advisory Committee of the Long Island Sound Study (LISS). The LISS, part of the National Estuary Program and charged with protecting and restoring LIS, had already developed a Comprehensive Conservation and Management Plan, formally approved in 1994 under requirements laid out in the Clean Water Act. But the science upon which the plan was based was approaching 20 years old and the management lexicon of the time didn't even include climate change. Much new research had been conducted, and since that time the National Estuary Program's faint whispers of ecosystem-based management using the best science in collaborative efforts that incorporate society and the economy had become bold calls embraced by scientists (McLeod et al. 2005) and adopted by policymakers (National Ocean Council 2012, New York Ocean and Great Lakes Ecosystem Conservation Act 2006).

Hobbie (2000) described synthesis as "the bringing together of existing information in order to discover patterns, mechanisms, and interactions that lead to new concepts and models." Kemp and Boynton (2012) defined synthesis as "the inferential process whereby new models are developed from analysis of multiple data sets to explain observed patterns across a range of time and space scales." They further emphasized the key need to "improve mutually supporting linkages between synthesis research and coastal management."

The purpose of this book then is to bring together existing information about LIS, to take stock of what we know, in order to help discover and explain its patterns and mechanisms. Six technical chapters sum our knowledge about its human history, geology, physical oceanography, geochemistry, pollutant history, and biology and ecology. It is expected that a more synthetic knowledge of the science of LIS will provide a firmer foundation for improving ecosystem-based management—particularly as we confront climate change. To test this, the last chapter attempts to identify the linkages between the scientific synthesis of LIS and environmental management.

While the book is largely targeted to a technical, scientific audience, it is our belief that science is too important to be left exclusively to scientists, planning too important to be left exclusively to planners, and environmental management too important to be left exclusively to managers. Successful ecosystem-based management of LIS will require that if the old walls between these fields must remain, they should at least have larger windows.

References

Carpenter et al (2009) Bioscience 59(8):699–701
Connecticut Department of Environmental Protection (1977) Long Island Sound: an atlas of natural resources. Hartford
Kemp WM, Boynton WR (2012) Synthesis in estuarine and coastal ecological research: What is it, why is it important, and how do we teach it? Estuaries Coasts 35:1–22

Koppelman LL, Weyl PK, Gross MG (1976) The urban sea: Long Island Sound. Praeger, New York, p 223

Hobbie JE (2000) Estuarine science: the key to process in coastal ecological research. In Hobbie JE (ed) Estuarine science: a synthetic approach to research and practice. Island Press, Washington, pp 1–16

McLeod KL, Lubchenko J, Palumbi SR et al (2005) Scientific consensus statement on marine ecosystem-based management. http://www.compassonline.org/sites/all/files/document_files/EBM_Consensus_Statement_v12.pdf. Accessed 18 Feb 2011

Milly PCD, Betancourt J, Falkenmark M et al (2008) Stationarity is dead: wither water management? Sci 319:573–574

Nassau-Suffolk Regional Planning Board (1978) The Long Island comprehensive waste treatment management plan. vol 1: Summary Plan

National Ocean Council (2012) Draft National Ocean Policy Implementation Plan. http://www.whitehouse.gov/sites/default/files/microsites/ceq/national_ocean_policy_draft_implementation_plan_01-12-12.pdf. Accessed 22 Feb 2012

New England River Basins Commission (1975) Final report of the Long Island Sound regional study. vol 1 Summary, Vol 2 Supplement

New York Ocean and Great Lakes Ecosystem Conservation Act (2006) Environmental Conservation Law. Article 14. http://www.oglecc.ny.gov/media/ECL_Article%2014.pdf. Accessed 22 Mar 2012

Valiela I, Martinetto P (2005) The relative ineffectiveness of bibliographic search engines. Biosci 55(8) pp 688–692

Acknowledgments

This project received support from the US Environmental Protection Agency under Cooperative Agreements LI-97183601 and LI-97186501 with the Connecticut Sea Grant College Program, University of Connecticut, and Cooperative Agreement LI-97241708 with the New York Sea Grant College Program, Research Foundation of SUNY. We wish to heartily acknowledge the support and perseverance of the following people and organizations. Without their help, this project would never have been accomplished.

Thomas Bianchi	External Peer Review	Texas A&M University
Jon C. Boothroyd	External Peer Review	University of Rhode Island
Robert Christian	External Peer Review	Eastern Carolina University
Sylvain DeGuise	Funding Logistics	CT Sea Grant
William Dennison	External Peer Review	University of Maryland
Johanna Hunter	Funding Logistics	EPA New England
Virge A. Kask	Scientific Illustration	UCONN Biology Central Services
Jason Krumholz	Special Editing	NOAA NMFS
Jim Kuwabara	External Peer Review	US Geological Survey
Natalie A. Naylor	External Peer Review	Hofstra University
Edna Nolfi	Special Editing	EPA Long Island Sound Office
Cornelia Schlenk	Funding Logistics	NY Sea Grant
Bonnie W. Stephens	Technical Editor	Stony Brook University
Kuo Wong	External Peer Review	University of Delaware

Contents

Contributors

W. Frank Bohlen Department of Marine Sciences, University of Connecticut, Groton, CT 06340, USA

Malcolm Bowman School of Marine and Atmospheric Sciences, Stony Brook University, Stony Brook, NY 11794, USA

Vincent T. Breslin Science Education and Environmental Studies, Southern Connecticut State University, New Haven, CT 06511, USA

Drew Carey Coastal Vision, Newport, RI 02840, USA

James T. Carlton Williams-Mystic Maritime Studies Program, Williams College, Stonington, CT 06355, USA

Robert Cerrato School of Marine and Atmospheric Sciences, Stony Brook University, Stony Brook, NY 11794, USA

J. Kirk Cochran School of Marine and Atmospheric Sciences, Stony Brook University, Stony Brook, NY 11794, USA

Daniel Codiga Graduate School of Oceanography, University of Rhode Island, Narragansett, RI 02882, USA

Carmela Cuomo Department of Biology and Environmental Sciences, University of New Haven, West Haven, CT 06516, USA

Hans Dam Department of Marine Sciences, University of Connecticut, Groton, CT 06340, USA

Rob DiGiovanni Riverhead Foundation for Marine Research and Preservation, Riverhead, NY 11901, USA

Chris Elphick Department of Ecology and Evolutionary Biology, University of Connecticut, Groton, CT 06340, USA

Todd Fake Department of Marine Sciences, University of Connecticut, Groton, CT 06340, USA

Diane B. Fribance Department of Marine Science, Coastal Carolina University, Conway, SC 29528, USA

Michael Frisk School of Marine and Atmospheric Sciences, Stony Brook University, Stony Brook, NY 11794, USA

Corey Garza Division of Science and Environmental Policy, California State University, Monterey Bay, Seaside, CA 93955, USA

Christopher Gobler School of Marine and Atmospheric Sciences, Stony Brook University, Stony Brook, NY 11794, USA

Lyndie Hice School of Marine and Atmospheric Sciences, Stony Brook University, Stony Brook, NY 11794, USA

Penny Howell Marine Fisheries Division, Connecticut Department of Energy and Environmental Protection, Hartford, CT 06106, USA

Adrian Jordaan University of Massachusetts Amherst, Amherst, MA 01003, USA

James S. Latimer Office of Research and Development, US Environmental Protection Agency, Narragansett, RI 02882, USA

Ralph Lewis Long Island Sound Resource Center, University of Connecticut, Groton, CT 06340, USA

Senjie Lin Department of Marine Sciences, University of Connecticut, Groton, CT 06340, USA

Sheng Liu Department of Marine Sciences, University of Connecticut, Groton, CT 06340, USA

Darcy Lonsdale School of Marine and Atmospheric Sciences, Stony Brook University, Stony Brook, NY 11794, USA

Glenn Lopez School of Marine and Atmospheric Sciences, Stony Brook University, Stony Brook, NY 11794, USA

Kamazima Lwiza School of Marine and Atmospheric Sciences, Stony Brook University, Stony Brook, NY 11794, USA

Anne McElroy School of Marine and Atmospheric Sciences, Stony Brook University, Stony Brook, NY 11794, USA

Maryann McEnroe School of Natural and Social Sciences, Purchase College, Purchase, NY 10577, USA

Kim McKown Bureau of Marine Resources, New York State Department of Environmental Conservation, East Setauket, NY 11733, USA

George McManus Department of Marine Sciences, University of Connecticut, Groton, CT 06340, USA

John R. Mullaney US Geological Survey, East Hartford, CT 06108, USA

James O'Donnell Department of Marine Sciences, University of Connecticut, Groton, CT 06340, USA

Rick Orson Orson Ecological Consulting, Branford, CT 06405, USA

Bradley Peterson School of Marine and Atmospheric Sciences, Stony Brook University, Stony Brook, NY 11794, USA

Chris Pickerell Cornell Cooperative Extension of Suffolk County, Riverhead, NY 11901, USA

Elizabeth Pillsbury Riverdale Country School, The Bronx, NY 10471, USA

Ron Rozsa Coastal Management Program, Connecticut Department of Energy and Environment Protection, Hartford, CT 06106, USA

Sandra E. Shumway Department of Marine Sciences, University of Connecticut, Groton, CT 06340, USA

Amy Siuda Sea Education Association, Falmouth, MA 02540, USA

Paul E. Stacey Great Bay National Estuarine Research Reserve, New Hampshire Fish and Game Department, Durham, NH 03824, USA

Kelly Streich Bureau of Water Management Planning and Standards, Connecticut Department of Energy and Environmental Protection, Hartford, CT 06106, USA

R. Lawrence Swanson School of Marine and Atmospheric Sciences, Stony Brook University, Stony Brook, NY 11794, USA

Stephanie Talmage School of Marine and Atmospheric Sciences, Stony Brook University, Stony Brook, NY 11794, USA

Gordon Taylor School of Marine and Atmospheric Sciences, Stony Brook University, Stony Brook, NY 11794, USA

Mark A. Tedesco Long Island Sound Office, US Environmental Protection Agency, Stamford, CT 06904, USA

Ellen Thomas Department of Geology and Geophysics, Yale University, New Haven, CT 06511, USA

Karl K. Turekian Department of Geology and Geophysics, Yale University, New Haven, CT 06511, USA

Margaret Van Patten Connecticut Sea Grant, University of Connecticut, Groton, CT 06340, USA

Johan Varekamp Earth and Environmental Sciences, Wesleyan University, Middletown, CT 06459, USA

Jamie Vaudrey Department of Marine Sciences, University of Connecticut, Groton, CT 06340, USA

Marilyn E. Weigold Dyson College of Arts and Sciences, Pace University, Pleasantville, NY 10570, USA

Michael Whitney Department of Marine Sciences, University of Connecticut, Groton, CT 06340, USA

Gary Wikfors Northeast Fisheries Sciences Center, National Oceanic and Atmospheric Administration, Milford, CT 06460, USA

Robert E. Wilson School of Marine and Atmospheric Sciences, Stony Brook University, Stony Brook, NY 11794, USA

Charles Yarish Department of Ecology and Evolutionary Biology, University of Connecticut, Groton, CT 06340, USA

Roman Zajac Department of Biology and Environmental Science, University of New Haven, New Haven, CT 06516, USA

Acronyms and Common Units

ADCP	Acoustic Doppler Current Profilers
ASP	Amnesic Shellfish Poisoning
ASR	Aquatic Surface Respiration
AVS	Acid-Volatile Sulfides
BBiom	Bacterial Biomass
BIS	Block Island Sound
BSi	Biogenic Silica
BNP	Bacterial Net Production
BOD	Biological Oxygen Demand
BTZ	Benthic Turbidity Zone
C&GS	Coast and Geodetic Survey
CB	Chesapeake Bay
CCE	Cornell Cooperative Extension of Suffolk County
CCMP	Comprehensive Conservation and Management Plan
CERCLA	Comprehensive Environmental Response, Compensation, and Liability Act
CFA	Carbonate Fluorapatite
Chl a	Chlorophyll a
CLIS	Central Long Island Sound
cm	Centimeters
cm/s	Centimeters Per Second
CMSP	Coastal Marine Spatial Planning
C/N	Carbon to Nitrogen Ratio
COD	Chemical Oxygen Demand
CORGM	Marine Organic Carbon
CORGT	Terrestrial Carbon
CORGWW	CORG flux from WWTFs
CR	Connecticut River
CSO	Combined Sewer Overflows
CTD	Conductivity, Temperature, and Depth
CTDEP	Connecticut Department of Environmental Protection
CTDEEP	Connecticut Department of Energy and Environmental Protection (formerly CTDEP)

CTDEP MFIS	Connecticut Department of Environmental Protection Marine Fisheries Information System
CWA	Clean Water Act
DDA	Dark DI^{14}C Assimilation
DDT	Dichlorodiphenyltrichloroethane
DIC	Dissolved Inorganic Carbon
DIN	Dissolved Inorganic Nitrogen
DNC	Dominion Nuclear Connecticut
DO	Dissolved Oxygen
DOC	Dissolved Organic Carbon
dpm/L	Disintegrations Per Minute Per Liter
DSi	Dissolved Silica
EBM	Ecosystem-Based Management
EDC	Endocrine Disrupting Compounds
EF	Enrichment Factors
ELIS	Eastern Long Island Sound
EMAP	Environmental Monitoring and Assessment Program
EOBRT	Eastern Oyster Biological Review Team
EOC	Extracellular Organic Carbon
EOF	Empirical Orthogonal Function
ERL	Effects Range Low
ERM	Effects Range Median
ESG	Eastern Sound Gyre
ESI	Environmental Sensitivity Index
ESID	EcoSpatial Information Database
EXRK	Execution Rocks
FIS	Fishers Island Sound
FOAM	Friends of Anoxic Mud
FVCOM	Finite-Volume Coastal Ocean Model
g/m^2/yr	Grams Per Square Meter Per Year
g/mole	Grams Per Mole
GESAMP	The Joint Group of Experts on the Scientific Aspects of Marine Environmental Protection
GIS	Geographic Information System
GNHS	Geological and Natural History Survey
GPP	Gross Primary Production
GSA	Gill Surface Area
gww	Grams Wet Weight
HAB	Harmful Algal Blooms
Hb	Blood Hemoglobin
HR	Housatonic River
I-95	Interstate 95
I-287	Interstate 287
JFM	January, February, and March
kg/d	Kilograms Per Day

kg/s	Kilograms Per Second
kg/m³	Kilogram Per Meter Cubed
kg/ha/yr	Kilogram Per Hectare Per Year
kg/day	Kilograms Per Day
LIE	Long Island Expressway
LILCO	Long Island Lighting Company
LIPA	Long Island Power Authority
LIRR	Long Island Rail Road
LIS	Long Island Sound
LISICOS	Long Island Sound Integrated Coastal Observing System
LISS	Long Island Sound Study
LISTS	CTDEEP LIS Trawl Survey
LOI	Loss on Ignition
MAB	Middle Atlantic Bight
MFIS	Marine Fisheries Information System
m/Pa	Meters (sea level rise) Per Pascal (unit wind stress)
m/s	Meters Per Second
m/y	Meters Per Year
m/yr	Meters Per Year
m³/d	Cubic Meters Per Day
m³/s	Meters Cubed Per Second
m³/yr	Meters Cubed Per Year
mg/l	Milligrams Per Liter
m/s²	Meters Per Second Squared
MHW	Mean High Water
mg/kg	Milligrams Per Kilogram
mg/kgww	Milligrams Per Kilogram Wet Weight
mg/L	Milligrams Per Liter
microgram/g	Micrograms Per Gram
Mkg/yr	Million Kilograms Per Year
mm/yr	Millimeters Per Year
mmol/m²/day	Millimoles Per Meter Squared Per Day
mol/L	Moles Per Liter
MOS	Margin of Safety
NADP	National Atmospheric Deposition Program
NAO	North Atlantic Oscillation
NASQAN	National Stream Quality Accounting Network
NAWQA	National Water Quality Assessment
NCA	National Coastal Assessment
NCE	Nitrogen Credit Exchange
NDBC	National Data Buoy Center
NEIWPCC	New England Interstate Water Pollution Control Commission
NEP	National Estuary Program
ng/g	Nanograms Per Gram
ng/gdw	Nanograms Per Gram Dry Weight

ng/gww	Nanograms Per Gram Wet Weight
NGVD	National Geodetic Vertical Data
NHANES	National Health and Nutrition Examination Survey
NMFS	National Marine Fisheries Service
NOAA	National Oceanic and Atmospheric Administration
NOS	National Ocean Service
NPP	Net Primary Production
NRC	National Research Council
NS&T	National Status and Trends
NURC	National Undersea Research Center
NURTEC	Northeast Underwater Research Technology and Education Center
NWC	Northwest Control
NYCDEP	New York City Department of Environmental Protection
NYSDEC	New York State Department of Environmental Conservation
OHO	Oxyhydr(oxides)
OM	Organic Matter
OMZ	Oxygen Minimum Zones
P	Grazing Percentage
PAH	Polycyclic Aromatic Hydrocarbons
PBDE	Polychlorinated Dibenzodioxins and Furans and Polybrominated Diethyl Ethers
PCA	Principal Component Analysis
PCBs	Polychlorinated Biphenyls
PE	Peconic Estuary
PFC	Perfluorinated Compounds
PFOA	Perfluorooctanoic Acid
PFOS	Perfluorooctane Sulfonate
P-I	Oxygen-Based Photosynthesis-Irradiance
POC	Particulate Organic Carbon
POM	Particulate Organic Matter
PPCPs	Pharmaceuticals and Personal Care Products
psu	Practical Salinity Units
QAC	Quaternary Ammonium Compounds
RCRA	Resource Conservation and Recovery Act
REMOTS	Remote Ecological Monitoring of the Seafloor
RFMRP	Riverhead Foundation for Marine Research and Preservation
ROV	Remotely Operated Vehicle
RPD	Redox Potential Discontinuity
S	Salinity
SPI	Sediment Profile Imaging
SoMAS	School of Marine and Atmospheric Sciences
SQUID	Sediment Quality Information Database
SSER	South Shore Estuary Reserve
SWEM	Systemwide Eutrophication Model

T	Temperature
TBT	Tributyltin
TDI	Total Dissolved Inorganic Nitrogen
TDIP	Total Dissolved Inorganic Phosphorus
TDML	Total Maximum Daily Load
TOC	Total Organic Carbon
TIN	Total Inorganic Nitrogen
TRS	Total Reduced Sulfides
μg/g	Micrograms Per Gram
μg/gdw	Micrograms Per Gram Dry Weight
μg/gww	Micrograms Per Gram Wet Weight
μg/L	Micrograms Per Liter
USACE	US Army Corps of Engineers
USGS	United States Geological Survey
WLIS	Western Long Island Sound
WWTF	Wastewater Treatment Facilities

Childe Frederick Hassam (1859-1935). The Mill Pond, Cos Cob, 1902; Oil on canvas, 26 ¼ x 18 ¼ in. Collection of the Bruce Museum, Greenwich, CT, Anonymous Gift, 94.25

William Sidney Mount (1807-1868). Crane Neck Across the Marsh, 1841. Oil on
panel, 12.875 x 17". The Long Island Museum of American Art, History & Carriages.
Gift of Mr. and Mrs. Carl Heyser, Jr., 1961

Chapter 1
Long Island Sound: A Socioeconomic Perspective

Marilyn E. Weigold and Elizabeth Pillsbury

1.1 Discovery and Early Settlements

Four centuries ago Captain Adriaen Block, a Dutch merchant, undertook multiple voyages to the New York area to engage in the fur trade. In the course of one of his three or four trips, he discovered the waterway that nineteenth century states-man Senator Daniel Webster called the "American Mediterranean." According to most accounts, the discovery was made in 1614 when Block and his crew, whose ship the *TIGER* was destroyed by fire, constructed a new vessel in Manhattan (Hart 1959) and sailed it up the East River through the swirling torrents of Hell Gate, and out into the open Sound (Stokes 1967). Some sources suggest that the discovery of the Sound occurred during the 1612 voyage Block made on another ship, the *FORTUYN,* or the 1612–1613 journey to New Netherland undertaken on the same ship. During that trip, Block may have sailed from east to west in Long Island Sound (LIS) in 1613 (Stokes 1967). Sailing the opposite way in 1614, Captain Block explored the Connecticut coastline, as well as the Connecticut River, landed at Montauk and possibly on Block Island (Varekamp 2006).

A decade later, the Dutch established a permanent settlement on Manhattan Island and a short-lived settlement consisting of only a few families near the mouth of the Connecticut River. The latter venture lasted only 2 years but Dutch traders continued to visit the area to secure furs from the native inhabitants. They were displaced, however, by the English who built Fort Saybrook at the mouth of the Connecticut River (Cave 1996). English settlement of the Sound shore of Connecticut proceeded rapidly after the Pequot War of 1637, which decimated the native peoples of eastern Connecticut (Eisler 2001). In 1638, the New Haven Colony was founded and within a few years

M. E. Weigold (✉)
Dyson College of Arts and Sciences, Pace University, Pleasantville, NY 10570, USA
e-mail: mweigold@pace.edu

E. Pillsbury
Riverdale Country School, Bronx, NY 10471, USA
e-mail: epillsbury@riverdale.edu

J. S. Latimer et al. (eds.), *Long Island Sound*, Springer Series on
Environmental Management, DOI: 10.1007/978-1-4614-6126-5_1,
© Springer Science+Business Media New York 2014

it spawned settlements in Branford, Guilford, and Stamford on the Connecticut coast and across the Sound in Southold on the North Fork of Long Island. By the mid-1600s, English settlers from Massachusetts and Connecticut moved to Huntington and other places along the Island's north shore, including the area from Oyster Bay west through Queens, a region originally included in the New Netherland Colony.

1.2 Transportation

1.2.1 The Sound as a Nautical Highway

Whether they dwelled on the mainland or on Long Island, inhabitants of the Sound area used the waterway as a nautical highway to transport locally produced foodstuffs to the markets of New York City and Boston. In the first half of the seventeenth century, the settlers of eastern Connecticut and eastern Long Island established strong ties with Boston, not only because of their proximity to the Puritan metropolis but because of the New Netherland colony's onerous trade regulations. Once the British assumed control, renaming the colony New York, Manhattan became a popular destination for trading vessels departing from ports up and down the Sound. Some ships venturing forth from LIS harbors participated in the coast-wide trade while others transported dried fish, lumber, and other items to the West Indies and brought back sugar and spices (Decker 1986). The harbor of Black Rock, a thriving maritime center long before the community was absorbed by neighboring Bridgeport, was home port for vessels engaged in the West Indian trade. Farther east on the mainland coast, New London assured its prominence in the West Indian trade by building sizable vessels capable of transporting considerable cargo. In recognition of both its location at the eastern end of the Sound and its growing trade, a customs commissioner was appointed for New London in the mid-seventeenth century (Caulkins 1852). In 1756, a customs house for the western part of the Sound was established in New Haven. Three decades later, with its thriving maritime activity centered around Long Wharf, New Haven was a major seaport on LIS (Maynard 2004). In addition to its trade with the Far East, Europe, and the West Indies, in the nineteenth century New Haven generated income from sealing and whaling. Although New London was Connecticut's premier whaling port, Mystic, Stonington, Bridgeport, East Haddam, and Norwich sent out whaling vessels, as did Cold Spring Harbor on the north shore of Long Island.

1.2.2 Overland Travel

Although trips to faraway places, whether on whaling ships or trading vessels, could be very lucrative, most inhabitants of the Sound shore area did not venture far from home. But they did have to stay connected with the outside world.

Consequently, as settlements expanded from the immediate coastline to the hinterland of the Sound, roads were needed to connect the interior with the wharves that were a key element in trade. The inhabitants of areas of Connecticut located on rivers flowing into the Sound could transport their farm products by boat downriver to coastal ports, but mainland colonists who were not so well situated had to rely upon roads, some of which were merely enhanced versions of Indian trails. Given Long Island's paucity of rivers, roads, such as they were, constituted the only viable way to get to the coast. In contrast, with the limited north/south roads leading to the Sound was the east/west North Country Road (Newsday 1998). While it extended from the western end of Long Island all the way out to Suffolk County, a trip on the road was something to be endured rather than enjoyed. A keen-eyed rider on horseback may have been able to spot and avoid the inevitable ruts in the road, but a fully loaded coach was not so maneuverable. Its passengers were jostled as the vehicle swayed from side to side and if it touched down in a crater, even momentarily, its human cargo could literally hit the ceiling.

Travelers on the mainland side of the Sound were no less immune to these hazards. The Boston Post Road, portions of which were laid out in the 1670s, could be challenging (Baird 1974). Frigid temperatures and snow in winter made for slow going and during the spring thaw the ribbon of highway linking the Sound shore communities turned into a sea of mud. Even in good weather, coaches leaving lower Manhattan for the long trip to Boston often journeyed no farther than Rye the first day. A libation from the tap room of Haviland's Inn, a good meal, and a night's rest, such as it was given the fact that colonial regulations stipulated that five men could occupy a bed, prepared the weary travelers for the next day's adventure on the Post Road. At different times, John Adams and George Washington passed this way and stayed at the widow Haviland's Inn (Baird 1974). Earlier in the eighteenth century Sarah Kemble Knight, an Englishwoman traveling between Boston and New York on horseback, forded streams, rode over rickety wooden bridges, and stayed at inns that were, in her view, mostly inferior. Now and again, such as in New Rochelle, she was pleased with her accommodations (Knight 1935). No matter how comfortable the lodgings along the way, getting to one's destination by traveling overland remained an ordeal. All of this changed in the early nineteenth century, however, with the introduction of steamboats on LIS.

1.2.3 The Era of Steam Navigation

The age of steam had commenced in 1807 with the successful voyage of Robert Fulton's *CLERMONT* on the Hudson River. The War of 1812 delayed the introduction of steamboats on the Sound but the year after the conflict ended, Captain Elihu Bunker piloted the aptly named steamboat *FULTON* 75 miles between New York City and New Haven in 11 h. The return trip stretched out to 15 h because of rough seas. This was, nevertheless, quite an achievement. Within a year, Captain Bunker had a new ship, *COMMONWEALTH*. All white, except for its vivid green trim, the vessel was anything but understated. But this was all to the good because the eye-catching *COMMONWEALTH* stimulated interest in steam navigation. By 1817, regular service between New York and New Haven was instituted and quickly increased from two trips per week to three. For a time, the all-water route was suspended because the New York State legislature had granted a monopoly to Robert Fulton's company. As the exclusive legal provider of steamboat service on the state's waterways, the Fulton interests were able to prevent others from plying New York waters. This led to a makeshift arrangement whereby Connecticut captains steamed down to Greenwich, on the New York border, where passengers boarded stagecoaches for the remainder of their journey to Manhattan (Baird 1974). To the delight of captains and travelers, this considerable inconvenience ended in 1824 when the Supreme Court of the United States, in the case of Gibbons v. Ogden, ruled the Fulton monopoly unconstitutional. This decision inaugurated an era of unfettered competition among steamboat companies as they vied with one another to attract customers by offering more comfortable ships, bargain fares, and faster trips.

To reach their destination in the shortest possible time, some captains did not stop to refuel but instead were supplied with fresh wood, the fuel used by the first generation of steamboats, by sloops near Fishers Island in the eastern Sound. In the 1830s, coal was substituted for wood. Steamboats were able to carry enough of it to make non-stop trips, even if excess coal was required to plough through heavy seas or to challenge a rival boat to a race. Some of these contests had disastrous results because a great deal of steam pressure was required to achieve maximum speed and this could lead to explosions. Following a race between the *PROVIDENCE* and the *NEW ENGLAND* in 1833, the latter vessel exploded when it reached its destination, Essex on the Connecticut River upstream from the Sound (Daily Herald 1833). Such accidents engendered fear in the traveling public but steamboat designers came up with a solution, namely to elevate steamship boilers by placing them above the deck. Passengers were relegated to the lower deck, which was deemed safer in the event of an explosion. The famous British author Charles Dickens traveled throughout the Sound on a boat of this design on an American tour in 1842 and found the vessel disagreeable from an aesthetic standpoint. More pleasing in appearance were traditionally designed steamboats, most of which operated quite safely. Now and again, however, there were horrific accidents and they were not always the result of racing.

Although the *LEXINGTON*, which had been built for Commodore Cornelius Vanderbilt, had engaged in a famous race with the *JOHN W. RICHMOND*, racing had nothing to do with the loss of the *Lexington*. Instead, it was improper placement of cargo on the top deck of the ship that set off a dreadful chain reaction on a frigid night in January 1840. Since the hold of the ship was already full, bales of cotton being shipped to New England textile factories were placed on the upper deck next to the smokestacks. Given the extremely cold weather and ice floes in the Sound, the ship's boilers were working overtime to propel the vessel through the angry waters. As the *LEXINGTON* made its way east, sparks from the smokestacks ignited the cotton. With the ship engulfed in flames, Captain George Child pointed the vessel toward the Eaton's Neck lighthouse off Huntington, but the tiller ropes burned and he was unable to steer. At that point, the lifeboats were launched, but only one made it to the north shore of Long Island. It was empty, but four people did survive by clinging to unburned bales of cotton. One person drifted 40 miles east, scaled the bluffs at Baiting Hollow, and walked nearly a mile to a house whose inhabitants took him in. In the aftermath of the sinking of the *LEXINGTON*, unscrupulous opportunists who had read accounts of the disaster in New York City newspapers converged upon the beaches of the north shore to peel through the wreckage in hopes of finding hard currency carried by businessmen traveling on the ship that night. This prompted New York State to dispatch guards to patrol a fifteen mile stretch of the north shore to safeguard bodies, baggage, and anything else that washed up in the aftermath of a disaster that claimed 120 lives (McAdam 1957).

A predictable but temporary falloff in steamboat travel occurred following the sinking of the *LEXINGTON,* but the introduction of larger, beautifully appointed vessels helped overcome passenger apprehension. The ships of the Fall River Line, founded in 1847, were especially popular (Covell 1947). The Narragansett Steamship Company, established by financiers Jay Gould and Jim Fisk in the 1860s, competed with the Fall River Line by offering beautiful décor and personnel outfitted in natty uniforms (Covell 1947). Accurate charts of LI waters produced by the US Coast Survey beginning in the 1830s contributed to a safe passage through the Sound. The Coast Survey's charts and LI maps were the federal government's "first major venture in scientific cartography" (Allen 1997). A half century after the first charts were produced, steamboats and other vessels plying the waters of LIS experienced an easier passage through Hell Gate, a treacherous stretch of swirling water in the vicinity of today's Robert F. Kennedy Bridge (Triborough Bridge), after the US Army Corps of Engineers used dynamite to eliminate underwater rocks in 1876. The blast was the world's largest up to that time (Jackson 1995). LIS steamboats were not the only vessels to benefit from this. Ships coming into New York Harbor from the Atlantic had to cross a sand bar at Sandy Hook. Ships drawing more than 23 feet of water could not navigate this obstacle and, as a result, New York Harbor's future was limited unless the federal government appropriated sufficient funds to rectify the problem. Absent a huge Congressional appropriation, the solution was to enhance the Hell Gate passage, enabling large ships to transit from the Atlantic to LIS and down to New York Harbor (Klawonn 1977).

In addition to vessels that passed through Hell Gate and plied the whole length of LIS linking New York with Boston and other New England destinations, there were vessels that transported New Yorkers to seaside resorts in Connecticut and Long Island. Some of these ships were owned by the Montauk Steamboat Company, which became a subsidiary of the Long Island Rail Road (LIRR) in 1899 (NYT, May 14, 1899, 3).

The LIRR had been conceived in the 1830s by Brooklyn business interests who spied an opportunity to make a fortune by transporting people between New York and Boston by way of Long Island. Absent a direct rail link between those two cities on the mainland, and gambling on the likelihood of none materializing in the near future because of the cost of bridging the rivers flowing into the Sound, the LIRR's investors funded the creation of a line from the western end of the Island, which was easily reached by ferry from Manhattan, to Greenport on the North Fork (Ziel 1965). There, Boston bound passengers took a steamboat across to Stonington, CT where they boarded a train for the remainder of the 11.5-h trip to Boston. When the LIRR made its debut as a major regional carrier in 1844, there was a gigantic celebration at the Greenport railroad station. But the merriment so evident on that occasion gave way to gloom and doom just 4 years later when the New York, New Haven, and Hartford Railroad completed a line from New York to New Haven where it linked up with an existing line running inland to Hartford and Springfield and across to Boston. This all-rail route soon became the preferred way to travel between Manhattan and New England, and the LIRR was reduced to being a local line serving an area whose population would be insufficient to support a railroad until the next century. Following bankruptcy in 1850, the LIRR gradually recovered and built feeder lines running to north shore communities that started to grow thanks to tourism and, in the twentieth century, suburbanization. In 1891, the LIRR attempted to revive its cross-Sound service by transporting railroad cars on the Oyster Bay line to Wilson Point in Norwalk, CT on special ferries. The cars were then connected to trains heading to different points in New England. It was a novel idea but it did not attract the anticipated ridership.

Far more popular were excursion and commuter steamboats. They did a lively business in the summer by transporting people on day trips to picnic groves along the Sound and by delivering New Yorkers to their summer residences. One of the most popular commuter vessels was the *SEAWANHAKA*, which remained a favorite of its regular passengers even after losing an exciting race in 1867 to the *JOHN ROMER* of the Greenwich and Rye Steamboat Company, whose president was William Marcy Tweed, the notorious boss of New York City's corrupt Tammany Hall political machine. More than a dozen years later, on a June day in 1880, the *SEAWANHAKA* caught fire as it headed up the East River en route to its regular stops in Roslyn and Sea Cliff on the north shore of Nassau County. The ship's quick-thinking captain, Charles Smith, beached the vessel on a shoal, enabling most of the passengers to escape. Despite the encroaching flames, he did not let go of the wheel until the boat came to rest (NYT, June 30, 1880, 1). Not far from where the *SEAWANHAKA* met its fate, the excursion steamer *GENERAL SLOCUM* burned and sank with the loss of over a thousand lives in June 1904 as it headed out to the Sound on a church-sponsored excursion. Three years later, the steamer *LARCHMONT* of the Joy Line went down off Watch Hill, RI with the loss of at least 189 lives. A final determination of the number who perished could not be made because the passenger list was aboard the sunken ship (NYT, Feb. 13, 1907, 2). There were four casualties in January 1935 when the *LEXINGTON* of the Colonial Line went down in the East River (NYT, Jan. 4, 1935, 44). That same year the Fall River Line, which had been acquired by the New Haven Railroad in an effort to regulate competition between land and sea transport, suspended its New London service. Other service suspensions followed, as did a strike. Combined with the line's growing deficit during the Great Depression, the work stoppage proved to be the final blow. The Fall River Line was phased out by the New Haven Railroad in 1937 and its famous sidewheel steamboats, the *COMMONWEALTH,* the *PRISCILLA,* the *PLYMOUTH,* and the *PROVIDENCE* were sold for scrap. Six years earlier steamboat service from Hartford to New York via the Connecticut River and LIS had been terminated (Delaney 1983).

1.2.4 Ferries

Just as it had been a factor in the demise of the steamboat business on LIS, the Great Depression curtailed but did not totally eliminate cross-Sound ferry service. One line that did vanish during the 1930s, the Rye-Oyster Bay ferry, had existed since 1739. In the colonial era, the ferry was an important economic stimulus. A store that was part of the ferry complex in Rye stocked an array of goods, including textiles, foodstuffs, guns, gunpowder, and rum (New York Mercury, April 30, 1759, Supplement, 2). A retail establishment of this type was able to thrive because its customers included not only residents who lived nearby but countless ferry travelers, including Westchester Quakers who during the summer months traveled to Long Island to worship with brethren at the Matinecock Friends Meeting House in Glen Cove (Fox 1919). Some people who traveled on the ferry moved permanently to the opposite side of the Sound and transported their farm implements and household possessions with them. The original price list for the Rye-Oyster Bay ferry noted the cost of transporting such items as beds, bedding, and tables (Weigold 1975). Besides the Rye ferry, in the eighteenth century there was cross-Sound service from New Rochelle to Hempstead and from New London to Orient Point.

Although colonial era ferries served an important purpose, they were utilitarian sailing vessels that simply transported people from one side of LIS to the other. The ferries of the early twentieth century, in contrast, were, in a sense, pleasure craft because their patrons often boarded, with or without automobiles, with the thought of a nautical outing, perhaps combined with a picnic on the other side of the waterway. This was especially so during the 1920s, which was the golden era of ferries on the Sound. During that decade, it was possible to cross the waterway at numerous points. At the western end of the Sound, there was a ferry linking Clauson Point in the Bronx with College Point in Queens. Moving a bit eastward, there was service between New Rochelle and Sands Point, Rye and Sea Cliff, and Rye and Oyster Bay. The ferries originating in Rye were operated by the same company that linked Greenwich and Oyster Bay. The Long Island terminus for these vessels was actually Bayville, in the Town of Oyster Bay.

The Greenwich-Oyster Bay service supposedly originated upon the recommendation of industrialist Andrew Carnegie, who at one point had to get from the mainland to Long Island without delay. The trip for Mr. Carnegie and his carriage cost $250. Theodore Roosevelt paid considerably less in 1917 when

the cost of transporting a car and driver on the Greenwich-Oyster Bay ferry was
$2. Still, the former president was not pleased with the trip because the ferry
departed a full hour behind schedule. Having to make six round trips a day at
the height of the summer season was simply too much for the weary crew. The
remedy proposed by Roosevelt was to eliminate the crossing with the least ridership. The management of the ferry company agreed and after that it was smooth
sailing. By the mid-1920s, the Greenwich-Oyster Bay ferry had some nearby
competition in the form of a new line linking Stamford with Huntington via a
vessel with a capacity of 40 cars and 1,000 passengers. The Greenwich-Oyster
Bay ferry responded by increasing the number of trips, thereby enabling it to
transport 700 cars per day. Farther east on the Connecticut coast, the Bridgeport-
Port Jefferson ferry, established in 1872, and the New London-Orient Point ferry,
which has been linking CT and LI since colonial times, transported vehicles and
passengers across the waterway. These ferries continue to operate, but all of the
other lines were phased out during the 1930s, in part because of the economic
impact of the decade-long Great Depression, but also because of the opening of
two new East River bridges, the Triborough in 1936 and the Bronx-Whitestone in
1939.

1.2.5 Highways

Bridges were part of the arterial highway system envisioned by master builder
Robert Moses who, starting in the 1920s, pursued an ambitious development program that included parks and parkways, as well as bridges. When Moses set out
to build parkways on Long Island, the Island was lagging behind the mainland
when it came to modern highways. Westchester County's first parkway, the Bronx
River, was completed in 1925. During the same decade, the Westchester County
Park Commission developed a vast network of public open spaces and recreational areas, including two on LIS: Glen Island in New Rochelle and Playland in
Rye. The Commission also began building parkways through the Saw Mill and
Hutchinson River valleys. Across the border in Connecticut, the Merritt Parkway
was the continuation of the "Hutch" and this seamless, bi-state highway through
the western portion of Sound Shore communities relieved congestion on the busy
Boston Post Road. Robert Moses envisioned similar relief for Long Island. The
route he chose for a parkway through towns bordering LIS, however, was opposed
by Nassau County estate owners. They lobbied to have New York State acquire
and modernize the Vanderbilt Motor Parkway, which stretched from Queens to
Lake Ronkonkoma in Suffolk County. This private highway had been built by
the multi-millionaire sponsor of the Vanderbilt Cup races. Spectator casualties
in the early 1900s, when the races were held on public roads, caused Vanderbilt
to formulate plans for an automobile racetrack at the Suffolk County terminus
of his parkway, but they did not come to fruition. Instead of acquiring the Motor
Parkway, the state accepted the route recommended by Moses, but to placate

opponents of this route, instead of going directly east, the Northern State Parkway was designed with a 2-mile southern bend that avoided the estates of its most vociferous opponents (Caro 1974).

When construction of the Northern State Parkway began in the 1930s, it seemed unimaginable that two decades later a six-lane expressway cutting through the center of Long Island would be needed to accommodate the Island's explosive growth in the period following World War II. Construction started on the much-maligned Long Island Expressway (LIE) in 1953. At that time, planners expected 80,000 vehicles a day to use the highway once it was completed in 1972. Yet, in 1966, the partially opened road was accommodating 140,000 cars a day on average, with tens of thousands more on summer weekends. No wonder the LIE was dubbed "the world's longest parking lot" (Kimmel 1983). With the addition of high occupancy vehicle lanes in the 1990s and the rebuilding of major stretches of the roadway in the first decade of the twenty-first century, the LIE became more user friendly. But some people traveling between the Island and the mainland, heading to the Whitestone or Throgs Neck Bridge, might yet gaze longingly at the exit for the Seaford-Oyster Bay Expressway, which was to have been the approach road to Robert Moses's last great project, the one that did not materialize: the Rye-Oyster Bay Bridge. Although the Throgs Neck Bridge, which opened in 1961, was initially underutilized, within 5 years 30.6 million vehicles crossed the bridge annually; another 25.8 million used the Whitestone Bridge each year (Moses 1970). From Moses's perspective, the combined traffic on the two spans called for action in the form of a cross-Sound bridge.

1.2.6 Bridging the Sound

The idea for such a span was not new. In 1938, the United States Senate's Commerce Committee discussed the feasibility of a bridge stretching nearly twenty miles from Orient Point to Plum, Gull, and Fishers Islands and from there to a landfall on the mainland at Groton, CT or Watch Hill, RI. Proponents of this Depression-era plan argued that the bridge would provide Long Island businesses with access to New England markets. A 1963 study commissioned by the Long Island Sound Tristate Bridge Committee, a private group lobbying for a new span from Orient Point to a terminus in the vicinity of the Connecticut-Rhode Island border, also touted the economic benefits of such a bridge. Robert Moses, however, thought he had a better idea, namely to create a crossing between Oyster Bay and Rye. A feasibility study for this route done by the engineering firm of Madigan-Hyland supported the idea. Moses presented the Madigan-Hyland study to the New York State legislature in 1965 in an effort to persuade the legislature to authorize the Triborough Bridge and Tunnel Authority, which Moses headed, to expand its jurisdiction beyond New York City in order to construct the proposed bridge (Traffic, Earnings and Feasibility of the Long Island Sound Crossing 1965). To Moses, this was a mere technicality although New York State Governor Nelson

A. Rockefeller noted that when facilities built by the Triborough Bridge and Tunnel Authority were fully paid for by toll revenues, the City of New York assumed ownership. How the city could own a bridge that was not within its borders puzzled the governor. Ultimately, it would be Rockefeller who would block construction of the bridge but not until howls of opposition were heard on both sides of LIS.

Anticipating negative reaction to his proposed span, Moses went on the offensive, denouncing critics who according to him resisted change and did not build anything (Traffic, Earnings and Feasibility of the Long Island Sound Crossing 1965). But the critics had their say, including citizens groups on both sides of the Sound, the Yacht Racing Association of LIS, which called for a tunnel instead of a bridge, and environmental groups. Citing the potential impact of the bridge and its approach roads on the Sound's aquatic life, wetlands, and vistas, the opposition was relentless. Then, in June 1973, before stepping down as governor to assume the vice presidency of the United States, Nelson Rockefeller announced that he was withdrawing his support for the bridge (NYT, June 21, 1973, 30). He subsequently signed a bill withholding authorization for the Metropolitan Transportation Authority, a regional entity that included the Triborough Bridge and Tunnel Authority, to build the bridge.

Had the Rye-Oyster Bay Bridge become reality, it would have funneled westbound traffic onto the Cross-Westchester Expressway (I-287) and the New England Thruway/Connecticut Turnpike (I-95) raising the possibility that those roadways would be quickly overwhelmed by increased traffic volume. Even without the bridge, both highways would soon be straining under mounting vehicular loads. In 1957, when the Connecticut Turnpike was completed save for the Byram River Bridge, the missing link that would join the New York and Connecticut portions of I-95, the turnpike was viewed as the longest urban highway in the nation. That may have been true of the road's western part, which cut through four major cities: Stamford, Norwalk, Bridgeport, and New Haven, but as Connecticut's governor, Abraham Ribicoff, observed, during a two-day preview of the turnpike in the fall of 1957, the area from New Haven east to the Rhode Island border had considerable open space and, with the exception of New London, limited urbanization and suburbanization. But this was about to change (NYT, Oct. 24, 1957, 35).

1.3 Land-Use Patterns

1.3.1 Pre-Contact and Contact Periods

The completion of the Connecticut Turnpike in 1958 signaled the beginning of a major transformation for the area east of New Haven but anyone traveling through that region in the late 1950s would have found nautical and pastoral vistas reminiscent of an earlier time when people on both sides of the Sound derived a living from the sea and the land. Long before fast-moving cars and eighteen-wheelers began barreling down the new interstate linking the Sound shore communities, native peoples sustained themselves not only by hunting, fishing, and gathering edible fruits and berries but also by farming. Female members of the various Algonquin tribes dwelling on both the mainland and the north shore of Long Island conducted a simple but effective form of agriculture, raising corn, beans, pumpkins, and squash, all of which supplemented the food brought home by peripatetic male hunters and fishermen (Newsday 1998). Indians lived lightly upon the land, using the resources they needed for food and other purposes. They felled trees to build their canoes and used wetlands as safe anchorage for the vessels. Wetlands also supplied the raw material for their baskets.

1.3.2 "Taming" the Land: Farms, Mills, and Vineyards

Like the Indians who had preceded them, whether the Siwanoys in coastal Westchester, the Paugussets, Quinnipiacs, Hammonassetts, Nehantics, Pequots, or Mohegans of Connecticut, or Long Island natives such as the Matinecocks, Nissequogues, Setaukets, and Corchaugs, European settlers lived lightly upon the land until their increased numbers necessitated more intensive use of the resources of both land and sea. Aside from farming, land was needed for settlers' animals. Pigs could forage in the forest and could hold their own against predators, but sheep and cows were best kept in protected pastures, close to their owners' dwellings. Thus more land was needed. As the population increased, vast acreage had to be cleared for efficient farming. This meant cutting down trees and, if possible, extracting their stumps. At the very least, the forest canopy had to be removed to permit sunshine to bathe the crops. In time, ready markets for felled trees were found. Tall trees were in demand as masts for the sailing ships of the colonial era and boats of all sizes were built in communities along the Sound and the Thames, Connecticut, Housatonic, and Quinnipiac Rivers (Walsh 1978). On both sides of the Sound, cordwood, used for myriad purposes, became an income-producing sideline for farmers. Small sailing vessels transported wood to New York from landings on the north shore of Long Island (Bayles 1962). Vessels arrived at landings with the tide and then beached themselves to enable cordwood-laden wagons to pull up alongside and transfer cargo to the waiting ships at high tide (Austen 1992). Some of the cordwood produced on the Island never made it to landings on the Sound, however. Wood obtained from Long Island's Pine Barrens, occupying a wide swath of land in the middle of the Island, was used as fuel for the early locomotives on the LIRR. In Connecticut, locally harvested cordwood was used in the early iron industry. It would not be until the nineteenth century, however, that industry emerged as a major component of the economy on the mainland side of the Sound.

In the meantime, most inhabitants of the Sound shore area made their living from the land, at first growing subsistence crops, but as settlements grew and markets for their crops expanded farmers began to specialize. Long Island's sandy soil and long growing season made it the breadbasket of New York. Before grain grown in the Midwest began reaching Eastern markets via the Erie Canal, which was completed in 1825, the Hempstead Plains was a major supplier of wheat. Rye and hay were also important in the Island's agrarian economy, as they were on the mainland. Vegetables and fruits destined for the New York market were raised on both sides of the Sound. Asparagus was the specialty crop in Oyster Bay (Weidman and Martin 1981). New Rochelle gained fame for its blackberries (Nichols 1938). No matter what they were growing, to achieve maximum yields per acre, farmers had to use fertilizer. The Sound yielded abundant fish for this purpose (Bayles 1962). As the nineteenth century progressed, agricultural production far beyond the Sound shore area benefited from the Atlantic menhaden (*Brevoortia tyrannus*) fertilizer produced in factories in the Norwalk Islands, Groton, Greenport, and elsewhere in the region. In the twentieth century, chemical fertilizers and pesticides, including Aldicarb, were used. They helped increase crop yields but they also seeped into the groundwater, necessitating the installation of special filters for drinking water when homes were built on former farm land. In the waning years of the twentieth century, on both sides of the Sound, some open space was spared from intensive development when farms were converted to vineyards. The transformation of portions of the LIS region into wine production areas commenced in 1973 when Louisa and Alex Hargrave purchased a potato farm in Cutchogue, on Long Island's North Fork, and planted wine grapes. Nearby, in Peconic, Dr. Herodotus Damianos opened Pindar Vineyards. Many others followed; by 2009 the North Fork had thirty wineries. On the mainland, vineyards were flourishing in Clinton and Stonington.

1.3.3 Industrialization

To some, the success of the vineyards was a reminder that despite all of the development that occurred in Sound shore communities, particularly in the latter half of the twentieth century, the area's agrarian roots were still visible. If one were to turn the clock back, however, to the first half of the twentieth century or the nineteenth century, it must have seemed as though agriculture was being supplanted by industry, especially on the mainland, where rivers flowing into the Sound provided power for the early textile mills. Going back to the colonial era, farmers had harnessed water power to run gristmills. Locales where streams and inlets merged with LIS were ideal spots for mills powered by the force of the changing tide, Kirby's Mill in Rye being but one such example (Baird 1974). This was a small operation in contrast with the type of mills that dotted the New England landscape in the nineteenth century. Utilizing mass production techniques pioneered in New Haven at the very end of the eighteenth century by Eli Whitney to produce 10,000 muskets for the United States Army, factories in Sound shore communities and on rivers flowing into the Sound were economic powerhouses turning out all sorts of goods (Maynard and Noyes 2004).

Always on the lookout for new opportunities, some industrial entrepreneurs specialized in more than one product. Oliver F. Winchester, for example, manufactured both firearms and shirts. The putting-out method was used in making the latter. The material was cut to size in his New Haven factory and then sewn in workers' homes not only in coastal Connecticut but also across the Sound on Long Island. The installation of sewing machines, made by Nathan Wheeler, who had a light-metal business in Bridgeport, ended the putting-out method and insured uniformity. The mass production techniques used to turn out shirts were also employed in Winchester's other New Haven business, which manufactured the famous Winchester rifle. Clocks were another major product, as were carriages. Many clocks from Chauncey Jerome's factory were loaded onto boats at Long Wharf to begin their journey through the Sound and out to the Atlantic on their way to customers in England. Carriages made in New Haven found a ready market in the South during the ante-bellum period and some were exported to Brazil. Padlocks were another item made in New Haven but the premier lock manufacturer along the Sound shore was the Yale and Towne Company in Stamford. In addition to locks, Stamford was known for woollen textiles, the production of which entailed the application of chemicals and dyes that ended up in the Sound (Feinstein 2002). For a time, Norwalk was a major producer of hats but increased mechanization of the industry was accompanied by lower wages for employees. Strikes ensued and they were sometimes marked by violence. Even when it did not result in strikes, dissatisfaction on the part of factory workers was evident. But given the steady influx of immigrants, native-born workers knew only too well that they could be easily replaced. As the nineteenth century progressed, immigrants became a more significant component of the workforce in such coastal cities as Bridgeport. Well known as a center of textile manufacturing, Bridgeport also turned out carriages, sewing machines, and corsets. In 1860, the city had nearly a hundred factories (Grimaldi 1986).

At the opposite end of the state, New London Harbor was the egress for textiles manufactured in towns along the Thames River (Anderson 2002). Connecticut's other river valleys were also centers of industry and sources of pollution. In Danbury, which was "the hat manufacturing capital of the world" in the nineteenth century, mercury used in the production of felt was discharged into a tributary of the Housatonic River (Varekamp et al. 2005). Along the Naugatuck River in Waterbury, waterpower, which had formerly been used to operate a gristmill, was harnessed to power the machinery in a plant manufacturing brass in the first decade of the nineteenth century. Copper waste from the plant was discarded in the river. The metal was carried downstream to the Housatonic River and from there transported to LIS. The Housatonic River watershed, along with the Connecticut River watershed, which was impacted by metal manufacturing in Hartford and Springfield, would emerge as the main sources of heavy metal contamination for LIS and its coastal zone (Varekamp et al. 2005). Limited at first by the actual volume of brass production, pollution was a major challenge by the end of the nineteenth century when the American Brass Company was established in Waterbury. Scovill was another major brass manufacturer in Waterbury (Grant 1974). Since brass was used to manufacture myriad products including clocks, weapons, carriages, and whale oil lamps, there was plenty of business to go around and small firms existed alongside the giants. The plains flanking the Naugatuck became known as Brass Valley because, in addition to the large-scale operations in Waterbury, which was truly Brass City, USA., brass was produced in more than a half dozen other communities in the valley (Anderson 2002).

Compared with the Naugatuck, industry near the Byram River was limited but the Abendroth Foundry had manufacturing facilities along the river in both Port Chester and Greenwich (Mead 1911). To the south, Mamaroneck produced textiles and carriages. Nearby New Rochelle also had a carriage company and a printing plant but industrial output along the Sound shore of Westchester County was minimal in comparison with that of coastal Connecticut. Less still was that of Long Island. Although shipbuilding occurred on the north shore, the sort of industrial activity evident in Connecticut was almost totally absent on the opposite shore. Glen Cove had a leather works (MacKay 1985) and the world famous Duryea starch factory whose product was awarded medals at international expositions (Ross 1902). Huntington produced pottery utilizing clay found in the immediate vicinity and clay deposits in the town of Southold were the raw material for brick making in Greenport (Bayles 1962). In western Suffolk County, Cold Spring Harbor turned out woolen textiles. Farther east, shipbuilding was occurring in Stony Brook, Setauket, and Port Jefferson. The Hand Shipyard in Setauket developed a worldwide reputation for its schooners (Adkins 1955). Port Jefferson was the shipbuilding capital of Long Island, with more than 300 vessels constructed between the late eighteenth century and the mid-1880s (Newsday 1998). Sand and gravel mining also generated income for Port Jefferson and Mt. Sinai. The industry, which reshaped the shoreline, thrived until the 1930s (Bone 1998). In Stony Brook, William Sidney Mount, the man regarded as the nation's greatest genre painter of the nineteenth century, was the catalyst for the dredging of the western end of the harbor. The goal was to boost the economy of the community and reduce the travel time for ships departing for New York City and Connecticut. An initial dredging project was undertaken in 1856, but within 2 years shoaling necessitated additional work. This time, Mount was unable to muster adequate financial resources and the idea was dropped (Swanson et al. 2005). Mount concentrated on his artistic career, depicting scenes of rural life on LI, which remained primarily agrarian. The situation was very different on the mainland where industrialization and increased population in urban centers along the Sound and its tributaries resulted in pollution.

1.3.4 Population Growth and Sewage Pollution

Whether as a result of indifference or ignorance, Daniel Webster's "American Mediterranean," through which he sailed in the 1800s when commuting between his home in Massachusetts and New Rochelle to court the woman who became his second wife, was becoming a gigantic waste receptacle. Besides factories, the houses and stables lining the Sound's tributaries emptied their waste into the water.

This was an immediate solution to the disposal problem but the water courses on the receiving end and the Sound itself, the downstream resting place of industrial byproducts and sewage, were subjected to degradation. Recognition of the causal relationship between contaminated water and disease led to the construction of sewers along many Manhattan streets 50 years before the consolidation of the five boroughs in 1898. This approach did nothing to prevent the pollution of wells from sewage seepage in areas not included in the sewer project. Moreover, waste from parts of Manhattan that did have sewers was funneled into surrounding waterways. It was a similar situation in the major population centers of Connecticut. With more than a dozen miles of sewers by the last quarter of the nineteenth century, Bridgeport was diverting waste from only part of the city and the waste handled by these sewers ended up in LIS (Anderson 2002). In the interior, sewage from as far away as Massachusetts and Hartford was deposited in the Connecticut River, eventually making its way south to the Sound. By the late 1800s, waste from the sewers of New Haven, which was one of the first cities in the United States to build a sewer system, did not have to travel far to reach LIS (Maynard and Noyes 2004). Stamford, too, had sewers, starting in the 1880s, but Westchester County lagged behind.

Residue from manufacturing in Westchester's Sound shore and Hudson River communities, along with sewage produced by a growing population, ended up in the waterways, one of which, the Bronx River, became a real blight on the landscape by the late nineteenth century. Joining forces with New York City, Westchester County undertook a massive project to clean up the Bronx Valley, eliminating factories, private residences, privies, billboards, and other manmade intrusions upon the once pristine area. Returning the land to its natural state was only the beginning. A roadway, originally conceived as a carriage drive, but redesigned at the start of the automobile age as a multi-lane parkway for cars, was constructed to parallel the river. At the same time, a trunk sewer line was built to convey a considerable amount of Westchester's waste, which otherwise would have gone into the river and been carried down to LIS, to a final resting place in the Hudson River. In the 1920s, sewage treatment plants were built on the Hudson in North and South Yonkers, and in Rye and Mamaroneck on the Sound (Westchester County Department of Environmental Facilities 1997). Like the facilities built in Port Jefferson and Glen Cove, these were primary treatment plants in which solids were trapped by a screen and the effluent was treated with chlorine gas. A different approach was taken at the Norwalk sewage treatment plant, which came online in the 1930s. There, the rate of the wastewater entering the facility was slowed down to allow solids to settle to the bottom of tanks. Incineration was used to dispose of this sewage while the water discharged into the Sound was treated with chlorine (Anderson 2002). Although innovative methods employed in some treatment plants had a beneficial effect upon the Sound, by the 1930s it was evident that more was needed. Pollution from sewage and other sources was not solely a local issue but a regional challenge. In an effort to address the problem, the Tri-State Sanitation Commission was established in 1936 by a Congressionally approved compact between New York and New Jersey to oversee and regulate water quality. Connecticut became part of the commission in 1941; in 2000 the commission's name was changed to Interstate Environmental Commission.

Its jurisdiction includes the portion of LIS west of Port Jefferson and New Haven (Interstate Environmental Commission 2011).

In the late 1930s, secondary treatment was employed at the new Ward's Island plant at Hell Gate (NYT, Dec. 8, 1936, 30). This facility, which treated New York City sewage, added an extra layer to the standard filtering out of solids and chlorinating wastewater by oxygenating organic sludge, thereby encouraging the growth of microorganisms that devour the organic material. This worked quite well in dry weather but New York and older urban areas in Connecticut had combined sewers through which sewage and rain flowed. During heavy downpours, to prevent sewage treatment plants from being flooded, the plants were bypassed and everything coursing through the sewers went directly into LIS. Adding to this problem in the second half of the twentieth century was the deterioration of sewers that had been installed decades earlier. As a result of seepage from older sewers, waste water found its way directly to the Sound. That this was still happening more than a decade after passage of the Clean Water Act of 1972 was attributable to foot dragging by municipalities and counties, and to the fact that in the 1980s the federal government pulled back from providing outright funding for upgrading sewage treatment plants and substituted loans to municipalities for this purpose. Although the Clean Water Act mandated a minimum of secondary treatment for sewage, several plants in Connecticut, Long Island and Westchester were slow to comply. By the late 1980s, however, all but the Mamaroneck plant had become secondary treatment facilities. Mamaroneck was finally upgraded in 1993 (NYT, Aug. 15, 1993, 7).

In the meantime, in the late 1980s, an environmental engineering firm was given the responsibility of running the Norwalk treatment plant. It had become a secondary facility in the 1970s but it experienced ongoing problems that resulted in the banning of shell fishing in Norwalk Harbor for extended periods (Anderson 2002). Farther down the Connecticut coast, the Stamford sewage treatment plant was using activated sludge treatment. This method employed carefully measured amounts of oxygen and sludge to provide a breeding ground for bacteria that consumed most of the organic matter (Anderson 2002). Another positive development was a joint state-federal agreement on a no-net-increase policy governing nitrogen (N) discharges into LIS. This agreement was the first phase of a multi-phase strategy to address the problem of hypoxia, low dissolved oxygen levels, in LIS (see Sect. 1.6.3). At the start of the new millennium, the Long Island Sound Restoration Act provided funding for various initiatives to improve the Sound (Long Island Sound Restoration Act 2005). Sewage plant upgrades were included, but the actual amount appropriated was a fraction of the annual $40,000,000 that had been anticipated.

1.3.5 Suburbanization, Exurbanization, and Corporate Relocation

A drop-in-the-bucket approach to funding the rehabilitation of LIS was clearly inadequate in view of the tremendous growth of the Sound area in the previous century. Aside from the increase in New York City's population, communities located within easy rail commuting distance of the city experienced a population surge in the early twentieth century. Even before World War I, the north shore of Nassau County, the Sound shore of Westchester, and coastal Fairfield County began attracting middle- and upper-middle-class newcomers. The housing boom in Westchester prompted *The New York Times* to declare, in 1913, that Westchester was waking up (NYT, April 13, 1913, 8). That same year, the New York State Department of Health judged New Rochelle to be the healthiest city in the state. Other attributes of the close-in county were highlighted in a real estate exhibit mounted in Grand Central Station. Featuring a model home outfitted with state-of-the-art appliances and paintings of Westchester vistas, including LIS at Oakland Beach in Rye, the exhibition was a major promotional effort. Alluring as Westchester was, however, it was regarded as an expensive place even back then. More affordable was a new development at Flushing Bay touted as being suitable for cultured urban workers with limited means who wished to live in the country (NYT, April 25, 1913, 10).

Following World War I, the proliferation of automobiles and the building of parkways contributed to a major wave of suburbanization in the 1920s. On Long Island's fabled Gold Coast, an area stretching along the Sound shore of Nassau County and extending into western Suffolk, great estates developed between the 1880s and World War I gave way to smaller suburban homes. The disappearance of large parcels of land once occupied by mansions, stables, garages, greenhouses, and other outbuildings caused some north shore municipalities to adopt master plans as a way of controlling suburban sprawl. Yet the erection of new homes, especially in communities within easy commuting distance of Manhattan, continued at a fairly rapid pace in the 1920s. In Manhasset, the Metropolitan Museum of Art sponsored Monsey Park, a development named for a benefactor who had bequeathed hundreds of acres to the museum. Across the Sound in Westchester new homes sprouted on and near the shore from Pelham to Port Chester. The county was seen as a natural outlet for city dwellers because getting there did not require a ferry (NYT, Feb. 17, 1924, 9). Connecticut communities enjoyed the same advantage and they, too, experienced a boom in suburban home building in the 1920s. In the aftermath of the stock market crash of 1929 and the onset of the Great Depression, residential construction slowed to a trickle but resumed after World War II.

In the late 1940s and 1950s, William Levitt's mass-produced housing transformed a wide swath of agricultural land in the town of Hempstead into instant housing, with entire streets of homes seeming to rise overnight. Despite critics' warnings that Levittown, which was initially a rental community, was a slum in the making, strict rules governing upkeep and the decision to sell rather than rent the homes ensured the community's future as a tidy suburb with well maintained, albeit at least in the early years, look-alike houses (Smits 1974). During the mid-twentieth century, suburban home building near LIS was concentrated in Nassau, Westchester, and lower Fairfield Counties, but as the twentieth century progressed, the insatiable demand for housing, combined with the building of superhighways on both sides of the waterway, resulted in the suburbanization of formerly exurban areas. Following the completion of the Connecticut Turnpike, areas east of New Haven experienced development. When the Long Island Expressway pushed into Suffolk County, reaching its terminus in Riverhead in 1972, a housing boom that had transformed western Suffolk continued unabated with the result that communities as far east as Riverhead witnessed not only population growth but increased taxes to support schools and infrastructure enhancements. LIS was impacted as well. More people translated into additional road runoff and sewage pollution. Yet, people had to live somewhere, whether in single-family homes on an acre or more of land or in clustered housing, a concept endorsed by the Nassau-Suffolk Regional Planning Board as a way of maximizing utilization of vacant land. That agency also called for the building of apartments to accommodate Long Island's growing population.

In the second half of the twentieth century, multi-family housing was constructed on both sides of the Sound. With funding from the Mitchell-Lama middle-income housing program, Co-op City rose on the shore of Eastchester Bay in the Bronx. Built on the site of the former Freedomland Amusement Park and completed in 1970, this community of 60,000 inhabitants has towering apartment buildings, schools, and stores. Throughout the Sound shore area, multi-family housing was a component of urban renewal efforts in the mid-twentieth century. Over in Nassau County, North Hempstead, Manhasset, and Roslyn secured federal funding to demolish older buildings, replacing them with multi-family housing (Smits 1974). Housing was also erected in Glen Cove along with the Village Square Shopping Center (Harrison 2008). On the mainland, New Haven was the beneficiary of massive federally funded revitalization in the 1960s. Deteriorated housing was demolished to make way for apartments and town houses for low- and moderate-income families and senior citizens. Abutting LIS, the Long Wharf area, once a bustling commercial hub, was transformed with the addition of a theater, manufacturing, and office space. In the 1990s, a four-year study was undertaken for an upscale shopping mall at Long Wharf but the idea was abandoned (NYT, June 1, 1993, A1). Several decades earlier, during the period of intense urban renewal, a new housing development rose across New Haven harbor in West Haven, on the site of the once famous Savin Rock Amusement Park (NYT, Dec. 29, 1963, 155). Housing was also the centerpiece of renewal efforts in Stratford, Milford, and New London. Apartment towers transformed Stamford as well in the second half of the twentieth century but the revitalization of this Sound shore community's downtown also included new hotels, the Town Center shopping mall, office buildings, the restored Palace Theater, the Stamford Center for the Arts, and a branch of the Whitney Museum of American Art. Beyond the downtown core, on a peninsula jutting out into the Sound, Stamford Landing, with its mix of apartments, townhouses, commercial buildings, and marina rose in the 1980s. Farther down the Sound, a section of Port Chester along the Byram River was redeveloped in the early years of the twenty-first century (NYT, July 31, 2005, WE1) (Fig. 1.1).

With the approach of the new millennium, New Rochelle, where mid-century urban renewal had included the construction of an enclosed shopping mall, office

Fig. 1.1 Urban redevelopment along the Byram River, Port Chester, NY, included Costco plus additional stores and a multiplex cinema

buildings, and hotel witnessed the rebirth of the former mall site as an entertainment center, with an adjacent hotel and upscale housing. Additional upscale apartments in high-rise towers were built by Avalon Bay, which also erected a mid-rise complex in Mamaroneck. In New Rochelle, the high-rise condominiums built by Donald Trump in conjunction with Louis Capelli transformed not only the appearance of the downtown area but that of the region as well. New Rochelle's skyscrapers, like Co-Op City and North Shore Towers erected in the 1970s on the Nassau-Queens border, were visible from miles away. More unobtrusive was the former Bloomingdale's department store building in New Rochelle that was converted to condominiums. Interestingly, this structure occupied the site of the LeRoy House, a nineteenth century hotel owned by the family of the woman Daniel Webster sailed down the Sound to court (NYT, Jan. 28, 1885, 8).

By the early twenty-first century, the shores of Webster's "American Mediterranean" had been radically transformed, not only by suburbanization and urban redevelopment but also by corporate relocation. In Connecticut, the backcountry of Greenwich became home to sprawling corporate campuses. Nearby Stamford attracted the Swiss Bank Corporation, which in the mid-1990s established its North American headquarters on a site where Morgan Stanley had planned to build before backing out. Norden, a division of United Aircraft, established its corporate headquarters in Norwalk and General Electric chose Fairfield for its headquarters. In Bridgeport, a different type of development was occurring. In the 1990s the Shoreline Star, a greyhound racing track, opened. Located not far from the Sound, this recreational venue was expected to funnel considerable revenue into the city's coffers at a time when the Connecticut economy was shaken by cutbacks in defense spending that had resulted in workforce reductions at Sikorsky Aircraft and the Electric Boat Division of the General Dynamics Corporation. Similar reductions were occurring at the Grumman Corporation on Long Island.

Although the greyhound track, like the jai alai fronton that had preceded it in the 1970s, closed, Bridgeport did not give up on gambling. Proposals were sought from Donald Trump and others for a major casino but the concept was stillborn because the Connecticut legislature refused to support it. Undaunted, Bridgeport officials adopted a plan calling for housing, retail, conference, and office space to be built at Steel Point on the Sound early in the new millennium (NYT, May 8, 1998, B8).

Meanwhile, farther east and inland, the Mashantucket Pequots opened a bingo parlor on their tribal land in Ledyard (Eisler 2001). From that small beginning, in 1986, the gigantic Foxwoods Resort Casino evolved. A dozen miles away, the Mohegans opened a casino in 1996 (Eisler 2001). The original facilities at both Foxwoods and Mohegan Sun expanded dramatically. The casinos were not only a factor in the economy of the state but they had an impact on the land they occupied. This was very evident when the Monhegan Sun Hotel opened in 2002, for it was the second tallest building in Connecticut (NYT, June 30, 2002, D2). The casinos impacted the land in another way, as well. Increased traffic on the once quiet roads leading to these gambling meccas meant more runoff. Across the Sound, the North Fork of Long Island also experienced increased traffic as casino-bound motorists headed to the ferry terminal at Orient Point. The expanded parking required to accommodate their vehicles became a major environmental concern on the North Fork. These issues became more pronounced after the Mashantucket Pequots funded the building of a high-speed passenger-only vessel capable of making a 45-minute crossing to New London, where buses waited to take people to the casinos.

New London was more than a transfer station for gamblers, however. In 1995, the city unveiled a plan for Ocean Quest, a project featuring a whaling village, hotel, retail space, and a theater at the mouth of the Thames River on a site previously occupied by a textile factory and, later, a printing plant and linoleum manufacturer. Nothing came of this, but New London persevered in its efforts to redevelop the waterfront. After the Pfizer pharmaceutical company opened a new research center in New London in 2001, the city proceeded with plans to redevelop the adjacent Fort Trumbull area for condominiums and a hotel and conference center. To acquire the property needed for this project, the city used eminent domain. Taking land for such public benefit projects as highways was one thing, but using eminent domain to secure sites for a private project, albeit one that would benefit the public through the additional tax revenue flowing to the municipality, was something else. One homeowner in the neighborhood surrounding the fort, which had been built during the Revolutionary War, decided to fight city hall. The case made it all the way to the US Supreme Court, which ruled in 2005 that state and local governments could transfer private property from its owner to a private developer in anticipation of creating additional revenue or jobs. The Supreme Court, in effect, considered such public uses to be public benefits (Benedict 2009). In the aftermath of this decision, many states either enacted legislation or adopted constitutional amendments aimed at safeguarding private property from the excessive use of eminent domain (Benedict 2009). Such action seemed less urgent following the announcement in November 2009 that Pfizer intended to withdraw from New London less than a decade after its controversial arrival. Leaving behind the city's largest office complex, Pfizer planned to move its 1,400 New London employees to Groton within 2 years (NYT, Nov. 13, 2009, 1). To some, the economic impact of Pfizer's decision was the twenty-first century equivalent of the decline of whaling. While it lasted, however, whaling and other fisheries-based activities had sustained not only New London but also much of the Sound shore region for centuries.

1.4 Commercial Fishing in Long Island Sound

1.4.1 Introduction and Overview

Located at the northern edge of the migratory range of fishes like summer flounder (*Paralichthys dentatus* also commonly called fluke or Northern fluke) and at the southern end of species like American lobster (*Homarus americanus*) and winter flounder (*Pseudopleuronectes americanus*), the Sound's open waters, harbors, and inlets provide shelter and food for a great diversity of marine life. And in turn, its fish and shellfish have long provided a source of sustenance, commerce, and recreation for the peoples living near its shores. Prior to European settlement, members of the Pequot and Quinnipiac tribes raked for the eastern oysters (*Crassostrea virginica*) that flourished within the Sound's estuarine harbors and rivers. Early European colonists used nets to capture the alewives (*Alosa pseudoharengus*) and shad (*Alosa sapidissima*) that migrated each spring up the Connecticut River and caught bluefish (*Pomatomus salatrix*), several species of flounder, and striped bass (*Morone saxitilis*) with hook and line on open waters. In the eighteenth century, farmers on Long Island and the Connecticut shore caught fish for market, for their own consumption, and for use as fertilizer on depleted fields.

By the nineteenth century, thriving fishing economies emerged in the region, and dealers sold fish and shellfish across the country and across the ocean. In western bays, self-titled "baymen" gathered oysters and hard clams (*Mercenaria mercenaria*) for market while capturing with hook and line and small nets migratory fish for sale locally. Fishermen in the 1830s began transplanting very young oysters called "seed" from Virginia and natural beds in Connecticut to cultivated bay bottoms that came to be known as oyster farms or plantations (Ingersoll 1881; Clark 1887). In doing so, they vastly increased the numbers of oysters produced in the region. By the 1850s, LIS oysters found ready market in places as far away as frontier towns in Wyoming and restaurants in London (Kurlansky 2006; Pillsbury 2007). Yet just as the Sound became a place for large-scale commercial fishing, it boomed as a center for recreation. Elite fishermen from the nearby cities of New York and Hartford set off in boats from fishing clubs, while urban industrial workers cast their lines from the shores and piers of Bridgeport and New Haven, seeking leisure as well as sustenance (Pillsbury 2008, 2009).

With so many interests in the waters, then, it is no surprise that many commercial and recreational fishermen became concerned about the future of the fisheries. They feared that a combination of factors ranging from overfishing to pollution and habitat destruction was devastating the region's natural abundance. Fishermen and an emerging cadre of natural scientists began advocating for laws and state and federal fish commissions to protect fish populations. Despite numerous federal and state fisheries investigations, little consensus emerged about how to protect fish populations among different groups of fishermen and scientists. Some fisheries biologists and fishermen advocated limiting netting in certain waters, while others sought to enact fishing seasons that would in theory protect fish when they spawned, but in practice often served to limit the fishing operations of their competitors. By the end of the nineteenth century, these different parties could agree on only one thing: that state and federal agencies could advance the interests of all fishermen by helping to supplement the coastal fisheries by cultivating and releasing fish eggs and fry in the Sound. By the mid-twentieth century, scientists determined that for most salt water fisheries this practice held little effect, but it proved popular among politicians and fishermen who saw it as an easy solution that did not require the imposition of fishing restrictions or the curtailment of habitat destruction (Allard 1978; McEvoy 1986; Stickney 1996; Taylor 1999).

As various fish stocks in the Sound declined during the course of the twentieth century, the question of how to ensure the continued health of fisheries and the fisheries industries remerged again and again. Only after coast-wide fish catches plummeted in the 1970s and 1980s did state and federal authorities finally enact fisheries management plans that actually held the power to effectively reduce the fishing effort (Atlantic States Marine Fisheries Commission 1991). These plans have helped to rebuild many of the region's fish populations to sustainable levels, though never again will LIS be the place of abundance encountered by European settlers.

1.4.2 From Whales to Menhaden

At the turn of the nineteenth century, whales, seals, and cod (*Gadus morhua*) comprised the most lucrative fisheries for those living along the ports of eastern Connecticut and Long Island. This meant that the most valuable fish traded on the Sound came from waters and fishing grounds far offshore. Connecticut fishing vessels joined those from Rhode Island, Massachusetts, and New Hampshire on the Georges Bank fishing for cod and halibut, and sold their catch fresh in New York City. Fishing vessels from New London sailed along the New England coast to capture mackerel, which they sold to Boston dealers, while captains who went farther to collect furs and oils in seas a great distance from the protected waters of LIS made and lost fortunes (Clark 1887). Whaling brought tremendous wealth to the coastal towns around the ELIS (Clark 1887; Jones 1976). In New London, for example, whaling firms employed over 2,500 men in the 1840s, and the industry touched on the lives of nearly every family in the city (Decker 1986).

In the 1860s, however, as global whale stocks dwindled, the region's industrial leaders began looking for new sources of oil closer to home. The development of new fishing and extraction technologies allowed companies to render menhaden— a small, oily fish that schools up and down the Atlantic Coast—into oil for market (Clark 1887). Menhaden oil became cheaper and less risky (for financial investments and human lives) than whale oil and allowed the exponential growth of the industry in the decades that followed (Franklin 2007). The "whitefish," "bony fish," and "mossbunker," as menhaden were called in New York and Connecticut, swam in the LIS from the late spring to the late fall before traveling to offshore wintering grounds. The waters were said to be "alive with them" (Goode 1887). From the Colonial era until the mid-nineteenth century, LIS's farmer-fishermen captured the fish using nets called haul seines. These fishermen would string a long net between two boats just offshore. They would then row the seine into the shore, forming a large "U" shape in the water and encircling schools of menhaden and numerous other species within their net (Goode 1887). They then drew in the net from the beach, gathering up the fish from the shallow waters, thus "hauling the seine" (Franklin 2007).

In the 1860s, fishermen began employing a new kind of net that freed them from the shores. The purse seine extended some 750–1,800 feet in length. When suspended in the water, the purse seine dangled between 75 and 150 vertical feet. Like a haul seine, the purse net was weighted at the bottom, with floats at the top and could form a "flexible wall" of net encircling the fish. Taken out by sailing schooners and later steamships, the nets were put in place by fishermen in two small dories, each laying out one half of the net moving in opposite directions around a school of fish. Once set in a full circle, the bottom edge of the net was then cinched together to form a large bag or "purse" that could entrap entire schools of fish. The fishermen then brought the net to the large fishing vessel to scoop the fish out of the top of the twine enclosure and into the watertight tank of the steamer. With an average mesh size of smaller than 2.5 inches, the seines could capture even the smallest fish (Goode 1879) (Fig. 1.2)

Purse-seining enabled the capture of menhaden as they swam in the open waters and not simply when they came into shore to feed, dramatically increasing the number of menhaden captured. Soon menhaden oil was incorporated into numerous industrial processes ranging from lubrication to tanning and soap making. Firms that had once traveled the high seas in search of sperm whales (*Physeter macrocephalus*) began going after what was sometimes called the "miniature whale" (Goode 1879). One haul of the purse seine frequently captured more oil in the form of the small fish than a multiyear whaling voyage. Within a decade of the purse seine's introduction, fishing companies had established dozens of factories in Connecticut, New York, and Rhode Island for the rendering of menhaden oil and fertilizer. By 1880, 630 Connecticut fishermen and industrial workers landed and processed more than 65 million fish, while New Yorkers caught almost 290 million in a single year. Towns that had been oriented around whaling now became centers for the capture of menhaden. Mystic, once home to 27 whaling ships, now laid claim to oil and guano factories, 14 large menhaden vessels, and

THE MENHADEN FISHERY.

School of menhaden surrounded with purse-seine and the fish striking the net. (Sect. v. vol. I, p. 339.)

From sketch by Capt. B. F. Conklin.

Fig. 1.2 Image of the menhaden fishery—school of menhaden surrounded by purse-seine

36 small boats, upon which 240 men found employment. New York experienced a similar transformation. By 1877, in the former whaling centers of the eastern end of Long Island, an estimated 540 men found employment on more than 60 vessels (Goode 1879).

The industry's rapid expansion had real consequences for the environment around the Sound. Some coastal residents complained of the noxious odors emitted by the factories on shore and at sea. Others protested the oil slicks that sometimes emitted from rendering factories on shore and at sea, while a growing number of recreational anglers, market fishermen, and conservationists feared the damage done by purse-seining to local fish populations. Many fishermen believed that aggressive seining depleted the menhaden stock that served as a critical food supply for game and market species like bluefish, mackerel, striped bass, and weakfish (*Cynoscion regalis*) (US Senate Committee on Fisheries 1884; Franklin 2007).

In response to these conservation concerns, state legislatures enacted laws that prohibited seining in certain waters. Connecticut and New York both limited the use of the purse seines in bays and harbors, and in the western waters of the Sound. Despite concerns raised up and down the seaboard, national scientific studies conducted at the time concluded that fishing in open waters should continue unfettered (US Senate Committee on Fisheries 1884). Despite the scientific community's optimistic predictions, seining alongside unchecked coastal development and pollution contributed to the decline of the Sound's once enormous menhaden populations. As fish stocks dwindled, the fishing effort increased up and down the coast. By the 1930s, the fishery around LIS collapsed, and for the next three decades, landings were variable and small. Factories shuttered and closed. By the 1970s, the only seines operating in the LIS were aboard vessels based in Rhode Island and New Jersey. No longer were the fishermen around the Sound employed in the large-scale capture of the once bountiful fish. Up and down the coast, the situation was similar. By 1980, vessels relied on spotter planes to locate the schools of fish and only eleven factories processed the fish. Today, only 11 vessels and one facility are in operation on the Atlantic Coast, rendering the fish into oil and meal used in animal feed and in countless products from orange juice to vitamin supplements for human consumption (Wise 1975; Atlantic States Marine Fisheries Commission 2007; Franklin 2007).

Fisheries management plans put in place in the early 1980s have helped improve menhaden stocks, but most of the fish today are taken as juveniles in southern waters. As the fish migrate to northern waters as they age, this means that in the twenty-first century a far smaller percentage of the menhaden population reach the LIS than they did in the nineteenth. Some scientists suggest that the loss of so many menhaden has done more than devastate the fishing industry and taken away the food of many larger game and market fishes. They suggest that the destruction of this fish, which naturally filters waters as it swims, has meant the loss of natural protection against eutrophication in the bays and harbors of the Atlantic Coast (Atlantic States Marine Fisheries Commission 2007; Franklin 2007).

1.4.3 Fishing in the Connecticut River

The concern about overfishing in the mid-to-late nineteenth century was not limited to menhaden. By the 1860s, many recreational and commercial fishermen lamented the depletion of the anadromous species such as shad, salmon (*Salmo salar*), and striped bass that migrated up the Connecticut River and numerous smaller rivers within the state. Already they had witnessed the steady depletion of these fish from overfishing, pollution, and dam construction, and they advocated for measures to prevent future losses (Pillsbury 2009).

In the Colonial era, European settlers considered the salmon that spawned in the Housatonic and Connecticut Rivers to be the most desirable of the region's fish species and perceived the more abundant shad a poor substitute for salt pork. Still, settlers from the mouth of the Connecticut River in Saybrook to Bellows Falls, Vermont—where the shad's run ended—relied on the fish for critical sustenance. Slowly shad gained in popularity and by the 1740s dealers in Massachusetts and Connecticut frequently advertised shad alongside salmon. In response, the farmers along the river took up fishing to meet this new market demand (McDonald 1894).

All along the river, fishermen employed simple netting technologies to capture large amounts of the abundant shad and less abundant salmon. Fishermen on the Connecticut River employed a variety of techniques as the fish entered the river in April and May and migrated upstream, including scoop nets with which a lucky fisherman beneath the Hadley Falls in Massachusetts might capture 2,000–3,000 fishes in a single day in the early nineteenth century. Yet at the same time that commercial interests in the fishery increased so too did interests in the river as a site for power production and pollution disposal. In 1798, a dam at Turner Falls, Massachusetts prevented fish from ascending further north. Almost 50 years later, a dam at Holyoke cut off 36 more miles to the fish, including many important spawning grounds. Fishermen all along the river documented declining catches after the dam's construction, raising apprehension among state leaders and politicians. In 1857, leaders in Vermont—a state which no longer saw shad or salmon migrating into its waters—became so concerned about the situation that they commissioned George Perkins Marsh to investigate what could be done to restore to state waters the fish that once provided critical nourishment for poor farmers and active sport for recreational anglers (McDonald 1894; McFarland 1911; Judd 1997; Marsh and Trombulak 2001; Taylor 1999).

The devastation caused by the dams was magnified by a more intensive fishing effort in Connecticut both within the river and along the LIS. In the 1850s, fishermen began using stationary nets called pound nets or trap nets. Pounds consisted of nets weighted to the bay bottom by heavy rings and held upright by stakes set every couple of feet. Once set, the pounds stood still while different kinds of fishes swam along the pound's long leader, into its heart-shaped center and finally into the netted pouch called a bowl where they remained until someone scooped them out. While expensive to construct, the pounds could capture huge numbers. As one fisheries expert at the time remarked, the pounds enabled the capture of "all kinds of fish, both large and small, whether swimming at the surface or at the bottom," with "no skill," day or night, "whether the fisherman be awake or asleep" (True 1887). Near the mouth of the Connecticut River, growing numbers of fishermen in the mid-nineteenth century used pounds and seines to capture huge numbers of shad before they even entered the river (Fig. 1.3).

Fishermen and conservationists throughout New England became so alarmed by the depletion of salmon and shad that they advocated for the establishment of fish commissions to address the problem. Massachusetts established a commission in 1856, New Hampshire and Vermont in 1865, and Connecticut in 1866. The commissioners from each of the New England states convened in 1866 and agreed that if each state met certain conditions the fishery might be rebuilt (United States Commission of Fish and Fisheries 1873). The commissioners decided that Connecticut would cease destructive netting at the mouth of the river, Massachusetts would build fishways over its dams, and Vermont and New Hampshire would stock the river with eggs and fry.

Despite assurances made by the commissioners during the meetings, no state lived up to their promises. In 1868, Massachusetts sued the Holyoke Water Power Company to build a fishway over its dam. Finally completed in 1873, the

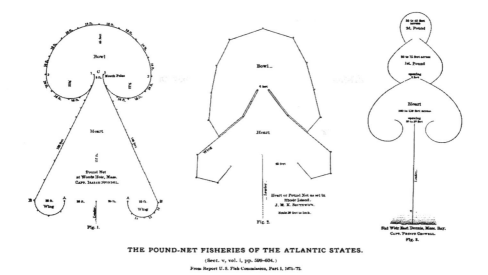

THE POUND-NET FISHERIES OF THE ATLANTIC STATES.
(Sect. v, vol. i, pp. 599–604.)
From Report U. S. Fish Commission, Part 1, 1871–72.

Fig. 1.3 Image of the pound net fisheries of the Atlantic States

fishway never functioned properly and was eventually allowed to deteriorate. New Hampshire and Vermont never provided the fish fry and eggs that they had agreed to supply. In 1871, when the Connecticut Legislature prohibited pound nets all along the coast, commercial fishermen reacted with outrage and successfully worked to have the law repealed within a year (Hudson 1873). Pollution destroyed the shad and salmon populations in the other rivers of Connecticut by the last decades of the nineteenth century. And making matters still worse, in 1885 a dam at Enfield, CT blocked some if not all of the shad migrating northward into Massachusetts along the Connecticut River. As fishermen continued to aggressively capture shad, the situation deteriorated rapidly. In 1879, fishermen captured more than 450,000 shad in Connecticut and Massachusetts, but by 1891 that number had dropped to 20,503 shad. None were captured north of the Connecticut border (United States Commission of Fish and Fisheries 1873; Stevenson 1899; Wise 1975; Judd 1997; Frisman 2002).

During the course of the twentieth century, the shad population on the Connecticut River never again reached the high catches of the mid-nineteenth century. Catches fluctuated dramatically, though they only occasionally reached lows similar to those observed in the early 1890s. In the 1950s, an effective fishway was finally established at the South Hadley dam allowing fish to return to northern spawning grounds. After the Enfield dam fell into disrepair in the 1970s, fishery experts with the Connecticut Department of Environmental Protection (CTDEP) stated that the number of migratory fish north of the dam "shot way up" (Frisman 2002). The removal of these obstacles to spawning coincided with the establishment of federal and state laws to improve water quality. And these improvements, coupled with fishing restrictions, have helped to reverse some of the historic damage done to the population. Today, fisheries experts believe the stock is stable, if not nearly as abundant as it once was. Thousands of anglers capture shad for recreation, while commercial fishermen on the river land between 30,000 and 87,000. Even salmon, once almost entirely gone from the river, have come back thanks to state and federal stocking efforts. Still, it remains difficult to imagine the abundance once described by those living along the banks of the Connecticut River (Wise 1975; Frisman 2002; National Marine Fisheries Service 2009).

1.4.4 Oysters and Clams

More than any other species of fish or shellfish found in the Sound, European settlers prized the oyster. So popular was the shellfish, that early settlers in New York and Connecticut noted depletion of certain oyster stocks as early as the seventeenth century. The Connecticut and New York State Legislatures as well as local town councils passed laws during the course of the eighteenth century to prevent continued depletion by prohibiting the harvest and sale of oysters during the summer spawning season and limiting the numbers of people who could gather oysters from the public commons. Yet, despite these measures, oystermen confronted an ever-diminishing number of oysters produced from the natural beds (Kirby 2004; Kurlansky 2006).

In response, oystermen sought ways to control and improve their harvests. Between the 1810s and the 1830s, LIS fish dealers began to import oysters from Virginia and New Jersey to sell to market. Following this practice, oystermen began to buy small oysters and then plant them in protected waters, "seeding" underwater plots of ground. Over the course of the spring and summer, the oysters could gain a fourth to a third in size, making them fat and ready for market in the fall. In the years that followed, oystermen in Connecticut and New York began transplanting young oysters from nearby natural beds in the Quinnipiac, the East River, and the Housatonic. Oyster planting became common practice by the mid-nineteenth century, and the majority of oysters brought to market from the waters of the LIS came from cultivated oyster beds rather than wild or natural oyster bars. By 1889, more than 1.4 million bushels of oysters were harvested from Connecticut's artificially planted grounds, while the natural beds yielded only 73,850 bushels (Ingersoll 1881; Collins 1891; Sweet 1941; McDonald 1894; Kochiss 1974; Kirby 2004).

As oyster planters demanded secure rights to their harvest, towns began selling and permitting private underwater oyster lots within their jurisdiction. The states followed suit, granting coastal towns the right to permit oyster lots in their waters and enacting protections for the natural beds. By the last decades of the nineteenth century, both states placed commissions in charge of shellfish regulation and developed state leasing and franchise schemes in the deeper waters of the Sound governed by the states (Ingersoll 1881; Collins 1891; Sweet 1941).

The notion that individuals could privately "own" lands beneath the common waters became hotly contested. It caused particular ire among independent baymen who plucked oysters and clams from natural beds and caught the many different species of fish that inhabit the LIS. The men who worked the natural beds protested vehemently when these areas were deemed no longer productive and sold to private individuals. As the private areas grew and the natural beds became less abundant and smaller in legally recognized size, those who had long worked the natural beds were squeezed between their ecological decline and the privatization of the bottom. It is perhaps then no surprise that they frequently responded not only through legal protests but also through surreptitious and illegal means, stealing oysters from large and productive private beds and occasionally resorting to violence. At the same time, those "natural growthers" who sold seed oysters to the planters raked and tonged the natural beds intensively, often in direct violation of laws meant to protect the beds. As one fisheries expert at the time observed, "the natural-growth beds have been almost incessantly and unscrupulously drawn upon for their products.... The private beds have thus been increased at the expense of the natural grounds" (Ingersoll 1881; Collins 1891; McCay 1998).

Unfortunately, overharvesting was but one of the threats against natural beds located in urban areas like New Haven and New York City. Oystermen confronted beds covered in garbage, mud, sewage, oil, and other refuse, which destroyed and contaminated the oysters below. By the 1920s, Americans began to perceive the mollusks as conveyors of diseases like typhoid. Thus, the oystermen of LIS confronted a dual problem during the course of the twentieth century: not only were they able to produce ever fewer oysters, but those oysters that were harvested consumers did not want to purchase. In 1910s and 1920s, Connecticut's natural beds produced fewer and fewer seed oysters or "sets," and in some years produced none at all. This destroyed not only the once booming seed industries of New Haven and Bridgeport, but also much of the planting industry as well, which no longer found Chesapeake oystermen willing to sell valuable seed oysters to northern planters (McDonald 1894; Sweet 1941; Wise 1975; Pillsbury 2007).

Oyster harvests plummeted. Between 1880 and 1910, the peak of the oyster planting industry in Connecticut, production averaged 10 million pounds of meats harvested. By the 1930s, only about five million pounds of meats were landed in Connecticut. Thirty years later, that figure had dropped to 133,000 pounds of meats.

Today, oysters and hard clams comprise the most valuable species harvested in the waters of LIS. Yet, rather than coming from the once abundant natural beds of New Haven Harbor and the bays of the western Long Island Sound (WLIS), the vast majority of these shellfish are carefully cultivated and groomed as crops before being sold to market. Federal and state authorities worked with oyster planters to better control predators, develop crops genetically resistant to disease, and add oyster shells, or cultch, to bay bottoms, all of which helped to improve harvests during the last decades of the twentieth century (Blackford 1885; Wise 1975; National Marine Fisheries Service 2009). These shellfish operations reflect a tradition of aquaculture begun in the Sound nearly 200 years ago. They also reveal how much has changed within the LIS ecosystem. No longer able to depend exclusively upon seed from natural oyster beds, shellfishermen carefully propagate disease-resistant shellfish to plant on vigilantly tended grounds. Even the recreational grounds from which thousands of coastal residents pluck shellfish for fun reflect this transition, as town officials purchase and plant each year oysters and clams upon the "public grounds." Ultimately, like the loss of the massive schools of menhaden in the Sound, many scientists wonder whether the disappearance of natural beds of shellfish contributes to the development of hypoxic conditions within the Sound each summer.

1.4.5 Lobster and Flounder: Market Fishing in Long Island Sound

Oysters, shad, and menhaden may have comprised the most valuable fisheries in the LIS, but by the mid-nineteenth century fishermen used a variety of shore-based nets, lobster and eel pots, and hooks and lines to pluck numerous species from local waters for sale in the markets of New London, New York City, and Hartford. In the eastern ports of Connecticut, many fishermen earned their living by fishing locally for market species. In 1880 in Mystic, CT, for example, 4 small smack vessels, a number of small boats, 3 haul seines, and about 45 fyke nets were used to capture black sea bass (*Centropristes striata*), cod, bluefish, and other species. Fishermen in larger vessels cruised from Montauk to Block Island to capture fish using hooks and lines. Fishermen with small boats in New London set lobster traps in Fishers Island Sound, while those with larger vessels traveled as far away as Block Island Sound. Fishermen from New London captured some "150,000 pounds of flounders, eels (*Anguilla rostrata*), tautog (*Tautoga onitis* also known as blackfish), smelts (probably mostly *Osmerus mordax*) and other species, and about 30,000 pounds of lobsters" in LIS in 1880 (Clark 1887).

Baymen in WLIS captured a variety of fish for sale in local markets, but few relied on the finfishes for their living. With oysters so easily cultivated and natural clams so easily gathered, migratory finfishes proved a far less desirable target. And it is not surprising that according to the United States Fish Commission's 1880 survey of the fisheries and fishery industries, there were "no important general fishing stations in Connecticut west of New Haven" (Clark 1887). In South Norwalk, the local supply of fish came from the markets in New York City, and there was no fishing there "except for sport." New Rochelle, NY was declared to offer sport to anglers but nothing from a "commercial point of view" (Clark 1887). And in Oyster Bay, the local baymen took approximately 6,000 bushels of hard clams, 75,000 bushels of oysters, 50,000 bushels of soft clams (*Mya arenaria*), $1,000 worth of menhaden, and a number of finfishes so small they were not deemed significant enough by the United States Fish Commission to record (Mather 1887).

During the course of the twentieth century, the fishermen in the Sound began to employ ever more efficient methods of capturing finfishes, bringing record catches and profits for species once considered of negligible economic importance. In 1905

and 1906, fishermen began using internal combustion engines, liberating boats from
the prevailing winds and allowing trawlers to operate more consistently. By the
1920s, the adoption of the otter trawl within the Sound enabled fishermen to capture
a wide variety of fishes, including winter flounder and scup (*Stenotomus chrysops*),
with far more ease than ever before possible. The otter trawl, like the purse seine
before it, dramatically increased the amount of fish captured in the Sound. Whereas
in 1890, Connecticut fishermen captured a total of 101,000 pounds of scup, they harvested more than 1.5 million pounds in 1950. In 1890, Connecticut fishermen took
443,000 pounds of different kinds of flounders, while from 1951 to 1971, the catch
of winter flounder alone ranged between 780,000 lbs to 1.7 million pounds (McHugh
and Conover 1986; Wise 1975; National Marine Fisheries Service 2009) (Fig. 1.4).

Since the moment of their introduction in local waters, many recreational fishermen and conservationists protested the use of the trawlers, which they feared
contributed to the overfishing of some species and damaged the benthic environment by scraping ocean bottoms. Their fears had merit. The rising catches enabled
by the otter trawls did not reflect the improving health of the fish populations, and
in fact indicated just the opposite. Fishermen in the mid-twentieth century captured ever greater numbers from smaller and smaller fish populations. In the 1970s
and 1980s, catches of winter flounder, scup, and many more species in southern
New England and the Northern Middle Atlantic began to decline. And state and
federal authorities began implementing fisheries management plans to rebuild
the stocks, in part by limiting the fishing effort (Wise 1975; Matthiessen 1986;
McHugh and Conover 1986; National Marine Fisheries Service 2009).

Unfortunately, for many of the remaining commercial fishermen of LIS, these
management plans represented not a new opportunity for future business growth
and stability but rather an additional blow to a struggling industry. Seasonal and
netting restrictions coupled with daily and seasonal limits for the commercial fisheries meant that it became ever more difficult to make a profit. This had a very
real effect for once vibrant fishing communities along LIS. By the 1990s, the
Stonington fishing fleet remained the only significant one in Connecticut with 18
draggers and 12 lobster boats (Matthiessen 1986; NYT, Dec. 4, 1994, 1).

For fishermen struggling to make ends meet during the 1970s and 1980s, lobstering provided the one bright spot in the troubled industry. Lobsters, which had

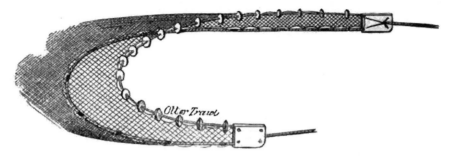

Fig. 1.4 Image of the otter trawl

once been considered so inferior that they were given to servants in Massachusetts and used to fertilize fields in Maine, gained in popularity during the eighteenth century and eventually formed the basis of a lucrative fishery. Yet, like so many other fisheries in the late nineteenth century, overfishing contributed to a decline in the lobster population, raising concerns that ultimately led to some measures that helped protect the valuable shellfish (Clark 1887; McFarland 1911).

In 1880, Connecticut fishermen captured more than 700,000 pounds of lobsters in pots, largely located within the waters of the Sound. Yet, observers of the fishery noted, that without protective laws like those in Massachusetts and New York that set a minimum length for lobsters sold, Connecticut fishermen did not make nearly as much as they might have otherwise. While lobster catches increased during the 1880s, by the 1890s catches declined and certain grounds were believed to be "totally exhausted from over-fishing." At the same time, the lobster gained in popularity as a luxury food item, dramatically increasing in price between 1880 and 1905. Connecticut passed laws to limit the lobster harvest and, along with the United States Fish Commission, released hundreds of thousands of lobster fry into local waters in order to help increase the stock (Clark 1887; McFarland 1911).

The Connecticut lobster catch continued to decline during the first part of the twentieth century. In 1956, it reached a historic low of 183,000 pounds. Yet, during the 1960s, the population naturally increased, encouraging ever more part- and full-time fishermen in Connecticut and New York to set pots in the waters of the Sound. Whereas in 1956, New York and Connecticut licensed 12,000 lobster pots in the LIS, in 1974 they licensed more than 40,000. For many part- and full-time fishermen on LIS in the 1970s and 1980s, lobstering provided a critical base of income, supplementing catches of finfish and collection of other shellfish. Lobster harvests peaked in 1996–1997 with more than 11 million pounds harvested annually. This meant that fishermen on the Sound were particularly vulnerable to fluctuations in the lobster stock. During the summer of 1999, lobsters in the western Sound suddenly began dying. While attributed to a wide range of factors from the spraying of pesticides to hypoxia to sudden temperature (T) changes at the bottom of the Sound, the lobster die-off meant a sudden halt to a still thriving lobstering industry. Some fishermen and scientists fear that as warming trends in the Sound continue, the lobster fishery may never fully recover. With lobsters so critical to their income, fishermen worry about the future of the Sound's fishing industry without a viable lobster fishery (Wise 1975; Wilson and Swanson 2005; National Marine Fisheries Service 2009).

1.5 Recreational Usage of Long Island Sound

1.5.1 Taking to the Water

Sharing Daniel Webster's "American Mediterranean" with commercial fishermen were increased numbers of recreational anglers. In the second half of the twentieth century, fishing became much more than a casual interval of surf casting or motoring a short distance off shore to cast a line. Residents of Sound shore communities continue to do both and thanks to cell phones, by the early twenty-first century, it was not uncommon to see locals fishing from the beach while using their phones to alert fellow anglers to the presence or absence of particular species. Accurate information about what was running was important to visitors as well because fishing is a component of tourism in Sound shore communities. Given the significance of recreational fishing in the tourism component of the Sound area's economy, maintaining the health of the waterway is vitally important. There have been times when either health concerns or fish die-offs have posed a genuine threat to fishing. Whether PCBs in bluefish or hypoxic events resulting in the absence of fish in portions of the Sound, the end result is that tourists and their dollars stay away. Similarly, beach closures have an economic impact. Whether the result of sewage pollution or oil spills, beaches have been periodically placed off limits. Fortunately, beach closures in prime tourist areas on both sides of the eastern Sound have occurred infrequently, and then only briefly. But as the twenty-first century progresses, one wonders whether subtle changes in the waterway's ecology may make a day at the beach less enjoyable. The proliferation of jellyfish in the eastern Sound in 2008, for example, may have kept away some of the locals as well as tourists from afar, and a recurrence may cause visitors to spend their vacation dollars elsewhere. On the other hand, a faltering economy led to more "staycations" for Sound area residents.

1.5.2 Recreational Fishing in Long Island Sound

Widespread sport fishing in LIS first boomed in the 1830s and 1840s as urban residents sought rural escapes and formed fishing and hunting clubs on nearby waters. On Long Island, most of these clubs were located on the south shore along Jamaica Bay and later the Great South Bay. In Connecticut and Westchester County, a number of hotels and clubs opened on the north shore of the Sound, including in the 1860s Boss Tweed's Americus Club on Greenwich's Round Island. In the East River near New York City, residents could hire yachts or rowboats from City Island to troll salt waters for sea bass, flounder, or bluefish. As a fishing guide from the 1840s described, an angler could hire a boat that would be "gently rowed along by a skillful oarsman, who rests his oars the moment a fish is struck, giving the angler full opportunity to play his fish with skill and care..." A fishermen trolling in this fashion might take a striped bass of "20, 30, and even 40 pounds" in the East River, "in the neighborhood of Hellgate" (Brown & Co 1846; Weigold 2004).

By 1848, the New Haven Railroad connected New York City to New Haven, while the LIRR began connecting people to places east in Brooklyn, Queens and later Suffolk County. Railroads helped encourage the development of suburban communities in Westchester County as New Yorkers sought summer retreats from the heat and pollution of the city. They also helped fuel the development of sport fishing on the Sound. By the 1870s, a well-to-do New Yorker in search of sport could, as one fisherman observed, "generally get good fun" off New London Harbor and take the Shore Line Railroad home "three times a day." Hotels and boarding houses along the Connecticut shore advertised their proximity to fishing grounds and the knowledge of their fishing guides (Hotchkiss 1878; Weigold 2004). The sport grew during the early twentieth century, booming in popularity during the Great Depression as people sought a form of escape as well as an alternative source of sustenance. By 1937, more than 120,000 people fished out of the LIRR's docks and piers at Fort Pond Bay and Montauk, NY. And many more uncounted sportsmen hired boats elsewhere, took out their own private boats, or cast their rods on public shores and piers (New York State Department of Conservation 1938; Judd 1997; Pillsbury 2009).

Despite the long existing tensions between the two groups, many commercial fishermen sought greater profits by catering to the new hobbyists by serving as fishing guides. In the 1930s, railroads began advertising "fishermen's specials" to bring residents from New York and other cities to the coast for fishing thrills. Party boats offered flat and affordable rates to working and middle class people who wanted a day of fishing, while charter boats were more expensive but allowed a more private adventure for special occasions. In Flushing and Whitestone, Queens and on City Island, party boats departed to troll the waters of the western Sound. Fleets of party and charter boats departed from Montauk and New London, bringing city residents to fish the waters of the Atlantic Ocean and ELIS for flounder, bluefish, and numerous other species (Long Island Railroad 1935; New York State Department of Conservation 1938).

Following the Second World War, these numbers grew even more. As one headline in the newly created *Saltwater Sportsman* magazine declared, "A Million Fishermen, Maybe More" resided in New York City. Armed with disposable income and greater leisure time than ever before, the residents of the growing suburbs of Connecticut and Long Island sought the Sound for sport. Their numbers skyrocketed, and while many fished freshwater rivers or lakes, others took to the abundant salt waters of the coast. By 1988, three-quarters of a million people were estimated to have fished in the Sound and more than 200,000 owned boats in the region. And in 1992, an economist commissioned by the Long Island Sound Study estimated that recreational fishermen on LIS generated more than 22 million dollars within the local economy each year (Evanoff undated; Altobello 1992; Weigold 2004).

Considering the vast numbers involved in the sport, it is no surprise that for certain species like winter and summer flounder and striped bass the sport catch rivaled and sometimes surpassed the commercial harvest. Thus, while sportsmen had once advocated limiting commercial practices, by the 1960s some sporting organizations advocated for the adoption of catch-and-release practices among fishermen not intending to consume their catch. While this helped instill a conservation ethic among many sportsmen, it was ultimately not enough. When fish stocks plummeted in the 1970s and 1980s, states implemented restrictions on the recreational catch as well as that of commercial operators, restrictions that helped some local stocks rebound.

1.5.3 Yachts and Excursion Steamers

Jellyfish notwithstanding, besides swimming in the Sound, tourists can opt to glide across the water in a sailboat or powerboat. Human beings have taken to the water in this fashion for hundreds of years. Whether native peoples in their canoes, colonists in their sailboats, or tycoons aboard their steam yachts, people have been navigating the waters of LIS for a very long time. Recreational boating, however, really emerged in the nineteenth century and along with it came yacht clubs. In 1871, the Seawanhaka Corinthian Yacht Club was established in Oyster Bay (Cooper 1977). On the opposite shore, the Larchmont Yacht Club was founded in 1880; although the club was a mecca for sailboats, one steamboat docked at the club on a regular basis. It was the *CRYSTAL STREAM,* which transported club members and their guests from lower Manhattan to Larchmont. The trip on this club-owned vessel took two hours. A short distance up the coast in Rye, the American Yacht Club was established in 1887 as the first club on the Sound devoted to steam yachts. Four years later, the Stamford Yacht Club was formed (Cooper 1977). In 1905, Port Washington had its own yacht club (Cooper 1977). For those who could not afford yachts, there were excursion steamboats that transported the masses to various destinations on both sides of LIS including Columbia Grove on Lloyd Neck and Locust Point in Huntington. One of the most popular spots was Glen Island. Located on several islands off New Rochelle, this grand amusement center was developed by John Starin, a former US Congressman, and owner of the Starin Line of steamboats that carried passengers and freight between

Fig. 1.5 Entrance to Westchester County's Playland Park on LIS, Rye, NY

New York and Connecticut. During the summer months, beginning in 1879, Starin's boats transported tens of thousands of daytrippers to Glen Island (Weigold 1984). Once there, they could enjoy a picnic, dine at a restaurant, visit a zoo, aviary and museum, swim, take boat rides on small craft, and in the evening view a fireworks display. In contrast with Glen Island, which was noted for its cleanliness and Disney-style family atmosphere, Paradise Park and Pleasure Park in Rye were less wholesome. In the 1920s, the Westchester County Park Commission acquired these parks, demolishing their rides and other attractions, and in 1928 opened the Playland Amusement Park. The 1920s was also the decade when the commission acquired Glen Island, filled in the areas separating the different islands, creating one large landmass, and built a bridge from the new Glen Island Park to the mainland. With the exception of the stone castle that had been the focal point of Starin's "Little Germany," an area developed to attract New York City's large German population, manmade structures were demolished and natural features were highlighted. So popular were the beach and picnic facilities at Glen Island that Westchester County imposed a residency requirement after World War II; until then non-residents, most of whom came from the nearby Bronx, had been admitted although they were charged a higher parking fee. Playland, however, which even today lacks a residency requirement, welcomed everyone (Fig. 1.5).

1.5.4 Parks: Connecticut and Long Island

Non-residents are also welcome at Connecticut state parks on the Sound: Sherwood Island in Westport, Hammonasset in Madison, and Rocky Neck in Niantic. Most municipal beaches, however, are for the exclusive use of residents. This policy was challenged in a lawsuit brought by a Stamford resident who had been denied access to a Greenwich town beach. The suit contended that Connecticut law required municipalities to hold beaches in trust for all residents of the state (NYT, Aug. 28, 2001, B6). The Connecticut Supreme Court upheld this interpretation but Greenwich discouraged outsiders by imposing a high user fee on non-residents seeking access to its beach. Pricing differentiation was employed by other municipalities as well as a way of banning outsiders. But on Long Island, as in Connecticut, state parks, including Sunken Meadow in Kings Park, Wildwood in Wading River, and Caumsett, the former Marshall Field estate in Lloyd Harbor, all of which are on the Sound, are open to everyone. The partially developed Caumsett State Park was purchased by the state in 1961. Four decades later, in 2005, New York State established the Jamesport State Park on Sound-front property that the Long Island Lighting Company (LILCO) had intended to use for a nuclear power plant (NYT, May 29, 2005, N9). Preservation of this open space located on the eastern fringe of Riverhead was vital because large sections of Riverhead experienced unprecedented development in the late twentieth and early twenty-first centuries. Condominiums had sprouted slightly to the west of the Jamesport State Park and developments of single-family homes sprang up on former farms in the scenic Sound Avenue corridor where the park is located. At the same time, Route 58, with its office buildings and shopping centers, including the gigantic Tanger Outlet Center, began to resemble Route 110 in western Suffolk County. The acquisition, in 1999, by the Town of Riverhead of 2,900 acres of the expansive US Navy property in Calverton where Grumman had built military aircraft meant further development (NYT, Feb. 7, 1999, LI17). Although a proposal to incorporate the existing Grumman runway into a jetport was rejected, a business incubator was established and plans were formulated for a mega-resort that would include an indoor ski slope.

1.5.5 Second Homes

The global economic crisis of 2008 and resulting recession had the effect of delaying, if not totally scrapping, ambitious development plans practically everywhere, including Riverhead. The sub-prime mortgage problem, which had sparked the crisis, led to a decline in real estate values, even in the highly desirable LIS area. Home values had increased substantially in the early and middle years of the first decade of the twenty-first century, enabling some people to parlay their existing residences into waterfront dream homes, often with the help of an adjustable rate mortgage. When the interest rates on those mortgages reset, some homeowners were in deep financial trouble. Short sales of properties for amounts less than the remaining balance on the mortgage proved to be the solution for some homeowners; at the same time, this method of disposing of real estate represented an opportunity for buyers. In the new recession era economic climate, prudent buyers took into account not only their ability to afford the purchase price and the real estate taxes but the cost of insuring a waterfront home. With some companies refusing to write new policies and even renew existing ones, homeowners found themselves in the position of having to choose from among fewer insurers and paying high premiums.

In theory, enhanced building codes for both new construction and renovations in Sound shore municipalities meant that there would be less damage in the event of northeasters and hurricanes. Still, a devastating storm like the category three hurricane of 1938, which wrecked havoc on the shore and for some distance inland in eastern CT, would have a huge impact, as would flooding in Connecticut's river valleys. In 1955, floods swept gasoline, vehicles, and entire buildings into Connecticut's rivers and eventually down to LIS. The potential for floods has caused owners of waterfront property on LIS or the rivers flowing into it to opt for flood insurance. National Flood Insurance became available in 1968 as a result of Congressional legislation and it has had the effect of encouraging development in areas that potentially could become inundated. Some communities, however, have prohibited building within the coastal erosion hazard zone. Southold, for example, allows improvements to be made to existing homes but prohibits the erection of new structures in the zone. In 2009, the new owner of a Sound-front lot, for which he had paid $1.1 million, sued the town because he had been denied a building

Fig. 1.6 Bostonian train in Stonington: hurricane of 1938

permit. However, environmentalists and some town board members held fast to their belief that allowing an exception in this case would lead to other exceptions that, taken together, would have a significant negative environmental impact (Suffolk Times 2009) (Fig. 1.6).

1.6 Managing and Enhancing Long Island Sound for Future Generations

1.6.1 Power for the People

On both sides of LIS, increased development, not only in the pre-recession boom years of the first decade of the twenty-first century but during the final two decades of the twentieth century, had far-reaching implications. More people meant more traffic, more sewage and garbage, more air pollution and increased use of fuel to generate the electricity required by a growing population. Going back to the mid-twentieth century, the utilities that served the region viewed nuclear power as the solution to the area's increasing energy requirements. Nuclear power was not something that was foreign and exotic in these parts. Since the 1950s, when the first nuclear powered submarine was christened in Groton, nuclear subs have been traversing the waters of the Thames River and LIS. These vessels are modern day reminders that warfare has periodically disrupted the placid lives of Sound area residents. During the Revolutionary War and the War of 1812, British vessels plied these waters, sometimes depositing soldiers in Sound shore communities. With Long Island occupied by the British from 1776 until the end of the Revolution, many people fled across the Sound to Connecticut. To the relief of people living on both sides of the waterway, the Sound region was not severely impacted by the War of 1812. The world wars of the twentieth century necessitated military patrols of the Sound and the Cold War required an enhanced naval capability, hence the submarine base at Groton. Nuclear powered subs coming and going in the eastern Sound were one thing, but having nuclear power plants on both sides of the waterway was quite another. In the 1960s, when LILCO announced plans for a nuclear power plant in Lloyd Harbor, the Lloyd Harbor Study Group was established to fight the proposal (Aron 1997). On the mainland, the Citizens League for Education about Nuclear Energy was formed to oppose Consolidated Edison's plan to build four nuclear reactors on David's Island, a former US Army base just off New Rochelle, which the utility purchased in the 1970s. Both the David's Island and Lloyd Harbor plans were dropped. The organization founded to oppose the latter was active in the unsuccessful struggle to prevent the construction of a nuclear power plant farther east along the north shore in Shoreham.

Work on the Shoreham plant commenced in 1973 simultaneous with the development of Suffolk County's emergency evacuation plans, which were to be implemented in the event of a major malfunction at the plant. Six years later when the Three Mile Island nuclear power plant in Pennsylvania overheated, the Nuclear Regulatory Commission imposed strict, new requirements for evacuation plans, yet plans were not in place in 1980 when low-level testing began at Shoreham. The previous year, Shoreham was the site of what was believed to be the world's largest protest against nuclear power. The demonstration, which took place on International Anti-Nuclear Day, attracted 15,000 people, 600 of whom were arrested for trespassing on plant property (NYT, June 4, 1979, A1). A principal concern of the protesters was evacuation in the event of a nuclear emergency. In 1982, Suffolk County held public hearings on the draft of an evacuation plan but county legislators concluded that evacuation was not feasible (McCallion 1995). Closing Shoreham was the only option. This task fell to the Long Island Power Authority (LIPA), a public utility agency established in the mid-1980s with the idea of providing electricity to the Island's homes and businesses at rates lower than LILCO was charging, thereby increasing Long Island's economic competitiveness (NYT, July 12, 1986, 31). In 1990, LIPA sought Nuclear Regulatory Commission approval to take over Shoreham and decommission the plant (NYT, May 25, 1990, B2). Since it derived most of its tax revenue from the plant, the Shoreham-Wading River School district initiated legal action against LIPA but the US Supreme Court rejected the motion, thereby allowing the decommissioning to move ahead. This process was extremely expensive because LILCO's low-level testing in the 1980s had resulted in radioactive contamination.

In the end, it took a thousand workers 2 years to complete the $200 million decommissioning of the plant. While the work of decommissioning the facility proceeded, LILCO attempted to recover hundreds of millions of dollars it had paid in property taxes on the grounds that the plant's assessment had been too high. A state court agreed and LILCO recovered part of what it had paid through the years (NYT March 20, 1999, B5). In 1998, in an effort to reduce the cost of energy and enhance Long Island's economic competitiveness, LIPA acquired LILCO's transmission lines while LILCO's power plants were sold to private companies (NYT, Feb. 25, 1998, B4). To ensure an adequate supply of electricity for the Island's growing population, LIPA erected several small conventional power plants and, as back-up, the Trans-Energie underwater cable built by Hydro-Quebec was laid between Connecticut and Shoreham in 2002 (NYT, Jan. 13, 2002, LI2).

The value of a cross-Sound cable had been demonstrated in 1996 when power from Long Island was transmitted through an earlier pipeline to supplement Connecticut's supply during a period when all three of the reactors at the Millstone Nuclear Power Plant had been taken off line because of safety considerations. Interestingly, when Northeast Utilities began building the Millstone facility in Waterford (near New London) in 1966, there was a noticeable absence of opposition. In time, some residents expressed concern when a third reactor was constructed but by and large the tax revenue pouring into the municipality from Northeast Utilities was viewed as a fair trade-off. By the early twenty-first century, however, opposition to the Millstone plants surfaced on eastern Long Island. Given its proximity to coastal Connecticut and the impossibility of evacuation in the event of a serious malfunction at the Waterford facility, the East End would be a dead end if a radioactive plume drifted across the Sound (NYT, March 2, 1999, B7).

Safety concerns notwithstanding, aging nuclear power plants are permitted to remain online because of the demand for the energy they produce. As the twenty-first century progresses, it is conceivable that more such plants will be constructed along the Sound. Of course there will be mass demonstrations aimed at preventing this from happening, especially in view of leakage from Japanese nuclear plants in the wake of the 2011 tsunami, just as there was considerable opposition to the plan to build a liquid natural gas terminal in LIS. Proposed by Broadwater Energy, a joint venture of TransCanada Corporation and Shell Oil, the seven-story-high floating terminal, with a length equivalent to three football fields, would have served as a processing facility where the liquefied product, brought in by ship, would have been turned into gas and piped to New York and Connecticut. Opposition by environmentalists, fishermen, and residents of the LIS area caused New York State to reject Broadwater Energy's proposal in 2008 (Environmental News Service, April 11, 2008). In April 2009, the US Commerce Department announced that it would not grant federal permits for the project (NYT, April 16, 2009, 28).

1.6.2 Dredging and Dumping

Even before the Broadwater decision, there was speculation about alternative sources of energy, including wind turbines proposed for the waters off Plum Island. Whether this or as yet unheard of methods of providing power to the LIS region, it would seem that nuclear power plants will continue to provide some of the area's energy, despite decades-old concerns about this method of generating electricity. Back in the 1970s, an in-depth study of the Sound by the New England River Basins Commission (NERBC) (NERBC 1975), a federal agency, noted that thermal pollution from both conventional and nuclear power plants was one of six major ecological threats to LIS. The other challenges were radioactive pollution from nuclear power plants, oil leaks and spills, sewage pollution, destruction of wetlands, and dredging and dumping. In one sense, all of these challenges related to increased population. More people needed more power, and the vessels delivering oil to electrical generating facilities or to home heating oil tank farms had occasional accidents. In 1969 and again in 1972, tankers hit Barlett's Reef in the eastern Sound. The 1972 spill covered nearly 50 square miles and oil washed ashore at Niantic where it coated waterfowl. To protect shellfish beds in the Niantic River, the State of Connecticut used containment booms to seal off the river from the Sound (NYT, March 23, 1972, 53). That same year, a brand new oil tanker split in two in Port Jefferson Harbor. Fortunately the oil had already been unloaded, but the accident made people wonder whether a proposed dredging project to deepen the harbor to permit larger tankers to access the LILCO power plant and the Consolidated Petroleum Company should go forward (NYT, Jan. 11, 1972, 1). In the last four decades, dredging has become controversial. Yet, there is an ongoing need to dredge approaches to power plants, harbors, and marinas accommodating pleasure craft whose numbers have continued to grow, recessions and fuel price spikes notwithstanding. The permitting process for these jobs is often lengthy and arduous because a host of potential environmental impacts must be taken into account, among them spawning and migration of different species of fish and nesting habits of shorebirds. The end result is increasingly narrow windows for dredging. An occasional frigid winter resulting in frozen inlets can further reduce the permissible time for dredging. Then there is the challenge of what to do with the dredged material.

When Thames River dredging was done in the mid-1990s to enable *SEAWOLF* submarines to access their base in Groton, the dredged material was dumped in LIS. Since the Sound is the only non-ocean governed by the 1972 Marine Protection, Research and Sanctuaries Act's strict testing regulations, New York State initiated legal action to halt the dumping because of the possible effect upon lobsters from toxins in the dredged material. Although the dumping was allowed to continue, stricter adherence to the regulations was required. This did not end the battle over dumping, however. For the next decade, Connecticut continued to assert that from an economic standpoint disposing of dredged materials in the Sound was the only way to keep the state's harbors open to commerce. All the while New York insisted that alternatives had to be sought. In 2007, an agreement brokered by the federal government made 2014 the deadline for implementing a new dredged material management plan. Without a satisfactory plan, dumping of dredged material in the Sound will cease that year (NYT, Jan. 13, 2008, LI14).

1.6.3 The LIS Studies

Dredging and dumping, as well as the other issues highlighted in the NERBC study of the 1970s, were revisited when the federal government and the states of

New York and Connecticut undertook a comprehensive study of the waterway as part of the National Estuary Program created by the Clean Water Act Amendments of 1987. Working with the Environmental Protection Agency, the states bordering the Sound sought to develop a plan for protecting water quality, fishing, and habitats. The resulting study concentrated on six issues: the impact of water quality on finfish and shellfish, hypoxia, toxic contamination, pathogens, land use, and floatable debris. In order to improve water quality and reduce hypoxic events that had caused fish to practically vanish from an area extending from the Throgs Neck Bridge to Greenwich in the summer of 1987, the study recommended reducing the nutrients being discharged into the Sound by sewage treatment plants (Long Island Sound Study 1994). Concerning toxic contaminants, whether toxic metals migrating to the Sound from the Thames, Connecticut, Quinnipiac, and Housatonic Rivers, or metals discharged into harbors from sewage treatment plants, landfills, and other sources, the study called for the monitoring and cleanup of harbors (LIS Study 1994). To deal with pathogens, the study recommended bacterial monitoring of harbors, abatement of combined sewers, elimination of illegal sewer connections, and designation of some parts of the Sound as no-discharge zones for waste from boats (Long Island Sound Study 1994). On the subject of land use, the study called for more education for local land use officials and open space preservation (Long Island Sound Study 1994). To counteract floatable debris, the study recommended combined sewer overflow abatement, stenciling of storm drains, beach cleanups, and programs to educate the public (Long Island Sound Study 1994). In the LIS area, such organizations as Save the Sound, SoundWaters, The Soundkeeper, and Clean Sound, Inc. have played an important educational role, as has Sea Grant, a university affiliated program. Established in 1966 by an act of Congress, Sea Grant became part of the National Oceanic and Atmospheric Administration (NOAA) in 1970.

Even with the help of organizations eager to educate the general public about LIS, and the presence on the shores of the Sound of the US Merchant Marine Academy in Kings Point, the Coast Guard Academy in New London, and SUNY's Maritime College in the Bronx, institutions renowned for their specialized maritime programs, attaining all of the goals set forth in the LIS plan would be expensive. In 1994, when New York, Connecticut, and the Environmental Protection Agency signed a formal agreement to rehabilitate LIS, it was estimated that as much as $700 million per year for 20 years would be needed for the New York part of the plan and a little over $100 million a year for the same period would be required for Connecticut's portion (NYT, Sept. 27, 1994, B7). In 1998, New York voters approved an environmental bond act providing $200 million for the Sound (NYT, Nov. 4, 1998, B12). That same year New York, Connecticut, and the federal government agreed to a $650 million 15-year plan to reduce by 60 % the amount of N going into the Sound and, within a decade, to restore at least 2,000 acres and a hundred miles of shoreline (NYT, Feb. 7, 1998, B4). The aforementioned Long Island Sound Restoration Act, passed in 2000, was another positive step but the annual appropriation fell far short of the projected amount in the early 2000s. Federal spending aimed at stimulating the economy following the global financial crisis of 2008, however, did provide assistance to regional wastewater infrastructure and Sound restoration efforts.

While financial strains on federal, state, and local budgets continue, solutions to the LIS region's twenty-first century challenges are in reach, and before the first century of new millennium ends, Daniel Webster's beloved "American Mediterranean" may once again be a clean, robust, and economically viable urban sea.

References

Adkins EP (1955) Setauket: the first three hundred years 1655–1955. David McKay Co, New York

Allard DC (1978) Spencer Fullerton Baird and US Fish Commission. Arno Press Inc, New York

Allen DY (1997) Long Island maps and their makers. Amereon House, Mattituck

Altobello MA (1992) The economic importance of Long Island Sound's water quality dependent activities. US Environmental Protection Agency Report

Anderson T (2002) This fine piece of water: an environmental history of Long Island Sound. Yale University Press, New Haven

Aron JB (1997) Licensed to kill? The Nuclear Regulatory Commission and the Shoreham plant. University of Pittsburgh Press, Pittsburgh

Atlantic States Marine Fisheries Commission (2007) Species profile: Atlantic menhaden, increased demand for omega-3 fatty acids may lead to increased harvest. ASMF Fisheries Focus. 16, no 1. http://www.asmfc.org. Accessed 1 Feb 2010

Atlantic States Marine Fisheries Commission (1991) 50 year history of the commission. Atlantic States Marine Fisheries Commission, Washington

Austen B et al (1992) Journey through time: the Riverhead Bicentennial, 1792–1992. Riverhead Bicentennial Commission, Riverhead

Baird CW (1974) Chronicle of a border town: history of Rye, Westchester County, New York, 1660–1870. Harbor Hill Books, Harrison

Bayles RM (1962) Historical and descriptive sketches of Suffolk County. IJ Friedman, Port Washington

Benedict J (2009) Little pink house: a true story of defiance and courage. Grand Central Publishers, New York

Blackford EG (1885) Report of the oyster investigation, 1884, 1885. In: Report of the commissioner of fisheries of New York State. State of New York, Albany

Bone F (1998) Sands of time: a history of the sand and gravel operations in Port Jefferson and nearby harbors. Three Village Historical Society, Setauket

Caro R (1974) The power broker: Robert Moses and the fall of New York. Knopf, New York

Caulkins FM (1852) History of New London, Connecticut from the first survey of the coast in 1612 to 1852. Case, Tiffany & Co., Hartford

Cave AA (1996) The Pequot War. University of Massachusetts Press, Amherst

Clark AH (1887) Connecticut and its fisheries. In: The fisheries and fishery industries of the United States. Government Printing Office, Washington

Collins JW (1891) Notes on the oyster fishery of Connecticut. United States Fish Commission Bulletin, V. 9, 1889. Government Printing Office, Washington

Cooper JN (1977) Land's end, Water's edge: Long Island Sound. Theo Gaus Ltd, Brooklyn

Covell WK (1947) A short history of the Fall River line. W King Covell, Newport

Daily Herald (1833) New Haven, Oct. 10, p 2

Decker RO (1986) The New London merchants: the rise and decline of a Connecticut port. Garland Publishing, Inc., New York

Delaney E (1983) The Connecticut River. The Globe Pequot Press, Chester

Eisler KI (2001) Revenge of the Pequots. Simon & Schuster, New York

Evanoff V (undated) A million fishermen! Salt water sportsmen. Long Island Collection, QueensBorough Public Library, Queens, New York

Feinstein E et al (2002) Stamford: an illustrated history. American Historical Press, Sun Valley

Fox G (1919) George Fox: an autobiography. Ferris & Leach, Philadelphia

Franklin HB (2007) The most important fish in the sea: Menhaden and America. Island Press/Shearwater Books, Washington

Frisman P (2002) Enfield Dam. Connecticut General Assembly, Hartford

Goode GB, Atwater WO (1879) A history of the menhaden. In: Baird S (ed) Annual report of the commissioner of fish and fisheries. Government Printing Office, Washington, pp 1876–1877

Goode GB, Clark AH (1887) Menhaden fishery. In: Goode GB (ed) Fisheries and fishery industries of the United States. Government Printing Office, Washington

Grant E (1974) Yankee dreamers and doers. Pequot Press, Chester

Grimaldi L (1986) Only in Bridgeport: an illustrated history of the Park City. Windsor Publications, Northridge

Harrison J (2008) Glen Cove. Arcadia Publishing, Charleston

Hart S (1959) The prehistory of the New Netherland Company. City of Amsterdam Press, Amsterdam

Hotchkiss C (1878) On the ebb: a few long lines from an old salt. Tuttle, Moreous & Taylor, New Haven

Hudson WM (1873) Letter from the State of Connecticut, Department of Fisheries. In: United States commission of fish and fisheries, Report on the condition of the sea fisheries of the south coast of New England in 1871 and 1872. Government Printing Office, Washington

Ingersoll E (1881) The oyster industry. Government Printing Office, Washington

Interstate Environmental Commission (2011) About us. http://www.iec-nynjct.org/about.who.htm. Accessed 24 Oct 2011

Jackson K (ed) (1995) Encyclopedia of New York City. Yale University Press, New Haven

Brown J & Co (1846) The American anglers guide. Burgess, Stringer & Co, New York

Jones RM (1976) Stonington borough: a Connecticut seaport in the nineteenth century. The City University of New York, New York

Judd RW (1997) Common lands, common people: the origins of conservation in northern New England. Harvard University Press, Cambridge

Kimmel D (1983) Traffic to flow on world's longest parking lot. IMSA 20(1):16–18

Kirby MX (2004) Fishing down the coast: historical expansion and collapse of oyster fisheries along continental margins. PNAS 101(35):13096–13099

Klawonn MJ (1977) Cradle of the Corps: a history of the New York district, US Army Corps of Engineers, 1775–1975. District Army Corps of Engineers, New York

Knight SK (1935) The journal of Sarah Kemble Knight. Peter Smith, New York

Kochiss JM (1974) Oystering from New York to Boston. Mystic Seaport. Wesleyan University Press, Middletown

Kurlansky M (2006) The big oyster: history on the half shell. Ballantine Books, New York

Long Island Railroad (1935) Fishing around Long Island. Long Island Railroad, New York

Long Island Sound Restoration Act: Aid to distressed communities (2005) Connecticut Bureau of Waste Management, Hartford

The Long Island Sound Study: Summary of the comprehensive conservation and management plan (1994) US Environmental Protection Agency. Long Island Sound Office, Stamford

MacKay RB, Rossano G (1985) Between ocean and empire: an illustrated history of Long Island. Windsor Publications, Northridge

Marsh GP, Trombulak SC (2001) So great a vision: the conservation writings of George Perkins Marsh. Middlebury College Press, Hanover

Mather F (1887) New York and its fisheries. In: Goode GB (ed) The fisheries and fishery industries of the United States, Section II: a geographical review of the fisheries industries and fishing communities for the year 1880. Government Printing Office, Washington

Matthiessen P (1986) Men's lives. Random House, New York

Maynard P, Noyes MB (eds) (2004) Carriages and clocks, carpets and locks: the rise and fall of an industrial city. University Press of New England, Hanover

McAdam R (1957) Salts of the sound. Stephen Daye Press, New York

McCallion KF (1995) Shoreham: the rise and fall of the nuclear power industry. Praeger, Westport

McCay BJ (1998) Oyster wars and the public trust: property, law, and ecology in New Jersey history. University of Arizona Press, Tucson

McDonald M (1894) Report of the United States commissioner of fish and fisheries for the year ending June 30, 1892. Government Printing Office, Washington

McEvoy AF (1986) The fisherman's problem: ecology and law in the California fisheries, 1850–1980. Cambridge University Press, New York

McFarland R (1911) A history of the New England fishes with maps. University of Pennsylvania, D. Apple, New York

McHugh JL, Conover D (1986) History and condition of food finfisheries in the middle Atlantic region compared with other sections of the coast. Fisheries 11:8–13

Mead S, Mead DM (1911) Ye historie of ye town of Greenwich. Knickerbocker Press, New York

Moses R (1970) Public works: a dangerous trade. McGraw Hill, New York

National Marine Fisheries Service (2009) Annual commercial landings statistics. NOAA, Washington

New England River Basins Commission (1975) People and the sound: a plan for Long Island Sound. NERBC, New Haven

New York State Department of Conservation (1938) Annual report for the year 1937. JB Lyon Company Printers, Albany

Newsday (1998) Long Island our story. Newsday, Melville

Nichols HB (1938) Historic New Rochelle. The Board of Education, New Rochelle

Pillsbury E (2007) Filthy waters, fattened oysters, typhoid veversfevers: The New York sewage battles, 1880-1925. Paper presented at the American Society for Environmental History, Baton Rouge

Pillsbury E (2008) Selling the bottoms, undermining the commons: understanding the oyster mapping efforts 1880-1920. Long Island Sound Fellow Reports, 29. Long Island Sound Study, Stamford, CT

Pillsbury E (2009) An American bouillabaisse: the ecology, politics and economics of fishing around New York City, 1870-present. Dissertation, Columbia University

Ross P (1902) A history of Long Island from its earliest settlement to the present time. The Lewis Historical Publishing Company, New York

Smits E (1974) Nassau, suburbia, USA. Friends of the Nassau County Museum. Syosset, New York

Stevenson CH (1899) The shad fisheries of the Atlantic Coast of the United States. In: Bowers GM (ed) Report of the commissioner of fish and fisheries for the year ending June 30, 1898. Government Printing Office, Washington

Stickney RR (1996) Aquaculture in the United States: a historical survey. Wiley, New York

Stokes IN Phelps (1967) The iconography of Manhattan Island. Arno Press, New York

Suffolk Times (2009) March 12, p 6

Swanson RL, Turano F, Amondolia J (2005) William Sidney Mount: His plans for the Stony Brook Harbor area. LI Hist J 19(1):110–122

Sweet G (1941) Oyster conservation in Connecticut: past and present. Geogr Rev 31(4):591–608

Taylor JE III (1999) Making salmon: an environmental history of the northwest fisheries crisis. University of Washington Press, Seattle

Traffic, earnings and feasibility of the Long Island Sound crossing (1965) Madigan-Hyland, New York

True FW (1887) The pound-net fisheries of the Atlantic states. In: Goode GB (ed) The fisheries and fishery industries of the United States, Section V: History and methods of the fisheries. Government Printing Office, Washington

United States Commission of Fish and Fisheries (1873) Report on the condition of the sea fisheries of the south coast of New England in 1871 and 1872. Government Printing Office, Washington

United States Senate Committee on Fisheries (1884) Fish and fisheries on the Atlantic Coast

Varekamp JC, Varekamp DS (2006) Adriaen Block, the discovery of Long Island Sound and the New Netherlands Colony: what drove the course of history. Wrack Lines 6(1):1–5

Varekamp JC, Mecray EL, Maccalous TZ (2005) Once spilled, still found: Metal contamination in Connecticut coastal wetlands and Long Island Sound sediment from historic industries. In: Whitelaw DM, Visgilio GR (eds) America's changing coasts. Edward Elgar, Northampton

Walsh JP (1978) Connecticut industry and the revolution. American Revolution Bicentennial Commission of Connecticut, Hartford

Westchester County Department of Environmental Facilities, Wastewater Division (1997) History of the Department. Westchester County Department of Environmental Facilities, White Plains

Weidman BS, Martin LB (1981) Nassau County, Long Island in early photographs, 1869-1940. Dover, New York

Weigold M (2004) The Long Island Sound: its people, places and environment. New York University Press, New York

Weigold M (1984) People and the parks: A history of parks and recreation in Westchester County. Westchester County Department of Parks and Recreation, White Plains

Weigold M (1975) Rye to Oyster Bay ferry. Rye Historical Society, Rye

Wilson RE, Swanson RL (2005) A perspective on bottom water temperature anomalies in Long Island Sound during the 1999 lobster mortality event. J Shellfish Res 24(3):825–830

Wise WM (1975) The fisheries and fishery resources of Long Island Sound. Dissertation, State University of New York at Stony Brook

Ziel R, Foster GH (1965) Steel rails to the sunrise. Duell, Sloan and Pearce, New York

Chapter 2
The Geology of Long Island Sound

Ralph Lewis

2.1 Introduction

This chapter is intended to provide an overview of our present understanding of the geology of the lowland that now contains both Long Island and Fishers Island Sounds: The Long Island Sound (LIS) Basin. A portion of the chapter is a grateful reminder of the contributions of previous investigators. This is followed by a discussion of the major geologic components of the Basin, and a synopsis of the region's geologic history and modern sedimentary processes. Implicit in these discussions is the knowledge that the morphology and composition of the Basin influence the physical, chemical, and biological processes that are at work in both Sounds. With this in mind, referrals to other related chapters of this book are included in the text. The LIS Basin lies at the juncture of the glacially modified, bedrock-controlled landscape to the north, and the Atlantic Coastal Plain to the south. As a result, the northern and southern flanks of the LIS Basin, and the coastlines that have developed on them, are compositionally and morphologically different (Fig. 2.1), an interesting combination when viewed in the broader context of the east coast of North America.

The Atlantic coast of the United States is sediment-dominated along the south shore of Long Island and from Staten Island southward to Florida. Excluding the glacial deposits that dominate the coastlines of western Rhode Island, Narragansett Bay, and Cape Cod, the coastline from Rhode Island northward through Maine is largely bedrock-dominated. Whether sediment- or rock-dominated, these coastlines are typically exposed to the large waves that are generated by winds blowing across the Atlantic Ocean. Locally, lower energy wave regimes prevail in protected harbors, inlets, coves, and embayments. Long Island

R. Lewis (✉)
Long Island Sound Resource Center, University of Connecticut, Groton, CT 06340, USA
e-mail: ralph.s.lewis@uconn.edu

J. S. Latimer et al. (eds.), *Long Island Sound,* Springer Series on
Environmental Management, DOI: 10.1007/978-1-4614-6126-5_2,
© Springer Science+Business Media New York 2014

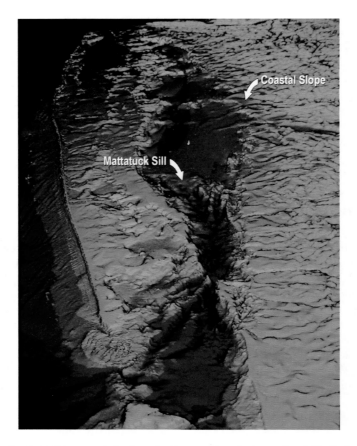

Fig. 2.1 A computer generated view looking westward down the LIS Basin. Aspects of this image have been enhanced to highlight the N-S grain of the glacially modified, bedrock-controlled landscape of the mainland and the E-W orientation of the moraines on Long Island. The muted topography of the Coastal Slope (slightly *darker green* coastal strip) can be seen along the I-95 corridor. The water has been removed and bathymetric features are enhanced to show the highly eroded eastern Sound in contrast to the depositional areas to the west of the unfortunately misnamed "Mattituck Sill". *Lightest blue* indicates shallowest depth and *darkest blue* indicates greatest depth

and Fishers Island Sounds are protected from the high wave energy of the Atlantic Ocean by Long Island and Fishers Island, and their low-energy coastlines reflect the geologic makeup of the region: sediment-dominated on Long Island and Fishers Island, and the longest stretch of low-energy, bedrock-dominated shoreline on the US Atlantic coast, to the north.

Along the north shore, and throughout the adjacent mainland, the distinct N-S grain of the present landscape (Fig. 2.1) results from a geologic legacy that began about 500 million years ago as the bedrock terranes of the Northeast were assembled from west to east, and subsequently nearly pulled apart. As less resistant components of the assembled bedrock units and roughly aligned N-S trending faults and fractures preferentially yielded to a few hundred million years of weathering and stream erosion, a well-developed, south-flowing bedrock drainage system evolved. The overall configuration of this pre-glacial drainage system, with its intervening N-S ridges, was not materially altered by glaciations or the deposition of layered sands and gravels as meltwater streams occupied emerging glacially modified bedrock valleys during the northward melt-back of the last glacier. Owing to the fact that the N-S grain of the bedrock still prevails, much of New England drains southward to LIS, and driving N-S in Connecticut is an easier option than driving E-W.

Along the somewhat muted topography of the coast (Figs. 2.1 and 2.2, the Coastal Slope), the bedrock still holds sway. From Throgs Neck eastward through Westchester County and nearly all of Connecticut, the natural shoreline is characterized by pocket beaches flanked by N-S trending crystalline bedrock promontories. The morphology of the bedrock surface determines where the south-draining stream valleys are and, since the melt back of the last glacier, the position of the stream valleys has largely controlled where beaches and marshes could form. Since both Sounds are fetch limited (O'Donnell et al., Chap. 3, in this volume), there is not enough wave energy in the system to overcome the influence that the bedrock has on the overall shape and character of the coast.

This is not the case along the sediment-dominated south shore, where, except for some outcrops in northwestern-most Queens, New York, crystalline bedrock is buried under Cretaceous Coastal Plain strata and/or younger glacial deposits. Wherever these deposits are thick enough to completely bury the underlying bedrock ridges and valleys, their depositional history has shaped their morphology. The E-W orientation of Long Island and Fishers Island (Figs. 2.1 and 2.2) is, for example, attributable to their morainal backbones. The semi-consolidated to unconsolidated deposits of the south shore are largely composed of boulders, gravels, sands, silts, and clays that are more susceptible to wave action than the crystalline bedrock of the north shore. On Long Island, this is particularly true east of Port Jefferson, where wave action has significantly eroded and straightened the shoreline (Fig. 2.3).

Just as the course of landscape development is influenced by a region's geologic history, the course of human endeavors is influenced by the character and composition of regional landscapes. The layout of transportation systems, development patterns, and the economic character of the land areas surrounding the LIS Basin are a good example of this. Early travel paths and the resulting road system are very much E-W on Long Island, but not so to the north. Driving E-W across Westchester County (New York) and throughout much of Connecticut often entails circuitous routes around and over hills, and across stream valleys (somewhat akin to traveling across the grain of a corrugated roof), but driving north–south usually involves more direct and flatter routes (with the grain) up and down valleys. As a result, Long Island developed from west to east, parallel to the moraines, while Connecticut developed up the valleys, with early E-W transport on LIS. Several railroads went bankrupt trying to build the hundreds of bridges needed to complete the shoreline route across the array of south-draining valleys that characterize coastal Connecticut.

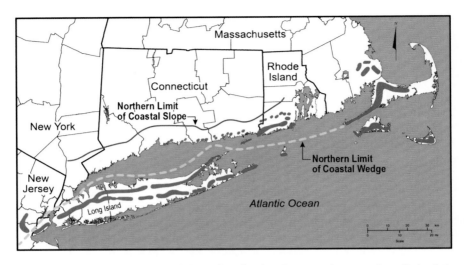

Fig. 2.2 An idealized map view of the region showing the approximate northern limit of the Coastal Slope (*brown line*) on the mainland, and the approximate northern limit of Cretaceous Coastal Plain strata (Northern Limit of Coastal Wedge, *dashed yellow line*) in Long Island, Block Island and Rhode Island Sounds, and along the southern margin of Buzzards Bay. For a more detailed map of Cretaceous Coastal Plain strata in LIS see sheet two of Stone et al. (2005). Moraine positions are approximated in *red* (for more detail see Fig. 2.7). Adapted from images courtesy of CTDEEP and Janet R. Stone

Bedrock control and a preponderance of fairly impermeable till favored surface runoff and hampered agriculture on the mainland, whereas the permeable glacial sediment of Long Island favored agriculture and greatly inhibited stream development. Early on, these differences determined where water power was available for industrialization and where farming would be most successful. Today, they relate to management issues such as how fresh water and its contaminant load are delivered to the LIS Basin. On the mainland, localized direct groundwater discharges to coastal waters exist but, at regional scales, ground water typically discharges to wetlands, lakes, ponds, rivers, and streams. Absent a large surface water component, ground water flowing through complicated and not always well understood glacial geology is the major contributor of fresh water along the north shore of Long Island (Cuomo et al., Chap. 4 and Varekamp et al., Chap. 5, in this volume).

Groundwater discharge from the Long Island side of LIS is the dominant source of fresh water from this area. In undeveloped areas on Long Island, about 50 % of the water that falls as precipitation recharges the groundwater system, and the other 50 % is lost through evapotranspiration. Groundwater discharges to LIS as it flows northward from the regional groundwater divide. Groundwater discharge from Long Island to LIS was estimated to include shallow (upper glacial aquifer—1.2×10^8 m^3/year) and deep groundwater (Magothy and Lloyd aquifers—2.3×10^8 m^3/year) (Buxton and Smolensky 1999; Scorca and Monti 2001).

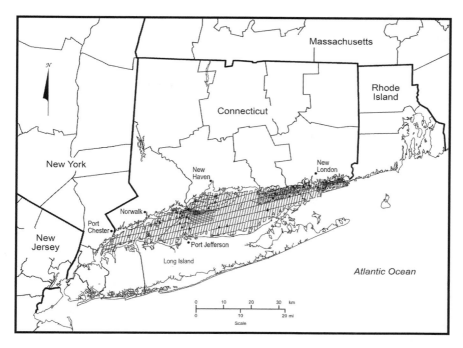

Fig. 2.3 Map of locations where seismic data (*black lines*) and cores (*triangles*) were collected in support of Phase 1 (1981–1990) of the GNHS/USGS Cooperative Mapping Program

Some ground water is also discharged to streams. The groundwater budget in more urbanized areas is affected by stormwater and wastewater management. Discharge of storm water may occur directly to surface water bodies and/or LIS, reducing recharge, and increasing direct runoff into the Sound. Other changes to the water budget under urbanization include excess recharge from stormwater infiltration, re-routing of water withdrawn by wells to sewage treatment plants, and on-site wastewater treatment systems. Groundwater discharge to the small number of streams on the north shore of Long Island is probably less than under pre-development conditions (Scorca and Monti 2001).

In Connecticut, the percent base flow of streams relatively unaffected by human activities is related to the percent of each watershed that is underlain by coarse-grained, stratified glacial meltwater deposits (Mazzaferro et al. 1979). Connecticut is underlain by about 18 % coarse-grained glacial meltwater deposits. This equates to a base-flow index of 46 % of total runoff, indicating that groundwater discharge is a significant part of the tributary flow in Connecticut watersheds. The estimated total runoff from the Connecticut part of the watershed for 1980–2010 is 0.66 m, or about 8.3×10^9 m^3/year (http://waterwatch.usgs.gov). Estimated groundwater discharge to rivers and LIS is about 3.8×10^9 m^3/year.

Direct discharge of ground water from Connecticut to tidally affected waters is probably limited to a small area adjacent to the shorelines and tidally affected sections of the major rivers, representing less than 500 km[2], or about 4 % of the area of the state.[1] The estimated direct groundwater discharge is on the order of 1.5×10^8 m[3]/year.

2.2 A Legacy of Geologic Contributions, 1870–2011

Speculation concerning the geologic origins of LIS began to appear in the scientific literature during the latter part of the nineteenth century. Although work conducted prior to World War II was largely rooted in extrapolations from on-land evidence, several pioneering investigators drew insightful conclusions from their observations. Dana (1870, 1890), Lewis (1877), Veatch (1906), and Fuller (1914) correctly suggested a fluvial origin for the LIS Basin, and Fuller (1905) reported that a water well had encountered granitic bedrock at about 86 m below Fishers Island. Hollick (1893), Antevs (1922), and Reeds (1927) offered evidence for a glacial lake north of Long Island. The lake that Reeds (1927) recognized in the vicinity of New York City is a portion of Glacial Lake Connecticut, which is now known to have occupied the entire LIS Basin (Stone et al. 1985) much as the Sound does today. Technological advances that occurred during and after World War II provided a means to test and augment the findings of earlier workers with seismic, magnetic, and gravity data that could be collected in and above the waters of Long Island, Fishers Island, and Block Island Sounds. Between 1950 and 1980, this fostered a series of topical investigations that cumulatively established a general understanding of the relationships of bedrock, coastal plain strata, and glacial deposits in the offshore of southern New England and southeastern New York. As is often the case, work conducted in one water body helped with the understanding of adjacent areas.

In 1951, Oliver and Drake published a depth-to-bedrock map of Long Island and Block Island Sounds. This was followed by Zurflueh's (1962) total magnetic intensity map of portions of LIS and reports of granite gneiss and schist at depths of -153 m and -488 m in wells drilled at Orient Point and at Brookhaven, respectively (de Laguna 1963; Pierce and Taylor 1975). Several of the bedrock valleys in and around LIS were discussed by Upson and Spencer (1964), and Bloom (1967) produced a report on the coastal geomorphology of the Connecticut coast. Tagg and Uchupi (1967) described the bedrock surface as low-relief topography overlain by unconsolidated sediments. From 1965 to 1973, researchers from the University of Rhode Island provided basic information regarding the subbottom in Block Island Sound (BIS) and ELIS (McMaster et al. 1968; McMaster and Ashraf 1973a, b, c). Grim et al. (1970) significantly advanced the geologic

---

[1] John R Mullaney of the USGS-Water Resources Division in East Hartford, CT has kindly supplied the summary of surface and groundwater discharges to LIS.

knowledge base for the LIS Basin with their insightful interpretation of marine seismic data from the central Sound. Their mapping provided solid evidence supporting the earlier observations of Veatch (1906) and Fuller (1914) that the LIS Basin was initially formed by fluvial erosion and later glacially modified. Their work also showed that the flanks of the Basin are primarily composed of bedrock, overlain by glacial deposits, to the north and Cretaceous Coastal Plain strata, overlain by glacial deposits, to the south. They also noted thick glacial deposits filling deep portions of the LIS Basin. Haeni and Sanders (1974) mapped a deep bedrock valley just offshore of New Haven, CT. Regional offshore bedrock characterizations were augmented by the aeromagnetic map of Harwood and Zietz (1977) and Dehlinger's 1978 free air gravity map of the region. The Dehlinger gravity map is available digitally on US Geological Survey Open-File Report 00-304 (Paskevich and Poppe 2000), and on the web (http://www.lisrc.uconn.edu/).

Investigations related to the glacial history of LIS and BIS progressed in a similar manner. Fuller (1905, 1914) conducted detailed studies of Fishers Island and Long Island, and Crandell (1962) worked on Plum Island. Suter et al. (1949) described the geologic formations and aquifers of Long Island. Goldsmith (1960), Schafer (1961, 1965), Donner (1964), Schafer and Hartshorn (1965), Isbister (1966), Mills and Wells (1974), and Sirkin (1971, 1976, 1982), studied the moraines of eastern Connecticut, Rhode Island, and Long Island. As part of his work, Schafer (1961) located glacial end-moraine deposits across the inner shelf on the basis of seafloor topography and bottom sediment characteristics described by McMaster (1960). Flint and Gebert (1976) traced a moraine offshore from Hammonasset State Park in Madison, CT.

After a century of debate, it appears that Myron Fuller's (1905, 1914) discussions regarding at least four glaciations of Fishers Island, Long Island, Martha's Vineyard, and Nantucket are gaining renewed traction in the modern literature. Stone (2012) has assigned middle to late Pleistocene ages to the deposits discussed by Fuller, and correlated them with four large Pleistocene deposits that have been recognized on Georges Bank (Lewis and Sylwester 1976; Lewis and Stone 2012).

Earlier discussions of the existence of a glacial lake in the LIS Basin (Hollick 1893; Antevs 1922; Reeds 1927) were revisited by Lougee (1953), and Newman and Fairbridge (1960) proposed the name "Glacial Lake Antevs" for their version of the LIS glacial lake. They correctly envisioned a series of lakes that traced from metropolitan New York through LIS, BIS and Rhode Island Sound, with the lakes being dammed by eastward extensions of the Long Island moraines. Later publication of "Long Island Sound: An Atlas of Natural Resources" (CTDEP 1977) unfortunately muddied the clear vision of Newman and Fairbridge (1960) with the incorrect assertion that the "Mattituck Sill," in east-central LIS, is a bedrock ridge that dammed a small glacial lake to its west.

Physical evidence of the occurrence of glacial lake clay in the vicinity of Block Island, Fishers Island, and ELIS was supplied by Frankel and Thomas (1966), who reported finding glacial lake-clay concretions in BIS, and Coch et al. (1974), who cored light red, laminated clay in Little Gull Channel at the eastern entrance of LIS. Bertoni et al. (1977) identified freshwater-lake deposits in central BIS on the basis of bottom samples, cores, and seismic profiles. Newman (1977) reported on glacial lake clay from borings in Flushing Bay and referred to these clays as the Flushing formation.

By the early 1980s, the composition and distribution of bottom sediments along the inner shelf of southern New England had been studied by numerous investigators (e.g., McMaster 1960; McCrone 1966; Savard 1966; Feldhausen and Ali 1976; Williams 1981). Trumbull (1972) and Schlee (1973) provided some information on the bottom sediments of Fishers Island Sound (FIS), and Akpati (1974) conducted the first comprehensive study of the mineralogy and sedimentology of eastern LIS (ELIS). Buzas (1965) and Dignes (1976) reported on the foraminifera of LIS and BIS. Bokuniewicz et al. (1976) outlined some of the issues surrounding the sediment budget for LIS, and Gordon (1980) and Bokuniewicz and Gordon (1980) discussed the sediment system of LIS. Bloom and Stuiver (1963) collected data related to sea level rise along the Connecticut coast.

In 1981, the State Geological and Natural History Survey of Connecticut (GNHS) worked cooperatively with the United States Geological Survey (USGS) to initiate a long-term, systematic geologic mapping program in BIS, FIS, and LIS. The first phase of this program (1981–1990) concentrated on mapping the sub-seafloor geologic components of the region and better defining its geologic history. During phase two (1991–present), the mapping program has concentrated on characterizing the seafloor of LIS. Several important partners have been involved in various aspects of the mapping program; they include the Bureau of Ocean Energy Management, Regulation and Enforcement (formerly known as the US Minerals Management Service-MMS Continental Margins Program), The Northeast Underwater Research Technology and Education Center (NURTEC) at UConn-Avery Point (formerly known as The National Undersea Research Center—NURC), NOAA's Atlantic Hydrographic Branch and students and faculty from the following colleges and universities: Boston University, Mount Holyoke College, Stony Brook University, US Coast Guard Academy, University of New Haven, University of Connecticut, University of Rhode Island, Vrije Universiteit (Amsterdam), Wesleyan University, and Williams College. As a result, more than 150 papers, articles, and map products have been published.

A seismic survey of BIS was conducted in 1981, and the results were published in 1984 (Needell and Lewis 1984). Goss (1993) used the 1981 seismic data from BIS to compare and integrate this information with ice marginal sedimentation and ice retreat models proposed for LIS (Lewis and Stone 1991) and Narragansett Bay (Peck and McMaster 1991).

Eighteen cruises involving seismic surveys, physical sampling, and remotely operative vehicle (ROV)/submersible dives were conducted in LIS and FIS between 1982 and 1990 (Fig. 2.3). NURTEC assisted this effort with ROV/submersible support and the Bureau of Ocean Energy Management, Regulation and Enforcement facilitated non-energy resource assessment work in Connecticut and New York waters for ten years beginning in 1984 (e.g., Neff and Lewis 1988; Kelly et al. 1989; Lewis et al. 1990, 1994, 1998). Using cores obtained during the

MMS-supported LIS sampling effort, Reimer (1986) produced a detailed report on varved (glacial lake) clays and Szak (1987) studied foraminifera to reconstruct a scenario of Late Quaternary events in ELIS.

Cruise reports, core locations, core logs, photographs of cores, trackline maps, and images of the seismic data for the years 1981–1990 are available on a DVD-ROM (Poppe et al. 2002b) and this information, as well as listings of publications from the mapping program, and the data and results from post-1990 work, can be found on web sites maintained by the Long Island Sound Resource Center (http://www.lisrc.uconn.edu/) and the USGS Woods Hole Field Center (http://woodshole.er.usgs.gov/project-pages/longislandsound/Pubs.htm). Results regarding the regional geologic framework of Long Island and Fishers Island Sounds have also been reported in paper form by Lewis and Stone (1991), Lewis and DiGiacomo-Cohen (2000), and Stone et al. (2005). This reporting benefitted from the legacy of geologic knowledge passed down from the previous investigators mentioned above and from a variety of workers who presented their findings as mapping progressed.

Related onshore knowledge was enhanced by the publication of the Bedrock Geologic Maps of Connecticut (Rodgers 1985) and Rhode Island (Hermes et al. 1994). A summary of the geology of New York was presented in Isachsen et al. (1991). Thompson et al. (2000) reported fault motion along the eastern border fault of Connecticut's Hartford Basin. Koteff and Pessl (1981) proposed a model for systematic ice retreat in New England, and Stone and Borns (1986) presented a Pleistocene glacial and interglacial stratigraphy for the area. Gustavson and Boothroyd (1987) cited the Malaspina Glacier in Alaska as a modern analog for their concept of the southeastern margin of the Laurentide ice sheet. Koteff et al. (1988) used studies of Glacial Lake Hitchcock to derive information regarding postglacial uplift. Stone et al. (1992) published the Surficial Materials Map of Connecticut and Cadwell (1991) produced the Surficial Geologic Map of New York.

The 1991 Lewis and Stone paper appeared in Special Issue No. 11 of the *Journal of Coastal Research,* which was edited by Paul Gayes, Ralph Lewis, and Henry Bokuniewicz and was dedicated to Walter Newman. LIS-related papers that were included in the special issue discussed estuarine paleoshorelines (Gayes and Bokuniewicz 1991), palynology (Shaw and van de Plassche 1991), geochemistry of mudflat and marsh sediments (Varekamp 1991; Varekamp et al., Chap. 5, in this volume), Late Holocene sea level rise (Thomas and Varekamp 1991; van de Plassche 1991; Patton and Horne 1991), and the stratigraphy of the Long Island platform (Sirkin 1991).

Between 1989 and 1998, the second phase of the GNHS/USGS Cooperative Mapping Program concentrated on characterizing the physical constituents of the sea floor and modern sedimentary processes. During this period, ten continuous sidescan sonar mosaics of selected areas (Fig. 2.4) in LIS and FIS were produced. Images of the ten sidescan sonar survey areas, as well as benthic ecological analysis, bathymetric, geochemical, and oceanographic information, maps of sedimentary environments, a surficial sediment database, and an extensive bibliography, were released on a CD-ROM as USGS Open-File Report 98-502

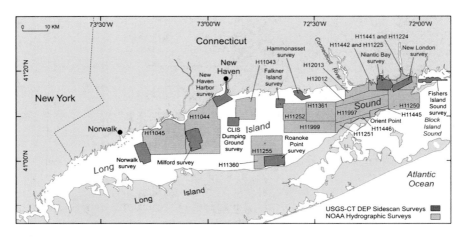

Fig. 2.4 Map of locations where data for the preparation of ten sidescan sonar mosaics (USGS-CTDEP) and nineteen multibeam hydrographic/bathymetric surveys (NOAA) have thus far been conducted during Phase 2 of the GNHS/USGS Cooperative Mapping Program. For downloadable images and reports related to this work visit the web sites maintained by the Long Island Sound Resource Center (http://www.lisrc.uconn.edu/) and the USGS Woods Hole Field Center (http://woodshole.er.usgs.gov/project-pages/longislandsound/Pubs.htm). Map courtesy of Larry Poppe

(Poppe and Polloni 1998). Cruise information, mosaic images, and listings of analytical and map products (typically USGS Geologic Investigation Series Maps, e.g., Poppe et al. 1998) that have been generated from the sidescan data are also available on web sites maintained by the Long Island Sound Resource Center (http://www.lisrc.uconn.edu/) and the USGS Woods Hole Field Center (http://woodshole.er.usgs.gov/project-pages/longislandsound/Pubs.htm).

Prior to the sidescan survey labeled Fishers Island Sound survey in Fig. 2.4, Poppe et al. (1994) used bottom samples to produce a map showing the distribution of surficial sediments in FIS. The remobilization of surficial sediment and the creation of furrows in the vicinity of the New Haven dumping ground (Central Long Island Sound Dumping Ground Survey, Fig. 2.4) were reported by Poppe et al. (2002a). NERTEC supported fieldwork related to the character of the sand sheet and westward mobility of the misnamed "Mattituck Sill" in ELIS was conducted between 1987 and 2003 (Fenster et al. 1990). Fenster et al. (2006) published data covering 16.1 years and have described the mechanism of westward movement of sand waves between 7 and 17 m in height.

A second volume of the *Journal of Coastal Research*, with a thematic section dedicated to LIS (Knebel et al. 2000), was published in 2000. This volume contained a new analysis of the sediment budget of LIS (Lewis and DiGiacomo-Cohen 2000), a regional overview of LIS seafloor environments (Knebel and Poppe 2000), a detailed map of LIS bottom sediment distributions (Poppe et al. 2000), and other papers relating to mercury (Varekamp et al. 2000), other contaminants (Mecray and Buchholtz ten Brink 2000; Cuomo et al., Chap. 4, in this volume), pollen, environmental change (Thomas et al. 2000), currents and sediment transport (Signell et al. 2000; O'Donnell et al., Chap. 3, in this volume), and a benthoscape analysis (Zajac et al. 2000; Lopez et al., Chap. 6, in this volume). USGS Open-File Report 00-304 (Paskevich and Poppe 2000) is a CD-ROM containing digital versions of much of the environmental and geologic mapping and information presented in the *JCR* thematic section. In addition to the contaminant mapping and indicators of environmental change chapters, there are digital versions of maps of the marine transgressive surface, the thickness of marine deposits, a gravity map, and maps of sedimentary environments and sediment distribution. The CD also has a chapter dedicated to photographs of various LIS seafloor settings. Figure 2.5 shows a sampling of computer-enhanced, subseafloor maps that originally appeared as black and white images in Lewis and Stone (1991), Lewis and DiGiacomo-Cohen (2000), and Paskevich and Poppe (2000). The Lewis and DiGiacomo-Cohen (2000) thickness of marine deposits map is also available in color on sheet two of the Quaternary Geologic Map of Connecticut and Long Island Sound Basin (Stone et al. 2005).

NOAA's Atlantic Hydrographic Branch began to share sidescan sonar data from their Branford survey (NOAA Survey H11043, Fig. 2.4) in 2001 (McMullen et al. 2008). Since then, sidescan sonar and multibeam data from the NOAA surveys shown in Fig. 2.4 have been provided to the GNHS/USGS cooperative program and a resulting series of digital USGS Open-File Reports (e.g., Poppe et al. 2006a, b, 2007a, b) and journal publications (e.g., McMullen et al. 2005, 2010; Poppe et al. 2008a, b, 2010b) have begun to appear. An effort to combine the multibeam bathymetric data from the NOAA ship THOMAS JEFFERSON surveys (Fig. 2.4) with new multibeam and near-shore LIDAR data to produce a continuous, geo-referenced bathymetric data set for northeastern LIS has also been completed (Poppe et al. 2010a, 2011). As of this writing, this cooperative mapping work is still under way, and a comprehensive, updated list of related publications can be found on the USGS Woods Hole Field Center web site (http://woodshole.er.usgs.gov/project-pages/longislandsound/Pubs.htm). Roger Flood and his students at Stony Brook University (Flood and Cerrato 2006; Kinney and Flood 2006; Cerrato and Flood 2008) are surveying Long Island harbors and bays using sidescan sonar and multibeam, while McHugh et al. (2007) and Vargas et al. (2008) have reported on their geophysical surveys in westernmost LIS. Onshore and offshore work in the northern part of Oyster Bay, New York (Stumm et al. 2004) reinforces earlier reporting by Foord et al. (1970) that thick glacial deposits may locally overlie extensively eroded segments of the coastal plain cuesta along the north shore of Long Island.

Peter Patton and James Kent authored a book published in 1992 entitled *A Moveable Shore, The Fate of the Connecticut Coast*. Although the geologic history chapter is now a bit outdated, this book provides a very nice overview of Bloom's (1967) mapping of the physical makeup of the north shore, and supplements this with a discussion of the role that coastal processes have played in the evolution of the coast. Information relating to sea level rise (e.g., Bloom and Stuiver 1963), and other process-oriented topics, is accompanied by a useful, coast-wide set of eight "Shoreline Erosion and Risk" maps. The Connecticut Department of Energy and

Fig. 2.5 A sampling of
computer-enhanced, color
versions of sub-seafloor maps
that have appeared in Lewis
and Stone (1991), Lewis and
DiGiacomo-Cohen (2000)
and Paskevich and Poppe
(2000) From *top* to *bottom*
they show: submerged end-
moraine deposits (*red*) and
ice-marginal lacustrine fan
deposits (*shades* of *green*)
of glacial Lake Connecticut;
deltaic (*solid* and *stippled
green* patterns) and varved
lake-bottom deposits
(*horizontal green line* pattern)
of glacial Lake Connecticut;
the channel system (*purple*)
cut into the top of glacial
Lake Connecticut deposits as
the lake slowly drained away
via the spillway at The Race;
structure contours drawn to
remaining portions of the
marine unconformity (mu,
Fig. 2.8) that was created
during the latest transgression
of the sea. Modern erosion
has removed most traces
of this unconformity in
the eastern third of LIS.
Computer enhancement
courtesy of Janet R. Stone,
adapted with permission
from The Journal of Coastal
Research

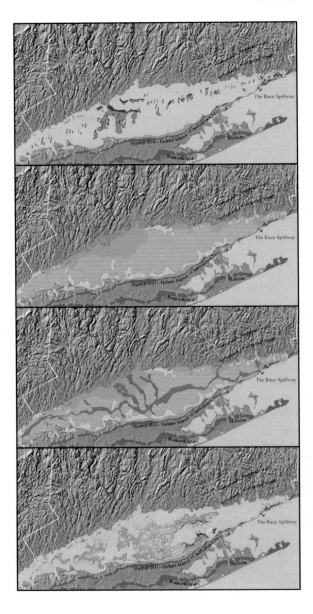

Environment Protection (DEEP) and The Nature Conservancy have created useful and informative web sites that relate to sea level rise and other coastal hazards.
The DEEP site can found at (www.ct.gov/dep/coastalhazards) and The Nature
Conservancy site address is (http://www.nature.org/ourinitiatives/regions/northam
erica/unitedstates/connecticut/explore/coastal-resilience-tool.xml).

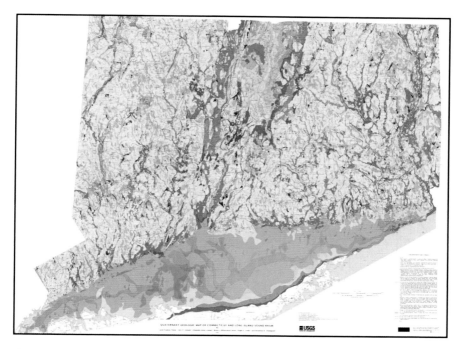

Fig. 2.6 A digital image of sheet one of the Quaternary Geologic Map of Connecticut and LIS Basin (Stone et al. 2005). The LIS Basin portion of this 1:125,000-scale map is a compilation of information from individual maps (such as the sampling shown on Fig. 2.5) and other sub-sea-floor data collected as part of the GHNS/USGS Cooperative Mapping Program. Image courtesy of Janet R. Stone

Building on the legacy of Bloom and Stuiver (1963), van de Plassche et al. (1989), and Patton and Horne (1991), studies related to sea level rise have continued (Stone and Lewis 1991; Nydick et al. 1995; Orson et al. 1998). Between 1989 and May of 2009, the late Orson van de Plassche had been a major contributor to this and related efforts, having reported findings from Connecticut marshes in Clinton (van de Plassche et al. 1989, 1998), Guilford (Edwards et al. 2004), Westbrook (Roe and van de Plassche 2005), and East Lyme (van de Plassche et al. 2006).

In addition to detailed mapping (1:125,000-scale) of the Quaternary geology of mainland Connecticut and the entire LIS-FIS Basin (Fig. 2.6), the Quaternary Geologic Map of Connecticut and LIS Basin (Stone et al. 2005) presents a scenario for the marine incursion into the Sound. Aspects of this scenario have been refined by Varekamp et al. (2006), and Varekamp and Thomas (2008) are continuing their work on paleo-environmental change in LIS (Varekamp et al., Chap. 5, in this volume). More recent work at the University of Rhode Island (Bryan Oakley's dissertation work and a Bryan Oakley-Jon C. Boothroyd manuscript entitled "Reconstructed

Topography of Southern New England Prior to Isostatic Rebound With Implications of Total Isostatic Depression and Relative Sea Level" (in press, *Quaternary Research*, as of January 2012) favors a thinner ice model with less isostatic depression (30 m) than the 80 m that was assumed by Stone et al. (2005). Less depression would impede the incursion of the sea more than the Stone et al. (2005) assumption would.

While it is clear that issues regarding the nature and timing of the marine transgression into the LIS Basin remain unresolved, several workers have helped to refine other related aspects of our knowledge of ice retreat. Balco and Schafer (2006) and Balco et al. (2009) have determined a [10]Be age of 20.2 ± 1.0 cal ka (thousands of calibrated calendar years before present) for the Ledyard moraine in southeastern Connecticut. Using the varve chronology of Ridge (2010), Janet R. Stone (personal communication, December 2011) has determined a calibrated calendar year age of 21 ka (calibrated NAVC, Ridge 2010) for the same locality. She is also hoping to correlate the 175 years of LIS varves reported by Reimer (1986) to the varve chronology (personal communication). Stone's varve chronology date, and the [10]Be work by Balco et al. (2009), would push the calendar year age of the Old Saybrook-Wolf Rocks moraine (20.5 ka as inferred by Stone et al. 2005) to before 21 ka because the Old Saybrook-Wolf Rocks moraine lies parallel to, and just to the south of, the Ledyard moraine (Fig. 2.7).

Information regarding the bedrock geology of eastern Connecticut, with implications for the offshore, has been published over the past few years. Aleinikoff et al. (2007), Walsh et al. (2007), and Wintsch et al. (2007) have outlined the latest thinking on a newly recognized bedrock terrane (Gander) in eastern Connecticut/ Rhode Island. A general interest book by Skehan (2008) provides an informative overview of the bedrock geology of western Connecticut, and Coleman (2005) discusses the pre-rift tectonic history of Connecticut's bedrock.

2.3 Overview of Geologic Framework

The discussion that follows presents an overview of the major geologic components of the Sounds (Fig. 2.8) from oldest to youngest. All radiocarbon dates have been converted to thousands of calibrated calendar years (cal ka) using the Fairbanks 0107 calibration curve (http://radiocarbon.ldeo.columbia.edu/cgi-bin/radcarbcal?id=0&fig=1&entry_typ e=0&add=1&age=21000&std=100) unless otherwise indicated.

2.3.1 Bedrock Components (~1.6 Billion to ~250 Million Years Old)

2.3.1.1 Mesoproterozoic (~1.6 to 1.0 Billion Years Old) and Paleozoic (~540 to 250 Million Years Old) Crystalline Bedrock

The crystalline bedrock foundation of southern New England and southeastern New York was assembled from west to east over the course of four plate tectonic

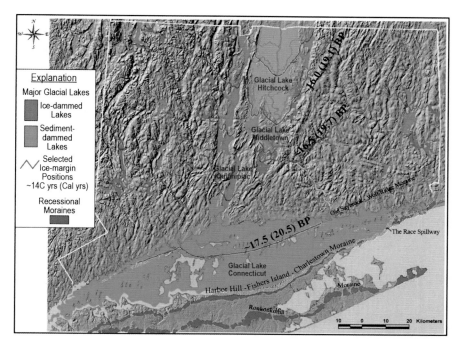

Fig. 2.7 A computer-enhanced image derived from sheet two of the Quaternary Geologic Map of Connecticut and LIS Basin (Stone et al. 2005). This image shows the extent of the major glacial lakes in Connecticut and the LIS Basin; mapped moraines; and inferred ~14Carbon and calibrated calendar year (cal yrs) ages (in thousands of years) for selected ice positions. The Ledyard moraine lies just north of the Old Saybrook-Wolf Rocks moraine. Image courtesy of Janet R. Stone

collisions (orogenies) that sequentially emplaced rock units representing the Mesoproterozoic Grenville (~1.6 to 1.0 billion years ago), and Paleozoic Taconic, Acadian, and Appalachian (~540 to 250 million years ago) mountain building events (Rodgers 1985; Isachsen et al. 1991; Hermes et al. 1994; Coleman 2005; Walsh et al. 2007). From Ravenswood-Astoria Queens, New York (Brock and Brock 2001) eastward to the Connecticut/Rhode Island line, the fabric of the crystalline basement is generally aligned N-S owing to the progressive eastward accretion of bedrock units representing the Grenville, Iapetos, Gander, Bronson Hill, and Avalonian terranes. Later Mesozoic faulting and fracturing also follow this general trend, so that differential weathering and erosion of the crystalline bedrock units of the mainland have produced the pattern of south-draining bedrock valleys, flanked by N-S trending bedrock promontories, that typifies the north shore of the Sound today.

Although reports of direct observation and/or sampling of the bedrock beneath Long Island and Fishers Island Sounds are lacking, existing well data from Long Island (de Laguna 1963; Pierce and Taylor 1975) and Fishers Island (Fuller 1905) and offshore seismic, magnetic, and gravity data support the inference that the crystalline bedrock terranes of the mainland extend seaward under the Sounds (Zurflueh 1962; McMaster et al. 1968; Grim et al. 1970; McMaster and Ashraf

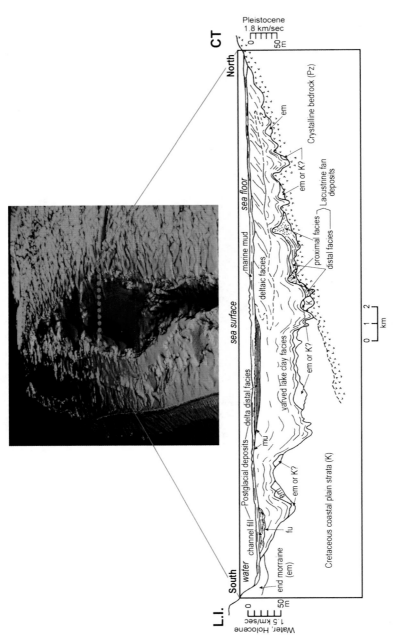

Fig. 2.8 Major components of the LIS Basin in an idealized, vertically-exaggerated, cross-sectional view. Images of the seismic data used to create this idealized view can be found on DVD-ROM (Poppe et al. 2002b) and on web sites maintained by the Long Island Sound Resource Center (http://www.lisrc.uconn.edu/) and the USGS Woods Hole Field Center (http://woodshole.er.usgs.gov/project-pages/longislandsound/Pubs.htm). Detailed mapping of the glacial deposits in LIS and FIS (with cross sections) can be found on the Quaternary Geologic Map of Connecticut and Long Island Sound Basin (Stone et al. 2005). *mu* marine unconformity. *fu* fluvial unconformity, other units labeled on cross section. Cross section adapted with permission from The Journal of Coastal Research

1973a, b, c; Haeni and Sanders 1974; Harwood and Zietz 1977; Dehlinger 1978; Lewis and Stone 1991; Lewis and DiGiacomo-Cohen 2000). Seismic mapping conducted in central Long Island Sound (CLIS) by Grim et al. (1970) and mapping throughout the LIS Basin (Fig. 2.3) by Lewis and Stone (1991) clearly delineate a submerged crystalline bedrock surface that very much resembles the glacially modified morphology of the adjacent Coastal Slope (Figs. 2.1 and 2.2). This surface dips southeastward (at about 2°) from exposures on the mainland (Pz, Fig. 2.8), and it retains recognizable aspects of the south-flowing drainage pattern that developed prior to deposition of Coastal Plain strata in Cretaceous time (Lewis and Stone 1991). From this we can infer that glacial modification of the crystalline bedrock surface under northern LIS was similar to the modification that occurred on the Coastal Slope. The overall N-S trend of bedrock interfluves was preserved as they were smoothed, and pre-Cretaceous drainage patterns survived in a muted form.

2.3.1.2 Mesozoic (~250 to 67 Million Years Old) Sedimentary and Igneous Bedrock

Assembly of the crystalline terranes was followed by rifting as the Atlantic Ocean began to form in the Mesozoic (~250 to 67 million years ago). On the mainland, this resulted in extensional faulting and emplacement of the sedimentary and igneous Newark terrane rocks of the Hartford Basin. Differential weathering and erosion over the ensuing years have preferentially removed less resistant bedrock and exploited the Mesozoic fault/fracture zones to create the N-S promontories and the intervening coves and embayments, which typify the north shores of LIS and FIS. A notable example of this is New Haven Harbor. This large embayment lies at the southern end of the Hartford Basin but is flanked on both sides by more resistant crystalline bedrock promontories.

Seismic evidence reported by Haeni and Sanders (1974) and Lewis and Stone (1991) shows that a deep southwestward-trending bedrock valley extends seaward from New Haven Harbor and an area of anomalously deep bedrock lies further offshore to the southwest (Fig. 2.9). Rodgers (1985) mapped a southwest-trending offshore extension of the "eastern border fault" of the Hartford Basin in the vicinity of the valley mapped by Haeni and Sanders (1974), and these features, as well as the area of anomalously deep bedrock to the southwest (Lewis and Stone 1991), lie within a magnetic low mapped by Zurflueh (1962). Unsubstantiated reports of sandstone in a Long Island well, and the discovery of Mesozoic basins south of The Island, have fueled speculation that the Hartford Basin extends under LIS. Seismic data show that the bedrock valley just offshore of New Haven is filled with something that overlies crystalline bedrock and underlies (and therefore predates) the glacial lake deposits. Given that there is presently no direct observation or sampling of this material, its age could potentially range anywhere from Mesozoic to Pleistocene. The work of Haeni and Sanders (1974) provides good evidence for Mesozoic extensional faulting in the crystalline bedrock under LIS,

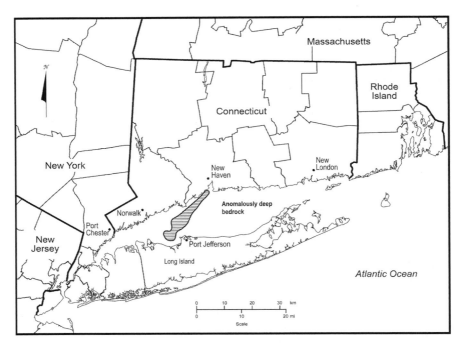

Fig. 2.9 Map showing approximate area of anomalously deep bedrock as reported in Lewis and Stone (1991)

but definitive answers regarding the presence of Newark terrane rocks or significant remnants of the Hartford Basin await further investigation through carefully planned drilling.

2.3.2 Coastal Plain Component (~140 to 66 Million Years Old) Cretaceous Strata

Unconsolidated to semi-consolidated gravels, sands, silts, and clays of Cretaceous age (~140 to 66 million years old), form the geologic foundation of Long Island (Fuller 1914; Sirkin 1991), and overlie crystalline bedrock (Lewis and Stone 1991) in southern LIS (northern limit of coastal wedge, Fig. 2.2). The Cretaceous coastal plain strata that underlie Long Island and southern LIS (Cretaceous coastal plain strata—K, Fig. 2.8) are the eroded remains of a seaward-thickening wedge of sediment that was deposited in Cretaceous and Tertiary time, as sediments eroded from the Appalachian Mountains built the Atlantic continental shelf. Strata of Tertiary age are not known to exist in the LIS Basin, and are absent on Long Island, but are found elsewhere on the continental shelf of the region. The reasons for their local absence remain unclear.

They may have existed and been selectively eroded away, or they may not have been deposited in the first place.

The fact that the bedrock surface along the Coastal Slope is slightly steeper and flatter (Fig. 2.1) than the bedrock surface of the rest of Connecticut has led to speculation that the thin landward margin of the coastal plain strata may once have extended far enough northward to have covered the Coastal Slope. In this scenario, the pre-Cretaceous bedrock surface of the Coastal Slope would have been covered and protected until the initiation of major stream erosion during periods of lowered sea level in the Tertiary. Subsequent stream action combined with later glacial modifications removed most of the thinner, landward portion of the sediment wedge, forming the LIS Basin and re-exposing bedrock in northern LIS and along the Coastal Slope. Seismic mapping conducted in CLIS by Grim et al. (1970) and more extensive coverage (Fig. 2.3) by the GNHS/USGS Cooperative Mapping Program clearly show that the Coastal Slope and the crystalline bedrock surface of the northern Sound have similar morphologies, and therefore probably share similar re-exposure histories.

Although there is little doubt that crystalline bedrock has been re-exposed and glacially modified in the northern Sound, the southern margin of the LIS Basin presents some interpretative problems related to glacial modification of the Coastal Plain remnant. These problems have been brought to light by offshore seismic and onshore well data from the vicinity of Oyster Bay, New York (Stumm et al. 2004) that are consistent with earlier well data from Smithtown, New York (Foord et al. 1970). These data indicate that a thick glacial mantle locally overlies highly eroded segments of the Coastal Plain remnant along the north shore. This may explain why Lewis and Stone (1991) noted a lack of internal reflectors in LIS seismic units inferred to represent Cretaceous strata. This lack of internal reflectors has long been troubling because in adjacent BIS strong, flat, gently south-dipping internal seismic reflectors, which are truncated along a north-facing cuesta scarp, typify the Cretaceous strata. The cuesta in BIS is inferred to have taken shape when fluvial erosion to the north removed interior portions of the Coastal Plain as the BIS Basin formed (Needell and Lewis 1984).

Mapping of the LIS Coastal Plain remnant by Lewis and Stone (1991) and the thick Cretaceous section shown under Long Island in Fig. 2.8 represent an interpretation of seismic data influenced by findings from BIS. This interpretation suggests that a prominent, north-facing cuesta, cut in Cretaceous strata, underlies southern portions of LIS and forms the southern flank of the LIS Basin. A Tertiary fluvial origin for the LIS Basin, with subsequent glacial modification, could be inferred from the position and orientation of the necks and bays west of Port Jefferson (Lewis and Stone 1991) if they cut deeply into in-place Cretaceous strata. If, on the other hand, the necks are mostly Pleistocene (Stumm et al. 2004) the story changes and the hydrogeology becomes more complicated.

In addition to the aforementioned troubling well data and the lack of internal reflectors in inferred Cretaceous strata, "Cretaceous" outliers north of the mapped cuesta scarp (K in Fig. 2.8) also arouse suspicion. In cross section, they do not have a break in slope where glacial material is inferred to overlie the Cretaceous

(Lewis and Stone 1991). This seems odd if the glacial material was deposited on an existing, glacially modified Coastal Plain remnant. It may very well be that there is more glacial material, and a much more muted cuesta, in southern LIS than previously thought. This would have important implications related to our understanding of how ground water moves and discharges along the north shore of Long Island. A suite of strategically placed onshore and offshore drill holes that penetrate to crystalline bedrock are needed to more definitively determine the extent of glacial modification that occurred in the southern Sound.

2.3.3 Glacial Component (~781,000 to 15,000 (~18.1 cal ka) Years Old) Middle Pleistocene and Wisconsinan Glacial Deposits

Although evidence for two glaciations (Illinoian and Wisconsinan age tills, Stone et al. 2005) is all that has been preserved in Connecticut, Stone's (2012) reaffirmation of Myron Fuller's work (1905, 1914) serves as a reminder that at least four ice sheets overrode the southern New England mainland and reached as far south as Long Island, Fishers Island, Martha's Vineyard, Nantucket, and Georges Bank. These ice sheets are inferred to have smoothed the bedrock surface of the mainland and helped in sculpting the Cretaceous remnant. The glaciers that overrode the LIS Basin appear to have modified a pre-existing lowland topography that was cut by streams during the periods of lowered sea level in the latter half of Tertiary time. As mentioned above, existing seismic mapping of the Cretaceous remnant under the southern Sound (Lewis and Stone 1991) assumes that glacial modification of the southern flank of the LIS Basin was not extensive and a prominent cuesta scarp exists there. There is mounting evidence that this inference is not entirely correct, and that a thick glacial section locally overlies a muted Cretaceous remnant. If this is true, glacial modification of the southern flank of the LIS Basin was more extensive than current mapping (Lewis and Stone 1991) indicates.

Direct evidence of the three pre-Wisconsinan glaciations that reached as far south as Long Island is lacking in the LIS Basin. The moraines, lacustrine fans, and lake deposits shown in Figs. 2.5, 2.6 and 2.7 are inferred to relate to the recession of the last (Wisconsinan) glacier as it sequentially melted back (northward) across the Basin (Stone et al. 2005). Around 19,000 radiocarbon years ago (~22.6 cal ka), the ice front stood atop the eroded northern margin of Cretaceous coastal plain strata, along what is now the north coast of Long Island (Fig. 2.7), and it deposited a series of moraine segments (Harbor Hill-Fishers Island-Charlestown Moraine, Fig. 2.7) that became the dam for Glacial Lake Connecticut. As the glacier melted out of the LIS Basin, Lake Connecticut expanded northward.

By 17,500 radiocarbon years ago (~20.5 cal ka), the ice front had melted out of southeastern-most Connecticut, and stood along a position marked by the Old Saybrook Moraine (Fig. 2.7). To the west, the glacier was still in the lake and the Norwalk Islands Moraine was being deposited along the ice margin west of

Bridgeport (Fig. 2.7). To the east, where the lake was deeper, a large lacustrine fan was emplaced south of New Haven (Fig. 2.5, top panel). These deposits represent the two types of material that accumulated along the base of the glacier as it paused at several recessional positions in Glacial Lake Connecticut (Stone et al. 2005). Lacustrine fans developed where meltwater was issuing into the lake from a portion of the ice front, and moraines developed elsewhere along the front (Stone et al. 2005).

The lake deposits that nearly fill the LIS Basin (Figs. 2.7 and 2.8) were deposited as the glacier stood at recessional positions in the lake, and across southern Connecticut. While the glacier was in contact with lake water, huge amounts of finely ground rock "flour" issued directly from the ice front as density underflows or settled to the lake bottom from suspension. This glacial lake clay coated and buried everything it landed on, and is locally up to 150 m thick in southern portions of the LIS Basin (Lewis and Stone 1991). Given a bit more time, Glacial Lake Connecticut deposits may well have completely filled the LIS Basin, and it is not a stretch to argue that the Wisconsinan glacier did more to fill in the LIS Basin than it did to carve it out (see varved lake clay facies and deltaic facies, Fig. 2.8).

As the glacier melted out of the lake, dry land began to separate the ice front from the lake. When this happened, meltwater streams reoccupied old bedrock valleys and began to carry glacial sediments southward to the lakeshore. Deltas of various sizes built into the lake at the mouths of these streams (Fig. 2.5, second from top and Fig. 2.8, deltaic facies). The sediments in these deltas generally become finer southward, where they grade into and become indistinguishable from the glacial lake clay. Wherever meltwater streams flowed, layered sand and gravel deposits eventually filled their valleys. The bedrock uplands were left blanketed by a thin veneer of till. As the melt back of the glacier proceeded and the sea began to flood the mainland, the bedrock uplands became promontories and islands, and the meltwater deposits choking the intervening valleys (dark green coastal deposits, Fig. 2.6) became the sediment source for pocket beaches and a platform for marsh development.

The Wisconsinan glacier was completely out of Glacial Lake Connecticut by 16,500 radiocarbon years ago (~19.7 cal ka) and out of Connecticut entirely a bit after 15,500 radiocarbon years ago (~18.7 cal ka). A much more detailed discussion of the glacial history of LIS can be found in the pamphlet that accompanies the Quaternary Geologic Map of Connecticut and LIS Basin by Stone et al. (2005).

2.3.4 Post Glacial Component (~15,000 Radiocarbon Years Ago = (18.1 cal ka) to the Present)

At its maximum, Glacial Lake Connecticut occupied roughly the same area that LIS now occupies (Fig. 2.7). There was also a glacial lake in BIS. These lakes existed because much of the world's water was tied up in glaciers, and sea level was as much as 100 m lower than it is today. The initial spillway for the lake in BIS was west of Block Island at the head of Block Channel (Needell and Lewis

1984), and the spillway for Lake Connecticut was across the moraine between Orient Point, Long Island, and Fishers Island, New York at The Race (Fig. 2.7). Over time, water exiting the lakes eroded their spillways. The lake in BIS had to begin to drain away before water levels in Lake Connecticut could begin to lower. Recent work at the University of Rhode Island indicates that during later stages of lake drainage, the spillway for the BIS lake shifted to the east side of Block Island at the "Mudhole" (Jon C. Boothroyd-Bryan Oakley personal communication, December 2011).

Direct evidence from the LIS Basin contradicts the popularly held belief that Glacial Lake Connecticut drained catastrophically (Stone et al. 2005). Deglaciation of the Connecticut coast proceeded from east to west, so the oldest Lake Connecticut deltas are found along the north shore of FIS, and related deltas get progressively younger westward toward New Haven. This westward progression of younger ages is accompanied by a westward lowering of the elevation of the topset/foreset contacts (indicators of lake levels) in the Lake Connecticut deltas, a clear indication that Lake levels were steadily dropping as deglaciation of the Connecticut coast proceeded. Another circumstance that argues against catastrophic lake drainage is the fact that, prior to rebound, the drainage system exiting Lake Connecticut and the adjacent lake in BIS had very little gradient and therefore little erosive power.

By about 15,500 (~18.7 cal ka) years ago, the lake in BIS and Glacial Lake Connecticut had completely drained away to the still lowered sea (Stone et al. 2005). In the LIS Basin, a river system was flowing eastward, across exposed lake deposits and through the former spillway at The Race (Figs. 2.5 and 2.6). This drainage system continued through the BIS Basin, and reached the sea through Block Channel. Stream channels were cut in exposed lake deposits at this time (channel fill, Fig. 2.8). These stream-cut channels contain sands and gravels that were carried across the exposed lake bottom as the stream system developed. Glacial meltwater from southern New England and elsewhere had begun to replenish the oceans, so sea level was rising as Lake Connecticut was draining.

At the same time, as the weight of the receding Wisconsinan glacier was removed, the land was rebounding. Early sea-level rise scenarios for LIS must take into account the fact that, until rebound began to wane (~9,000 radiocarbon years ago = 0.1 cal ka), there was a race between sea level rise and rebound. Stone et al. (2005) present a curve that shows a possible pre-rebound incursion of the sea followed by a regression as rebound kicked in. This was followed by the final incursion of the sea (Fig. 2.5, bottom panel), inferred by Stone et al. (2005) to have begun about 15,000 years ago (~18.1 cal ka). Varekamp et al. (2006) argue for an early incursion in deep stream channels at ~15,000 years ago (~18.1 cal ka), followed by a broader incursion about 5,000 years later. Work by Bryan Oakley and Jon C. Boothroyd at University of Rhode Island would also argue for a later incursion (personal communication, December 2011). More work is needed if we are to better understand this complicated phase of LIS history.

The Stone et al. (2005) scenario holds that about 500 years after Lake Connecticut had completely drained, the sea had risen enough to begin to enter the eastern end of the LIS Basin. It did so by initially moving up the channels that had been cut during the draining of Lake Connecticut. This began the creation of the LIS estuary that we know today. As the channel system filled with seawater, a long, narrow estuary extended westward through the middle of the LIS Basin. Varekamp et al. (2006) suggest that much of the LIS Basin may have been a dry, tundra-like basin at this time. As sea level rose further, the channels were overtopped and the estuary widened. Wave action planed across the lake deposits creating a wave-cut surface (ravinement), which was subsequently buried by marine deposits (Fig. 2.5, bottom panel and Fig. 2.8, marine unconformity-mu, delta distal facies, marine mud). As the estuary widened and deepened, waves and currents began to influence the distribution and character of these deposits (Poppe et al. 2000; Knebel et al. 2000). Widening of the estuary continues today as the rising sea advances northward up the river and stream valleys of the Connecticut coast and southward up the bays and inlets of Long Island's north shore. East of Port Jefferson, New York, the transgression of the sea is exemplified by retreating bluffs and straightening of the shoreline.

Along the mainland to the north, the deltas that formed in Glacial Lake Connecticut now choke the mouths of most bedrock valleys, and they are being drowned and eroded as the sea migrates landward, through coves and up rivers. Most of the sediment available to build beaches comes from the eroding glacial deltas (dark green coastal deposits, Fig. 2.6), so the distribution and size of natural beaches are largely controlled by the distribution and size of the glacial deltas that are their primary sediment source. This is why most natural mainland beaches are pocket beaches confined by rocky headlands. The drowned eroded remnants of the glacial deltas also provide a good platform for marsh development, so most larger marshes are also found in association with drowned valleys.

In the Sound's protected waters, wave energy is not sufficient to overcome the influence of the intervening bedrock promontories, which project seaward enough to inhibit major longshore sediment transport. Since large amounts of sediment are not available to move along the coast, barrier spits remain relatively small and commonly do not completely close off the mouths of inlets. Excluding the protected waters of Clinton Harbor, all of the larger barrier spits that have developed along the Connecticut shore have grown from east to west (e.g., Bluff Point/Bushy Point Beach, Griswold Point, and Point No Point/Long Beach). This may reflect the influence of nor'easters, and it is a clear indication that, over time, the net energy regime along exposed portions of the north shore of LIS favors westward longshore sediment transport (Patton and Kent 1992).

Offshore, a tremendous amount of sediment redistribution has occurred since tides developed in LIS. Although the Mattituck Sill is not a sill in the sense of a structural high (it is underlain by Glacial Lake Connecticut deposits), it does mark the transition from the highly eroded eastern Sound to the depositional environments to the west (Fig. 2.1). Its morphology is strictly the result of erosion and transport of Lake Connecticut deposits out of the east and to the west, and over time the "sill" is moving westward (Lewis and Stone 1991; Fenster et al. 2006). Recognition that Lake Connecticut sediments had been eroded from the eastern Sound and transported and redistributed westward allowed Lewis and

DiGiacomo-Cohen (2000) to construct a sediment budget that balanced Holocene sediment inputs with the volume of marine sediment in LIS, and addressed earlier concerns that all of the marine sediment in LIS could not be accounted for (Bokuniewicz et al. 1976). An excellent regional overview of LIS seafloor sedimentary environments is presented by Knebel and Poppe (2000), and a detailed map of LIS bottom sediment distributions was prepared by Poppe et al. (2000).

Present day LIS is unique in many respects. Owing to the extensive glacial lake deposits that nearly fill its basin, the Sound is fairly shallow, averaging only 65 feet. It parallels the mainland coast and has two openings to the sea, where classic estuaries are more or less perpendicular to the coast, with one opening to the sea. The Sound's orientation relative to the mainland gives rise to two distinct coastlines, one rock-dominated and the other sediment-dominated. The rock-dominated coastline to the north is protected from the large waves of the Atlantic Ocean by Long Island. This protection creates the longest stretch of low-energy, rocky shoreline on the east coast of the US.

2.4 Management Implications

While we understand these and many other aspects of Long Island Sound's unique geologic character, several deficiencies in our present knowledge have management implications. Resolving questions concerning the presence or absence of faults/deposits associated with the Mesozoic Hartford Basin would help address concerns about seismicity and energy infrastructure in LIS. The possibility that significant glacial deposits overlie highly eroded Coastal Plain strata along the north shore of Long Island runs counter to many current geologic portrayals and, if true, would require work to redefine how ground water flows and discharges to the Sound. Unresolved questions related to the nature and timing of the marine incursion into the LIS Basin have implications for our understanding of sea-level rise and climate change. While current efforts to better define the character of the seafloor in LIS are proceeding, they are not receiving the coordinated support necessary to gain the type of holistic understanding of the benthic ecology that is necessary for proper stewardship.

References

Akapati BN (1974) Mineral composition and sediments in eastern Long Island Sound, New York. Marit. Sediments 10:19–30

Aleinikoff JN, Wintsch RP, Tollo RP, Unruh DM, Fanning CM, Schmitz MD (2007) Ages and origins of rocks of the Killingworth dome, south-central Connecticut: implications for the tectonic evolution of southern New England. Am J Sci 307(1):63–118

Antevs E (1922) The recession of the last ice sheet in New England. Am Geograph Soc Res Series 11:120 pp

Balco G, Schaefer JM (2006) Cosmogenic-nuclide and varve chronologies for the deglaciation of southern New England. Quat Geochronol 1:15–28

Balco G, Briner J, Finkel RC, Rayburn JA, Ridge JC, Schaefer JM (2009) Regional beryllium-10 production rate calibration for late-glacial northeastern North America. Quat Geochronol 4:93–107

Bertoni R, Dowling JJ, Frankel L (1977) Freshwater-lake sediments beneath Block Island Sound. Geology 5:631–635

Bloom AL (1967) Coastal geomorphology of Connecticut. Final report. Geography Branch, Office of Naval Research, Contract No 401(45), 79 pp

Bloom AL, Stuiver M (1963) Submergence of the Connecticut coast. Science 139:332–334

Bokuniewicz HJ, Gordon RB (1980) Sediment transport and deposition in Long Island Sound. Adv Geophys 22:69–106

Bokuniewicz HJ, Gebert J, Gordon RB (1976) Sediment mass balance of a large estuary, Long Island Sound. Estuar Coast Mar Sci 4:523–536

Brock PC, Brock PWG (2001) Geologic map of New York City. School of Earth and Environmental Sciences, Queens College, Flushing, New York

Buxton HT, Smolenski DA (1999) Simulation of the effects of development of the ground-water flow system of Long Island, New York. US Geological Survey Water-Resources Investigations Report 98-4069, 57 pp

Buzas MA (1965) The distribution and abundance of foraminifera in Long Island Sound. Smithsonian Misc Coll 4604(149):1–89

Cadwell DH (1991) Surficial geologic map of New York. New York State Museum, Map and Chart Series No 40, 5 sheets (1:250,000-scale), Albany, New York

Cerrato RM, Flood RD (2008) Benthic habit mapping in Long Island harbors and bays. Ninth Biennial Long Island Sound research conference: program and abstracts (Connecticut College). Long Island Sound Foundation, UConn-Avery Point, Groton, CT, 15

Coch CA, Cinquemani LJ, Coch NK (1974) Morphology and sediment distribution in the Little Gull Channel, western Block Island Sound, New York. Geol Soc Am Abstracts with Programs 5(1):149

Coleman ME (2005) The geologic history of Connecticut's bedrock. State Geological and Natural History Survey of Connecticut Special Publication No 2. Connecticut Department of Environmental Protection, Hartford, CT, 30 pp

Connecticut Department of Environmental Protection (1977) Long Island Sound: an atlas of natural resources. Connecticut Department of Environmental Protection, Hartford

Crandell HC (1962) Geology and groundwater resources of Plum Island, Suffolk County, New York. US Geological Survey Water Supply Paper 1539-X, 35 pp

Dana JD (1870) Origin of some of the topographic features of the New Haven region. Trans Conn Acad Sci II:42–112

Dana JD (1890) Long Island Sound in the Quaternary Era, with observations on the submarine Hudson River channel. Am J Sci (Third Series) 40:425–437

de Laguna W (1963) Geology of Brookhaven National Laboratory and vicinity, Suffolk County, New York. US Geol Surv Bull 1156-A:35 pp

Dehlinger P (1978) Marine gravity. Elsevier Scientific Publishing Company, Amsterdam-Oxford-New York, p 322 pp

Dignes TW (1976) Latest Quaternary benthic foraminiferal paleoecology and sedimentary history of vibracores from Block Island Sound. Dissertation, University of Rhode Island, 110 pp

Donner JJ (1964) Pleistocene geology of eastern Long Island, New York. Am J Sci 262:355–376

Edwards RJ, van de Plassche O, Gehrels WR, Wright AJ (2004) Assessing sea-level data from Connecticut, USA, using a foraminiferal transfer function for tide level. Mar Micropaleontol 51:239–255

Feldhausen PH, Ali SA (1976) Sedimentary environmental analysis of Long Island Sound, USA with multivariate statistics. In: Merriam DF (ed) Quantitative techniques for the analysis of sediments: an international symposium. Pergamon Press, Oxford, pp 73–98

Fenster MS, Fitzgerald DM, Bohlen WF, Lewis RS, Baldwin CT (1990) Stability of giant sand waves in eastern Long Island Sound. Mar Geol 91:207–225

Fenster MS, Fitzgerald DM, Moore MS (2006) Assessing decadal-scale changes to a giant sand wave field in eastern Long Island Sound. Geology 34(2):89–92

Flint RF, Gebert JA (1976) Latest Laurentide ice sheet: New evidence from southern New England. Geol Soc Am Bull 87:182–188

Flood RD, Cerrato R (2006) Benthic habit mapping in Long Island harbors and bays. 8th Biennial Long Island Sound research conference, proceedings 2006 (US Coast Guard Academy). Long Island Sound Foundation, UConn-Avery Point, Groton, CT, 76

Foord EF, Parrott WR, Ritter DF (1970) Definition of possible stratigraphic units in north central Long Island, New York, based on detailed examination of selected well cores. J Sediment Petrol 40(1):194–204

Frankel L, Thomas HF (1966) Evidence of freshwater lake deposits in Block Island Sound. J Geol 74:240–242

Fuller ML (1905) Geology of Fishers Island. New York Geol Soc Am Bull 16:367–390

Fuller ML (1914) The Geology of Long Island, New York. US Geological Survey Professional Paper 82, 231 pp

Gayes PT, Bokuniewicz HJ (1991) Estuarine paleoshorelines in Long Island Sound, New York. J Coastal Res, Special Issue No 11:1–23

Goldsmith R (1960) A post Harbor Hill-Charlestown moraine in southeastern Connecticut. Am J Sci 258:740–743

Gordon RB (1980) The sedimentary system of Long Island Sound. Adv Geophys 22:1–39

Goss MC (1993) High-resolution seismic and ice-marginal sedimentation in Block Island Sound and adjacent southern Rhode Island. Dissertation, State University of New Jersey, Rutgers, 131 pp

Grim MS, Drake CL, Hertzler JR (1970) Sub-bottom study of Long Island Sound. Geol Soc Am Bull 8:649–666

Gustavson TC, Boothroyd JC (1987) A depositional model for outwash, sediment sources, and hydrologic characteristics, Malaspina Glacier, Alaska: a modern analog of the southeastern margin of the Laurentide Ice Sheet. Geol Soc Am Bull 99:187–200

Haeni FP, Sanders JE (1974) Contour map of the bedrock surface, New Haven-Woodmont Quadrangles, Connecticut. US Geological Survey Miscellaneous Field Studies Map, MF-557A. Scale 1:24,000

Harwood DS, Zietz I (1977) Geologic interpretation of an aeromagnetic map of southern New England. US Geological Survey Geophysical Investigations Map GP-906

Hermes OD, Gromet LP, Murray DP (1994) Bedrock geologic map of Rhode Island. Office of the Rhode Island State Geologist, Kingston, RI, Rhode Island Map Series No 1, (1:100,000-scale)

Hollick A (1893) Plant distribution as a factor in the interpretation of geologic phenomena, with special reference to Long Island and vicinity. New York Acad Sci Trans 12:189–202

Isachsen YW, Landing E, Lauber JM, Rickard LV, Rogers WB (eds) (1991) Geology of New York: a simplified account. The State Education Department, New York State Museum, Geological Survey, Educational Leaflet No 28, Albany, New York, 284 pp

Isbister J (1966) Geology and hydrology of northeastern Nassau County, Long Island, New York. US Geological Survey Water Supply Paper 1825, 87 pp

Kelly WM, Albanese JR, Aparisi MP, Rodgers WB (1989) Assessment of aggregate and heavy-mineral resources potential in New York coastal waters. In: Hunt MC, Doenges S, Stubbs GS (eds) Proceedings, second symposium on studies related to continental margins—a summary of year-three and year-four activities. Bureau of Economic Geology, The University of Texas at Austin, Austin, pp 161–168

Kinney J, Flood RD (2006) Multibeam bathymetry reveals a variety of sedimentary features in the Peconics potentially significant to management of the system. 8th Biennial Long Island

Sound research conference, proceedings 2006 (US Coast Guard Academy). Long Island Sound Foundation, UConn-Avery Point, Groton, CT, 20–25

Knebel HJ, Lewis RS, Varekamp JC (guest eds) (2000) Regional processes, conditions, and characteristics of the Long Island Sound sea floor. Thematic Section, J Coastal Res 16(3):519–662

Knebel HJ, Poppe LJ (2000) Sea-floor environments within Long Island Sound: a regional overview. Thematic Section, J Coastal Res 16(3):533–550

Koteff C, Pessl F (1981) Systematic ice retreat in New England. US Geological Survey Professional Paper 1179

Koteff C, Stone JR, Larsen FD, Ashley GM, Boothroyd JC, Dincauze DF (1988) Glacial Lake Hitchcock, postglacial uplift, and post-lake archaeology. In: Field Trip Guidebook AMQUA 1988: University of Massachusetts Department of Geology and Geography Contribution No 63, Holyoke, MA

Lewis E (1877) Water courses on Long Island. Am J Sci III (xii), 142

Lewis RS, DiGiacomo-Cohen ML (2000) A review of the geologic framework of the Long Island Sound basin with some observations relating to postglacial sedimentation. Thematic Section, J Coastal Res 16(3):522–532

Lewis RS, Neff (Friedrich) NE, McMaster RL (1990) Non-energy resources, Connecticut and Rhode Island coastal waters. In: Hunt MC, Doenges S, Stubbs GS (eds) Proceedings, second symposium on studies related to continental margins—a summary of year-three and year-four activities, Bureau of Economic Geology, The University of Texas at Austin, pp 169–181

Lewis RS, Neff (Friedrich) NE, Stone JR (1994) Stratigraphic and depositional history of Long Island Sound. In: Dellagiarino G, Masterson AR, Miller LA (eds) Proceedings, third symposium on studies related to continental margins—a summary of year-five and year-six activities, Bureau of Economic Geology, The University of Texas at Austin, pp 120–127

Lewis RS, DiGiacomo-Cohen ML, Neff (Friedrich) NE, Hyde R (1998) Non-energy resources, Connecticut Coastal Waters, year-nine and-ten activities (Abstract). In: Dellagiarino G, Miller LA, Doenges S (eds) Proceedings, fourth symposium on studies related to continental margins—a summary of year-nine and year-ten activities, Bureau of Economic Geology, The University of Texas at Austin, p 29

Lewis RS, Stone JR (1991) Late Quaternary stratigraphy and depositional history of the Long Island Sound Basin; Connecticut and New York. J Coastal Res, Special Issue No 11:1-23

Lewis RS, Stone BD (2012) Multiple glacial development of upper Georges Bank. Geol Soc Am Abstracts with Programs 44(2):71

Lewis RS, Sylwester RE (1976) Shallow sedimentary framework of Georges Bank. US Geological Survey Open-File Report 76-874, 14 pp

Lougee RJ (1953) A chronology of postglacial time in eastern North America. Sci Monthly 7:259–276

Mazzaferro DL, Handman EH, Thomas MP (1979) Water resources inventory of Connecticut, Part 8—Quinnipiac River Basin. Connecticut Water Resources Bull 27:88 pp

McCrone AW (1966) Sediments from Long Island Sound (New York), physical and chemical properties reviewed. J Sediment Petrol 36:234–236

McHugh CM, Cormier MH, Marchese P, Zheng Y, Stewart G, Acosta V et al (2007) Late Quaternary depositional history and anthropogenic changes of western Long Island Sound, New York. EOS Trans AGU 87(52):Abstract OS31B-1638

McMaster RL (1960) Sediments of the Narragansett Bay System and Rhode Island Sound, Rhode Island. J Sediment Petrol 30:249–274

McMaster RL, Ashraf A (1973a) Subbottom basement drainage system of inner continental shelf off southern New England. Geol Soc Am Bull 84:187–190

McMaster RL, Ashraf A (1973b) Drowned and buried valleys on the southern New England Continental Shelf. Mar Geol 15:249–268

McMaster RL, Ashraf A (1973c) Extent and formation of deeply buried channels on the continental shelf off southern New England. J Geol 81:374–379

McMaster RL, La Chance TP, Garrison LE (1968) Seismic-reflection studies in Block Island and Rhode Island Sounds. Am Assoc Petrol Geol Bull 52:465–474

McMullen KY, Poppe LJ, DiGiacomo-Cohen ML, Moser MS, Christman EB (2005) Surficial geology of the sea floor in west-central Long Island Sound as shown by sidescan-sonar imagery. Northeast Geol Environ Sci 27(1):60–70

McMullen KY, Poppe LJ, Schattgen PT, Doran EF (2008) Enhanced sidescan-sonar imagery, north-central Long Island Sound. US Geological Survey Open-File Report 2008-1174 (DVD-ROM)

McMullen KY, Poppe LJ, Danforth WW, Blackwood DS, Schaer JD, Ostapenko AJ et al (2010) Surficial geology of the sea floor in Long Island Sound offshore of Plum Island, New York. US Geological Survey Open-File Report 2010-1005, CD-ROM

Mecray EL, Buchholtz ten Brink MR (2000) Contaminant distribution and accumulation in the surface sediments of Long Island Sound. Thematic Section, J Coastal Res 16(3):575–590

Mills HC, Wells PD (1974) Ice-shove deformation and glacial stratigraphy of Port Washington, Long Island, New York. Geol Soc Am Bull 85:357–364

Needell SW, Lewis RS (1984) Geology of Block Island Sound, Rhode Island and New York. US Geological Survey Miscellaneous Field Studies Map MF-1621 (1:125,000- scale)

Neff (Friedrich) NE, Lewis RS (1988) Non-energy resources, Connecticut and Rhode Island coastal waters. In: Hunt MC, Ratcliff DC, Doenges S, Condon C (eds) Proceedings, first symposium on studies related to continental margins—a summary of year-one and year-two activities. Bureau of Economic Geology, University of Texas at Austin, pp 228–237

Newman WS (1977) Late Quaternary paleoenvironmental reconstruction: some contradictions from northwestern Long Island, New York. Ann NY Acad Sci 288:545–570

Newman WS, Fairbridge RW (1960) Glacial lakes in Long Island Sound. Geol Soc Am Bull 71:1–36

Nydick KR, Bidwell AB, Thomas E, Varekamp JC (1995) A sea-level rise curve from Guilford, CT. Mar Geol 124:137–159

Oliver JE, Drake CL (1951) Geophysical investigation in the emerged and submerged Atlantic Coastal Plain, Part VI—the Long Island area. Geol Soc Am Bull 62:1287–1296

Orson RA, Warren RS, Niering WA (1998) Interpreting sea level rise and rates of vertical marsh accretion in a southern New England tidal salt marsh. Estuar Coast Shelf Sci 47:419–429

Paskevich VP, Poppe LJ (2000) Georeferenced sea-floor mapping and bottom photography in Long Island Sound. US Geological Survey Open-File Report 00-304 (CD-ROM)

Patton PC, Horne GS (1991) A submergence curve for the Connecticut River estuary. J Coastal Res, Special Issue No 11:181–196

Patton PC, Kent JM (1992) A moveable shore, the fate of the Connecticut coast. Duke University Press, Durham, p 143 pp

Peck JA, McMaster RL (1991) Stratigraphy and geologic history of Quaternary sediments in lower West Passage, Narragansett Bay, Rhode Island. J Coastal Res, Special Issue No 11:25–37

Pierce DS, Taylor PK (1975) Geotechnical considerations at Shoreham Nuclear Power Station. In: Wolff MP (ed) Guidebook, 47th meeting. New York State Geological Association, Hofstra University, Hempstead, NY, pp 157–176

Poppe LJ, Polloni C (1998) Long Island Sound environmental studies. US Geological Survey Open-File Report 98-502 (CD-ROM)

Poppe LJ, Lewis RS, Quarrier S, Zajac R, Moffet AM (1994) Map showing the surficial sediments in Fishers Island Sound, New York, Connecticut, and Rhode Island. US Geological Survey Miscellaneous Investigation Series Map I-2456 (1:24,000-scale)

Poppe LJ, Lewis RS, Denny JF, Parolski KF, DiGiacomo-Cohen ML, Tolderlund DS (1998) Sidescan sonar image, surficial geologic interpretation, and bathymetry of the Long Island Sound sea floor in Niantic Bay and vicinity, Connecticut. US Geological Survey Geologic Investigations Series Map I-2625, 3 sheets

Poppe LJ, Knebel HJ, Mlodzinska ZJ, Hastings ME, Seekins BA (2000) Distribution of surficial sediment in Long Island Sound and adjacent waters: texture and total organic carbon. Thematic Section, J Coastal Res 16(3):567–574

Poppe LJ, Knebel HJ, Lewis RS, DiGiacomo-Cohen ML (2002a) Processes controlling the remobilization of surficial sediment and formation of sedimentary furrows in north-central Long
Island Sound. J Coastal Res 18(4):741–750

Poppe LJ, Paskevich VF, Lewis RS, DiGiacomo-Cohen ML (2002b) Geological framework data
from Long Island Sound, 1981-1990: a digital data release. US Geological Survey Open-
File Report 02-002 (DVD-ROM)

Poppe LJ, Ackerman SD, Doran EF, Beaver AL, Crocker JM, Schattgen PT (2006a) Interpolation
of reconnaissance multibeam bathymetry from north-central Long Island Sound. US
Geological Survey Open-File Report 2005-1145 (DVD-ROM)

Poppe LJ, Ackerman SD, Doran EF, Moser MS, Stewart HF, Forfinski NA et al (2006b) Geologic
interpretation and multibeam bathymetry of the sea floor in southeastern Long Island Sound.
US Geological Survey Open-File Report 2006-1059 (DVD-ROM)

Poppe LJ, DiGiacomo-Cohen ML, Doran EF, Smith SM, Stewart HF, Forfinski NA (2007a)
Geologic interpretation and multibeam bathymetry of the sea floor in the vicinity of The
Race, eastern Long Island Sound. US Geological Survey Open-File Report 2007-1012
(DVD-ROM)

Poppe LJ, Denny JF, Williams SJ, Moser MS, Forfinski NA, Stewart HF et al (2007b) The geology of Six Mile Reef, eastern Long Island Sound. US Geological Survey Open-File Report
2007-1191 (DVD-ROM)

Poppe LJ, McMullen KY, Williams SJ, Crocker JM, Doran EF (2008a) Estuarine sediment transport by gravity-driven movement of the nepheloid layer, Long Island Sound. Geo-Marine
Lett 28(4):245–254

Poppe LJ, Williams SJ, Moser MS, Forfinski NA, Stewart HF, Doran EF (2008b) Quaternary
geology and sedimentary processes in the vicinity of Six Mile Reef, eastern Long Island
Sound. J Coastal Res 24(1):255–266

Poppe LJ, Danforth WW, McMullen KY, Parker CE, Lewit PG, Doran EF (2010a) Integrated
multibeam and LIDAR bathymetry data offshore of New London and Niantic, Connecticut.
US Geological Survey Open-File Report 2009-1231(CD-ROM)

Poppe LJ, McMullen KY, Ackerman SD, Blackwood DS, Irwin BJ, Schaer JD et al (2010b)
Sea-floor geology and character offshore of Rocky Point, New York. US Geological Survey
Open-File Report 2010-1007 (DVD-ROM)

Poppe LJ, Danforth WW, McMullen KY, Parker CE, Doran EF (2011) Combined multibeam and
LIDAR bathymetry data from eastern Long Island Sound and westernmost Block Island
Sound. US Geological Survey Open-File Report 2011-1003 (DVD-ROM)

Reeds CA (1927) Glacial lakes and clays near New York City. Nat Hist 27:55–64

Reimer GE (1986) Sedimentology and interpretation of environment of deposition of rhythmi-
cally-laminated silt and clay facies of LISAT #6 core, Long Island Sound. Report to the
State Geological and Natural History Survey of Connecticut, Hartford, Connecticut, Project
86-993, 47 pp with slides

Ridge J (2010) North American Glacial Varve Project. http://geology.tufts.edu/varves/default.asp

Rodgers J (1985) Bedrock geological map of Connecticut. Connecticut Natural Resources Atlas
Series, State Geological and Natural History Survey of Connecticut, Hartford, Connecticut,
2 sheets (1:125,000-scale)

Roe HM, van de Plassche O (2005) Modern pollen distribution in a Connecticut saltmarsh:
implications for studies of sea-level change. Quatern Sci Rev 24:2030–2049

Savard WL (1966) The sediments of Block Island Sound. Dissertation, University of Rhode
Island, Kingston, Rhode Island, 66 pp

Schafer JP (1961) Correlation of end moraines in southern Rhode Island. US Geological Survey
Professional Paper 424-D, pp D68–D70

Schafer JP (1965) Surficial geologic map of the Watch Hill Quadrangle Rhode Island-
Connecticut. US Geological Survey Geologic Quadrangle Maps of the United States, Map
GQ-410 (1:24,000-scale)

Schafer J, Hartshorn J (1965) The Quaternary of New England. In: Wright J, Frey D (eds) The
Quaternary of the United States. Princeton University Press, Princeton, pp 113–127

Schlee J (1973) Atlantic continental shelf and slope of the United States—sediment texture of the northeastern part. US Geological Survey Professional Paper 529-L, 64 pp

Scorca MP, Monti J Jr (2001) Estimates of nitrogen loads entering Long Island Sound from ground water and streams on Long Island, New York. US Geological Survey Water-Resources Investigations Report 00-4196, 29 pp

Shaw J, van de Plassche O (1991) Palnology of late Wisconsinan/early Holocene lake and marsh deposits, Hammock River Marsh, Connecticut. J Coastal Res, Special Issue No 11:85–103

Signell RP, List JH, Farris AS (2000) Bottom currents and sediment transport in Long Island Sound: a modeling study. Thematic Section, J Coastal Res 16(3):551–566

Sirkin LA (1971) Surficial glacial deposits and postglacial pollen stratigraphy in central Long Island, New York. Pollen Spores 13:93–100

Sirkin LA (1976) Block Island, Rhode Island—evidence of fluctuation of the late Pleistocene ice margin. Geol Soc Am Bull 87:574–580

Sirkin LA (1982) Wisconsinan glaciation of Long Island, New York to Block Island, Rhode Island. In: Larson GJ, Stone BD (eds) Late Wisconsinan glaciation of New England. Kendall/Hunt, Dubuque, Iowa, pp 35–60

Sirkin L (1991) Stratigraphy of the Long Island platform. J Coastal Res, Special Issue No 11:217–227

Skehan JW (2008) Roadside geology of Connecticut and Rhode Island. Mountain Press Publishing Company, Missoula Montana, p 288 pp

Stone BD (2012) Contrasts in early and late Pleistocene glacial records in northeastern USA. Resulting from topographic relief, climate change and crustal depression. Geol Soc Am Abstracts with Programs 44(2):86

Stone BD, Borns HW (1986) Pleistocene glacial and interglacial stratigraphy of New England, Long Island and adjacent Georges Bank and Gulf of Maine. In: Sibrava V, Bowen DQ, Richmond GM (eds) Quaternary glaciations in the northern hemisphere. Pergamon Press, Oxford, pp 39–52

Stone JR, Lewis RS (1991) A drowned marine delta in east-central Long Island Sound: evidence for a −40-m relative sea level at >12.3 ka. Geol Soc Am, Abstracts and Programs 23(1):135

Stone JR, Stone BD, Lewis RS (1985) Late Quaternary deposits of the southern Quinnipiac-Farmington lowland and Long Island Sound Basin: their place in a regional stratigraphic framework. In: Tracy RJ (ed) Guidebook for fieldtrips in Connecticut and adjacent areas of New York and Rhode Island, Guidebook No 6. State Geological and Natural History Survey of Connecticut. State of Connecticut, Hartford, CT, pp 535–575

Stone JR, Schafer JP, London EH, Thompson WB (1992) Surficial materials map of Connecticut. US Geological Survey Special Map, 2 sheets (1:125,000-scale)

Stone JR, Schafer JP, London EH, DiGiacomo-Cohen ML, Lewis RS, Thompson WB (2005) Quaternary geologic map of Connecticut and Long Island Sound Basin. US Geological Survey Scientific Investigations Map 2784, 2 sheets (1:125,000-scale) and pamphlet, 71 pp

Stumm F, Lang AD, Candela JL (2004) Hydrogeology and extent of saltwater intrusion in the northern part of the Town of Oyster Bay, Nassau County, New York: 1995-98. US Geological Survey Water-Resources Investigations Report 03-4288, 55 pp

Suter R, de Laguna W, Perlmutter NM, Brashears Jr ML (1949) Mapping of geologic formations and aquifers of Long Island, New York. New York Water Power and Control Commission Bulletin GW-18, 212 pp

Szak C (1987) The nature and timing of late Quaternary events in eastern Long Island Sound. Dissertation, University of Rhode Island, 83 pp

Tagg AR, Uchupi E (1967) Subsurface morphology of Long Island Sound, Block Island Sound, Rhode Island Sound, and Buzzards Bay. US Geological Survey Professional Paper 575C, pp 92–96

Thomas E, Varekamp JC (1991) Paleo-environmental analyses of marsh sequences (Clinton, Connecticut): evidence for punctuated rise in relative sea level during the latest Holocene. J Coastal Res, Special Issue No 11:125–158

Thomas E, Gapochenko T, Varekamp JC, Mecray EL, Buchholtz ten Brink MR (2000) Foraminifera and environmental changes in Long Island Sound. Thematic Section, J Coastal Res 16(3):641–655

Thompson WG, Varekamp JC, Thomas E (2000) Fault motions along the eastern border fault, Hartford Basin, CT, over the past 2,800 years. EOS Trans AGU 81(19):S311

Trumbull JVA (1972) Atlantic continental shelf and slope of the United States—sand-size fraction of bottom sediments, New Jersey to Nova Scotia. US Geological Survey Professional Paper 529-K, pp K1–K45

Upson JE, Spencer CW (1964) Bedrock valleys of the New England coast as related to fluctuations in sea level. US Geological Survey Professional Paper 454-M, 44 pp

van de Plassche O (1991) Late Holocene sea-level fluctuations on the shore of Connecticut inferred from transgressive and regressive overlap boundaries in salt-marsh deposits. J Coastal Res, Special Issue No 11:159–179

van de Plassche O, Mook WG, Bloom AL (1989) Submergence of coastal Connecticut 6000-3000 ^{14}C-years BP. Mar Geol 86:349–354

van de Plassche O, van der Borg K, de Jong AFM (1998) Sea-level climate correlation during the past 1400 yr. Geology 26(4):319–322

van de Plassche O, Erkens G, van Vliet F, Brandsma J, van der Borg K, de Jong AFM (2006) Salt-marsh erosion associated with hurricane landfall in southern New England in the fifteenth and seventeenth centuries. Geology 34(10):829–832

Varekamp JC (1991) Trace element geochemistry and pollution history of mudflat and marsh sediments from the Connecticut coastline. J Coastal Res, Special Issue No 11:105–123

Varekamp JC, Thomas E (2008) Long Island Sound: A thousand year perspective. Ninth Biennial Long Island Sound research conference: program and abstracts (Connecticut College). Long Island Sound Foundation, UConn-Avery Point, Groton, CT, 1

Varekamp JC, Buchholtz ten Brink MR, Mecray EL, Kreulen B (2000) Mercury in Long Island Sound sediments. Thematic Section, J Coastal Res 16(3):613–626

Varekamp JC, Thomas E, Lewis RS (2006) Early history of Long Island Sound. Eighth Biennial Long Island Sound research conference: proceedings (US Coast Guard Academy, New London). Long Island Sound Foundation, UConn-Avery Point, Groton, CT, 27–32

Vargas W, Cormier MH, McHugh C (2008) High resolution geophysical survey of western Long Island Sound offshore of New York: an estuary floor shaped by bottom currents and human activity. Ninth Biennial Long Island Sound research conference: program and abstracts (Connecticut College). Long Island Sound Foundation, UConn-Avery Point, Groton, CT, 26

Veatch AC (1906) Underground water resources of Long Island, New York. US Geological Survey Professional Paper 44, pp 19–32

Walsh GJ, Aleinikoff JN, Wintsch RP (2007) Origin of the Lyme Dome and implications for the timing of multiple Alleghanian deformational and intrusive events in southern Connecticut. Am J Sci 307(1):168–215

Williams SJ (1981) Sand resources and geological character of Long Island Sound. Coastal Engineering Resource Center, US Army Corps of Engineers, Technical Paper No 81-3, Fort Belvoir, Virginia, 65 pp

Wintsch RP, Aleinikoff JN, Walsh GJ, Bothner WA, Hussey II AM, Fanning CM (2007) SHRIMP U-Pb evidence for a Late Silurian age of metasedimentary rocks in the Merrimack and Putnam-Nashoba terranes, eastern New England. Am J Soc 307(1):119–167

Zajac RN, Lewis RS, Poppe LJ, Twichell DC, Vozarik J, DiGiacomo-Cohen ML (2000) Relationships among sea-floor structure and benthic communities in Long Island Sound at regional and benthoscape scales. Thematic Section, J Coastal Res 16(3):627–640

Zurflueh EG (1962) A magnetic map of the Long Island Sound and the southward continuation of geological units in Connecticut. Trans of the AGU 43:435

Chapter 3
The Physical Oceanography of Long Island Sound

James O'Donnell, Robert E. Wilson, Kamazima Lwiza, Michael Whitney,
W. Frank Bohlen, Daniel Codiga, Diane B. Fribance, Todd Fake,
Malcolm Bowman and Johan Varekamp

3.1 Introduction

The ecology and geochemistry of Long Island Sound (LIS) are strongly influenced
by the physical processes that determine the spatial structure and temporal evolution of the temperature (T), salinity (S), and DO concentrations and the distribution of the sediments. Much has been learned about these processes in the last
decade through a combination of theoretical work, process studies, and the analysis of archived observations. The first wide-ranging summary of physical processes
in LIS was published in 1956 by the Peabody Museum of Natural History at Yale
University. Though the collection of papers mainly addressed biological oceanography, the chapter by Riley (1956) described the basic structure of the hydrography, tides, and currents, and much of his insight remains relevant. For the more
general reader, Koppelman et al. (1976) provided a broad overview of the geological origin and geomorphology of LIS and summarized major characteristics of
weather and climate. There has not been an extensive review of the literature since
then. Reviews of the physical processes in neighboring water bodies are available.

J. O'Donnell (✉) · M. Whitney · W. F. Bohlen · T. Fake
Department of Marine Sciences, University of Connecticut, Groton, CT 06340, USA
e-mail: james.odonnell@uconn.edu

R. E. Wilson · K. Lwiza · M. Bowman
School of Marine and Atmospheric Sciences, Stony Brook University, Stony Brook,
NY 11794, USA

D. Codiga
Graduate School of Oceanography, University of Rhode Island, Narragansett,
RI 02882, USA

D. B. Fribance
Department of Marine Science, Coastal Carolina University, Conway, SC 29528, USA

J. Varekamp
Earth and Environmental Sciences, Wesleyan University, Middletown, CT 06459, USA

J. S. Latimer et al. (eds.), *Long Island Sound*, Springer Series on
Environmental Management, DOI: 10.1007/978-1-4614-6126-5_3,
© Springer Science+Business Media New York 2014

In particular, the hydrography and circulation of the Middle Atlantic Bight are summarized by Mountain (2003) and Lentz (2008). A review of Block Island Sound (BIS) and Rhode Island Sound physical oceanography is given by Codiga and Ullman (2010) and corroborated and augmented with recent observations by Ullman and Codiga (2010). The physical processes in the Hudson River are summarized by Geyer and Chant (2006).

We begin with a description of the characteristics of the phenomena that have significant influence on the hydrography and circulation: the freshwater discharge, the wind and wave climate, and the tides. In Sect. 3.2, we then describe the evolution (on time scales longer than tidal) and structure of the T, S, and density distributions in the Sound using a broad array of datasets. The magnitude and structure of the residual circulation are summarized in Sect. 3.3 and evidence for seasonal variations is summarized in Sect. 3.4. In Sect. 3.5, we discuss the effects of wind on sea level and circulation. In Sect. 3.6, we review recent observational and theoretical work on the physical processes that influence the duration and extent of hypoxia. Hypoxia, low concentrations of DO, develops in the deeper waters of the western Sound every summer and the impacted region spreads eastward as summer progresses. The data records resulting from water quality monitoring programs are now several decades long and in Sect. 3.7, we summarize some characteristics of long-term changes that can be detected. In Sect. 3.8, we outline the characteristics and magnitude of the response of the Sound to severe storms and comment on the likely impact of climate change. In the final section, we summarize and comment on the main results of the review.

3.1.1 Freshwater and Saltwater Sources

LIS is often described as an estuary because it is significantly fresher than the adjacent shelf water of southern New England. But it is an atypical mid-latitude estuary in several ways. First, it is unusual that the largest freshwater tributary discharges into the estuary near the main connection to the ocean. Gay et al. (2004) summarized the long-term average discharge of the seven rivers entering the Sound (see Table 3.1). The Connecticut River is the dominant source of fresh water and contributes 75 % of the total gauged discharge. Gay et al. (2004) also pointed out that there is a distinct seasonality to the flow rates. Since the extensive watershed of the Connecticut River includes the mountains in New Hampshire and Vermont, which accumulate precipitation as snow and ice throughout the winter, there is a large freshet in the spring. Figure 3.1 shows that the mean monthly flow in March, April, and May all exceed the annual average flow and the mean April discharge is more than twice the average.

The mechanisms that lead to dispersal of the effluent from the Connecticut River are quite complex. This was the subject of a series of reports by Garvine (1974, 1975 and 1977). With ship surveys and drogue studies, he showed that a large area of ELIS was covered by a thin (2 m) surface layer of brackish water that

Table 3.1 Principal rivers entering LIS as Gauged by the USGS

USGS site name	USGS site number	Gauged area, square miles	1931–2003 mean flow (m³/s)	1995–2001 mean flow (m³/s)
Housatonic River at Stevenson, CT	01205500	1544	84	79
Naugatuck River, Beacon Falls, CT	00196500	115	14	15
Quinnipiac River at Wallingford, CT	01184000	9660	7	6
Connecticut River, Thompsonville, CT	01127500	89	545	508
Yantic River at Yantic, CT	01127000	404	5	5
Quinebaug River at Jewett City, CT	01122500	713	37	37

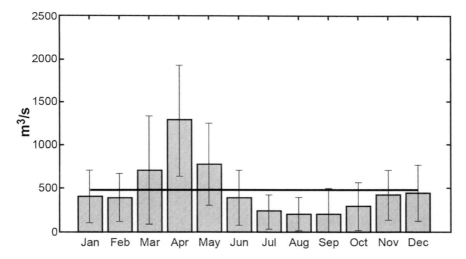

Fig. 3.1 Mean and standard deviation of the mean monthly discharge in the Connecticut River measured at Thompsonville, CT. The horizontal line shows the long-term mean

was separated from the Sound water by a line of foam and detritus, referred to as a front. This river plume was observed to be strongly influenced by the tidal flow in the Sound. On the ebb tide, the front was located to the east of the river mouth and to the west on the flood. During high flows, the front could be tracked almost to The Race. O'Donnell (1997) and O'Donnell et al. (1998, 2008a, b) have made detailed measurements of the structure and motion of water near the front and the mixing rates between the plume and Sound. It is likely that these near field processes affect the distribution and fate of the river water, but the small scales and transient character have made their inclusion in Sound-scale models impossible.

There is substantial inter-annual variation in the discharge of the river that is driven by regional scale meteorological variability, and Whitney (2010) has shown that the Connecticut and several other rivers of the eastern United States correlate with the North Atlantic Oscillation (NAO) index (Hurrell 1995). Using the same record, O'Donnell et al. (2010) showed that the day of the year by which 50 % of the annual discharge has passed the gauge, the "center of volume flow," has become earlier at a rate of 9 ± 2 days/century. It is consistent with the analysis of unregulated New England rivers and streams reported by Hodgkins et al. (2003) that the fresh water stored in the winter ice and snow pack arrives at the ocean through the Connecticut River earlier than in the past and is, therefore, likely to be a consequence of regional meteorological fluctuations rather than changes in watershed management.

A second unusual feature of the Sound is its connection to the estuary of the Hudson River by a tidal strait known as the East River. This causes the Sound to have two connections to the ocean and allows Hudson River water to enter the Sound. The magnitude and variability of the volume and salt fluxes through the East River to the Sound are not well established. The characteristics of the barotropic tidal flow have been described by Jay and Bowman (1975) who showed that the flow in the East River is determined by the phase and amplitude of the sea level imposed at the ends by the ocean and LIS tides, and by frictional processes at the seabed. Wong (1991) considered a broader range of frequencies and showed that though the East River filtered out tidal frequency oscillations, meteorologically forced motions at low frequencies propagated with little damping.

Blumberg and Pritchard (1997) combined direct current measurements and the results of a sophisticated three-dimensional numerical model to estimate the volume flux through the East River. They reported a net westward (toward New York Harbor) mean volume flux of $310 \text{ m}^3/\text{s}$ that was the difference between an eastward transport of surface water of $260 \text{ m}^3/\text{s}$ and lower layer flow of $570 \text{ m}^3/\text{s}$. The model results (see Fig. 3.2) also showed that these upper and lower layer transports had substantial variation ($150 \text{ m}^3/\text{s}$) with maxima in the winter. The variations in the two layers are negatively correlated with the result that the net westward transport only varies by $50 \text{ m}^3/\text{s}$.

Gay et al. (2004) developed a salt budget model for LIS that exploited an extensive archive of S measurements to estimate salt fluxes through the East River and The Race that were consistent with the S data and the Blumberg and Pritchard (1997) results. They concluded that salt was transported out of LIS into New York Harbor at a rate of 11,000 kg/s. This tends to freshen the western end of the Sound. Crowley (2005) also developed a circulation model that assimilated S and T observations in the Sound and used it to estimate the salt flux and freshwater flux. Her estimate of the average salt flux over the period April through September 1988 was 16,000 kg/s and the average freshwater flux was $150 \text{ m}^3/\text{s}$. This is a substantial flux, comparable to that of the major rivers except the Connecticut.

The model and analyses of Crowley (2005) also further our understanding of the relative importance of mechanisms contributing to the longitudinal flux of salt in LIS. She found, for example, that westward salt flux associated with the Stokes transport dependent on the phase relationship between tidal surface elevation and

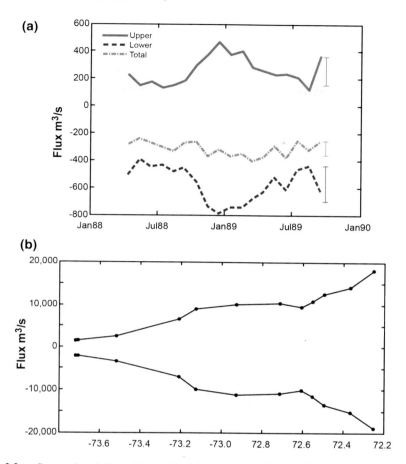

Fig. 3.2 **a** Seasonal variation of the volume fluxes through the East River from the model of Blumberg and Pritchard (1997). *Blue* shows the lower layer transport to the east and the *red* shows the counter flow in the surface layer. *The green* line shows the net transport. **b** The mean (April–September) volume exchange as a function of distance along the Sound. Redrawn from Crowley (2005)

tidal current made a substantially larger contribution to the total westward salt flux than the gravitational circulation. In a typical estuary, the upstream Stokes transport tends to be balanced by a downstream barotropic transport. In LIS, because of the unique boundary condition at the western end, that is not the case.

The largest volume exchange between LIS and surrounding water bodies occurs at the eastern boundary of the Sound through The Race. Riley (1956) used salt and drift pole observations to argue that the volume exchange in the eastern Sound was approximately 15,000 m³/s. This estimate ignored shear flow dispersion by tidal currents. Gay et al. (2004) allowed both dispersion and advection, and found that the exchange flow required to balance the salt budget was 25 % of Riley's estimate. Using 2 years of direct current observations from a vertically profiling acoustic current meter mounted on a ferry that crossed the eastern Sound approximately eight times a day during a 3-year interval (see Codiga 2007), the annual mean nontidal volume exchange was estimated by Codiga and Aurin (2007) to be 22,700 ± 5,000 m³/s. Codiga and Aurin (2007) suggested the low value of Gay et al. (2004) was a consequence of their parameterization of the variation of the transport coefficients using a power law dependence on freshwater inputs, or the inadequate representation of the lateral variations. The along-Sound variation of the positively and negatively directed volume fluxes in the model of Crowley (2005) is shown in Fig. 3.2b. At the western end of the domain, the fluxes are comparable to those of Blumberg and Pritchard (1997). At the eastern end, the values are much larger, 17,000 m³/s which is closer to the Riley (1956) estimate than that of Gay et al. (2004). Figure 3.2b also shows that along-Sound volume fluxes decrease rapidly from east to west in the eastern Sound and are only ~10,000 m³/s at −72.6°. Mau et al. (2008) and Hao (2008) recently developed models that followed the approach of Crowley (2005) and assimilated S and T measurements to predict the subtidal currents, fluxes, and values at the eastern end of the Sound. These also show general agreement with the Riley (1956) and Codiga and Aurin (2007) estimates. However, since the observations of Codiga and Aurin (2007) are in a region of rapid spatial variation in volume transport, and since there is an absence of salt flux estimates with which to evaluate the assimilative models, much remains to be learned about the flow and rates of exchange of materials between the Sound and adjacent waters.

Additional uncertainties in the freshwater budget of the Sound arise from lack of information about the rates of evaporation and precipitation. The maps of annual evaporation and precipitation rates in New England created by Krug et al. (1990) by integrating a diverse array of measurements suggest that there is a net delivery of fresh water at a rate of 0.5 m/year in coastal Connecticut. Assuming this rate applies over LIS leads to an approximate volume flux of 50 m³/s. Using well level measurements and a groundwater flow model, Buxton and Smolensky (1999) estimated that the delivery of fresh water to the Sound from deep and shallow aquifer systems was 109 m³/s. The magnitudes of these two sources are comparable to the discharge from the Housatonic River and 10–20 % of that of the Connecticut River. In their salt budget, Gay et al. (2004) assumed that the precipitation-evaporation fluxes were approximately in balance and that the groundwater flux was small. Though uncertainties of this magnitude may be tolerable for some purposes, the waters of New York Harbor, the New England shelf, and subsurface aquifers contain solutes (e.g., nitrate) at much different concentrations than those in the Sound and the volume exchanges consequently play a much larger role in those budgets. The improvement of estimates of these fluxes through direct measurement deserves attention.

3.1.2 Wind and Waves

Stress generated by wind at the surface of the ocean has a major influence on the circulation and rate of vertical mixing. Winds also generate surface gravity waves that influence sediment resuspension and the magnitude of the bottom stress on lower frequency motions (Signell et al. 2000; Wang et al. 2000). LIS lies in the westerlies but there is a distinct seasonality to the statistical characteristics of winds (Isemer and Hasse 1985). Using all long-term land station data archived by the National Climatic Data Center (NCDC, see http://www.ncdc.noaa. gov/oa/ncdc.html), Klink (1999) showed that the monthly mean surface wind velocity vectors in southern New England were directed to the southeast in winter and to the northeast in summer with much lower speeds. Lentz (2008) analyzed records of wind observations from five National Data Buoy Center (NDBC) buoys and six bottom attached towers and masts spanning the Middle Atlantic Bight (MAB). He reported seasonal trends in velocity that were consistent with Klink's continent-scale pattern. Additionally, his analysis of the wind stress revealed a MAB-wide annual mean of 0.02–0.03 Nm^{-2} directed toward the southeast with an annual amplitude cycle 0.03–0.04 Nm^{-2} in the same direction. This reinforces the mean across-shelf stress in the winter and opposes it in the summer. He also showed that the stresses at the buoys were more than twice those estimated at the coastal sites.

In the area of LIS, there are few long series of wind measurements. The longest record available in the NCDC archive is at Bridgeport (Sikorsky Memorial Airport), which has been in operation since 1942 and is located close to the shore and near the center of LIS (see Fig. 3.3a). Most analyses of the response of the Sound to winds have used this dataset or measurements at another regional airport. They assumed that the velocity can be extrapolated over water, and then estimated the wind stress with a bulk formula like that of Large and Pond (1981). Recently, over water observations have become available from the Long Island Sound Integrated Coastal Observing System (LISICOS) (http://lisicos.uconn.edu) at the Execution Rocks (EXRK), western and central Long Island Sound (WLIS and CLIS) buoys. The locations of these buoys are shown in Fig. 3.3a. Howard-Strobel et al. (2006) presented wind roses using the CLIS buoy observations. Here we summarize a more extensive record and provide a comparison of the statistical character of wind forcing over the central and western Sound.

Figures 3.3b and c show the annual mean wind stress vector components computed from hourly measurements of velocity at Sikorsky Airport (Bridgeport) and the WLIS and CLIS buoys for the period January 1, 2004 to December 30, 2009. The red symbols show the mean and 68 % confidence interval at the CLIS buoy assuming a three-day autocorrelation time scale (see Emery and Thompson 1997). The blue and green symbols show the components at the WLIS buoy and Bridgeport. The seasonal cycle is clear in the components of all three records. The monthly mean wind stress is weak from April to September and has a maximum magnitude in February of 0.07 Nm^{-2} toward the southeast at the CLIS buoy. Note that the amplitude of the cycle at the CLIS site is 0.04 Nm^{-2}, in agreement with the intra-annual variations in the MAB estimated by Lentz (2008). The means and the amplitude of the annual cycle at the WLIS buoy and Bridgeport are approximately 60   and 30 %, respectively, of the value at the CLIS buoy. The annual means, shown in Fig. 3.3a by the red arrows, are dominated by the winter conditions, which are shown in blue. Note that the orientations of the vectors (~125 *T*) at the three sites are consistent. They are also consistent with the southeasterly directed means computed throughout the MAB by Lentz (2008). However, the CLIS and

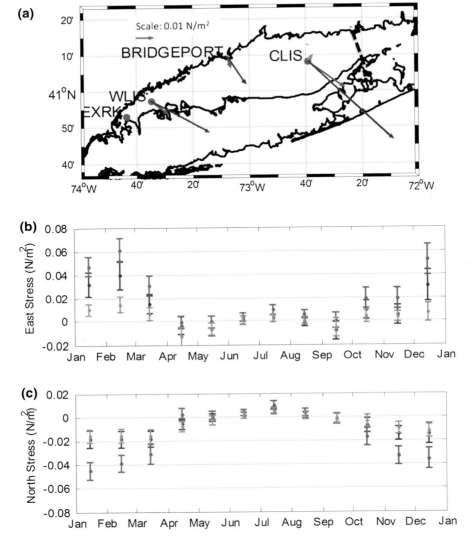

Fig. 3.3 **a** A map of the coastline of LIS showing the locations of the EXRK, WLIS and CLIS buoys, and Bridgeport Airport. The *red*, *blue*, and *green* vectors show the annual, the winter, and summer mean stress vectors. The *dashed line* shows the approximate track of the Bridgeport-Port Jefferson Ferry. **b** and **c** show the monthly mean east and north stress components for the period January 2004–December 2009 together with their 68 % confidence interval at Bridgeport, CT (*green*), WLIS (*blue*), and CLIS (*red*)

WLIS magnitudes, 0.026 and 0.012 Pa, are significantly different. It appears that the wind forcing of the central Sound at seasonal time scales is equivalent to that of the MAB shelf. It is also clear that the waters of the western Sound experience less stress as estimated by the parameterizations developed for ocean conditions.

The spatial scale over which the boundary layer adjusts to the surface stress imposed by the water remains unclear since there are few direct measurements of stress variation. Lentz (2008) noted that the stations closer to shore in the MAB showed reduced mean wind velocities and suggested that scale for the influence of land was less than 10 km. Our finding that the energy in the across-Sound stress at the buoy further from shore (CLIS) is much larger than that at the WLIS buoy is consistent with his result.

Mooers et al. (1976) demonstrated that synoptic scale weather systems dominate the variability in the wind, pressure, and precipitation in the MAB and that they can drive significant currents in LIS. Figure 3.4a and c show the winter and summer autospectra, $S_{\tau_{xr}}$, of the along-Sound (70° T) projection of the wind stress vector, τ_{xr}, estimated from velocity measurements at the CLIS (red) and WLIS (blue) buoys for the period 2004–2009 using the parameterization of Large and Pond (1981). The spectra, $S_{\tau_{yr}}$, of the across-Sound wind stress component, τ_{yr}, in winter and summer are shown in Fig. 3.4b and d. The spectra were computed using the Welch method (see Emery and Thompson 1997) with a Hanning window

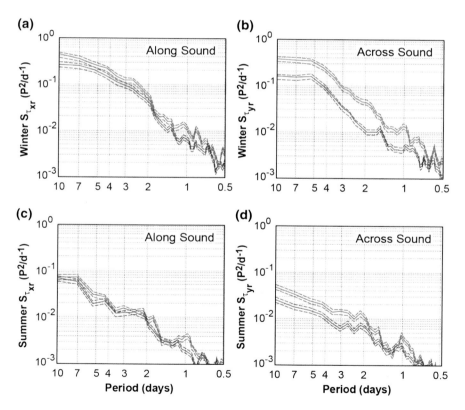

Fig. 3.4 The spectra of along- and across-Sound stress components. **a** and **b** Show the winter time spectra for the along-Sound components. **c** and **d** Show the same quantities for summer. In all graphs, *blue* represents the WLIS and *red* the CLIS spectra

and all possible 512 h nonoverlapping segments with less than 5 % data loss and no data gaps longer than 2 h. Short gaps were filled by linear interpolation. The spectra have a bandwidth of 0.47 days^{-1} and 62° of freedom.

All spectra show that energy in the 2–10 day (synoptic) band is an order of magnitude higher than at shorter periods as is typical in the mid-latitudes. By comparing Fig. 3.4a–d, it is clear that the magnitude of $S_{\tau_{xr}}$ at CLIS (red) and WLIS (blue) during both summer and winter is more alike than for $S_{\tau_{yr}}$. The width of the Sound appears to have a greater effect on the across-Sound wind stress. The stress fluctuations at the WLIS buoy are generally a factor of two to four less energetic than those at the CLIS buoy, with the exception of $S_{\tau_{xr}}$ in the summer (Fig. 3.4c) when they are of comparable magnitude. In the synoptic band, the coherence between the WLIS and CLIS buoys (not shown) is also high, greater than 0.6, as is to be expected since the synoptic weather systems are much larger scale than the separation of the buoys. At shorter periods, the coherence drops to less than 0.3. We conclude that though the mean and long-period fluctuations in the wind stress over the Sound are spatially coherent with larger amplitudes near CLIS, a large fraction of the variance at and below the daily period is not. The study of processes that are sensitive to high frequency variations in the wind (e.g., sea breezes) and their consequences will require local observations to resolve the spatial variability in this frequency range.

Figure 3.5 shows the frequency distribution of wind direction observed at LaGuardia Airport, New York during July and August using data from 1960 to 2010. There are three distinctive summertime wind direction regimes. The predominant direction is centered on approximately 180°. This is associated with

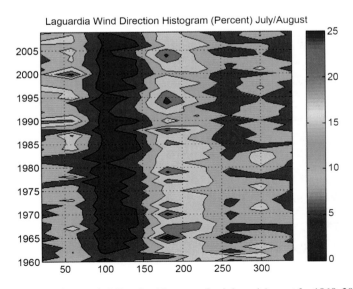

Fig. 3.5 LaGuardia Airport wind direction histogram for July and August for 1960–2010

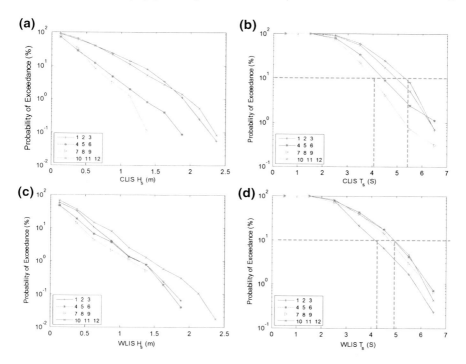

Fig. 3.6 Seasonal variation in the probability of exceedance, expressed as a percentage, for significant wave heights and periods at the CLIS, **a** and **b**, and WLIS sites, **c** and **d**. Data from January to March is shown by *blue lines* and '+' symbols; October–December by *magenta* with 'x' symbols; April-June by *red lines* and '*' symbols; and July to September by *green lines* and *triangles*

summertime ridge and trough systems. These winds are interrupted by intervals of wind from the northeast and northwest. Analyses of available NOAA historical daily weather maps indicate that northeast winds over the western Sound in the summer are typically associated with hybrid systems with low pressure to the southeast of the Sound and a high to the northwest. Northwest winds arise when a low is to the northeast of the Sound and a high is to the southwest.

There are few published observations of the characteristics of the wave height and period in LIS. Bokuniewicz and Gordon (1980a and b) and Signell et al. (2000) have made estimates of the wave-induced bottom currents using the predictions of fetch limited wave models to interpret patterns of sedimentation, but direct measurements with which to evaluate the performance of the models have been limited. Rivera Lemus (2008) provided a brief summary of the wind and wave climatology using a year-long interval of data from the LISICOS CLIS buoy. However, a much longer record has been established and we describe the observations from 6 years of observations of a three-axis directional wave gauge at the CLIS buoy and a single axis sensor at WLIS. Here we briefly summarize statistics of the observations and outline a simple explanation for the structure and variability.

To summarize the data, we define $f_H\left(H_{s,j}; M_i\right)$ as the fraction of hourly significant wave height observations obtained during the set of months $M_{i \in [1:12]}$, in the interval $H_{s,j} - \delta H < H_s < H_{s,j} + \delta H$. We then estimate the probability that an observation will exceed $H_{s,j}$ as $P\left(H_{s,j}; M_i\right) = 1 - \sum_{k=1}^{j} f_H\left(H_{s,k}; M_i\right)$. The probability that the significant wave period will exceed is computed in an analogous manner. The probability of exceedance, $P\left(H_{s,j}; M_i\right)$ for $M_{i=\{1,2,3\}}$, expressed as a percentage, for significant wave height during January–March at CLIS is shown by the blue line in Fig. 3.6a. The distribution closely follows that for $M_{i=\{10,11,12\}}$, October–December, which is plotted in magenta. The spring and summer distributions at CLIS are also similar to each other. Note that even with 6 years of data, there are few estimates of high H_s values and extrapolation is frequently employed to estimate the probability of very large values for engineering design applications. It is clear from the data, however, that the frequency of large waves is much higher in the October–March period. In fact, for much of the range of H_s, the probability of exceeding a particular value is a factor of 10 higher in the winter. $P\left(H_{s,j}; M_i\right)$ for WLIS is shown in Fig. 3.6c. It does not display as strong a seasonal variation as observed at CLIS though the probability of exceeding a particular H_s is at least a factor of two larger in October–December than in June–August.

The probability of exceedance of the significant wave period, $P\left(T_{s,j}; M_i\right)$, is shown in Fig. 3.6b and d for CLIS and WLIS, respectively. At CLIS, the seasonal variation is similar to that displayed in the H_s graph. In the summer, June–August, T_s is less than 4 s in 90 % of observations and in winter, January–March, the periods generally are longer with 90 % of observations less than 5.3 s. At WLIS (see Fig. 3.6d) only $P\left(T_{s,j}; M_i\right)$ for January–March is anomalous and wave periods are shorter than in the summer.

A short example of the time series of wind stress vector components and H_S from the LISICOS buoys during December 2007 is shown in Fig. 3.7a and b. The black line in Fig. 3.7b shows the evolution of H_S at CLIS and the red line shows the variation of H_S at WLIS. Examination of the wave heights series in Fig. 3.7b reveals high correlation. However, there are intervals when the wave heights are almost identical. Some examples are indicated by arrows in Fig. 3.7b. At other times H_s at WLIS is only half of that at CLIS. As has been shown previously, the components of the wind stress at WLIS and CLIS are clearly coherent with each other at longer periods, so it is not immediately obvious why there are differences in the wave heights at the two buoys. The gray points in Fig. 3.8a show the relationship between the mean significant wave height, $\bar{H} = (H_{SW} + H_{SC})/2$, and the along-Sound component of the wind stress at CLIS, τ_{xr}, using all observations between January 2004 and December 2009. The error bars show the median and 68th percentile range of $\bar{H}$ in 0.05 Nm^{-2} intervals of stress. Though there is considerable scatter in the data, the trend is for $\bar{H}$ to increase symmetrically with the magnitude of τ_{xr}. Figure 3.8b shows the dependence of the difference in wave height, $\Delta H = H_{SC} - H_{SW}$, on τ_{xr}. As in Fig. 3.8a, the data are shown by the gray dots,

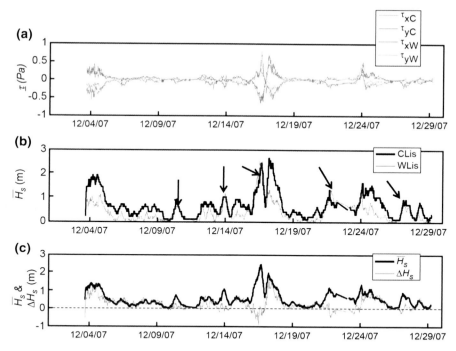

Fig. 3.7   a Time series observations of along (*red*) and across (*green*) stress at the CLIS buoy. The *magenta* and *blue lines* $\bar{H}_s = (H_{SW} + H_{SC})/2$ how the corresponding quantities at the CLIS site. **b** The significant wave height at CLIS (*black*) and WLIS (*red*). **c** The average (*black*) and difference (*red*) in the wave heights at CLIS and WLIS

and the median and 68th percentile intervals are shown by the error bars. Though positive along-Sound wind stress observations are more common, it is clear in Fig. 3.8b that wave heights are larger at CLIS when stress is positive (i.e., the wind is from the west southwest). During periods of strong winds, the waves in CLIS can be 1 m larger than at WLIS. In contrast, when the stress is negative (i.e., winds from the east northeast), the waves are the same. Figure 3.8c shows the influence of the across-Sound wind stress, τ_{yr}, on ΔH. Strong winds from the south are infrequent in southern New England, so most of the data in Fig. 3.8c is for $\tau_{yr} < 0$ It is clear, however, that $|\Delta H|$ increases with $|\tau_{yr}|$ and that the wave height at CLIS increases relative to WLIS with the magnitude of the across-Sound stress component.

Like the wind at short periods, the wave climate in WLIS is quite different from that in the central Sound. There is also significant seasonal variation that has higher amplitude in the central Sound. Waves generally have larger amplitude and longer period in the central Sound, especially in winter. In the western Sound, waves are largest when the winds are easterly (with a negative, or westward, along-Sound stress component) and H_s is comparable to that in the central Sound. As pointed out by Bokuniewicz and Gordon (1980a, b), (Signell

Fig. 3.8 **a** The *gray* points
show mean of the significant
wave height at WLIS and
CLIS, dependence on the
along-Sound component
of the wind stress at CLIS,
τ_{xr}, using all observations
between January 2004 and
December 2009. The error
bars show the median and
68th percentile range of $\bar{H}$
in 0.05 Nm^{-2} intervals of
stress. **b** The dependence
of the difference between
the significant wave height
observations at CLIS and
WLIS, $H_{SC} - H_{SW}$,
on the mean along-Sound
component of the wind stress.
c The dependence of the
$H_{SC} - H_{SW}$ on the mean
across-Sound component of
the wind stress at CLIS

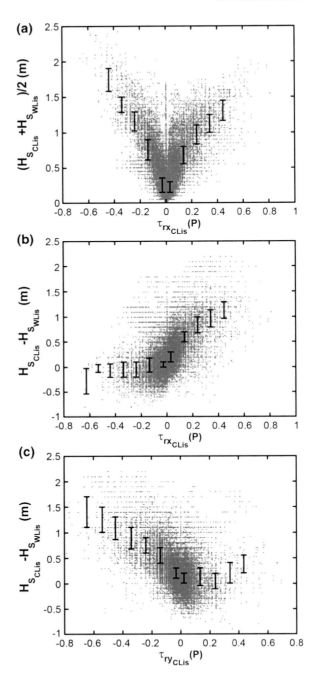

et al. 2000), and Rivera Lemus (2008), the dependence of waves on wind in LIS is largely consistent with the fetch limitation theory of Sverdrup and Munk (1946) and Bretschneider (1952). The difference between the response of wave in the western and central Sound is also consistent with this idea. When the wind is from the east, the fetch at both buoys is large and the wave statistics are similar. For all other directions, the fetch at the WLIS buoy is much smaller than at CLIS and so are the waves.

3.1.3 Tides

The dominant cause of motion of water in the Sound is the barotropic pressure gradient force created by sea level fluctuations on the adjacent continental shelf. Observations of sea level made by Le Lacheur and Sammons (1932), Redfield (1950), and Koppleman et al. (1976) showed that the amplitude of sea level fluctuations at the semidiurnal frequency increased by a factor of three from The Race in the east to Kings Point in the west. A simple mathematical model developed by Redfield (1950) suggested that this was because the length of the Sound was close to a quarter of the wavelength of the semidiurnal tidal wave and, therefore, the basin was near resonant at the semidiurnal frequency. The first comprehensive description of the tides in LIS was summarized by Swanson (1976) and his results were broadly consistent with the simple theory. Ianniello (Ianniello 1977a and b) extended the model approach adopted by Redfield (1950) to include the effects of the main characteristics of the basin geometry and predicted the along-Sound variation and vertical structure of the laterally averaged current and sea level amplitudes. Using the model, he investigated the impact of several plausible assumptions about the structure of the vertical eddy viscosity on the structure of the tidal currents, but the data available were inadequate to eliminate any of the possibilities. However, Ianniello's model showed that the mean flow driven by periodic tides might have comparable magnitude to that expected to be driven by baroclinic pressure gradients. This approach was extended to include the effects of lateral depth variations on the vertically averaged flow by Li and O'Donnell (1997 and 2005). Recently, Winant (2007) unified these theories in a concise framework that describes the three-dimensional circulation driven by barotropic tides.

   Swanson's (1976) description of the variation of phase in the western Sound and the East River, and the demonstration of the effect of winds on sea level predictions using harmonic analysis, prompted the development of the much more sophisticated two-dimensional numerical models of Murphy (1979) and Kenefick (1985). Though these models were very successful in simulating sea level variations, current observations were only available in a few locations and were inadequate to critically assess the predictions of the spatial structure of the velocity amplitudes. The numerical approach to the solution of the vertically averaged equations was extended by Bogden and O'Donnell (1998) to include the capability to assimilate current meter observations to predict tidal and meteorologically forced circulation in a segment of the central Sound. Using models and observations, they demonstrated the existence of significant across-Sound structure in the vertically averaged circulation that they attributed to wind forcing. Practical demands for predictions of the transport and fate of materials in the Sound, together with the development of acoustic Doppler current profilers (ADCPs), prompted the development of three-dimensional circulation models (see Valle-Levinson and Wilson 1994a and b; Valle-Levinson et al. 1995; Schmalz 1993; Blumberg et al. 1999; and Signell et al. 2000).

Recently, Hao (2008) extended the study of Crowley (2005) and Wilson et al. (2005) who implemented the Regional Ocean Modeling System (ROMS) (see Shchepetkin and McWilliams 2005) for the Sound and reported the results of a comprehensive analysis of the dynamics of the tidal circulation. Hao (2008) applied scaling arguments based on the three parameters introduced by Winant (2007) in his barotropic theory for frictional tidal flow in basins with topography: the basin aspect ratio $\alpha = \frac{B}{L}$, a friction parameter $\delta = \left(\frac{2K}{\omega H^2}\right)^{1/2}$, which is the ratio of the amplitude of the periodic boundary layer thickness to water column depth, and the ratio of basin length to tidal wavelength $\kappa = \frac{\omega L}{\sqrt{gH}}$. Using these parameters and scales appropriate for LIS, she argued that the longitudinal and lateral tidal-period momentum balances were fundamentally different. The lateral tidal-period balance was between acceleration, Coriolis force associated with the vertically sheared longitudinal current, and barotropic pressure gradient. This leads to tidal-period lateral circulation that is clockwise on flood and anti-clockwise on ebb as shown in Fig. 3.9. Since this circulation pattern may result in a transport of DO from the shallow, nearshore areas to the hypoxic deep water of the western Sound, these predictions deserve careful comparison with observations in the near future.

Bennett et al. (2010) reported observations of the vertical structure of the principal tidal constituents in LIS using long deployments of ADCPs at 11 sites and the analysis approach of Pawlowicz et al. (2002). They identified the tidal constituents that were dominant at each station and described their vertical structure. The results were presented in tables that should be used in the future to evaluate models and guide improvements to realistic simulations. Figure 3.10 shows the locations of the current meters and the near surface and near bottom major and minor axes of the M2 (lunar semidiurnal) tidal current ellipses, the largest harmonic at all sites. In the eastern Sound (see Fig. 3.10a), the observation reveals strong M2 major axis amplitudes oriented along the axis of the Sound at the surface and parallel to isobaths near the bottom. At the ADCP, labeled m03 in Fig. 2.10a, the near surface major axis amplitude of the M2 was 0.9 m/s. The other semidiurnal constituents, N2, S2, and L2, were 0.19, 0.12, and 0.08 m/s, respectively. The superposition of these harmonics leads to a large spring-neap variation, which Valle-Levinson and Wilson (1994a) described at a site slightly to the south of station M3. They also proposed that the relative importance of turbulent stresses in the tidal dynamics was substantially different at spring and neap stages.

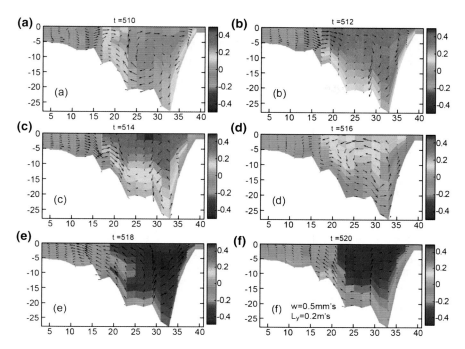

Fig. 3.9 Evolution of predicted along-Sound velocity component (*color shading*) and the velocity vector components in the across-Sound plane (*arrows*) in the central Sound at spring tide. **a** shows the flow at simulation hour and **b–f** show the structure at 2 hour intervals. The velocity units are m/s and the scale for the vectors is shown in panel (**f**)

At the stations shown in Fig. 3.10a, the phase variation in the M2 constituent in the horizontal is very small with currents at the shallower sites leading those in deeper water, a consequence of the increased relative importance of bottom friction. The variation in the M2 phase among the near surface records at stations M02, M03, and M06 is only 6°. However, at station M05 in the shallow water north of Long Sand Shoal, the M2 phase is advanced at all depths by 20° or 40 min. The observed weak phase variation along the Sound is consistent with the idea that the tide is almost a standing wave.

In the western Sound, the M2 constituent is also the dominant harmonic, and the major and minor axes of the near surface and bottom M2 ellipses are shown in Fig. 3.10b. Note that the scale is different from Fig. 3.10a and that the current amplitude is much smaller. At the ADCP labeled fb03, for example, the major axis amplitudes for M2 and N2 at the surface are 0.28 and 0.06 m/s. Bennett (2010) showed that this reduction in the amplitude of the semidiurnal constituents is largely consistent with a simple vertically averaged, frictional linear long wave model with channel convergence that represents the major geometric characteristics of the Sound. The maximum phase difference in the M2 constituent at the surface between the ADCP at location bw, in CLIS, and the western stations shown in Fig. 3.10b is only 40° or 80 min.

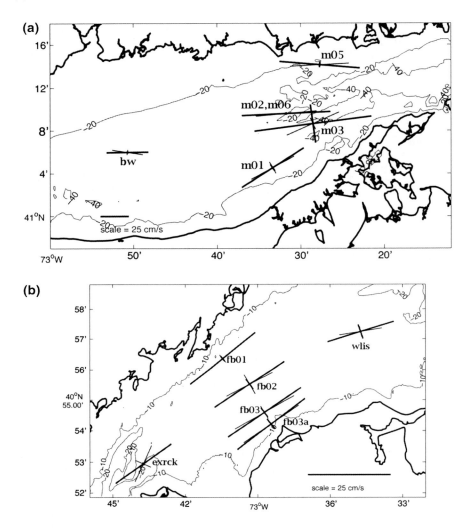

Fig. 3.10 Map of the coastline and bathymetry of **a** eastern and **b** western LIS showing the major and minor axes of the M2 tidal current ellipse at the surface (*thick lines*) and near the bottom (*thinner lines*). Note that the velocity scale is shown in the lower left corner of **a** and in the lower right of **b** and the bathymetric contours are at 10 and 20 m

The vertical structure of the principal harmonics was also described by Bennett et al. (2010) and they compared their results to the predictions of the model of Ianniello (1977a) using several alternate formulations of the vertical structure of the eddy diffusion coefficient. They found all were inconsistent with the observations. Codiga and Rear (2004) found similar results at sites in the approaches to BIS. They demonstrated that the most plausible explanation for the vertical structure was a modification of the planetary vorticity by horizontal shear in the residual currents. Whether it requires an unsteady closure model, a more realistic geometry, or inclusion of residual vorticity effects to simulate the vertical structure of the tides, is an important and unresolved question.

Figure 3.11a–d show the vertical structure of the M2 tidal ellipse parameters (semi-major axis amplitude, phase, orientation, and ellipticity, respectively) as defined by Pawlowicz et al. (2002) at the WLIS station, computed using four different deployment intervals. The profiles computed from data acquired during the more highly stratified seasons, when the surface to bottom density difference exceeded 1 σ_T, are shown in red and those from less stratified conditions for are shown in black for comparison. Though the magnitudes of the uncertainties are comparable to the differences between the more and less stratified amplitudes, it is evident that the vertical gradient of the major axis amplitude is significantly enhanced during the more stratified intervals. The profiles in Fig. 3.11b also show changes to the vertical structure of the phase. During the less stratified periods, the phase difference between the top and bottom bin averages approximately 20°, and the gradient is almost uniform. During the more stratified intervals, however, the phase changes approximately 25° in the lower half of the water column, whereas it remains uniform in the upper half. Figure 3.10c and d show that during times of higher stratification, the orientation is closer to zero in the lower water column and that the ellipticity increases substantially at the bottom. These characteristics are also evident at other stations. The theory for the structure of tidal currents that are influenced by turbulent friction, the Coriolis acceleration, and stratification has been reviewed by Soulsby (1990) and shown to be useful in areas far from shore by, for example, Codiga and Rear (2004). The most comprehensive analytic theory for the three-dimensional structure of tides in channels was recently developed by Winant (2007); however, Bennett (2010) found that its predictions were inconsistent with her observation in LIS. Since the Winant (2007) theory is based on

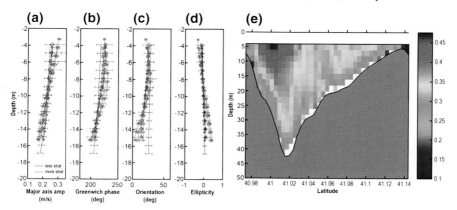

Fig. 3.11 The vertical structure of the M2 ellipse characteristics at the WLIS site computed using data from four separate deployments. Shown in panel **a** is the major axis amplitude, **b** the phase relative to Greenwich, **c** the orientation of the ellipse, and **d** the ellipticity (minor axis amplitude/major axis amplitude). The *red lines* show results from measurements acquired during more highly stratified, a surface to bottom density difference in excess of 1 σ_T, intervals. **e** the lateral and vertical structure of the M2 amplitude at the Bridgeport-Port Jefferson Ferry section

constant eddy coefficient turbulent closure and simple channel geometry, it seems likely that the complexity of the real geometry must be resolved and a sophisticated closure adopted as in the model of Hao (2008) for the tidal flow to be fully understood.

Observations from the vessel-mounted 600 kHz ADCP aboard the Bridgeport to Port Jefferson Ferry (see map in Fig. 3.24) provide a description of the structure of both longitudinal and lateral tidal currents in a lateral section in the central basin. Figure 3.11e shows the structure of amplitude of the M2 constituent of the component of current normal to the section for 2 m bins obtained from a harmonic analysis of a 22-day record segment; the center of the first bin is at approximately 5 m. Currents are surface intensified with significant lateral shear associated with depth variations. The corresponding structure of phase of the M2 constituent (not shown) indicates clear phase advance near the bottom and in the shallows on the channel flanks.

Tidal period fluctuations in the magnitude of vertical stratification in estuaries can result from differential advection of the mean along-Sound horizontal density gradient by vertically sheared along-Sound tidal currents. Simpson et al. (1990) referred to this effect as tidal straining and pointed out its influence on vertical mixing. A combination of current profile observations and density structure measurements has revealed that this mechanism is active in the eastern Sound. Figure 3.12a and c show the vertical structure of the M2 phase average of

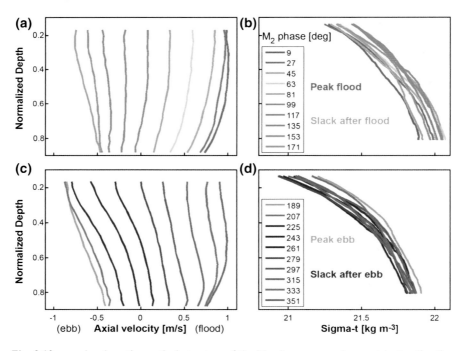

Fig. 3.12 **a** and **c** show the vertical structure of the M_2 phase averaged current in the direction 260° at station R08 in Fig. 3.23a. Zero phase is at maximum flood and colors represent phase values listed in the legends. Depths are normalized to the water depth. **b** and **d** show phase averaged density profiles

multiple years of current observations projected in the direction 260°. This direction best matches the alignment of tidal current ellipses. The measurements were obtained by Codiga (2007) using a ferry-mounted current profiler in an area midway between New London and Orient Point (near station R08 in Fig. 3.23a) where the water depth is 59 m. M2 phase averaged density profiles of several weeks of hourly ascents of a moored CTD profiler deployed by Codiga et al. (2002) at a site 46 m deep and approximately 4 km east of R08 are shown in Fig. 3.12b and d.

The velocity profiles show that the high amplitude tidal currents (~1 m/s) result in a sheared bottom boundary layer reaching as far as 20–30 m off the seafloor, about half the total water depth. The red lines in Fig. 3.12a and c show that during flood, the velocity is weakly sheared in the upper water column, and there is a subsurface velocity maximum below which the shear is strongest. In contrast, during and after ebb (blue-shaded lines), the velocity is sheared throughout the water column, in the opposite sense from deep flood conditions, with peak shears in the upper water column. The density profiles (Fig. 3.12b and d) show higher values after flood as is to be expected due to advection of the axial gradient. The slopes of the density profiles demonstrate that, while stratification in the deeper part of the water column varies modestly over the tidal cycle, the upper water column stratification is significantly enhanced during and after ebb (Fig. 3.12d) relative to the other half of the cycle (Fig. 3.12b). This is clear evidence that tidal straining is active. The quantitative implications of these cycles in shear and stratification on vertical fluxes and residual horizontal transport are very important topics for future investigation.

A completely unanticipated and unexplained result of the work of Bennett et al. (2010) was the discovery of high amplitude overtides in the western part of the basin. For example, near Execution Rocks (location exrck in Fig. 3.10b), the M6 was found to be approximately 25 % of the M2 amplitude. Figure 3.13 shows the along-Sound distribution of the ratio of the amplitude of the N2 and S2 to the M2 as black and red lines, respectively. Throughout the Sound, they remain at approximately 20 %. Since these constituents are close in frequency, friction and geometric effects should affect them in the same way and the ratios should, therefore, be constant. The ratio of the M4 to the M2 amplitude is shown in blue. It is also almost uniform and though it can be modified by a variety of nonlinear dynamical processes, either their effects cancel out or are small. The ratio of the M6 to M2 is similar to that of the M4 to the M2 in the eastern Sound, but it is significantly amplified at the three western Sound stations. This difference in the distribution of the characteristics of the overtides is likely due to the spatial variation in the magnitude of the nonlinear terms in the estuarine dynamics and kinematics, and their simulation is therefore a very sensitive metric of the performance of models. Bennett (2010) evaluated the predictions of the theory proposed by Parker (1984) for the consequences of quadratic bottom friction and advection of momentum in the dynamic balance, and of finite amplitude elevation changes in the conservation of volume equation, on the generation and propagation of overtides in the long wave model. The model was unable to simulate the observed structure.

Evidently, the recent observations of the structure of characteristics of the tides in LIS have posed new modeling challenges. Can models simulate the horizontal and

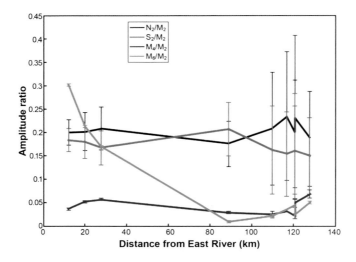

Fig. 3.13 The along-Sound structure of the amplitudes of the vertically averaged tidal current constituents scaled by the M2 amplitude. The green line shows the M6 and the blue line shows the M4

vertical variation in the characteristics of tides and the influence of stratification? If so, can they then predict the structure of the overtides, which are likely to be sensitive to the more subtle mechanisms in the models? It is important to improve our understanding of tidal currents by critical comparison of models and observations because their amplitudes far exceed those of the residual circulation, particularly in ELIS, such that tidal asymmetries may be an important influence on subtidal flow.

3.2 The Structure and Variability of the Hydrography

As in all estuaries, the hydrography and the circulation in the Sound are intrinsically coupled. The structure of the density field modifies the pressure gradient field set by sea level variations and strongly influences the circulation, particularly at low frequencies. The density-driven circulation generally leads to a transport of river water to the ocean and salt water toward land. Vertical gradients in density inhibit vertical mixing rates of heat, salt, and momentum and thereby modify the rates of transport. Since the Sound is shallow, pressure effects on density can be neglected and so the density variations are controlled by T and S. T is influenced by heat energy exchange with the atmosphere, solar radiation, and exchange with the adjacent ocean. Note that the magnitude of the exchange with the shelf at seasonal and annual scales is uncertain. Salinity (S) is controlled by the discharge of rivers and exchange with the ocean, and though precipitation-evaporation and inputs from aquifers have been assumed to play a secondary role, this is uncertain. Both the T and S budgets are modulated at seasonal scales, and these are manifested in observations of the distributions in the Sound.

The first comprehensive characterization of the structure and evolution of the S and T fields in LIS was reported by Riley (1952 and 1956). Using several years of ship surveys, he established the basic spatial trends and the range of seasonal variability. The State of Connecticut's Department of Environmental Protection (CTDEP) beginning in 1991 built on the surveys reported by Welsh and Eller (1991) and developed an extensive dataset with which to describe the seasonal and inter-annual variability of S, T, and the density fields. Kaputa and Olsen (2000) describe the program and review the first decade of results. Henceforth, we refer to this as the CTDEP dataset. It can be accessed at http://lisicos.uconn.edu/dep_ portal.php. Subsequently, Gay et al. (2004) used the S data to develop a salt budget and estimate residence times and exchange rates with BIS. Lee and Lwiza (2005) characterized the inter-annual variability in T and S, and Whitney (2010) used the data to examine the influences of river discharge fluctuations.

Figure 3.14 shows the distribution of station locations in the CTDEP dataset. The green symbols show the locations of the stations sampled at monthly intervals throughout the year and at twice that frequency in the summer. Sampling at

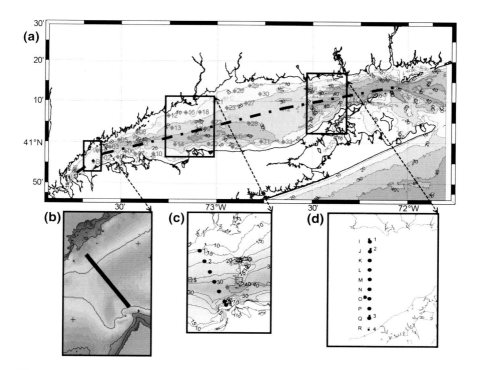

Fig. 3.14 a Map of LIS showing the bathymetry (depths in m) and the locations of the CTDEP hydrographic survey stations. **b** The coastline and cruise track used in the observation program described by Bennett (2010). **c** and **d** Show the stations used by O'Donnell and Bohlen (2003). The *thick dot-dash line* shows the along-Sound location for the vertical sections shown in Figs. 3.14, 3.15, 3.16, and 3.33

the seven stations, shown in green, began in 1991. The westernmost station in the East River, A2, was terminated in 1995. Year round sampling at the stations shown in yellow began in 1994. Another station, N3, not shown in the figure, was occupied in BIS from 1995 to 2002. Though there is inconsistency in the sampling interval and the high frequency fluctuations in the Sound are aliased, this dataset reveals the broad character of mean structures and the spatial and seasonal cycles of the S, T, and density, σ_T, distributions. The application of objective analysis, often referred to as OA (see Wilkin et al. 2002), to the estimation of the seasonal cycles from this data archive requires approximation of the space–time correlation function and the measurement noise for each variable. These statistics may depend upon location and time, and so are very difficult to estimate. Ad hoc approximations for the characteristics of the correlation functions are common. Lee and Lwiza (2008), for example, mapped the monthly evolution of the DO concentration in LIS with OA and used the sample station spacing as an estimate of the width of the correlation function. This is a common approach. If the sampling plan was developed to resolve the field with a limited number of stations, then choosing the correlation scale as the separation would be consistent. Further, the general appearance of contour maps is often insensitive to the details of the shape of the correlation function, though the characteristics of small-scale structures and estimates of gradients can be very sensitive to the assumptions adopted. The central value of the OA method is that it provides uncertainty estimates as well as the mapped field. The uncertainty estimates are very sensitive to the correlation function and noise levels chosen. Additional effort to refine the estimates of these statistics is, therefore, warranted. In the contour maps we present here using the CTDEP dataset, we follow the ad hoc approach of Lee and Lwiza (2008). Specifically, we assume that the covariance function at $\vec{x_0}$ is the form

$$C\left(x,y,z; \vec{x_0}\right) = \left\{ n^2 I\left(x,y,z\right) + \sigma^2 \exp\left\{ -\left(\frac{x-x_0}{L_x}\right)^2 - \left(\frac{y-y_0}{L_y}\right)^2 - \left(\frac{z-z_0}{L_Z}\right)^2 \right\} \right\}$$

where s^2 and n^2 are the signal and noise variances; x,y, and z are lags in the along-Sound, across-Sound, and vertical directions; and $L_{\{x,y,z\}}$ are the decorrelation length scales. The function $I(x,y,z)$ models the correlation structure of the noise and we assume $I\left(|x| > 0, |y| > 0, |z| > 0\right) = 0$ and $I(0,0,0) = 1$, i.e., noise is uncorrelated in space and time. We then assume $L_x = 25$ km, $L_y = 5$ km, and $L_z = 5$ m, and $n^2 = 100$. Note that the data are averaged by month at each station when more than one profile is available, which implicitly assumes that the correlation in time is a step function with a half width of 15 days. This is an assumption that should be reconsidered.

The vertical and along-Sound patterns of the monthly averaged S during April, August, and December are shown in Fig. 3.15. The section follows the black dashed line through stations A4, C1, I2, and M3. Note that the three-dimensional OA procedure allows data from all available stations to contribute to the structure. The horizontal axis is shown as distance along the dashed line from Execution Rocks (A4). Throughout the year, the lowest S is in the western Sound (25.0–27.0) and the

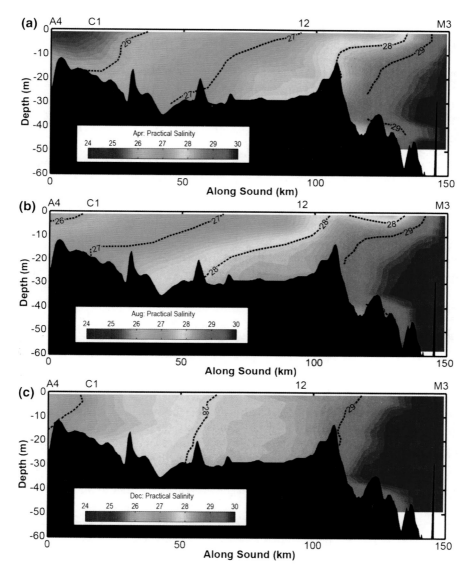

Fig. 3.15 The structure of the salinity distribution along the *dot-dash line* in Fig. 3.14 computed from the CTDEP survey data using data collected in **a** April, **b** August, and **c** December

highest values are at the eastern boundary (29.5–31.0). The amplitude of the annual cycle of S variations is approximately 1.5 at all stations. Throughout the Sound, the minimum S occurs in May, and the highest values are in November. In addition to this simple yearly oscillation, there are more subtle variations in the vertical and horizontal gradients in S. Comparison of the locations of the 26 and 29 contours in the April and August, and August and December S distributions of Fig. 3.15 demonstrates that the horizontal gradients are largest in April and lowest in December.

In the spring, the freshwater that discharges from the Connecticut River and the Hudson River appears to freshen the whole Sound. Though the S in eastern Sound is reduced, the western Sound is reduced slightly more and the horizontal gradient is enhanced. Between April and August, S increases throughout the Sound and the 27 and 28 contours move westward. The slope of the isohalines in the western Sound indicates that salt stratification persists there throughout the year though in December they have a steeper slope as a consequence of increased mixing.

The evolution of the monthly average T field is shown in Fig. 3.16. Note that the color scales are different in each frame since the magnitude of the annual cycle (22 °C) is much greater than the range due to the spatial structure. In January (not shown), the average T is near zero. By April, the water warms to between 5 and 7 °C with the warmest water in the western Sound surface waters. The near surface waters warms through the summer to more than 22 °C throughout much of the Sound and a thermocline develops approximately 10 m below the surface. In August, the bottom waters reach 20 °C. A region of high horizontal T gradients develops near the Mattituck Sill (at 100 km in Fig. 3.16) where the deeper waters are influenced by exchange with BIS and the continental shelf. Insolation is reduced in the fall and the wind stress associated with synoptic scale atmospheric patterns becomes greater, leading to cooling and vertical mixing in the Sound. By December, the T of the Sound falls to 7–10 °C and the isotherms are almost vertical. Note that the coldest water is in the western Sound.

The net effect of the fresh water and heating on the evolution of the density field, σ_T, is displayed in Fig. 3.17. Comparison of the location of the $\sigma_T = 21$ isopycnal in December and April shows that it moves between 40 and 60 km eastward in the spring due to warming and increased river discharge. By August, significant vertical stratification develops throughout most of the Sound, as shown by the reduced slope of the isopycnals, and density reaches a minimum with the $\sigma_T = 21$ isopycnal restricted to the eastern part of the Sound. The vertical structure of density shows a broad pycnocline centered at 10 m depth.

It is evident that both the vertical and horizontal density gradients are enhanced in August; however, the magnitude of the changes is not clearly revealed in the cross sections. In Fig. 3.18, we show the evolution of a finite difference estimate of the monthly mean stratification at 10 m below the surface, $\Phi(-10) = -\frac{\partial \rho}{\partial z}\Big|_{z=-10} = \frac{\rho(z=-15)-\rho(z=-5)}{10}$, at stations A4, C2, F3, H4, I2, and K2 which lie along the central axis of the Sound (see Fig. 3.14). Stratification is at a minimum at all stations in October–January. It begins to increase slowly in February at stations A4, C2, F3, H4, and I2, but accelerates in April and reaches a maximum in July. The highest stratification, $\Phi(-10) = 0.15 \, \text{kg/m}^4$, occurs at H4 with values closer to 0.1 kg/m^4 elsewhere. At station K2, in the eastern Sound, the rapid development of stratification begins in February, and the peak is reached in April. The behavior at M3 is similar (not shown) though the maximum is only half of that at K2.

If the equation of state for seawater is approximated in a layer $z_L \leq z \leq z_T$ as $\rho(S,T,P) = \rho\left(\bar{S},\bar{T},0\right) + \alpha\left(\bar{S},\bar{T}\right)\left(T - \bar{T}\right) - \beta\left(\bar{S},\bar{T}\right)(S - \bar{S})$ where $\bar{S}(t)$ and $\bar{T}(t)$ are the average S and T in the layer for a particular month, then the stratification

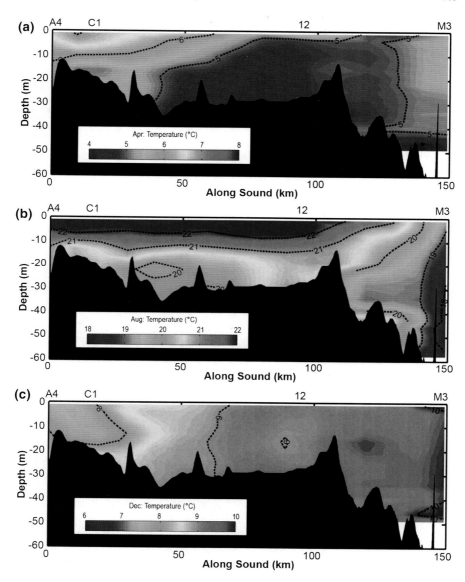

Fig. 3.16 The structure of the temperature distribution along the *dot-dash line* in Fig. 3.14 computed from the CTDEP survey data using data collected in **a** April, **b** August, and **c** December

can be approximated $\Phi(z) = -\alpha\left(\bar{S},\bar{T}\right)\left.\frac{\partial T}{\partial z}\right| + \beta\left(\bar{S},\bar{T}\right)\left.\frac{\partial S}{\partial z}\right|$ where α and β are the thermal expansion and haline contraction coefficients estimated following McDougall (1987). Using the notation $\langle\bar{T}\rangle$ and $\langle\bar{S}\rangle$ to represent the annual averages of the layer average T and S, then $\Phi(z)$ can be approximated as the sum of the thermal, Φ_T, and haline, Φ_s, effects as $\Phi = \Phi_T(T) + \Phi_s(S) = -\alpha\left(\langle\bar{S}\rangle,\bar{T}\right)\left.\frac{\partial T}{\partial z}\right|_z + \beta\left(\bar{S},\langle\bar{T}\rangle\right)\left.\frac{\partial S}{\partial z}\right|_z$.

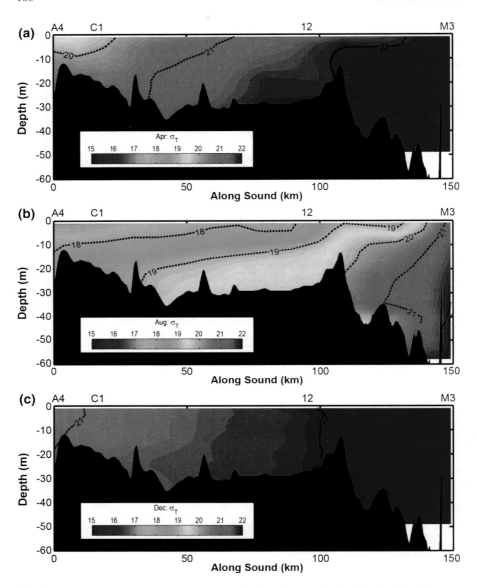

Fig. 3.17 The structure of the density (σ_T) distribution along the *dot-dash line* in Fig. 3.14 computed from the CTDEP survey data using data collected in **a** April, **b** August, and **c** December

Numerical experiments demonstrate the error due to the neglect of the T dependence of the haline expansion coefficient, and the S dependence of the thermal expansion coefficient is $< 5\%$.

To reveal the roles of warming and freshwater inflow variations on stratification at stations A4, C2, F3, H4, I2, and K2, Fig. 3.18 shows the annual cycle of the Φ_T as green lines and Φ_s as red lines. Since the red lines overlie the black lines from January

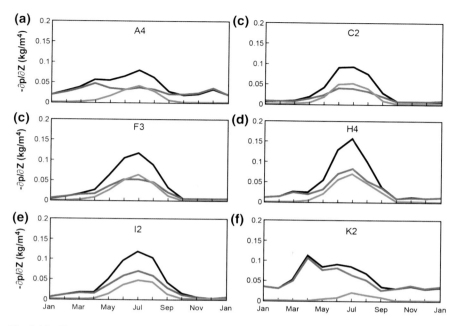

Fig. 3.18 Time evolution of the vertical stratification (*black line*) at CTDEP stations **a** A4, **b** C2, **c** F3, **d** H4, **e** I2, and **f** K2. The *red* and *green lines* show the approximate contributions of salinity and temperature to variations to the stratification, respectively

to April at each station, it is clear that it is the increase in freshwater discharge (see Fig. 3.1) that is responsible for the initial stratification enhancement. At Station A4, at the western end of the Sound near the East River, the freshwater effect peaks in April and then diminishes through the rest of the year. A similar pattern occurs at stations K2 and M3, which are located in the eastern Sound near the Connecticut River. At the other stations, the peak of the freshwater-induced stratification (Φ_s) is delayed until July, likely as a consequence of the time required to transport it. During the May to September interval, the contribution of the thermal effect on stratification is significant at all stations. It peaks in July when the magnitude of the effect equals that of salt at all stations except K2 and M3, which are deeper and, presumably, more influenced by exchange with BIS and the continental shelf. In late summer and fall, both the effects of salt and heat drop and by October stratification reaches its minimum level, which is solely due to the S structure. This situation persists until January.

Pritchard (1956) proposed that the modification to the pressure field created by spatial gradients in density was a key component of the force balance that determines the circulation in estuaries. Wilson (1976) applied the basic idea to estimate the vertical structure of the axial circulation in LIS using observations obtained during a series of cruises. The horizontal baroclinic pressure gradient at level z can be expressed $H_B(z) = \frac{g}{\rho_0} \int\limits_{z}^{0} \frac{\partial \rho}{\partial x} dz$ where $g = 9.8$ m/s is the acceleration of gravity at the surface of the earth and $\rho_0 = 1{,}000 \, \text{kg/m}^3$ is a reference density. Note that the force is $-\rho H_B$ and is directed toward the region of lower density.

Since the vertical structure of the horizontal gradient is weak, the temporal pattern is similar at all depths and magnitude of H_B increases linearly with distance below the surface. The solid lines in Fig. 3.19 show the annual evolution of $H_B(z = -20 \, \text{m})$ estimated using a finite difference approximation between stations A4 and C1 (blue), C1 and I2 (red), and I2 and M3 (magenta). All values are positive, indicating the baroclinic pressure gradient force is directed to the west, and have a magnitude of order $10^{-5} \, \text{m/s}^2$. The mean values in the eastern and western ends of the Sound are comparable and approximately four times larger than in CLIS, between C1 and I2. Wilson (1976) found a similar pattern in his more limited dataset. However, the seasonal evolution of H_B in the east and west is quite different. Between A4 and C1 (solid blue line), there is an enhancement of the gradient in April, a minimum in July–August, and an increase until January. In contrast, the gradient between I2 and M3 (solid magenta line) at the eastern end of the Sound varies with a comparable amplitude but with a maxima in June–August and minima in the winter. In CLIS, between C1 and I2, the seasonal cycle appears to be weaker and in phase with the variations in ELIS.

To assess the cause of the variability, we divide the variations in the baroclinic pressure gradient into two components, $H_B(z) = H_{BT}(z) + H_{BS}(z)$, where $H_{BS}(z) = \frac{g}{\rho_0} \int_z^0 \beta\left(\bar{S}, \langle \bar{T} \rangle\right) \frac{\partial S}{\partial x} dz$ is the contribution resulting from variations in S and $H_{BT}(z) = -\frac{g}{\rho_0} \int_z^0 \alpha\left(\langle \bar{S} \rangle, \bar{T}\right) \frac{\partial T}{\partial x} dz$ is the effect of T. In Fig. 3.19, the dashed lines show $H_{BS}(z = -20 \, \text{m})$ for the three sections using the color code defined above. The

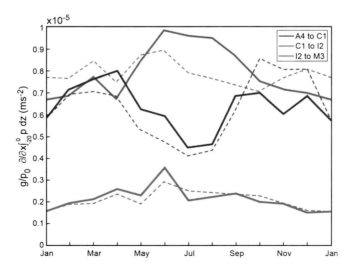

Fig. 3.19 The annual evolution of the horizontal baroclinic pressure gradient between CTDEP stations A4 and C1 (*blue*), C1 and I2 (*red*), and I2 and M3 (*magenta*). The *dashed line* shows the approximate contribution of salinity variations

differences between the solid and dashed lines represent the effect of T. During intervals in which the dashed line falls above the solid line on the graph, the horizontal T gradient acts to oppose the effects of the S gradient. Since the lines of the same color are close together, it is clear that the baroclinically driven flow is dominantly controlled by the S variations. Though the vertical T gradient is largest during the summer in the central Sound, the T effect on $H_B(z)$ is modest (see red lines). In ELIS (magenta lines), the dashed line is above the solid line between September and May when atmospheric cooling reduces T in the central Sound more than in the deeper waters of the eastern Sound (see Fig. 3.16). A similar behavior is evident in the western Sound (blue lines) during September–January where the shallow waters cool more rapidly.

There are few reports describing observation of the across-Sound structure of the S, T, density, and pressure gradients, though Riley (1956) noted the lateral gradients in surface properties and Hardy (1972) showed hydrographic casts along a north to south section in the eastern Sound in which significantly fresher water occupied an upper water column layer that deepened toward Long Island. Even early numerical models show substantial lateral shear in the mean flow (Murphy 1979). Recent theoretical work (e.g., Kasai et al. 2000 and Valle-Levinson 2008) suggests that the magnitude of vertical friction and the width of the estuary control the degree of lateral variation in estuaries, so periods of weak vertical stratification in the wider areas of the Sound are likely to exhibit significant across-Sound gradients.

To remove the influence of tidal frequency variations from cross-Sound surveys, several sections must be observed for at least a whole tidal cycle. This approach has been employed by O'Donnell and Bohlen (2003) and Bennett (2010) at the three locations shown in Fig. 3.14b, c, and d. Bennett (2010) describes the observations and analysis of data collected during two 25 h intervals between March 5 and 8, 2005 and July 31 and August 1, 2005 along the line indicted in Fig. 3.14b. In the March experiment, they visited four equally spaced stations with a profiling conductivity, T, and depth (CTD) sensor, and in the summer they employed an undulating towed vehicle and achieved much higher spatial resolution. Figure 3.20a and b show the mean vertical structure of the S and T fields observed in July when viewed looking eastward with the coast of Connecticut on the left and Long Island on the right. The S exhibits a maximum of 26.2 at the center of the section at the bottom and a top to bottom difference of 0.8. The isohalines slope slightly downward to the south. The pattern of the isotherms in Fig. 3.20b shows a similar structure with a maximum of 23 °C at the surface and a 5 °C vertical variation. There is no appreciable slope to the isotherms. The winter situation is illustrated in Fig. 3.20c and d. Here the maximum S is 26.1 psu, and it is located at the bottom on the northern side of the section. The minimum is 25.5 psu and is at the surface near the Long Island shore. The steeply sloping isohalines show that the vertical gradients are weak but the horizontal gradients are relatively large. Bennett (2010) reported a similar structure in March 2006.

At a section further to the east, O'Donnell and Bohlen (2003) surveyed the S, T, and velocity fields during three two-day sampling programs along a section with seven stations near Stratford Shoals (see Fig. 3.14b) during the interval July 20 to August 6, 1995. The mean structure of the S, T, and density (σ_T) is shown in Fig. 3.21 a–c. The horizontal isolines indicate that the across-Sound gradients in

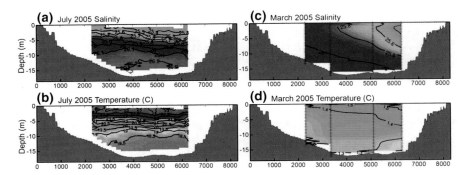

Fig. 3.20 Across-Sound sections of the **a** salinity and **b** temperature observed along line shown in Fig. 3.13b during July 2005 by Bennett (2010). **c** and **d** Show the same properties observed in March 2005. Adapted from Bennett (2010)

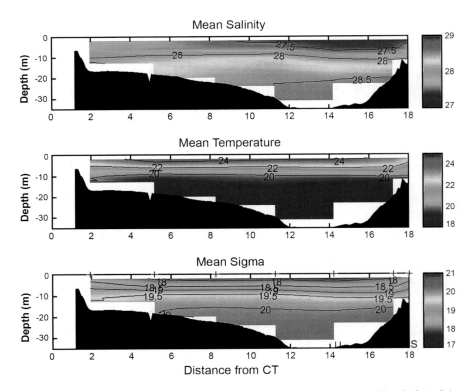

Fig. 3.21 The vertical and across-Sound structure near Stratford Shoals (see Fig. 3.13c) of the mean salinity (*top frame*), temperature (*middle frame*), and density (σ_T) during the interval, July 20–August 6, 1995. Note that the Connecticut coast is on the left and the '+' symbols at the top of 18c show the station locations. Adapted from O'Donnell and Bohlen (2003)

all three properties appear to be very weak. This is consistent with summer observation of Bennett (2010) shown in Fig. 3.20a and b.

In a survey earlier the same year (April 26 to May 6, 1995) that visited 10 stations along a section near Hammonassett Point (see Fig. 3.14d), O'Donnell and Bohlen (2003) found the mean structure of the S, T, and σ_T fields shown in Fig. 3.22 a–c. These show significant across-Sound gradients. Though the mouth of the Connecticut River, the largest source of fresh water to the Sound, is only 10 km to the east of the northern end of the section, the freshest water (S ~ 21) is found on the southern end. Unfortunately, observations were inadequate to determine whether this structure persists throughout the year or whether it is only a feature of the spring hydrography as suggested by the Bennett et al. (2010) observations.

The first persistent observations to reveal the mean and seasonal variability of the lateral structure of the near surface S field in the eastern Sound were obtained by the New London to Orient Point Ferry observation system described by Codiga (2007) and made available at http://www.po.gso.uri.ed u/~codiga/foster/getdata.htm. Salinity and other properties were measured in water pumped from an intake in the vessel hull approximately 2.5 m below the water surface and recorded every 30 s together with the ship's position. When

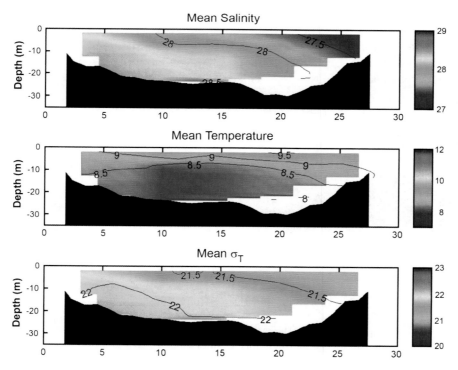

Fig. 3.22 The across-Sound structure near Hammonassett Point (see Fig. 3.13d) of the mean salinity (*top frame*), temperature (*middle frame*), and density (σ_T) during the interval April 26–28, 1995. Adapted from O'Donnell and Bohlen (2003)

operating, the ferry transits the Sound approximately eight times a day between 0700 and 2300. In Fig. 3.23a, we show the coastline and bathymetry of the eastern Sound in the vicinity of the ferry's track. The dashed line shows a convenient across-Sound coordinate and the solid line indicates the boundary of bins Codiga and Aurin (2007) used to average their measurements. These are labeled R01-R17, with the numbers increasing to the south. The southernmost segment, in Plum Gut, is labeled G01. Using s01 and s17 to represent the surface S in bins R01 and R17, we compute the time series of the five day average across-Sound S difference as $\delta s_{\{01-17\}} =< s01 - s17 >$, and the standard deviation as $\sigma s_{\{01-17\}} =< (s01 - s17)^2 >^{1/2}$. In Fig. 3.23b, we show the evolution of $\delta s_{\{01-17\}} \pm \sigma s_{\{01-17\}}$ for the interval July 2005 to September 2006. Note that there is a data gap between January and March of 2006 and that $\delta s_{\{01-17\}}$ is less sensitive to the instrument drift that makes long-term measurements of S difficult. The long-term mean monthly discharge, together with the 68th percentile interval, in the Connecticut River is displayed in Fig. 3.23c. The discharge during the interval of the ferry surveys is also shown. Note that the Connecticut River discharge is well above the long-term average during October 2005 and May–July 2006. The National Climatic Data Centers monthly precipitation index for southern New Hampshire (http://www7.ncdc.noaa.gov/CDO/cdo) shows that this is a consequence of anomalous precipitation in the watershed.

The principal result of this analysis is that $\delta s_{\{01-17\}} > 0$, i.e., the freshest water was at the southern shore throughout the fall of 2005 and in the spring of 2006 when the discharge in the Connecticut River was low, less than $500 m^3/s$. This suggests that the observations of low S water at the southern end of the Hammonassett Point

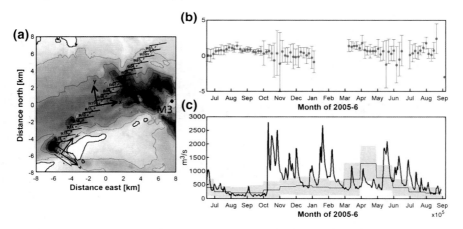

Fig. 3.23 **a** The coastline and bathymetry of eastern LIS showing boundaries of averaging intervals for the salinity observations shown in panel (**b**), from Codiga and Aurin (2007). **b** shows the evolution of the salinity difference across the section, $\delta s_{\{01-17\}} \pm \sigma s_{\{01-17\}}$ for the interval July 2005 to September 2006. The daily average discharge in the Connecticut River at Thompsonville is shown in (**c**) by the *solid black line* for the same period. The long-term mean discharge is shown by the histogram with one standard deviation shown in *gray*

section by O'Donnell and Bohlen (2003) were not anomalous, but a robust feature of the eastern Sound hydrography. When the discharge was anomalously high in the fall of 2005 and summer of 2006, $\sigma s_{\{01-17\}}$ became very large and, therefore, the sign of the gradient is uncertain. The aerial surveys of the river plume front of the Connecticut River by Garvine (1974) showed that the plume could extend to the eastern end of the Sound during ebb tide and high discharge conditions. It is therefore plausible that the ferry schedule leads to aliasing of the tidal frequency fluctuations in the S field during periods of high discharge in the Connecticut River. The observation of the variability of the subsurface hydrographic structure and the development of a quantitative understanding of the response of the hydrography to discharge variability are important tasks for the future.

The evolution of the vertical structure of the S, T, and density fields at sites in the central (between stations 19 and 20 in Fig. 3.14) and eastern Sound (near K2 in Fig. 3.14) has been described by Codiga et al. (2002) using a CTD system that profiles most of the water column at hourly intervals. They found that the density profile took a broad range of shapes, commonly being weakly stratified throughout the water column or having stronger stratification in a layer 10–20 m deep overlying weaker stratification. Less commonly, it consisted of a well-mixed layer overlying stratification. At the CLIS site, during a spring 2002 deployment, the transition from nearly vertically uniform to stratified conditions occurred in a relatively smooth fashion over several weeks, while in contrast the breakdown of stratification during the summer-fall deployment occurred in a series of abrupt mixing episodes associated with storm events. During the simultaneous summer-fall deployments in the central and eastern Sound, the density stratification was effectively eliminated within hours at both sites in response to certain strong storm events, but restratification was complete within about a day at the eastern site and within about 2 days at the central site. The response to other wind events was different at the two sites, consistent with the earlier demonstration that local winds at different sites in the Sound are not coherent on short time scales.

The important dynamic influence of the S and T fields arises through their influence on density and horizontal pressure gradients. The magnitude of the acceleration due to the along-Sound baroclinic pressure gradient in both the western and eastern ends of the Sound, shown in Fig. 3.19, is on the order of $0.5 \times 10^{-5} \mathrm{m/s^2}$. The magnitudes of the corresponding across-Sound quantities are at least comparable. These are large values. If, as in the model of Hansen and Rattray (1965), the divergence of the vertical turbulent momentum flux was required to balance H_B in a layer of 20 m depth, then using an eddy viscosity magnitude of $10^{-3} \mathrm{m^2/m^2/s}$, an along-estuary residual velocity at the surface of approximately $H_b \frac{D^2}{A_z} = 10^{-5} \times 10^3 \times \frac{20^2}{5} \sim 10 \mathrm{cm/s}$ would be required. Since H_B varies by 50 % over the year, the magnitude of the seasonal variation in the residual circulation should be 5 cm/s. If the springtime across-Sound gradient were to be near geostrophic balance, then the lateral variation in the along-Sound current in the spring would have to be on the order of 1 m/s. Since this would be evident in the most primitive current measurements, the barotropic pressure gradient and frictional stresses must also be significant in the lateral dynamic balance.

3.3 Residual Circulation

The residual circulation in LIS was first characterized by Riley (1952) by summarizing drift bottle observations and measurements made by the US Coast and Geodetic Survey using current poles and current log lines (Le Lacheur and Sammons 1932). He created a general view of residual currents near the surface and described large-scale gyres in each of the three basins, augmented by many smaller-scale features. In CLIS, he observed a general clockwise rotation, and he found counterclockwise surface currents in the eastern and western basins. These observations were augmented by those of Larkin and Riley (1967), Gross and Bumpus (1972), Hollman and Sandberg (1972), Gordon and Pilbeam (1975), and Paskausky (1976) to include near bottom drift estimates in the central Sound. These revealed a two-layer east–west exchange flow consistent with the estuarine circulation described by Pritchard (1956), except nearshore in less than 10 m of water where mixing occurred throughout the water column. Large-scale nontidal events were found to correlate with maximum daily wind values. These events were typically on the order of 5–20 days in the winter and less than 3 days in the summer, and so they concluded that a 10-day record in the summer and a 20-day record in the winter should be sufficient to characterize the mean flow. General trends included westward flow mid-Sound, northward flow in shallower waters off the Connecticut coast, and some southward flow closer to the Long Island shore.

Even now, observations of the long-term mean circulation in the Sound are very limited, so the theoretical models are necessary to sensibly interpolate what is available. There is a hierarchy of complexity in theories. The earliest and simplest ones have focused on two driving mechanisms, the density gradients (Hansen and Rattray 1965; Wilson 1976) and the nonlinearity of the tidal flow (Ianniello 1977a; Murphy 1979). Though a large fraction of the fresh water in LIS comes from the Connecticut River and is delivered to the northern side of the Sound, the early theories for the residual circulation in LIS (Wilson 1976 and Ianniello 1977a) assume that the estuary is laterally homogenous. Only recently have measurements been available that resolve lateral variations and motivated the need for more sophisticated models. For completeness, we comment on the early models and then describe the more recent model results that yield a comprehensive view of the circulation that is largely consistent with the available observations.

Ianniello (1977a) developed an analytic model of the residual circulation in an estuary with weak density gradients to evaluate the magnitude and distribution of the tidally driven residual circulation. His model predicted that the Lagrangian residual circulation had a strong vertical structure: westward at the surface and eastward at depth. Note that this pattern of circulation is counter to the traditional notion of estuarine circulation in which the baroclinic pressure gradient is balanced by turbulent friction. In a complementary study, Wilson (1976) developed a numerical model of the buoyancy-driven circulation in LIS that adopted the baroclinic-friction balance of Hansen and Rattray (1965) but specified the density field using observations from east–west transects along the center of LIS.

The predictions of the model showed westward motion at depth and eastward motion at the surface as expected and with magnitudes that were consistent with the data available. Murphy (1979) expanded on the dynamics used in Ianniello's horizontally averaged model and studied the effect of more realistic geometry and bathymetry on the across-Sound structure of the circulation using a numerical model. He neglected the density-driven motion. Murphy found an extremely complex Eulerian residual circulation pattern with several large eddies in the eastern Sound. At the time, there was little observational evidence with which to discriminate the relative value of these models.

Schmalz (1993) developed a complex three-dimensional model that included both density variations and nonlinear effects. The prediction of the distribution of the mean velocity at the surface showed similar patterns and orders of magnitude velocities to those of Murphy (1979). Though complex, the dominant pattern in both models showed a westward motion on the north side of the Sound and eastward motion along the Long Island shore with a large cyclonic (counterclockwise) gyre dominating the central basin as was proposed by Riley (1956). The near bottom circulation was predicted to be to the west everywhere except in the vicinity of the Mattituck Sill and Stratford Shoals where topography complicated the flow. Valle-Levinson and Wilson (1998) studied the dynamics in the vicinity of the Mattituck Sill in more detail with a similar model and Valle-Levinson et al. (1995) noted the role of the interaction of the exchange flow through the East River, wind, and tides on the rate of vertical mixing in WLIS. Blumberg et al. (1999) developed a three-dimensional model to assist in the prediction of water treatment plant upgrades on water quality and Signell et al. (2000) developed a similar model that also included the effect of surface gravity waves on the bottom stress in order to understand the distribution of bottom sediments in the Sound.

The most comprehensive model prediction of the circulation in the Sound available to date is that of Crowley (2005) and Wilson et al. (2005). They implemented a data assimilating version of the ROMS three-dimensional circulation model for LIS. They performed four 1–3 month simulations using meteorological and sea level observations collected in 1988 to include the effects of coupling to the atmosphere and adjacent ocean. T and S observations from an intensive campaign in 1988 were incorporated through data assimilation to simulate the effects of river discharge fluctuations and seasonal heating rate cycles. This enabled examination of seasonal variability in the circulation. The model demonstrated the amplitude and phase of the principal tidal harmonics well and predictions also compared favorably with the S and T observations. They also provided a qualitative comparison to the moored current meter program reported by Vieira (2000), which included approximately 10 sensors on six transects across LIS between March and October 1988. The locations are shown in Fig. 3.24. These observations were not synoptic. Each transect was occupied for approximately 30 days and then the instruments were recovered and redeployed on the next transect to the east. Transects 5 and 6 were occupied simultaneously in both June and September. Figure 3.24 shows a very complex current pattern in both speed and direction. The near surface flow, shown in red, is generally eastward and the deeper flow is westward as expected from the results of the simple baroclinic pressure gradient

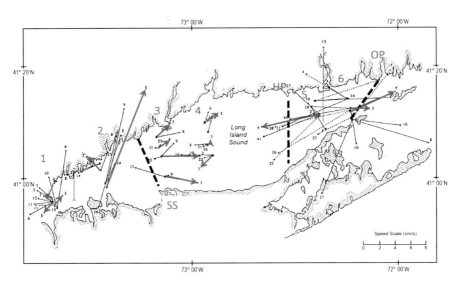

Fig. 3.24 A map the coastline of LIS showing the mooring locations and the mean flow estimates reported by Vieira (2000). The *dashed vectors* show the results of the June deployments. Sections are labeled 1–6 increasing to the east and the small numbers at the tip of the arrows indicate the depth of the measurement. Measurements above 5 m are shown in *red*. The *dashed lines* labeled SS and HP show the approximate locations of O'Donnell and Bohlen (2003) measurements and the OP shows the ferry transect described by Codiga and Aurin (2007). Section 3 is similar to that of Bennett (2010). Modified from Vieira (2000)

balancing turbulent friction theories of Hansen and Rattray (1965) and Wilson (1976). However, there are very large across-channel velocity components at all depths and a coherent pattern is difficult to extract. The extent to which this is due to the nonsynoptic sampling plan is unclear. For this reason, a model that is consistent with these and other measurements, and also with our understanding of the dynamics, is likely to yield a better estimate of the circulation and its variability.

The Crowley (2005) predictions of the mean along-Sound velocity at the Vieira sections labeled 3, 4, and 6 in Fig. 3.24 are shown in Fig. 3.25. The general structure of the velocity component distributions at Section 3, Fig. 3.25a and b, have some similar characteristics. The model predicts in the deeper parts of the cross-section that near the surface (above 15 m), the flow is eastward (positive) and westward below that and in the shallower areas near the Connecticut shore. The observations in Fig. 3.25b are sparse but do show the surface layer moving eastward. The magnitudes appear to be half of that predicted. At Section 4, shown in Fig. 3.25c, the eastward flowing layer in the model is much shallower than at Section 3 across most of the section, but fills almost the entire water column near the Long Island shore. The observations shown in Fig. 3.25d were only obtained in the region of slow westward flow, but are largely consistent with the predictions. At Section 6, the model predicts a very large eastward flow above 20 m in the southern half of the section (see Fig. 3.25e) and a strong westward flow elsewhere.

Only six moorings were available on Section 6; however, Fig. 3.25f shows that they do reveal the strong westward motion in the deep area and substantial lateral and vertical shears that are similar to that predicted. As at Section 3, the magnitude of the observed eastward motion is substantially less than that predicted.

Observations from the vessel mounted 600 kHz ADCP aboard the Bridgeport to Port Jefferson Ferry resolve the lateral structure of longitudinal residual currents along Section 3 of Vieira (2000) and Crowley (2005). As Fig. 3.25 shows, this is slightly to the east of Section SS of O'Donnell and Bohlen (2003). Figure 3.25g shows the mean east velocity component observed between August 6 and September 28, 2004. Westward flow (blue) is confined to the main channel with a zero crossing at approximately 7 m. Eastward flow (red) is confined to the water column above approximately 7 m and to the shallow northern flank of the section. Note that the center of the first bin is at approximately 5 m. The structure of the flow is remarkably consistent with that predicted by Crowley (2005) and shown in Fig. 3.25a.

Hao (2008) has reported more detailed analysis of aspects of the dynamics of the tides and residual circulation in the Crowley (2005) and Wilson et al. (2005a) model. Figure 3.26a and b show maps of the coastline of the Sound with the predicted mean flow vectors near the surface and at mid-depth, respectively. Some very large velocities appear in the model predictions that are likely to be associated with abrupt topographic and coastal irregularities, and are therefore very sensitive to model resolution.

The near surface circulation (Fig. 3.26a) shows a vigorous flow in the eastern Sound that is very similar to that predicted by the barotropic nonlinear layer model of Murphy (1979). To the east of the Hammonasset Point (HP in Fig. 3.26b) section, there is a strong (5–10 cm/s) current along the north shore of Long Island to the east and a westward flow of comparable magnitude on the north side of the Sound. These two features of the residual circulation appear to be connected by a southward flowing current to the east of the HP section to form the eastern Sound gyre (ESG). To the west of the HP section, the northeastward flowing surface waters from the central Sound are predicted cross the HP section near its center at approximately 10 cm/s, and then converge with the southward flowing waters on the western side of the ESG.

The across-Sound survey campaign of O'Donnell and Bohlen (2003) included ship-mounted acoustic Doppler current profiler measurements that allowed estimation and removal of the tidal period currents from their measurements. The red arrows in Fig. 3.26c show their estimates of the residual circulation at 5 m below the surface on April 26–28, 1995 at the HP section. Slow westward currents are present on the southern half of the section, and the strong eastward flow predicted by the model is evident in the central part of the section. The agreement between the location, width, and magnitude of this feature is excellent. O'Donnell and Bohlen (2003) show two other sets of observations from the HP section in April 1995 that display the same structure but with variation in the magnitude of the maximum currents of approximately 30 %.

Figure 3.26b shows Hao's (2008) prediction for mid-depth, approximately 10–15 m over most of the eastern Sound. To the east of the HP section, the general pattern of the predicted circulation is flow to the south and west though there is a northeastward motion between the HP section and the Connecticut River. On the southern half of the HP section, the model predicted mid-water flow is westward

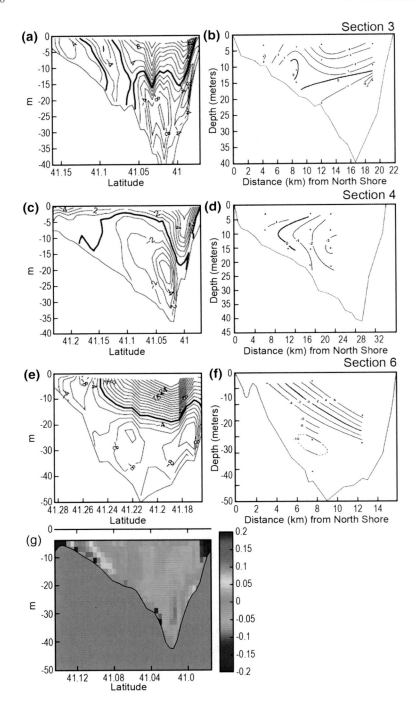

◄ **Fig. 3.25** Comparison of the mean along-Sound flow at sections 3, 4 and 6, (shown in Fig. 3.24) predicted by Crowley (2005) with the currents measured by Vieira (2000). **a**, **c** and **e** show the observations at sections 3, 4 and 6. These should be compared to the observations shown in **b**, **d**, and **f**. **g** shows the easterly component of velocity at, approximately, section 3 from the Bridgeport Ferry ADCP observations

at approximately 5 cm/s. The blue arrows in Fig. 3.26c show the O'Donnell and Bohlen (2003) observations at 15 m are consistent with the model predictions across most of the section. However, the observations indicate a westward motion at the northern end of the section that does not appear in the predictions. Other observations reported by O'Donnell and Bohlen (2003) suggest that this feature is more variable than the westward flow at the southern end of the section. Of course it may also be a consequence of wind or hydrographic variability not captured in the model.

Between the Stratford Shoals (SS in Fig. 3.26b) and HP sections, the near surface flow pattern predicted by the model (see Fig. 3.26a) is eastward along the southern shore with weak southwesterly flow elsewhere. A similar pattern appears at mid-depth in Fig. 3.26b. A more vigorous across-Sound circulation appears slightly to the east of the SS section, where the bathymetry shoals and the Sound narrow. At the southern end of the SS section a strong near surface current to the east is predicted

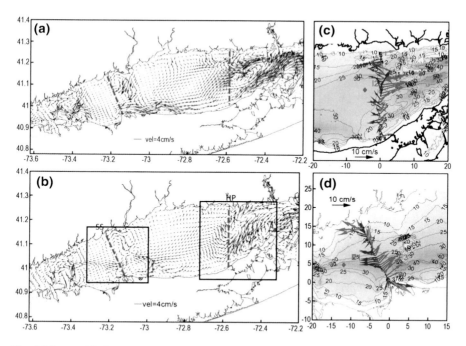

Fig. 3.26 **a** and **b** Show the near surface and mid-depth mean current vectors from the model of Hao (2008). The *black rectangles* and *dashed lines* show the location of the observations displayed in **c** and **d**. **c** and **d** Show the mean currents estimated by O'Donnell and Bohlen (2003). *Red shows* the vectors at 5 m and *blue* shows those at 15 m

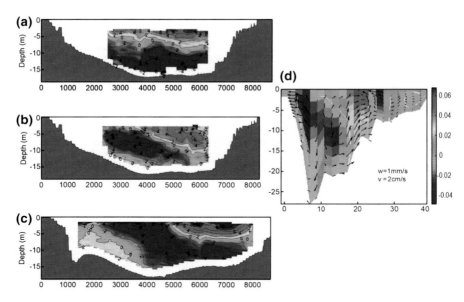

Fig. 3.27 Residual along-Sound velocity components at the LISICOS section (near Vieira's Sect. 3.1 in Fig. 3.21) observed in **a** March 2005, **b** July 2005, and **c** March 2006. Positive (*red*) values indicate eastward along-Sound flow and units are cm/s. The Connecticut coastline is on the left. Section (**c**) is located 3 km to the east of (**a**) and (**b**) where the Sound is wider. **d** shows predictions of Hao's (2008) model slightly to the east of Sect. 3.1. The *arrows* show the across-channel and vertical components. **a–c** are from Bennett (2010) and **d** is adapted from Hao (2008)

with a counter flow at mid-depth. Comparison of the red arrow in Fig. 3.26d with 3.26a, and the blue arrows in 3.26d with 3.26b, demonstrates that the model is reasonably successful in simulating the flow structure. Again there are discrepancies between the observations and predictions at the northern end of the section.

Bennett (2010) also used a ship-mounted ADCP to measure the velocity field during the hydrographic surveys described in Fig. 3.20. The along-Sound component of the mean flow for three intervals, (a) March 2005, (b) July 2005, and (c) March 2006, are shown in Fig. 3.27. The velocity components in Fig. 3.27c were obtained 3 km to the east of those in Fig. 3.27a and b where the Sound is slightly wider. In all three sections, the red colors show an eastward flow at 4–8 cm/s that is confined to a near surface layer adjacent to the Long Island shore. A westward flow (blue) fills the rest of the section. Though there are significant differences in the structure of the hydrography observed during these three transects, the structure of the velocity fields is quite consistent.

3.4 Seasonal Variability in the Circulation

The observations of O'Donnell and Bohlen (2003) and Bennett (2010) show that there is a tendency for eastward flow on the southern side of the Sound. However, their measurements were obtained only during light winds and for periods of a few days at a time. More sustained observations of the lateral and vertical structure of the velocity field are available from two ferry-mounted ADCP systems. Codiga and Aurin (2007) report the analysis of the observations obtained from the New London to Orient Point Ferry along the section labeled OP in Fig. 3.24 at the eastern end of LIS. Stony Brook University operates a similar system on a ferry between Bridgeport, CT and Port Jefferson, NY in the central Sound along line 3 in Fig. 3.24. Using approximately eight sections per day during the period November 2002 to January 2005, Codiga and Aurin (2007) computed the average nontidal current in the lateral bins shown in Fig. 3.23a to reveal the seasonal variation of the nontidal flow shown in Fig. 3.28. In these figures, the coast of Connecticut (north) is on the right side, and the shaded areas indicate westward flow into the Sound. Note that the arrows show the across-channel velocity components and that the values above 7 m are not sampled by the ADCP because of the deep draft of the vessel.

There is a distinct lateral-vertical structure of the ELIS exchange flow, consisting of an eastward outflow that is concentrated in the south and near the surface together with a westward inflow that is concentrated near the bottom and toward the north (Codiga and Aurin 2007). This lateral-vertical structure is consistent with the predictions of the models of Kasai et al. (2000) and Valle-Levinson et al. (2003) in which the Coriolis acceleration modifies the exchange flow driven by the horizontal pressure gradient and resisted by bed stress. Codiga and Aurin (2007) address the dynamics in this area.

Subtle seasonal fluctuations in the along-Sound velocity components are evident in Fig. 3.28. The maximum inflow and outflow velocities of 25 cm/s are observed between May and August. In the winter, the maximum falls to 20 cm/s. This suggests that the rate of volume exchange between LIS and BIS is seasonally modulated; however, the areas of the inflow and outflow also vary, so Codiga and Aurin (2007) integrated the measured velocity component over the area of the section where it was directed inward and obtained 25,000 m^3/s from May to August, and 13,000 m^3/s from November to February. Since the bottom 10 % of the water column is not accessible to ship-mounted ADCPs, they extrapolated the measurements to the bottom and estimated the unmeasured transport. This adds approximately 5,000 m^3/s to the inflow flux estimate. The best estimate of the annual mean volume inflow is then 23,300 m^3/s. The model of Lee (2009) is consistent with this. The annual mean outflow must be comparable since the mean flux of fresh water from rivers and through the East River is an order of magnitude smaller. A very large fraction of the outflow occurs in the unmeasured near surface layer, so the net exchange remains uncertain. It is clear, however, that the volume flux may be much larger than inferred from the salt budget of Gay et al. (2004). These recent volume transport estimates must now be reconciled with the heat and salt budgets of the Sound.

Due to sampling limitations, observations alone do not provide an adequate description of the mechanisms that control the seasonal evolution of bottom T and S. Models provide an important adjunct to data interpretation and guide prioritization of additional measurements. For example, Crowley (2005) noted the presence

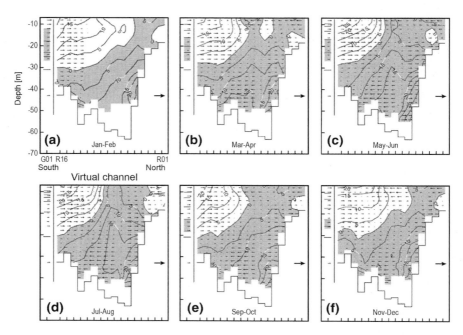

Fig. 3.28 Seasonal variation of the along-Sound residual flow observed by Codiga and Aurin (2007) averaged in two-month intervals. Note that the southern coast of the Sound is on the left and that the *shaded areas* indicate flow into the Sound (westward). Values above were not measured. The *vertical line* on the left side of each frame represents Plum Island and flow through Plum Gut, and segment G01 in Fig. 3.23a, appears to its left

of a pool of cold bottom water in the central Sound between Mattituck Sill and Hempstead Sill in her analysis of the CTDEP hydrographic archive. Figure 3.29a and b show the mean spring (April 12 through June 8) T and S distributions along the thalweg of the Sound (see Fig. 3.29c). Note that Fig. 3.16 shows that the cool pool warms through the summer but remains cold relative to the waters over the Mattituck Sill to the east and Hempstead Sill to the west. Crowley's (2005) model points to the importance of horizontal advection of warm saline bottom waters from Mattituck Sill to the warming of this pool and to the seasonal evolution of both bottom T and bottom S in the Sound. Crowley's (2005) Fig. 3.20 (not shown) emphasizes that this intrusion accelerates in July.

To examine the processes further, we analyze the seasonal variation in longitudinal density gradient across the sill from model results at stations on either side of the Mattituck Sill (Sect. 200 and 240 in Fig. 3.29c). Figure 3.30a shows the evolution of the haline and thermal contributions to the baroclinic pressure gradient across the sill. We see a period of maximum density gradient in July (days 180–200). The haline contribution (shown in green) is dominant and is associated primarily with an elevation in eastern boundary S. The thermal contribution to the baroclinic pressure gradient begins to increase through July as the basin warms

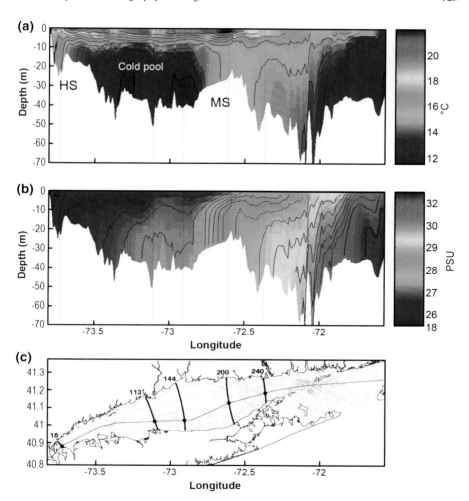

Fig. 3.29 Sections of seasonally averaged values of **a** temperature and **b** salinity along the thalweg of LIS for spring. MS and HS show the locations of Mattituck Sill and Hempstead Sill. **c** shows the coastline of the Sound and the location of the thalweg (from Crowley 2005)

due to surface heating (see red line in Fig. 3.30a) and it significantly augments the salt-induced stratification. Between days 100 and 200, stratification over the sill increases in June (days 160–180) as a consequence of heating (see red line in Fig. 3.30b). Though stratification is strongly modulated by monthly variations in the tidal mixing rate, it is clear that stratification increases significantly in July because the thermal-induced stratification is augmented by haline stratification. A more complete analysis of the interaction of the circulation and stratification can be pursued following the approaches of Holt and Proctor (2003) and Horsburgh et al. (2000) by computing from numerical solutions the rate of heat delivered by horizontal advection and that due to vertical turbulent diffusion. This calculation

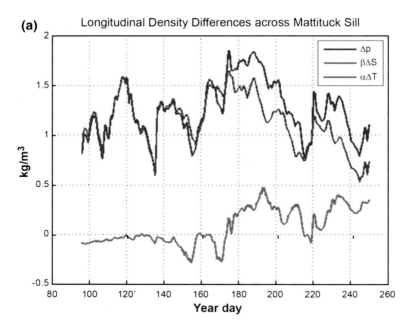

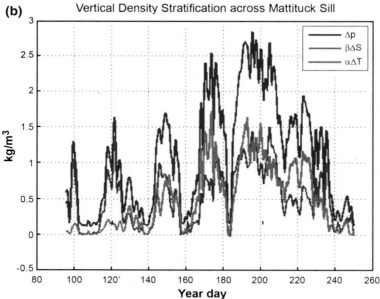

Fig. 3.30 **a** Contributions to difference in depth mean density at channel stations Sections 240 and 144, and **b** Contributions to surface to bottom density difference in density at the thalweg station on Sect. 200 (see Fig. 3.29c)

confirms that advection makes the dominant contribution to changes in sub-pycnocline T in the late summer. A similar analysis of the terms in the S budget also confirms the dominance of the advective contribution to the changes in the lower water column S.

It is important to note that the influence of topographic structures of 20–40 km scale, like the Mattituck Sill, can have an important role on the evolution of the Sound as a whole. Though the extensive archive of hydrographic observations upon which we relied for a description of the seasonal cycles in hydrography (Sect. 3.1.2) is very valuable, it is inadequate to resolve these important smaller scale structures. In Fig. 3.31a, we show a map of the bathymetry of the eastern Sound with the locations of seven stations occupied by a cruise in August 2002 and sampled in approximately a 6 h period. Figure 3.31b and c show the vertical structure of the S and T distribution observed along the cruise track. There is clearly an abrupt surface S front over the eastern side of the shoal with a S change of 1 over 5 km. A similar gradient is also evident near the bottom. The T field is similar on the eastern side of the shoal but there is also a near bottom T front on the western side between the waters of the central Sound cold pool and the warmer waters over the shoal. Codiga et al. (2002) observed similar behavior. The pressure gradients created by these fields are likely to drive significant across-Sound circulation and may play an important role in the enhancement of dispersion in the eastern Sound.

3.5 Synoptic-Scale Wind Forced Circulation

As in other estuaries, sea level varies due to winds blowing locally over the estuary (local effects) and due to wind-induced variations along the continental shelf (remote effects). Beardsley and Butman (1974) analyzed current and water level response to winter extra-tropical cyclones in the section of the Mid-Atlantic Bightfrom Nantucket Shoals to Sandy Hook, NJ. They showed that for cyclones passing to the south of Long Island, the current and sea level response could be described as a relatively simple Ekman response. Northeasterly winds contributed to an onshore Ekman transport and a rise in coastal sea level, and alongshore wind-driven currents were in approximate geostrophic balance with the pressure gradient associated with the across-shore sea level slope. Winds blowing axially along the estuary generate the largest local effects, while winds blowing along the continental shelf coast create the largest remote effects (Garvine 1985). During westerly winds in LIS, sea level is lower at the head and the surface is inclined upward toward the mouth (east) due to local effects and sea level is lowered throughout the estuary and along the shelf coast due to coastal upwelling dynamics (Bokuniewicz and Gordon 1980b; Wong 1990). During easterly winds, sea level is higher at the head and the surface is inclined upward toward the head (west) due to local effects and sea level is increased due to coastal downwelling dynamics along the shelf. Consequently, coastal flooding can occur when setup from easterly wind events coincides with astronomical high tides. This was

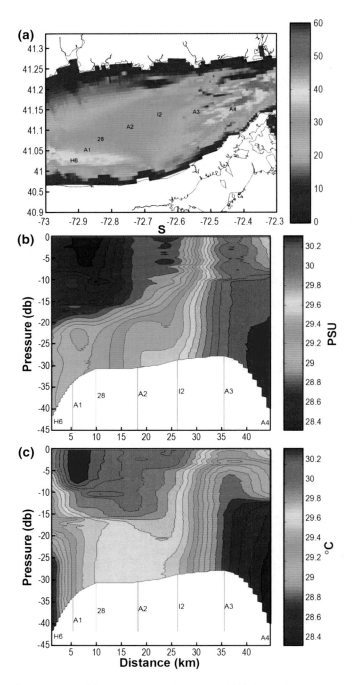

Fig. 3.31 **a** Bathymetry and Station positions for August 2002 SoMAS cruise across Mattituck Sill and the vertical cross-section of **b** salinity and **c** temperature

clearly demonstrated by Wong (1990) who used a year of observations of wind and sea level at stations in LIS to show that more than 80 % of the subtidal sea level variance was correlated with the east–west component of wind stress. Strong easterly and westerly winds (15 m/s, 0.37 Pa) associated with a winter storm passage generated sea level set-up and set-down with approximately 1 m amplitude (Bokuniewicz and Gordon 1980b). The relative importance of remote and local effects can be assessed by comparing sea level at Montauk Point on the shelf to the difference between sea levels at LIS stations and Montauk Point. Wong (1990) found the remote effects (with 0.12 cm standard deviation) are 2–3 times larger than the local effects, which are smaller in C LIS than WLIS (with 0.04 and 0.07 cm standard deviations, respectively). A section-averaged analytical model with wind stress, an along-estuary sea level gradient, and bottom friction compares favorably to observations and indicates the remote effect has a spatially constant amplitude, while the local response increases almost linearly toward the estuary head (Wong 1990).

It has been considerably more difficult to isolate the current response to wind events in LIS. These difficulties are due to the horizontal and vertical variability in currents, other subtidal signals (e.g., gravitational circulation and tide-generated residuals), the limited spatial coverage of moored observations, and the limited temporal coverage of shipboard observations. Gordon and Pilbeam (1975) found no correlation between near bottom currents from multiple locations in CLIS and wind speed and direction, though testing correlation with wind stress components may have helped. Ullman and Wilson (1984) concluded that subtidal currents through a lateral section in WLIS (south of Norwalk, CT) are significantly correlated with the along-estuary wind stress and sea level gradient; each variable is individually correlated with 36 % of the variance of the first spatial mode. Their analysis indicates a westward (up-estuary) flow at depth in response to the along-estuary pressure gradient generated by eastward (down-estuary) winds. Subsequent observations over the deepest area of this section (Schmalz 1993) also indicate an upwind flow response for near bottom currents. For this observational record, 62 % of the subtidal variance is correlated with along-Sound wind velocities and a 1 m/s wind corresponds to a 0.01 m/s current response (Signell et al. 2000).

Recently, Whitney and Codiga (2011) have used several years of data from the ferry-mounted ADCP on the section OP (see Fig. 3.24) to reveal the wind-event velocity anomaly fields. They identified period of westward and eastward wind with stress in excess of 0.1 Pa and averaged the de-tided and de-meaned along-Sound current components during those times to depict the effect of wind directions in the circulation patterns. Figure 3.32a shows the anomaly during intervals of eastward winds. Westward current anomalies (shaded blue in the figure) with 0.19 and 0.03 m/s maximum and mean values occupy most of the section. There is also evidence for downwind anomalies in the shallow areas on the south side of the section. The response to westward stress is shown in Fig. 3.32b. Again upwind current anomalies of similar magnitudes are evident in the center of the section, and these are flanked by near-surface downwind flow anomalies on both sides. Though there are no observations in the upper 7 m of the water column, it seems likely that current anomalies are

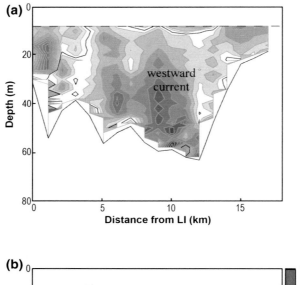

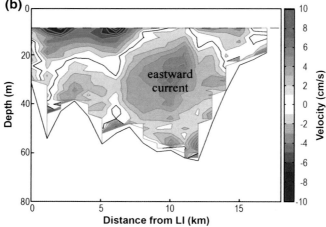

Fig. 3.32 Cross-section of observed wind-driven along-estuary velocity along the eastern LIS ferry transect: **a** wind-event velocity anomaly during westerly winds, and **b** wind-event velocity anomaly during easterly winds. The south side of the estuary is shown on the left. Eastward velocities are *shaded red* and westward velocities are *shaded blue*. The *dashed line* indicates the surface-most level observed

downwind close to the surface. The magnitude of the wind-driven anomalies is several times smaller than the total subtidal velocities at this location (see Fig. 3.28). The wind response tends to augment the mean subtidal flow and increase exchange with BIS during westerly winds and partially counters the background flow and reduces exchange during easterly winds (Whitney and Codiga 2011).

Numerical simulations are useful for studying the spatial structure of wind response throughout the LIS and diagnosing the controlling dynamics. Whitney and Codiga (2011) analyzed model runs with tides, river discharge, and 0.1 Pa eastward and westward wind events to isolate the spatial structure of current vector anomalies. In Fig. 3.33a, we show that the current anomalies during eastward winds are downwind at the surface in most areas and fastest along each coast. Maximum magnitudes vary along the estuary from 0.15 to 0.42 m/s. The near-bottom velocity anomaly field is shown in Fig. 3.33b and is consistent with previous modeling results (Signell et al. 2000) that show downwind flow along each coast flanking upwind anomalies that are strongest in the deepest areas. The relatively weak wind response in the CLIS explains why Gordon and Pilbeam (1975) were unable to identify the wind response in observations, while the stronger response to the west accounts for the significant correlations in WLIS (Ullman and Wilson 1984; Signell et al. 2000).

The spatial structure of wind response is qualitatively consistent with analytical models including wind stress, sea level gradients, and friction developed by Csanady (1973), applied to estuaries (e.g., Wong 1994), and modified to include rotation by Winant (2004). The main features of surface-intensified downwind flow along the coasts flanking a deeper central upwind flow are present in analytical models and the wind-event anomalies in numerical simulations

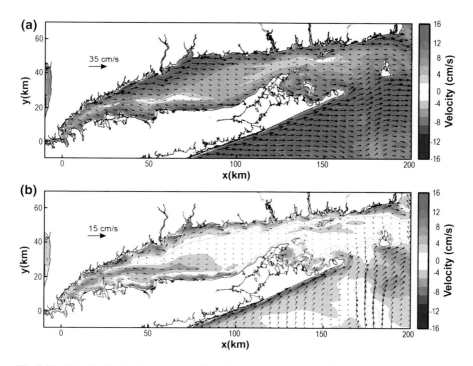

Fig. 3.33 Simulated wind-event anomalies during eastward (westerly winds) at the surface (**a**) and near the bottom (**b**). Velocity vectors are shown as arrows and the East–West velocity component is color-contoured with eastward velocities in *red* and westward velocities in *blue*

and observations. Simulation diagnostics of the section-averaged along-estuary momentum balance indicate the barotropic pressure gradient has the largest wind response. The response in bottom stress is significant, but responses in the Coriolis acceleration, advection, and the baroclinic pressure gradient (all typically neglected in analytic solutions) are equally important. The ratio of the maximum depth across a lateral section to the Ekman depth ranges from 1.7 to 4.8 in LIS, suggesting moderate importance of rotational effects (Winant 2004; Whitney and Codiga 2011). The cross-estuary flow in simulation results is due to this rotational influence (Fig. 3.33). The Wedderburn number (W) measures the relative magnitude of the along-estuary baroclinic pressure gradient and the wind stress (Imberger and Parker 1985). The estuary averaged W magnitude (0.5) indicates that dynamics are strongly wind influenced for 0.1 Pa events that are frequently experienced in the LIS (Whitney and Codiga 2011).

The mean structure of the density field and stratification are significantly modified by wind-induced advection, stirring, and isopycnal straining (Whitney and Codiga 2011). The Sound is characterized by large spatial variations in the sign and magnitude of horizontal density gradients and the intensity of stratification because of the location of rivers, the bathymetry, the subtidal circulation patterns, and variations in tidal mixing conditions. In many parts of LIS, wind-driven stirring, advection of stratification, and isopycnal straining (the change of vertical stratification by vertically sheared currents in a fluid with horizontal density gradients) work together to reduce stratification. In other areas, advection and/or straining work against stirring to limit the stratification reduction or actually increase stratification during wind events. The interplay between these factors changes considerably with wind direction and leads to asymmetries in the response to wind and makes the character of the response highly site-specific. It is likely, therefore, that the response described by O'Donnell et al. (2008a, b) in WLIS area will not be the same throughout the estuary. Additional observational and modeling research is needed to resolve the wind response throughout the LIS and to study wind influence on Lagrangian particle transport and exchange pathways.

3.6 Physical Processes Influencing Hypoxia

Hypoxia, or very low concentrations of DO, occurs in the bottom waters of the western Sound and East River in the summer (Welsh and Eller 1991; Kaputa and Olsen 2000). Two metrics have been adopted by LIS water quality managers to quantify the severity of hypoxia: the duration and the areal extent of hypoxic conditions. Observations from the survey program described Sect. 3.2 and Fig. 3.14 are used to monitor their trends. CTDEP (2009) has described sampling protocols, instrumentation, and methods for defining the two metrics from the data and the metric time series themselves. The New York City Department of Environmental Protection (NYCDEP) has executed a similar program and their station E10, the easternmost sampling location in the deepest part of the channel in the East

River (see Fig. 3.14), exhibits severe summertime hypoxia. Figure 3.34 shows the evolution of the mean May–October DO concentration computed using the CTDEP dataset. The 3 mg/l and 5 mg/l contours are shown. In April and May, the DO begins to drop below saturation in the waters deeper than 10 m and by May (Fig. 3.34a), the signal is evident between A4 and I2. The mean June DO is significantly depressed below 10 m in the western half of the Sound and the 5 mg/l contour then moves eastward to approximately 80 km in August. The bottom waters between A4 and C1 are below 3 mg/l. In the record of CTDEP observations, the earliest date of onset of hypoxia (DO less than 3.5 mg/l) in LIS was June 20 and the latest end date was September 26. The mean start and end dates were July 4 and September 9. Observations by the NYCDEP in the East River show similar start dates for the hypoxic period and a slightly longer duration.

To reduce the extent and duration of hypoxia, the States of Connecticut and New York developed a plan to reduce the amount of nitrogen (N) delivered to the Sound by wastewater treatment facilities (WWTFs) and agricultural run-off, which the US Environmental Protection Agency (USEPA) approved in 2001. Lee and Lwiza (2008) studied the CTDEP measurements obtained during the interval 1993–2004 and found that long-term changes in the extent and duration of seasonal hypoxia in LIS were not statistically significant. Since N discharges had been reduced, they proposed that they had not yet reached levels during late winter and spring that limit primary production. However, they did find that the inter-annual variability of hypoxic volume (the volume of water below the hypoxic threshold) was weakly correlated with variations in summer wind speed, spring total N, spring Chlorophyll a, and maximum springtime river discharge. The spring values for total N and Chlorophyll a were used because that is when maximum phytoplankton biomass concentrations occur and these were thought to determine the size of the organic carbon pool. The maximum river discharge was used to represent the fluctuations in vertical density stratification. Using multivariable linear regression, Lee and Lwiza (2008) found that these variables S, had a strong correlation with hypoxic volume fluctuations, but since the springtime total N explained the lowest fraction of the variance, they questioned the potential effectiveness of the N reduction strategy.

Torgersen et al. (1997) outlined a simple DO budget for the hypoxic region of the Sound and argued that there must be a substantial vertical or horizontal flux of DO to explain the difference between measured respiration rates and the rate of decline of near bottom DO over the summer. McCardell and O'Donnell (2009) used buoy observations to infer the magnitude of the vertical flux of heat and DO across the pycnocline and concluded that it was comparable to the vertically integrated water column respiration rates estimated by Goebel and Kremer (2007). O'Donnell et al. (2006 and 2008) also examined DO, and T series from moored instruments at the EXRK and WLIS sites shown in Fig. 3. These are near CTDEP stations A4 and C1 shown in Figs. 3.14 and 3.34, which are in the region of persistent summertime hypoxia. These records showed that the seasonal decline of DO in the bottom waters was intermittently interrupted by increases that they termed ventilation events. By correlating these periods with wind stress, they noted that

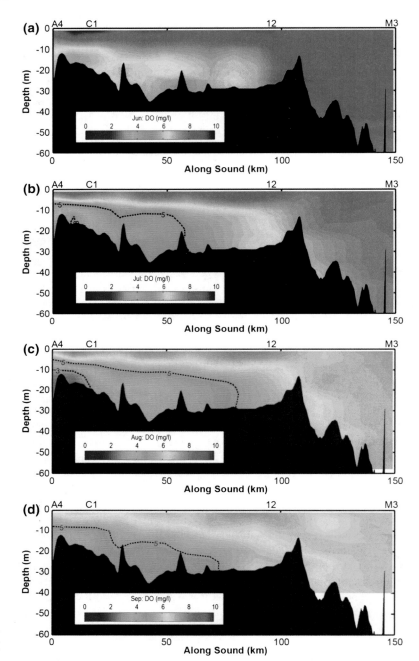

Fig. 3.34 Along-Sound cross sections of the distribution of dissolved oxygen concentration along the *dot-dash line* in Fig. 3.14 during **a** June, **b** July, **c** August, and **d** September computed by monthly averaging the CTDEP dataset and objective analysis

winds from the northeast generally coincided with ventilation. They proposed that the northeast winds modulated the buoyancy-driven circulation that brings fresh water from the East River into LIS to maintain the stratification that is constantly being reduced by tidal shear induced mixing. Scully et al. (2005) noticed the influence of wind on stratification in another estuary and referred to it as wind-induced straining. Contemporaneously, Wilson et al. (2008) used DO observations in the East River and a numerical model like that of Crowley (2005) to investigate the sensitivity of the stratification and DO in the western Sound to wind direction and also demonstrated quantitatively that straining was potentially important.

More recently, Scully (2010a) has demonstrated that inter-annual variations in the summertime volume of hypoxic water in the Chesapeake Bay are positively correlated with the duration of westerly winds and negatively correlated with the duration of southeasterly winds. Scully (2010b) then used a numerical model to investigate the mechanisms through which the extent of hypoxia in the Chesapeake is influenced by wind direction. He found that interactions between vertical mixing over shoal areas and lateral flows driven by winds from the south dominate the supply of DO to the hypoxic regions, whereas winds from the west are the least effective. Lateral advection was shown to be the most important mechanism producing ventilation rather than simple longitudinal or lateral straining suggested by O'Donnell et al. (2008a, b) and Wilson et al. (2008) as the important mechanism in LIS.

Wilson et al. (2008) used the NYCDEP observations at station E10 at the eastern end of the East River (see Fig. 3.14) to compute the time that bottom DO was below 3.5 mg/l using linear interpolation between samples. This is the hypoxia duration metric used by the CTDEP. The CTDEP summer hypoxia duration times series from 1991 to 2009 and the E10 duration time series when Winkler titrations were employed (1988–2006) were both 19 years in length with an overlap of 16 years. The correlation coefficient between the CTDEP hypoxia duration estimates for stations in CLIS and WLIS for 1991–2009 and the percent of hourly summer winds (in $40°$ bins) from different directions are shown in Fig. 3.35a. Experimentation showed that using wind data from June 1 to September 10 for each year and wind speeds above 3.25 m/s maximized correlations and minimized p-values, the measure of statistical significance (see Wunsch 2006). Using wind data from this period each year produced the same directional relationships as data from the extreme or mean range described above but with higher correlations. The high negative correlation for winds from the northeast ($40–60°$) is remarkable. For winds in the $40°$ bin centered on $50°$, the correlation with hypoxia duration is -0.83 with a $\log_{10}(p$ value$)$ of -5.03. High positive correlations and low p-values for winds from the northwest are evident in Fig. 3.35a. For winds in a $40°$ bin centered on $300°$, correlation with hypoxia duration is $+0.74$ with a $\log_{10}(p$ value$)$ of -3.53. Correlations with winds from other quadrants are not significant.

The correlation between hypoxia duration for 1988–2006 at NYC DEP station E10 and winds from different directions (Fig. 3.35b) affords a comparison between directional response at this western Sound station and that inferred from the distributed CTDEP stations (Fig. 3.35a). For this single station in WLIS,

Fig. 3.35 Correlation
between **a** CTDEP hypoxia
duration, **b** NYCDEP E10
hypoxia duration, and **c**
CTDEP hypoxia areal
extent, and percent of winds
in 40° bins from different
directions for 1991–2009

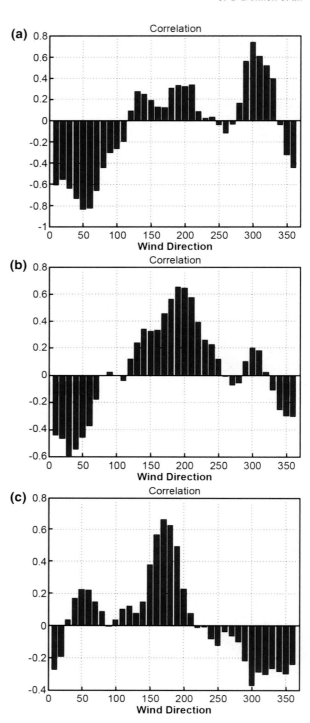

results showed that using wind data from July 5 to September 1 for each year and wind speeds above 3.0 m/s maximized correlations and minimized p-values (increased statistical significance). High negative correlations and low p-values were found for winds from the north-northeast (20–40°). For winds in the bin centered at 30°, the regression coefficient was –0.60 with a $\log_{10}(p$-value) approximately –2.20. Very high positive correlations and low p-values were found for winds from the south. For winds from approximately 185° correlations are approximately +0.69 with a $\log_{10}(p$-value) of approximately –2.96. Correlations with winds from other quadrants are not significant.

The correlation between hypoxia areal extent for 1991–2009 for the distributed CTDEP stations and winds from different directions (Fig. 3.35c) affords description of the directional response for this metric. Analyses showed that using wind observations from June 20 to August 20 for each year and wind speeds above 3.0 m/s maximized correlation with areal extent and minimized p-values. High positive correlations and low p-values are found for winds from the south. For winds from approximately 165° correlations are approximately +0.67 with a $\log_{10}(p$-value) of approximately –2.75. Correlations with winds from other quadrants are not significant.

Correlations in Fig. 3.35a point to a statistically significant relationship between inter-annual variations in summertime hypoxia duration at the CTDEP stations in CLIS and WLIS and wind direction. Winds in a 40° (all compass directions are true north) bin centered on 50° are strongly negatively correlated with variations in duration and so influence the ventilation of bottom waters. At the NYCDEP station E10 in the far western Sound, winds in a bin centered on 30° are strongly negatively correlated with variations in duration (Fig. 3.35b). To interpret the variations in directional response, it is useful to consider the orientation of the thalweg developed from gridded bathymetry used in numerical simulations by Crowley (2005) and Hao (2008). The thalweg heading varies between 60° and 70° with the west central Sound, and it decreases rapidly toward the far western Sound. Winds from 50° are rotated counterclockwise approximately 15° from the thalweg in the west central Sound; winds from 30° have the same orientation relative to the thalweg near E10. The directional response is not inconsistent with wind-induced straining of a longitudinal density gradient, and the 15° deflection could indicate an influence of rotation causing some transport to the right of the wind. The 15° deflection could also indicate that straining of a lateral S gradient is important or that lateral advection as described by Scully (2010b) contributes to ventilation of bottom waters. It is useful to note that exactly the same directional relationships are obtained if only the overlapping 16-year period of 1991–2006 between the CTDEP and NYCDEP data is considered.

At NYCDEP station E10, winds in a bin centered on 190° are strongly positively correlated with inter-annual variations in duration (Fig. 3.35b). This is consistent with the straining of a longitudinal density gradient, although at this station winds are again deflected counterclockwise relative to the thalweg orientation. Unlike the response at E10, Fig. 3.35a indicates that for the CTDEP stations, winds in a bin centered on 300° are strongly positively correlated with variations in duration and so presumably contribute to the rate of restratification. In the more open waters represented by the CTDEP duration metric, winds from the northwest contribute to the maintenance of stratification. This is consistent with straining of a longitudinal density gradient, but winds from 300° do have a significant component normal to the thalweg. This directional response in these open waters is more distinct from that in the constricted waters of the far western Sound.

The inter-annual variations in areal extent of summertime hypoxia are positively correlated with winds from the south-southeast (Fig. 3.35c). Because of longitudinal variations in the orientation of the channel, these winds would contribute to straining of the longitudinal density (S) gradient in the western narrows near E10. In the west central Sound, winds from this direction would contribute to straining of the lateral density (S) gradient.

Figure 3.35a shows that hypoxia duration in the more open waters of the west central Sound is reduced when synoptic winds are from the northeast. Winds from the northwest are associated with an increase in duration. In the far western Sound (E10), winds from the northeast reduce hypoxia duration, while winds from the south contribute to increased duration. Southerly winds also contribute to increased hypoxia areal extent in the west central Sound. Inter-annual variations in synoptic period variability have significant influence on both duration and areal extent.

Further insight into the response of water column structure and mixing to synoptic events, specifically to summertime northeasters and northwesters, can be obtained by analyses of three-dimensional hind-cast numerical model simulations. The model used is described by Crowley (2005), Wilson et al. (2005a), and Hao (2008). We focus here on a description of transient water column structure in a single lateral section in the western Sound centered on −73.51°. This section is close to that of Bennett (2010), shown in Fig. 3.14b. At this section, the orientation of the thalweg is approximately 67 °T. To the west, the basin becomes narrower and the thalweg orientation becomes more northerly.

The summertime northeaster described by Wilson and Swanson (2005) is known to have produced significant mixing. Hindcast simulations for the wind forcing shown in Fig. 3.36 illustrate the evolution of water column structure during the event (Fig. 3.37). The most distinctive feature is downward tilting of the isopycnals on the south side (left) of the transect and the advection of deep water from the channel up into the shallows on the north side of the transect where it is mixed vertically. The low-pass filtered lateral velocity (not shown) is consistent with this advection of bottom waters toward the north side of the transect. This is consistent with mechanisms described by Scully (2010b). At the same time, the low-pass filtered longitudinal velocity (not shown) shows clear evidence of a destratifying strain with bottom velocities directed toward the east and surface velocities directed toward the west. This combination of processes contributes to very significant water column destratification and produces strong lateral density gradients that then adjust under gravity. So winds from the northeast can promote both lateral advection and a destratifying longitudinal strain.

Analyses of hindcast simulations for a number of northwesters show a similar evolution of water column structure. The northerly component of wind stress again

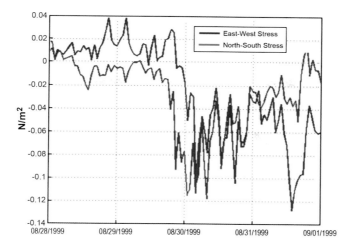

Fig. 3.36 Wind stress components for summertime northeaster simulation

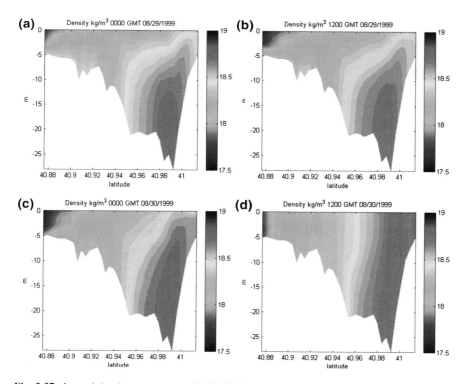

Fig. 3.37 Lateral density structure at **a** 0000 GMT 08/29/1999, **b** 1200 GMT 08/29/1999, **c** 1200 GMT 08/30/1999, and **d** 1200 GMT 08/30/1999. Note that the low density fluids and the south shore of the Sound are on the left of the figure

contributes to the lateral advection of the bottom waters in the channel toward the north side of the transect. The westerly component of wind stress apparently contributes to a stratifying longitudinal strain so the water column doesn't undergo the destratification that occurs under a northeaster. Hindcast simulations for periods with southerly winds characterized by diurnal sea breeze show relatively little response of water column structure to this type of forcing. This prediction must be assessed by comparison with observations and the response of embayments must be investigated.

3.7 Climate Change and Inter-Annual Variability

We have reviewed our understanding of the processes that influence the circulation and structure of LIS and affect hypoxia. It is clear that there are long-term variations in the large scale forcing mechanisms. However, much of our observation database is short-term. It is, therefore, very important that we consider the impact of climate change and inter-annual variability on our understanding.

Recently Whitney (2010) showed that annual cycles in S at the eastern end of LIS do not have the same characteristics as other LIS stations. Lee and Lwiza (2005) also commented on this. For example, there is a similar trend of S variability between stations F3 and M3 (see Fig. 3.14) on a longer time scale (>2 years), but their characteristics on shorter time scales (i.e., subseasonal) are different. This is explained very well in Gay and O'Donnell (2007 and 2009) who argue that the runoff distribution and the geometry of LIS drive a net flux of salt toward the west, which explains the negative curvature of the S profile at the western end of LIS. Lee and Lwiza (2005) have shown that on an inter-annual scale, direct precipitation in LIS does not significantly affect the S in LIS.

The river runoff, on the other hand, can influence the S in LIS with most of it coming from Connecticut River discharge, as it comprises 70 % of the total freshwater discharge into the Sound (see Fig. 3.1). Compared to depth-averaged S, the surface S of LIS is more influenced by river discharge (see Whitney 2010). Salinity records in LIS are too short to draw an unequivocal conclusion about its relationship to climatic processes that operate on longer time scales, decadal to multidecadal such as the NAO. Preliminary evidence from the Chesapeake Bay seems to indicate there might be some correlation to NAO. It is not yet clear whether NAO is manifested through its weak relation with river discharge or by directly affecting the adjacent shelf water. Whitney (2010) found that positive correlations between shelf and LIS surface S are stronger than with rivers, and can explain up to 82 % of the low-passed data with time lags of 0–2 months.

T in LIS is mainly dominated by seasonal variation with the minimum T occurring in February and the maximum in September. The annual range of the minimum and maximum T is higher in the western Sound (~0.7°–22.6 °C) than in the eastern part (1.0°–20.5 °C). The depth-averaged T contains larger inter-annual variability in winter than in summer. For example, the variance of depth-averaged T in the CLIS is 1.9 °C^2 in February and 0.2 °C^2 in September (Lee and Lwiza 2005).

The magnitude and the timing of the inter-annual variations of T in ELIS do not appear to be correlated with those in WLIS. For example, negative anomalies from the summer of 1992 to early 1994 seen at station F3 are larger than at station M3, and strong negative anomalies at station M3 in the winter of 1996 are hardly noticeable at station F3. This uncharacteristic differential heating has yet to be fully explained. Based on a seasonal time scale, Crowley (2005) showed that the estuarine circulation makes a relatively large contribution to the net heat transport during the stratified period and thus the exchange heat transport represents a loss of heat in LIS. Recently Lee (2009) used a numerical model, ROMS, to demonstrate the importance of horizontal advection, which cools the water column during spring-summer. The model results also indicate that spatial patterns of heat flux due to horizontal advection are driven by the longitudinal mean current whose momentum balance is between pressure gradient and advection, agreeing with the results of Hao (2008). In addition, Lee's (2009) results indicate that the net surface heat flux is the major mechanism that controls the heat storage during fall-winter. Evidently, further research is required to understand the mechanisms that control the inter-annual variability and longer time scales of evolution of T in LIS.

Since T in LIS is mainly affected by surface heat fluxes in fall-winter (Lee 2009), then the winter warming trend that has been reported (e.g., Stachowicz et al. 2002) is most likely driven by the regional climate change. Horizontal advection plays a minor role transporting the relatively warm water from BIS into LIS. During spring–summer, the role of horizontal advection is reversed (heat loss) and becomes important in the variability of T is possibly associated with adjacent shelf sea water. Long-term T changes also tend to affect trophic structures of the ecosystem. Although there has not been a systematic analysis done in the whole of LIS, evidence from neighboring habitats seems to point in that direction. In Narragansett Bay, taxonomic diversity has increased over time as the community shifted from fish to invertebrates of several phyla. The shifts in species composition correlate most strongly with spring-summer sea surface T, which increased 1.6 °C between 1959 and 2005 (Collie et al. 2008). Species composition is also correlated with the winter NAO index and the chlorophyll concentration, which has declined since the 1970s. Triggered primarily by rising T, these decadal changes have altered the trophic structure of the nekton community, resulting in a shift from benthic to pelagic consumers. Keser et al. (2005) have shown that the T in ELIS increased by 1 °C between 1979 and 2002, favoring the growth of knotted wrack (*Ascophyllum nodosum*, also commonly called bladder wrack)—an intertidal brown alga. However, their study also indicates that if the T continues to rise past 25 °C knotted wrack will die. Invertebrates are also affected by T variability; for example, the population of the tunicate *Diplosoma listerianum* only thrives if winter T stays above 4 °C. Recruitment failure occurs when T falls below 4 °C.

Long-term changes in surface T in LIS differ markedly between winter and summer. While winter temperatures have been shown to increase, the change for summer has been almost negligible (Lee and Lwiza 2005). Wilson et al. (2008) reported that the increase in thermal stratification in the WLIS during summer months (July and August) from ~0.5 to 2 °C between 1946 and 2006 was mainly due to the decrease in bottom T (see Fig. 3.38). It is proposed that the bottom T decrease is associated with the change in the wind regime. The change in the dominant wind direction during summer time has increasingly favored wind-induced straining as described by Scully et al. (2005) and O'Donnell et al. (2006). Since 1946, the mean wind direction has gradually been approaching 203 °T, which is the optimal direction for producing vertical stratification.

Light availability plays a critical role in triggering the initiation of spring blooms in marine ecosystems (Townsend et al. 1994; Iriarte and Purdie 2004). The inter-annual variations of surface irradiance are influenced by atmospheric conditions, especially cloud cover. Recent literature shows that there is significant inter-annual and decadal variability of the Earth's albedo, which is influenced by changes in cloud location, amount, and thickness (Palle et al. 2006). Figure 3.39a shows the 21-year average of the mean cloud cover over the continental US and the surrounding oceans during the winter months of January, February, and March (JFM) from 1984 to 2004. The Great Lakes effect on cloud cover can be clearly seen, and the northwest Atlantic region is characterized by high cloud cover of approximately 90 %, with a sharp gradient toward the coast. As the cold continental air moves over the warm sea surface, it rises to form the stratocumulus clouds over the Atlantic. The rest of the northeast region follows the continental climatology.

Figure 3.39b shows the sum of the three leading modes of the empirical orthogonal function (EOF) of cloud cover, which represents inter-annual variability. The Middle Atlantic States, and particularly the southern New England region, exhibit

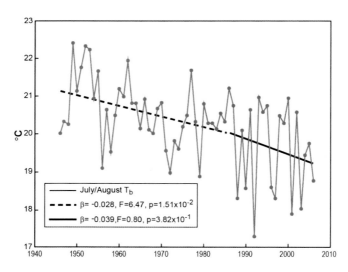

Fig. 3.38 Time series for bottom temperature at NYCDEP E10 averaged over July and August (Wilson et al. 2008)

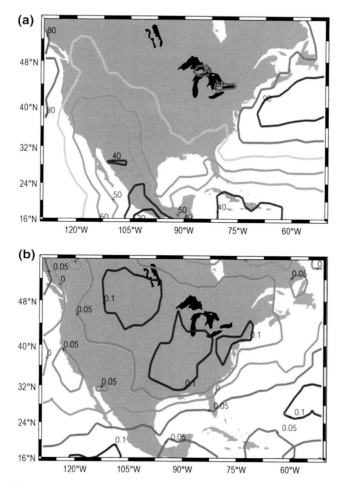

Fig. 3.39 **a** Winter mean cloud cover (%) for the months of Jan–Mar (1984–2004). Note the high cloud cover off the northeast US where there is a permanent stratocumulus cloud and **b** Sum of the three leading modes of the EOF of cloud cover shown in (**a**). Data source: ISCCP from NASA

high variability, which is likely to influence the amount of shortwave radiation reaching the surface. Shortwave radiation can be reduced by as much 16 % during winter months with high cloud cover (Fig. 3.39b). Such a significant decrease will affect the amount of solar radiation available for photosynthesis, particularly in the period of the spring bloom. One might expect that increases in shortwave radiation result in increases in surface air T. However, analysis of long-term datasets tends to suggest the opposite response. Warm winters lead to high cloud cover and low shortwave radiation, e.g., 1971, 1974, 1982, and 1988.

3.8 The Influence of Severe Storms on LIS

Extra-tropical cyclones (northeasters) are the major weather makers in New England. Most extra-tropical cyclones develop as low-pressure systems in the southern States that are then drawn to the northeast by the jet stream. The northeast track brings the system along the east coast past the mid-Atlantic and New England coastal states. Divergence in the upper atmosphere disburses the rising air at a faster rate than it is replaced at the surface, which, in conjunction with the Coriolis effect, causes the cyclone to intensify. In the lower atmosphere, the cyclonic winds bring warm, moist oceanic air over land where it meets colder, drier continental air. The greater the T differences between these two air masses, the more severe the storm can become. Though they occur throughout the year, they are usually more intense in fall and winter because of the large T differences between the converging tropical and continental air masses.

To guide analyses of the oceanic response to such storms, a model of the evolution of the meteorological forcing fields during an east coast extra-tropical winter cyclone has been developed by Mooers et al. (1976) as a composite of 34 extra-tropical cyclones affecting the area during the 1972–1973 and the 1974–1975 winter seasons. The model cyclone passes across the east coast between Cape May and Kiptopeake Beach. It then travels over open water toward the northeast at a speed of approximately 11.8 m/s; the central pressure drops from 1,002 mb to 995 mb over a period of 36 h and the size of the system increases.

The highest wind speeds observed in the LIS area are associated with hurricanes, or tropical cyclones. These form in the tropical North Atlantic and drift westward in June to October. Occasionally, they intensify in the Caribbean Sea, Gulf of Mexico, or over the Gulf Stream off the Carolinas, and then track northeastward across New England. Figure 3.40 shows the tracks of hurricanes identified by Colle et al. (2010) that led to surges of between 0.6 and 1 m at The Battery, NY. Only 13 named hurricanes have tracks that crossed LIS in the last 50 years so the structure and evolution of the wind fields and the sea level responses are only well resolved for a few major events. However, Phadke et al. (2003) have developed a parametric representation of hurricane wind fields that is useful for model sensitivity studies and engineering design applications. It requires specification of the storm track, forward speed, center pressure, and the radius at which maximum wind occurs. This allows simulation of the ocean response to infrequent major storms.

The most damaging aspects of tropical and extra-tropical cyclones are the flooding resulting from the combination of storm surge, freshwater accumulations, and wind wave set-up. Of these, the storm surge is usually the major culprit. Storm surge is a rise in water level along the coast due to the action of wind stress in combination with an inverted barometer effect and, as outlined in Sect. 3.5, this is sensitive to regional coastline geometry and bathymetry. The extent of flooding is also sensitive to storm timing of the storm surge relative to high tide. Much of the developed coastline around LIS, New York, and New Jersey is less than 3 m

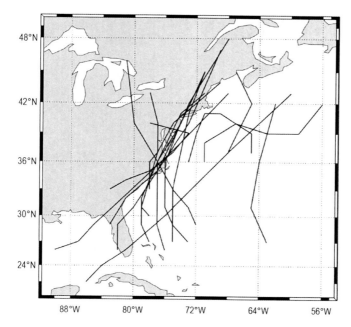

Fig. 3.40 The 48 h tracks of tropical cyclones that led to surge heights of 0.6–1.0 m at the battery between 1959 and 2007. From Colle et al. (2010)

above mean sea level making the region highly vulnerable to storm surge flooding from both extra-tropical and tropical storms. For example, the yellow shade in Fig. 3.41 shows the area of land in the altitude range 0–3 m above mean high water. Clearly the densely populated region of the western Sound is very vulnerable to inundation.

Zhang et al. (2000) have analyzed historical storm activity associated with both extra-tropical and tropical cyclones as it is reflected in storm surge records in century-long sea level series at stations along the east coast of the USA. This analysis is relevant because it includes both Sandy Hook and The Battery. Records are analyzed for storm count, duration, and a measure of intensity. The study affords a description of both seasonal and inter-decadal variability. More recently, Colle et al. (2010) have produced an updated climatology of the characteristics of both extra-tropical and tropical cyclones that produce surge events in New York Harbor as represented by sea level at The Battery. They addressed the complex question of how wind speed and direction evolve in the vicinity of New York Harbor for both weak-surge and moderate-surge events. They also defined the cyclone positions and tracks that favor storm-surge events for the harbor area.

Colle et al. (2010) defined minor surges to be between 0.6 and 1 m above the predicted tide level and showed that there was substantial inter-annual variability in the number of these surges each year and highlighted evidence of a reduction in the frequency of these events. However, the sea level record at The Battery shows

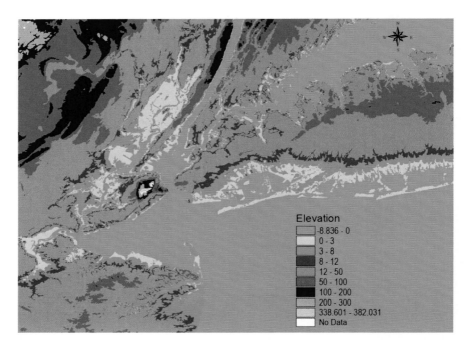

Fig. 3.41 Topographic contour map for the New York metropolitan region; elevations in meters (from Bowman et al. 2005)

that flooding, as measured by the total water elevation above mean high water, has increased. When the effect of long-term sea level rise, 2.8 mm/year at The Battery, was subtracted from the data record, the increasing trend in flooding was eliminated. This shows that a relatively small change in the water level can significantly influence flooding statistics.

The response of sea level in the New York Bight and WLIS to major storm events has been studied recently by Bowman et al. (2005) and Zheng (2006) using numerical simulation. Zheng's analyses confirmed the importance of direct set-up in WLIS, as described by Bokuniewicz and Gordon (1980b), and the coastal Ekman response described by Beardsley and Butman (1974). She found that coastal sea level was highly correlated with the component of wind stress toward 240°, which is essentially coast parallel with zero time lag; a linear relationship between Sandy Hook surge and wind stress toward 240° has a correlation coefficient of 0.89 and a sea level increase per unit wind stress of 1.5 m/Pa. She also found that sea level at Kings Point was highly correlated with the component of wind stress toward 270° with a time lag of approximately 4 h; a linear relationship between Kings Point surge and lagged wind stress toward 270° has an R of 0.92 and a gain of 4.5 m/Pa. A northeaster with this track did not produce any significant westerly winds during its transit and observations showed no evidence of a set-down at Kings Point. Consistent with the response described by

Bokuniewicz and Gordon (1980b), the surge at Montauk lagged that at Kings Point by approximately 2 h. Zheng found that because the Kings Point surge lagged that at The Battery and had significantly greater magnitude, a subtidal sea level slope was maintained across the East River for approximately 12.5 h, producing a subtidal volume flux toward the harbor with peak magnitude of approximately 2×10^5 m³/s and a total volume transported into New York Harbor by the storm of approximately 4.5×10^9 m³.

Hurricane Floyd (September 7, 1999) was fast moving with a track directly across western Long Island that was analyzed through hindcast simulations by Bowman et al. (2005). Atallah and Bosart (2003) described the extra-tropical transition of this storm. The sea level response to such a fast moving tropical cyclone exhibits differences both within the Sound and on the open coast from that of a slower moving extra-tropical cyclone with track to the south and approximately parallel to the island. Time series for lowpass filtered sea level at The Battery, Kings Point, and Montauk (Fig. 3.42a) show the character of the surge response during passage of this storm beginning on September 16, 1999. At the beginning of the storm, strong easterly winds created a set-up of approximately 0.6 m. Subsequently strong westerly winds led to a set-down. Maximum surge levels occurred at Kings Point (green line) in the western Sound.

The directional response at Kings Point to LaGuardia winds (Fig. 3.42b) is very well defined; surge level is highly correlated ($R = 0.83$) with the component of wind stress toward 270°, but with a time lag of approximately 7 h. A linear relationship between Kings Point surge and lagged wind stress has a gain of 2.86 m/Pa. The Battery, as representative of coastal sea level, was most highly correlated with winds toward 274° with a time lag of approximately 6 h; a linear relationship between The Battery surge and lagged wind stress has a gain of 2.24 m/Pa. One major difference with surge response for the extra-tropical storm is that coastal sea level is correlated with winds directed more in the across-shore direction. An important caveat in comparing these results on surge response for Floyd with those for the 2002 northeaster relates to the fact that Zheng (2006) based her results on MM5 model winds and not observed coastal winds such as those at LaGuardia.

Forecasts of flooding at the scale of municipalities that include the effects of rivers and stormwater management infrastructure require very high spatial resolution and a bathymetry and topography that have been merged to a common datum. The model of Chen et al. (2006), for example, has the capability to resolve complex coastal geometry, allows flooding and drying in grid cells, and can support several different levels of different resolution (grid nesting). Figure 3.43a shows a large domain with variable grid spacing that has very high resolution (~0.25 km) nearshore. Figure 3.43b shows a segment of this grid in high resolution near the eastern end of the Sound. The red point is the location of the New London tide gauge. Note that the grid has been extended over land up to the 30 m contour. This is the coarsest resolution at which inundation can be simulated. All surge models require meteorological forcing and forecasts have errors. Even hindcast fields are uncertain, especially during the major events. These are generally not very accurate during major events like the hurricane of 1938 that devastated much of the eastern Connecticut shoreline.

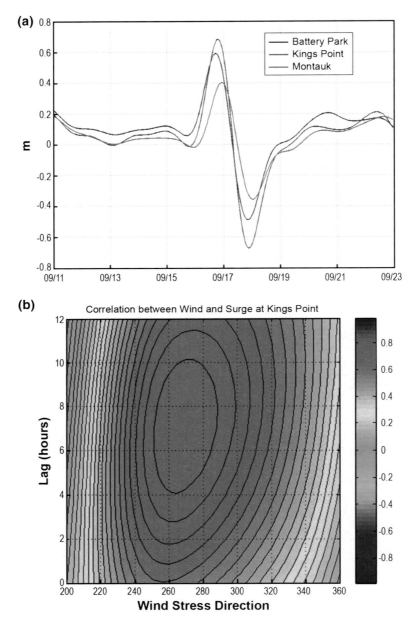

Fig. 3.42 **a** Low-pass filtered sea level at Battery Park, Kings Point and Montauk, NY associated with the storm surge during the passage of Hurricane Floyd (1999) and **b** The correlation between low-pass filtered sea level Kings Point, NY and wind stress estimated from winds at LaGuardia, NY for the storm surge associated with the passage of Hurricane Floyd, September 1999

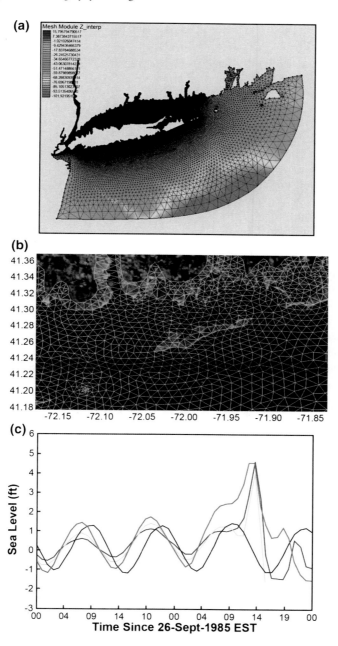

Fig. 3.43 a FVCOM model domain and bathymetry. **b** High resolution grid in the eastern Sound. Note the grid extends over land. **c** Sea level observed at New London (*red*), three alternative simulations are shown in *blue*, *black*, and *yellow*

To demonstrate the sensitivity of surge forecasts to the estimate of the wind field and the bathymetry, we compare the results of three simulations to the observations at New London during Hurricane Gloria in September 1985 (Fig. 3.43c). We use the parametric model of Phadke et al. (2003) to represent the wind and pressure field evolution. This allows studies of the sensitivity of solutions to errors in the track, speed, and intensity of the storm. Figure 3.43c, the red line shows the observations of sea level during the storm. The black line shows the solution using the wind field parameterization values suggested by Phadke et al. (2003). This solution produced a substantial error in surge and phase. We then modified the wind field parameters to better match the observations over land and show the sea level solution in yellow. This yielded a much more accurate surge forecast. Further improvement to the model bathymetry resulted in the best peak surge forecast result shown in blue. All three simulations did well in predicting the height of the maximum surge at 4.5 ft. But all had large errors in the evolution of the sea level. This model did not include river flow fluctuations very well and did not have realistic winds. However, diagnosing which of the many weaknesses is the most critical to the improvement of the model skill will require additional high resolution wind and water level measurements during major storm events.

Some important practical aspects of simulating the surge response within LIS have been summarized by Bowman et al. (2005) based on hindcast simulations of surge within the New York metropolitan area to a strong winter northeaster (December 2002) and to Hurricane Floyd. Among these are issues related to the development of high quality topographic and bathymetric databases including the need for:

- seamless transition from bathymetric to topographic datasets
- up-to-date topography and bathymetry
- accurate translation of available bathymetric datasets (NOS, USGS, USACE, and SoMAS) to a common datum (National Geodetic Vertical Data, NGVD)

3.9 Discussion and Conclusions

Physical oceanography is generally used to describe scientific study of the characteristics of waves, currents, and the density distribution in the ocean. These are closely linked to the dynamics and thermodynamics of the atmosphere and, in the coastal ocean, the hydrologic processes in the watershed of the tributaries. A complete description of the structure and variability of currents and water properties requires extensive and frequent sampling since variations occur across scales that range from seconds to decades and microns to hundreds of kilometers. Since this is not practical, we must combine available observations and theoretical models and seek to define the dominant characteristics of the important properties, quantifying the uncertainty in our knowledge.

Wind velocity, wave height, and period observations in the Sound have been measured for more than a decade at a few locations. However, comparison of the available buoy records at the sites labeled WLIS and CLIS in Fig. 3.3 to those at coastal stations whose records are much longer shows that the seasonal mean vectors are consistent with those over the shelf as summarized by Lentz (2008). The coastal station at Bridgeport, CT (also shown in Fig. 3.3) underestimates the mean over water stress at the WLIS buoy by a factor of two, and by a factor of three at the CLIS buoy. Approximately the same ratios apply to the monthly mean winds. Examination of the frequency spectra of the stress component fluctuations shows that the variance in the along-Sound components is similar at the WLIS and CLIS buoys at all frequencies. However, the across-Sound fluctuations are substantially smaller at the WLIS buoy. The buoy observations allow a first view of the spatial coherence of over water winds. We find that at frequencies higher than approximately 1/day, the winds at the two sites have low coherence and at lower frequencies the coherence is high. This suggests that the local observations will be required in applications in which short time scale meteorological events are important.

Though it is clear that we have not observed the whole range of possible conditions, we present the first statistical summary of long-term observations of buoy measured wave height and period statistics. Using 6 years of data, we show seasonal variation in the probability of exceedance for significant wave heights and periods at the CLIS and WLIS sites. By examining the time series of stress and wave parameters during large events, and the dependence of the differences in the observations of significant wave height at WLIS and CLIS, we find that when the wind is from the east, the wave fields at the two sites are similar. When the winds are from the west, the waves at the CLIS site are much larger. This is validates the idea that waves in LIS are fetch limited as proposed by Bokuniewicz and Gordon (1980b), Signell (2000), and Rivera Lemus (2008).

Sea level fluctuations in LIS due to tides have been well characterized by observations and models. However, there have been few direct current observations with which to evaluate the models. We summarize recent results and since the comparison of observations with simple models show they are not capable of representing the vertical structure, we recommend more detailed evaluations of the predictions of more complex models. Based on analyses of moored hydrographic profiler time series observations and ferry-based current observations, Codiga et al. (2002) and Codiga (2007) have developed strong evidence that straining by tidal currents is active in ELIS. It is, therefore, likely that the relative phase of tidal variations in shear and stratification are important to both horizontal dispersion rates and vertical turbulent diffusion. Bennett et al. (2010) recently reported the existence of large amplitude overtides in the western Sound. These are thought to be due to nonlinear dynamics and replicating the observations will pose a stringent test of the veracity of the simulations.

LIS is an estuary in the sense that it contains seawater that is significantly diluted by the fresh water from rivers. But it is more complex than most estuaries because there are major freshwater sources at each end of the Sound and a mean volume and salt flux through the basin. The Connecticut River enters at the eastern end of the Sound near the connection through which the tide propagates and ocean water intrudes. The western end is freshened by waters originating in the Hudson River watershed that are transported through New York Harbor to the East River by both natural processes and engineered water systems (WWTFs). We present a view of the seasonal variation in the along-Sound vertical structure of S,T, and density fields based on the data archive developed by the CTDEP survey program (see Kaputa and Olsen 2000). We also show the seasonal variation in vertical stratification and the horizontal baroclinic pressure gradients. The lateral structure is much less well resolved, and we summarize available observations. It appears that the lateral structure can be quite significant, and this needs to be better characterized by observations. Using ferry-based measurements in the eastern Sound where the influence of the Connecticut River discharge is very significant, we highlight the difficulty in observing the structure of the S field in the presence of high gradients and large tidal currents. However, we summarize convincing observational and model evidence that water is persistently fresher in the southern portion of ELIS. In addition, there is evidence that the inflow of brackish water to LIS from the East River is found on the southern side of the Sound.

The nontidal currents in the Sound are also difficult to measure and the data archive is limited. We summarize the available observations and model predictions and provide qualitative comparisons. The three-dimensional models of Blumberg and Pritchard (1997), Crowley (2005), and Hao (2008) have provided estimates of the along-Sound volume fluxes at several across-Sound sections and recent ship survey programs by O'Donnell and Bohlen (2003), Codiga and Aurin (2007), and Bennett (2010) have provided measurements that are generally consistent with the volume flux estimate of the models. However, detailed comparisons remain to be conducted. Gay et al. (2004) inferred exchange rates through The Race at the eastern end of the Sound that turned out to be significantly smaller than those of Codiga and Aurin (2007) at a section a few kilometers to the west. Attention should be paid to resolving this discrepancy through model analyses and additional measurements.

The Codiga and Aurin (2007) measurements from the New London to Orient Point Ferry also provide an estimate of the seasonal variation of the circulation in the eastern Sound and demonstrate a significant increase in the exchange volume flux in the summer. Using a model based on that of Crowley (2005), we provide evidence that there is also seasonal variation in the exchange across the Mattituck Sill that arises from the suppression of vertical friction by thermal stratification, which then allows the intrusion of salty water from the eastern Sound. This further increases stratification, reduces friction, and allows the baroclinic exchange. Near synoptic surveys of the late summer hydrographic structure demonstrate that resolving these mechanisms will require high resolution surveys, and seasonal scale transport fluctuations require persistent observations that resolve the lateral variation. Since the fluxes influence the character of the T and S distributions, they may have substantial influence on ecosystem characteristics.

We summarize the development of our understanding of the fundamental response of the LIS sea level and currents to synoptic period winds. The combination of the direct set-up in the Sound and the Ekman response over the shelf makes LIS sea level sensitive to winds with a large east component. Recent work by Whitney and Codiga (2011) has described the influence of the bathymetry on the circulation and we summarize their predictions, which include down wind currents in the near surface waters and higher downwind velocities near shore. Near bottom they find upwind currents in the channels. Using conditional averaging of observation from a ferry-mounted current profiler to extract the velocity response in the eastern Sound during strong eastward and westward wind, they found a strong subsurface flow counter to the wind direction. These transient currents may contribute significantly to the mean circulation.

In addition to achieving a more highly resolved view of the temporal and spatial variability of the circulation in LIS, there have also been marked advances in our understanding of the basic seasonal and basin scale features of both hydrography and circulation since the descriptions by Riley (1956). For example, exchange flows that control basin flushing are now understood in light of both models and observations to decrease from approximately 20,000 m^3/s at Orient Point to approximately 10,000 m^3/s at Mattituck Sill, implying substantial re-circulation of waters entering from BIS. Within the central basin west of Mattituck Sill, they remain at approximately 10,000 m^3/s until the vicinity of Hempstead Sill, where they begin to decrease rapidly to less than 2,000 m^3/s in the western narrows. This decrease in exchange flow again implies significant re-circulation of waters entering WLIS from the central basin.

The correlation between increased stratification and the development of hypoxia in the western Sound is well established, and the fundamental DO budget is described by Torgersen et al. (1997). Recently, the buoy observations of O'Donnell et al. (2006 and 2008a, b) revealed synoptic period variations in the near bottom DO concentration in the western Sound that occurred during periods of wind from the northeast. They conjectured that this was a consequence of modification of the rate of restratification by straining of the along-Sound S gradient by the exchange flow through the East River. Wilson et al. (2008) used a sophisticated circulation and mixing model and showed the sensitivity of mean mixing rates to wind directions. We summarize these results and present an analysis of the correlation between wind direction (measured at LaGuardia Airport) and the duration of hypoxia in the western Sound and in the East River. The role of the wind on the straining rate theory remains viable since the orientation of the topography is different at the two locations. New observations in the Chesapeake Bay suggest that lateral winds may play a large role in modulating the seasonal DO trend and we also present results of preliminary model calculations that explore this possibility in LIS. We find a strong response in the S and T field but the effect on the DO budget remains to be assessed.

The availability of long-term datasets now allows an assessment of decadal scale changes in the environment. We summarize the results of several studies including Lee and Lwiza (2005), Gay and O'Donnell (2007 and 2009), Wilson et al. (2008), and Whitney (2010) who examined T and S variations and looked for links to global scale forcing. Wilson et al. (2008) reported that the increase in thermal stratification in the WLIS during summer of between 0.5 and 2 °C between

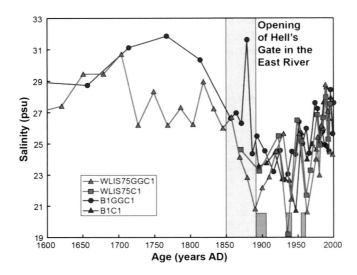

Fig. 3.44 The salinity in western LIS derived from oxygen isotope ratio measurements in shell fossils in cores obtained in western LIS near Execution Rocks

1946 and 2006 was mainly due to the decrease in bottom T (see Fig. 3.38). It is proposed that the bottom T decrease is associated with the change in the wind regime. However, Palle et al. (2006) have analyzed the cloudiness statistics and these show substantial inter-annual variability. The local effects of global changes remain uncertain and additional research is required.

Humans have greatly modified LIS. The development of the water delivery and disposal system for New York City and the surrounding municipalities has diverted water from the Hudson to the East River. Dredging of the Hell Gate in the East River in the late 19th century to make navigation safer is also likely to have increased the exchange with the Hudson and may have changed the S in the Sound. Varekamp (personal communication) has sampled and dated cores obtained near the EXRK buoy (see Fig. 3.3a) in the western Sound and analyzed the oxygen isotope ratio in fossil shells. Since the rate of accumulation of heavy isotopes depends on S, the history of S in the area when the shells were being formed can be inferred. Figure 3.44 shows the time series of "paleosalinity" between 1,600 and 2,000, and there appears to be a large decline between 1,800 and 1,900. While the uncertainty in these estimates is unknown at the moment, it is clearly important to attempt to understand in a quantitative way the impact of historical engineering work on the western Sound.

Currently, coastal development around LIS is very sensitive to severe weather. We summarize the characteristics of storms that influence the region with particular emphasis on sea level set-up and flooding. Colle et al. (2010) have provided a concise summary of observation and have highlighted recent trends in storm character and the impact of even modest sea level rise. Predictions about future flooding require simulation and we summarize some recent work in that area. It is clear that to make useful forecasts at the scale of individual buildings, circulation models must allow flooding of land, and topography and bathymetry maps must be aligned to a common data and merged. Models must then be tested at these scales and this will require high-resolution sampling in both the atmosphere and ocean.

In all aspects of this review a central theme resonates. Understanding of LIS requires sustained observations and the complementary development of mathematical models. Together these allow the quantitative evaluation of theoretical links between processes and variables, and the development of forecasts.

Acknowledgments We are grateful to the USEPA LIS Office and the LIS Study for motivating this article. During the preparation of the manuscript, the lead author was supported by the US IOOS office through a grant to the Northeast Regional Association of Coastal Ocean Observing Systems (NERACOOS) and the Connecticut Sea Grant College Program. REW acknowledges support from NY Sea Grant LIS project R/CE-30-NYCT.

References

Atallah EH, Bosart LF (2003) The extratropical transition and precipitation distribution of Hurricane Floyd (1999). Monthly Weather Rev 131:163–1881

Beardsley RC, Butman B (1974) Circulation on the New England continental shelf: response to strong winter storms. Geophys Res Lett 1:181–184

Bennett DC (2010) The dynamical circulation of a partially stratified, frictional estuary: Long Island Sound. PhD Dissertation, University of Connecticut, Groton, p 168

Bennett DC, O'Donnell J, Bohlen WF, Houk AE (2010) Tides and overtides in Long Island Sound. J Mar Res 68(1):21–35

Blumberg AF, Pritchard DW (1997) Estimates of the transport through the East River, New York. J Geophys Res 102(C3):5685–5703

Blumberg AF, Khan LA, St. John JP (1999) Three-dimensional hydrodynamic simulations of the New York Harbor, Long Island Sound and the New York Bight. J Hydraulic Eng 125:799–816

Bogden PS, O'Donnell J (1998) Generalized inverse with shipboard current measurements: Tidal and nontidal flows in Long Island Sound. J Mar Res 56(5):995(3)

Bokuniewicz HJ, Gordon RB (1980a) Sediment transport and deposition in Long Island Sound. Adv Geophys 22:69–106

Bokuniewicz HJ, Gordon RB (1980b) Storm and tidal energy in Long Island Sound. Adv Geophys 22:41–67

Bowman MJ, Colle B, Flood R, Hill D, Wilson RE, Buonaiuto F, Cheng P, Zheng Y (2005) Hydrologic feasibility of storm surge barriers to protect the metropolitan New York-New Jersey region. Technical Report, Marine Sciences Research Center, Stony Brook University, Stony Brook, NY, 98 pp

Bretschneider CL (1952) Revised wave forecasting relationships. In: Proceedings of second conference on coastal engineering, Chapter I:1–5

Buxton HT, Smolensky DA (1999) Simulation of the effects of development of the ground-water flow system of Long Island, New York. In: US geological survey water-resources investigations report 98-4069, 57 pp

Chen C, Beardsley RC, Cowles G (2006) An unstructured grid, finite-volume coastal ocean model (FVCOM) system. Oceanography 19(1):78–89

Codiga DL (2007) FOSTER-LIS gridded data products: observed current profiles and near-surface water properties from ferry-based oceanographic sampling in eastern long Island sound. Graduate School of Oceanography, University of Rhode Island, Report 2007–2001,

Narragansett, 14 pp http://www.po.gso.uri.edu/~codiga/foster/files/FOSTERLISGriddedDat aProductsMay2007.pdf

Codiga DL, Rear LV (2004) Observed tidal currents outside block Island sound: offshore decay and effects of estuarine outflow. J Geophys Res 109, C07S05. doi:10.1029/2003JC001804

Codiga DL, Aurin DA (2007) Residual circulation in eastern long island sound: observed transverse-vertical structure and exchange transport. Cont Shelf Res 27:103–116

Codiga DL and Ullman DS (2010) Characterizing the physical oceanography of coastal waters off Rhode Island, Part 1: Literature review, available observations, and a representative model simulation. Technical report 2, Appendix to Rhode Island Ocean special area management plan, 169 pp

Codiga DL, Waliser DS, Wilson RE (2002) Observed evolution of vertical profiles of stratification and dissolved oxygen in Long Island Sound. In: Proceedings of the New England estuarine research society/long island sound research Conference Groton, CT, 7–12

Colle BA, Rojowsky K, Buonaiuto F (2010) New York City storm surges: climatology and an analysis of the wind and cyclone evolution. J Appl Meteorol Climatol 49:85–100

Collie JS, Wood AD, Jeffries HP (2008) Long-term shifts in the species composition of a coastal fish community. Can J Fish Aquat Sci 65:1352–1365

Connecticut Department of Environmental Protection (2009) Long Island Sound hypoxia season review 2009. Hartford, 23 pp

Crowley H (2005) The seasonal evolution of thermohaline circulation in Long Island sound. PhD Dissertation, Marine Sciences Research Center, Stony Brook University, Stony Brook, NY, 142 pp

Csanady GT (1973) Wind-induced barotropic motions in long lakes. J Phys Oceanogr 3:429–438

Emery WJ, Thompson RE (1997) Data analysis methods in physical oceanography. Elsevier, San Diego 400

Garvine RW (1974) Physical features of the Connecticut river outflow during high discharge. J Geophys Res 79:831–846

Garvine RW (1975) The distribution of salinity and temperature in the Connecticut river estuary. J Geophys Res 80:1176–1183

Garvine RW (1977) Observations of the motion field of the Connecticut river plume. J Geophys Res 82:441–454

Garvine RW (1985) A simple model of estuarine subtidal fluctuations by local and remote wind stress. J Geophys Res 90(C6):1945–1948

Gay P, O'Donnell J, Edwards CA (2004) Exchange between long Island sound and adjacent waters. J Geophys Res 109:C06017. doi:10.1029/2004JC002319

Gay PS, O'Donnell J (2007) A one-dimensional model of the salt flux in estuaries. J Geophys Res 112:C07021. doi:10.1029/2006JC003840

Gay PS, O'Donnell J (2009) Comparison of the salinity structure of the Chesapeake Bay, the Delaware Bay and long Island sound using a linearly tapered advection-dispersion model. Estuaries Coasts 32:68–87. doi:10.1007/s12237-008-9101-4

Geyer RW, Chant R (2006) The physical oceanography processes in the Hudson River estuary. In: Levinton JS, Waldman JR (eds) The Hudson River Estuary, Cambridge University Press, New York, p 472

Goebel NL, Kremer JN (2007) Temporal and spatial variability of photosynthetic parameters and community respiration in Long Island sound. Mar Ecol Prog Ser 329:23–42

Gordon RB, Pilbeam CC (1975) Circulation in central long island sound. J Geophys Res 80:414–422

Gross MG, Bumpus DF (1972) Residual drift of near bottom waters in long island sound. Limnol and Oceanogr 11:636–638

Hansen DV, Rattray DM (1965) Gravitational circulation in straits and estuaries. J Mar Res 23:104–122

Hao Y (2008) Tidal and residual circulation in Long Island Sound. PhD Dissertation, Marine Sciences Research Center, Stony Brook University, Stony Brook, NY, 70 pp

Hardy CD (1972) Hydrographic data report: long island sound, 1970, Part II. Marine Sciences Research Center, State University of New York, Stony Brook, p 20

Hodgkins GA, Dudley RW, Huntington TG (2003) Changes in the timing of high river flows in New England over the 20th century. J Hydrol 278:244–252

Hollman R, Sandberg GR (1972) The residual drift in eastern long island sound and block island sound. New York Ocean Sci Lab Tech Rept 15, 19 pp

Holt JT, Proctor R (2003) The role of advection in determining the temperature structure of the Irish Sea. J Phys Oceanogr 33:2288–2306

Horsburgh KJ, Hill AE, Brown J, Fernand L, Garvine RW, Angelico MMP (2000) Seasonal evolution of the cold pool gyre in the western Irish Sea. Prog in Oceanogr 46:1–58

Howard-Strobel MM, Bohlen WF, Cohen DR (2006) A year of acoustic doppler current meter observations from central long island sound. In: 8th Biennial Long Island Sound Research Conference Proceedings 2006, pp 26–31

Hurrell JW (1995) Decadal trends in the North Atlantic oscillation regional temperatures and precipitation. Science 269:676–679

Ianniello JP (1977a) Non-linearly induced residual currents in tidally dominated estuaries. PhD Dissertation, The University of Connecticut, 250 pp

Ianniello JP (1977b) Tidally induced residual currents in estuaries of constant breadth and depth. J Mar Res 35(4):755–785

Imberger J, Parker G (1985) Mixed layer dynamics in a lake exposed to a spatially variable wind field. Limnol and Oceanogr 30:473–488

Iriarte A, Purdie DA (2004) Factors controlling the timing of major spring bloom events in an UK south coast estuary. Est Coast Shelf Sci 61:679–690

Isemer H-J, Hasse L (1985) Observations Vol 1. The bunker climate atlas of the North Atlantic Ocean. Springer, London 218

Jay DA, Bowman MJ (1975) The physical oceanography and water quality of New York Harbor and western Long Island Sound. Technical Report 23, Ref # 75-7, Marine Sciences Research Center, State University of New York, Stony Brook

Kaputa NP, Olsen CB (2000) State of Connecticut department of environmental protection, long Island sound ambient water quality monitoring program: summer hypoxia monitoring survey '91-'98 Data Review, CTDEP Bureau of Water Management, 79 Elm Street, Hartford, CT 06106-5127, p 45

Kasai A, Hill AE, Fujiwara T, Simpson JH (2000) Effect of the Earth's rotation on the circulation in regions of freshwater influence. J Geophys Res-Oceans 105:16961–16969

Kenefick AM (1985) Barotropic M2 **tides** and tidal currents in **Long Island** Sound: A numerical model. J Coast Res 1:117–128

Keser M, Swenarton JT, Foertch JF (2005) Effects of thermal input and climate change on growth of Ascophyllum nodosum (Fucales, Phaeophyceae) in eastern Long Island Sound (USA). J Sea Res 54:211–220

Klink K (1999) Climatological mean and interannual variance of United States surface wind speed, direction, and velocity. Int J Climatol 19:471–488

Koppelman LL, Weyl PK, Gross MG (1976) The urban sea: long Island sound. Praeger, New York 223 pp

Krug WR, Gebert WA, Graczyk DJ, Stevens DL, Rochelle BP, Church MR (1990) Map of mean annual runoff for the northeastern, southeastern, and mid-Atlantic United States water years 1951–1980. US Geological Survey Water Resources Investigations Report 88–4094, p. 11

Large WG, Pond S (1981) Open ocean momentum flux measurements in moderate to strong wind. J Phys Oceanogr 11:324–336

Larkin RR, Riley GA (1967) A drift bottle study in long Island sound. Bull Bingham Oceanogr Coll 19:62–71

Lee YJ (2009) Mechanisms controlling variability in Long Island Sound. PhD Dissertation, School of Marine and Atmospheric Sciences, Stony Brook University, New York, 147

Lee YJ, Lwiza K (2005) Interannual variability of temperature and salinity in shallow water: Long Island Sound, New York. J Geophys Res 110 doi:10.1029/2004JC002507

Lee YJ, Lwiza KMM (2008) Characteristics of bottom dissolved oxygen in Long Island Sound, New York. Estuar Coast Shelf Sci 76:187–200 doi:10.1016/j.ecss.2007.07.001

Le Lacheur EA, Sammons JC (1932) Tides and currents in long island and Block Island sounds. US coast and Geodetic survey. Special Publication, 174

Lentz SJ (2008) Seasonal variations in the circulation over the Middle Atlantic Bight continental shelf. J Phys Oceanogr 38:1486–1500

Li CY, O'Donnell J (1997) Tidally driven residual circulation in shallow estuaries with lateral in shallow estuaries with lateral depth variation. J Geophys Res 102:27915–27929

Li C, O'Donnell J (2005) The effect of channel length on the residual circulation in tidally dominated channels. J Phys Oceanogr 35:1826–1840

Mau J-C, Wang D-P, Ullman DS, Codiga DL (2008) Model of the long Island sound outflow: comparison with year-long HF radar and Doppler current observations. Cont Shelf Res 28(14):1791–1799. doi:10.1016/j.csr.2008.04.013

McCardell GM, O'Donnell J (2009) A novel method for estimating vertical eddy diffusivities using diurnal signals with application to western long Island sound. J Mar Syst 77:397–408

McDougall TJ (1987) Neutral surfaces. J Phys Oceanogr 17:1950–1964

Mooers CNK, Fernandez-Partagas J, Price JF (1976) Meteorological forcing fields of the New York Bight. University of Miami (RSMAS) Technical Report TR76-8, Miami, 151 pp

Mountain DG (2003) Variability in the properties of shelf water in the Middle Atlantic Bight, 1977–1999. J Geophys Res Oceans 108(C1):3014

Murphy DL (1979) A numerical investigation into the physical parameters which determine the residual drift in Long Island Sound. PhD Dissertation, Dept of Marine Sciences, The University of Connecticut, 362 pp

O'Donnell J (1997) Observations of near surface currents and hydrography in the Connecticut River plume with the SCUD array. J Geophys Res 102:25021–25033

O'Donnell J, Bohlen WF (2003) The structure and variability of the residual circulation in Long Island sound. Final Report, Connecticut Department of Environmental Protection, Hartford, CT. Grant CWF 325-R, 303 pp (http://www.lisrc.uconn.edu/DataCatalog/DocumentImages /pdf/Odonnell_Bohlen_2003.pdf http://www.lisrc.uconn.edu/DataCatalog/DocumentImages /pdf/Odonnell_Bohlen_2003.pdf)

O'Donnell J, Marmorino GO, Trump CL (1998) Convergence and downwelling at a river plume front. J Phys Oceanogr 28:481–1495

O'Donnell J, Bohlen WF, Dam HG (2006) Wind stress and the ventilation of the hypoxic zone of western Long Island Sound. In: Proceedings of the 8th Biennial Long Island Sound Research Conference, CT Sea Grant Program, New London

O'Donnell J, Dam HG, Bohlen WF, Fitzgerald W, Gay PS, Houk AE, Cohen DC, Howard-Strobel MM (2008) Intermittent ventilation in the hypoxic zone of western Long Island Sound during the summer of 2004. J Geophys Res 113 doi:10.1029/2007JC004716

O'Donnell J, Ackleson SG, Levine ER (2008b) On the spatial scales of a river plume. J Geophys Res-Oceans 113:C04017. doi:10.1029/2007JC004440

O'Donnell J, Morrison J, Mullaney J (2010) The expansion of the Long Island Sound Integrated Coastal Observing System (LISICOS) to the Connecticut River in support of understanding the consequences of climate change. Final Report to the CTDEP, LIS License Plate Fund, 20 pp

Palle E, Goode PR, Montañes-Rodriguez P, Koonin SE (2006) Can Earth's albedo and surface temperatures increase together? EOS Trans 87(4):37

Parker BB (1984) Frictional effects on the tidal dynamics of a shallow estuary. PhD Dissertation, The Johns Hopkins University, 304 pp

Paskausky DF (1976) Seasonal variation of residual drift in Long Island Sound. Est Coast Mar Sci 4:513–522

Pawlowicz R, Beardsley B, Lentz S (2002) Classical tidal harmonic analysis including error estimates in MATLAB using t_tide. Comput Geosci 28:929–937

Phadke AC, Martino CD, Cheung KF, Houston SH (2003) Modeling of tropical cyclone winds and waves for emergency management. Ocean Eng 30(4):553–578

Pritchard DF (1956) The dynamic structure of a coastal plain estuary. J Mar Res 1:33–44

Redfield AE (1950) The analysis of tidal phenomena in narrow embayments. Pap Phys Oceanogr Meteorol XI(4):36

Riley GA (1952) Hydrography of the long island and Block Island sounds. Bull of the Bingham Oceanographic Collection 13, Article 3, Peabody Museum of Natural History, Yale University, New Haven

Riley GA (1956) Oceanography of Long Island Sound: 1952–1954. II. Phys Oceanogr Bull of the Bingham Oceanographic Collection 15, Peabody Museum of Natural History, Yale University, New Haven

Rivera Lemus ER (2008) Wind waves in central Long Island Sound: a comparison of observations to an analytical expression. Masters Dissertation, Department of Marine Sciences, The University of Connecticut, 77 pp

Scully ME (2010a) The importance of climate variability to wind-driven modulation of hypoxia in Chesapeake Bay. J Phys Oceanogr 40:1435–1440

Scully ME (2010b) Wind modulation of dissolved oxygen in Chesapeake Bay. Estuaries Coasts 33:1164–1175

Scully M, Friedrichs C, Brubaker J (2005) Control of estuarine stratification and mixing by wind-induced straining of the estuarine density field. Estuaries 28(3):321–326. doi:10.100 7/BF02693915

Schmalz RA (1993) Numerical decomposition of Eulerian circulation in long island sound. In: Proceedings of the 3rd International Estuarine and Coastal Modeling Conference. ASCE, Chicago, pp 294–308

Shchepetkin AF, McWilliams JC (2005) Regional ocean model system: a split-explicit ocean model with a free-surface and topography-following vertical coordinate. Ocean Model 9:347–404

Signell R, List J, Farris A (2000) Bottom currents and sediment transport in Long Island Sound: a modeling study. J Coast Res 16:551–566

Simpson JH, Brown J, Matthews JP, Allen G (1990) Tidal straining, density currents and stirring in the control of estuarine stratification. Estuaries 12:129–132

Soulsby RL (1990) Tidal-current boundary layers. In: Le Mehaute B, Hanes DM (eds) The sea, ocean engineering science 9A. Wiley-Interscience, New York, pp 523–566

Stachowicz JJ, Terwin JR, Whitlatch RB, Osman RW (2002) Linking climate change and biological invasions: Ocean warming facilitates nonindigenous species invasions. Proc Nat Acad Sci US Am 99(24):15497–15500

Sverdrup HU, Monk WH (1946) Empirical and theoretical relations between wind, sea and swell. Trans Am Geophys Union 27(6):828–836

Swanson RL (1976) Tides. MESA New York Bight Atlas Monograph, 4, New York Sea Grant Institute, Albany, pp 34

Torgersen T, DeAngelo E, O'Donnell J (1997) Calculations of horizontal mixing rates using 222Rn and the controls on hypoxia in western Long Island Sound. Estuaries 20:328–343

Townsend DW, Cammen L, Holligan PM, Campbell DE, Pettigrew NR (1994) Causes and consequences of variability in the timing of spring phytoplankton blooms. Deep Sea Res I 41:747–765

Ullman DS, DL Codiga DL (2010) Characterizing the physical oceanography of coastal waters off Rhode Island, Part 2: New observations of water properties, currents, and waves. Technical Report 3, Appendix to Rhode Island Ocean Special Area Management Plan, 108 pp

Ullman DS, Wilson RE (1984) Subinertial current oscillations in western Long Island Sound. J Geophys Res 89:10,579–10,587

Valle-Levinson A (2008) Density-driven exchange flow in terms of the Kelvin and Ekman numbers. J Geophys Res 113:C04001. doi:10.1029/2007JC004144

Valle-Levinson A, Wilson RE (1994a) Effects of sill bathymetry, oscillating barotropic forcing and vertical mixing on estuary ocean exchange, J Geophys Res 99 (C3):5194–5169

Valle-Levinson A, Wilson RE (1994b) Effects of sill processes and tidal forcing on exchange in eastern Long Island Sound. J Geophys Res 99 (C6):12667–12681

Valle-Levinson A, Wilson RE, Swanson RL (1995) Physical mechanisms leading to hypoxia and anoxia in western Long Island Sound. Environ Int 21(5):657–666

Valle-Levinson A, Wilson RE (1998) Rotation and vertical mixing effects on volume exchange in eastern Long Island Sound. Estuarine Costal and Shelf Science 46:573–585

Valle-Levinson A, Reyes C, Sanay R (2003) Effects of bathymetry, friction, and rotation on estuary-ocean exchange. J Phys Oceanogr 33:2375–2393

Vieira MEC (2000) The long-term residual circulation in Long Island Sound. Estuaries 23(2):199–207

Wang YH, Bohlen WF, O'Donnell J (2000) Storm enhanced bottom shear stress and associated sediment entrainment in a moderate energetic estuary. J Oceanogr 56:311–317

Welsh BL, Eller FC (1991) Mechanisms controlling summertime oxygen depletion in western Long Island Sound. Estuaries 14:265–278

Whitney MM (2010) A study on river discharge and salinity variability in the Middle Atlantic Bight and Long Island Sound. Cont Shelf Res 30:305–318

Whitney MM, Codiga DL (2011) Response of a large stratified estuary to wind events: Observations, simulations, and theory for Long Island Sound. J Phys Oceanogr 41:1308–1327

Wilkin JL, Bowen MM, Emery WJ (2002) Mapping mesoscale currents by optimal interpolation of satellite radiometer and altimeter data. Ocean Dyn 52:95–103

Wilson RE (1976) Gravitational circulation in Long Island Sound. Estuar Coast Mar Sci 4:443–453

Wilson RE, Swanson RL (2005) A perspective on bottom water temperature anomalies in Long Island Sound during the 1999 Lobster Mortality event. J Shellfish Res 24:825–830

Wilson RE, Crowley HA, Brownawell BJ, Swanson RL (2005) Simulation of transient pesticide concentrations in Long Island Sound for late summer 1999 with a high resolution coastal circulation model. J Shellfish Res 24:865–875

Wilson RE, Swanson RL, Crowley HA (2008) Perspectives on long-term variations in hypoxic conditions in western Long Island Sound. J Geophys Res 113:C12011. doi:10.1029/200 7JC004693

Winant CD (2004) Three-dimensional wind-driven flow in an elongated, rotating basin. J Phys Oceanogr 34:462–476

Winant CD (2007) Three-dimensional tidal flow in an elongated, rotating basin. J Phys Oceanogr 37:2345–2362

Wong K-C (1990) Sea level variability in Long Island Sound. Estuaries 13:362–372

Wong K-C (1991) The effect of the East River on the barotropic motions in Long Island sound. J Mar Res 49:321–337

Wong K-C (1994) On the nature of transverse variability in a coastal plain estuary. J Geophys Res 99:14209–14222

Wunsch C (2006) Discrete inverse and state estimation problems with geophysical fluid applications. Cambridge University Press, Cambridge 371

Zhang K, Douglas BR, Leatherman SP (2000) Twentieth-century storm activity along the US East Coast. J Climate 13:1748–1761

Zheng Y (2006) Diagnosis of extra-tropical storm surge response in New York Harbor. Masters Dissertation, Marine Sciences Research Center, Stony Brook University, Stony Brook, 84 pp

Chapter 4
Geochemistry of the Long Island Sound Estuary

Carmela Cuomo, J. Kirk Cochran and Karl K. Turekian

4.1 Introduction

The complexity of Long Island Sound's physical oceanography, coupled with its natural geology, and its history of anthropogenic inputs exerts strong controls on the chemistry of both the water column and sediments of Long Island Sound (LIS). These, in turn, affect and are further affected by the organisms that live in these environments. Recognizing that the chemistry of LIS is too complex for a single chapter in a volume, the authors of this chapter, nevertheless, have attempted to describe, as succinctly and completely as possible, the present state of knowledge of the chemical processes, the influences on them, and their effects upon the larger LIS system. In order to accomplish this task, the chapter has been divided into several main topic areas, including processes (e.g., sediment diagenesis, bioturbation and bioirrigation,

Carmela Cuomo and Kirk Cochran dedicate this chapter to the memory of our co-author and colleague, Karl K. Turekian. It was both a privilege and pleasure to have been associated with Karl over the years, initially as graduate students at Yale University and then as colleagues, most recently collaborating on this chapter. Karl was a pioneer in recognizing the importance of Long Island Sound as a model laboratory to study both estuarine and, more broadly, ocean processes. Karl's group of students and postdoctoral scholars at Yale, together with those of Robert Berner, Donald Rhoads, and Robert Gordon, produced insightful studies that lead to a better understanding of Long Island Sound and paved the way for subsequent research that continues to the present and is embodied in this volume.
[†]Karl K. Turekian is deceased.

C. Cuomo (✉)
Biology and Environmental Sciences Department, University of New Haven,
West Haven, CT 06516, USA
e-mail: ccuomo@newhaven.edu

J. K. Cochran
School of Marine and Atmospheric Sciences,
Stony Brook University, Stony Brook, NY 11794, USA

K. K. Turekian
Department of Geology and Geophysics, Yale University, New Haven, CT, USA

J. S. Latimer et al. (eds.), *Long Island Sound*, Springer Series on
Environmental Management, DOI: 10.1007/978-1-4614-6126-5_4,
© Springer Science+Business Media New York 2014

and redox-sensitive metal cycling), associations (e.g., microbial communities), and inventories (e.g., radionuclides, metals). Although much of this chapter is focused on the sediments, the geochemical cycling discussed in this chapter involves the water column. Thus, we include geochemical processes in the water column as appropriate. It is the hope of the authors that this chapter will present an accurate description of the present state of knowledge of the basic chemical processes that are occurring in LIS, elucidate their connections to the greater LIS ecosystem, and identify the gaps that exist in the present understanding of LIS sediment geochemistry.

In his review of the conceptual understanding that underpins the majority of models dealing with sediment diagenesis within modern marine muds, Aller (2004) points out that while the models themselves are quite robust, they may not accurately portray the processes occurring within certain muddy marine environments, especially those occurring at the intersection of continents and the ocean. Sediment diagenesis in these systems, rather than being controlled primarily by the net accumulation rate of sediments, appears to be controlled by other sediment dynamics. It is true that Aller (2004) was discussing primarily mobile muds—those deposited episodically in high-energy environments with high sedimentation rates (e.g., the Amazon Basin). However, in examining some of the geochemical studies conducted on the LIS system within the past 25 years (Aller 1994a, b; Arias et al. 2004; Cuomo et al. 2005; Green and Aller 2001; Green et al. 1992, 1993, 1998; Hannides et al. 2005; Ingalls et al. 2000; Skoog et al. 1996; Sun et al. 1991, 1993, 1994, 1998, 1999, 2002; Wu et al. 1997), it appears that other sedimentary dynamics, rather than net sedimentation alone, may exert primary controls on the nature and magnitude of elemental and nutrient fluxes at the sediment–water interface in LIS—especially in the western and central basins. Furthermore, it also appears that the properties (e.g., temperature (T), stratification, dissolved oxygen (DO)) of the water column within the Sound interact in a dynamic fashion with the elemental and nutrient fluxes crossing the sediment–water interface. To this already dynamic exchange, add the seasonal influences of biotic processing, and the true nature of the complexity of sediment diagenesis in LIS becomes apparent. In this section, the authors endeavor to present an overview of the sedimentary geochemical environment in LIS as it relates to the organic carbon content of the sediments. A full discussion of anthropogenic inputs is beyond the scope of this chapter and will be considered in other chapters in this volume. The purpose of this section is to provide the reader with an understanding of the basic components of, and the processes that affect, sediment oxygen demand within the LIS system.

4.2 Long Island Sound Sediments

4.2.1 Sediment Conditions

4.2.1.1 Grain Size and Organic Carbon

The distribution of sediments within LIS varies from east to west. Finer-grained sediments predominate in the western and west-central portions of the Sound while the east-central and eastern ends of the Sound are dominated by coarser-grained sediments and bedrock (Figs. 4.1 and 4.2) (Knebel and Poppe 2000). The organic carbon content of surficial LIS sediments (Fig. 4.3) varies inversely with sediment grain size; therefore, sediments with high organic carbon content are primarily found in western LIS (WLIS), as well as in some of the harbor areas throughout the central and eastern basins. LIS sediments with the highest organic carbon content (~3 % by weight) are found just north of Hempstead Harbor, in WLIS (Poppe et al. 2000). The lowest sediment organic carbon values (<0.1 %) occur within the eastern basin of LIS, along the south shore (Poppe et al. 2000). It should also be noted that the sediment organic carbon content, especially in central LIS (CLIS) and WLIS, displays a significant amount of small-scale spatial variability (Cuomo et al. 2005; Poppe et al. 2000; Reid et al. 1979).

The higher total organic carbon (TOC) values in WLIS are generally attributed to excess phytoplankton productivity resulting from the presence of high nutrients (Anderson and Taylor 2001; Poppe et al. 2000), the adsorption of organic matter onto the surface of charged clay particles, and the fact that finer particles tend to accumulate in the western basin (Poppe et al. 2000). These processes, coupled with severe seasonal stratification and hypoxia, result in a steady introduction of labile organic carbon into the sediments of WLIS (Poppe et al. 2000).

Labile organic carbon, however, is not the only carbon present in the sediments of LIS. A large amount of the TOC (and N) stored in LIS sediments is in the form of refractory proteinaceous material derived from the decomposition of algal and plant material (Pantoja and Lee 1999; Westrich and Berner 1984). Work by Lugolobi et al. (2004) and others (Meyers 1994; Premuzic et al. 1982) has shown that a portion of the sedimentary organic carbon present in sediments along the Sound's north shore near Norwalk Shoals has a larger terrestrial carbon signature than sediments elsewhere in the Sound. Such organic matter, while not as labile as phytoplankton, can certainly contribute to an elevated sediment oxygen demand (SOD) in the sediments of LIS. It is possible, therefore, that long-term deposition of terrigenous debris into the sediments of LIS, and its subsequent anaerobic

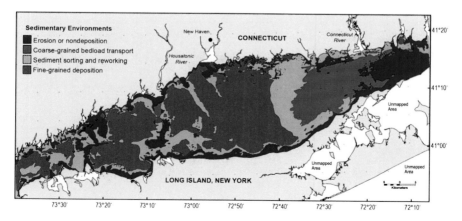

Fig. 4.1 Sedimentary environments of Long Island Sound (taken from Knebel and Poppe 2000)

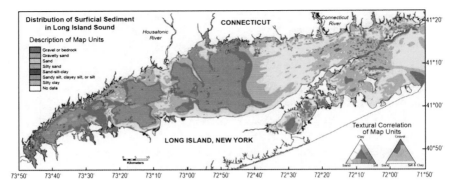

Fig. 4.2 Distribution of surficial sediments in Long Island Sound (taken from Poppe et al. 2000)

breakdown, may represent an additional source of organic carbon to the Sound's sediments. Recent work in Smithtown Bay (Cuomo et al. 2011) has revealed that a portion of the organic carbon contained in the sediments of the Bay is derived from marsh debris rafted out from the north shore of Long Island and entrained in the gyre of the Bay. Kristensen and others (Kristensen 2000; Kristensen et al. 1991) have also suggested that tubes of burrowing anemones (e.g., *Cerianthiopsis americanus*) and other macrofauna may actually comprise a moderate to high proportion of the organic matter present in marine sediments, including those of the Sound.

Moderate to high sediment carbon contents do not necessarily pose a problem in temperate estuarine and marine regions. Such carbon serves as an important food source for deposit-feeding benthos and is often associated with a rich and thriving benthic and nektonic community. If the amount of carbon being added to a sedimentary system exceeds the ability of that system to process it in the short term, however, then a substantial sediment organic carbon inventory can result. Under certain physical–chemical conditions (e.g., high temperatures) microbial processes occurring in such sediments can lead to the development of anoxic

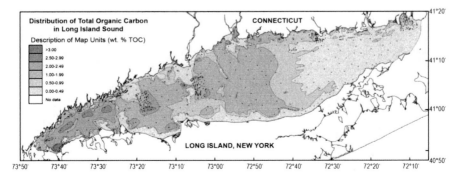

Fig. 4.3 Total organic carbon distribution within Long Island Sound sediments (taken from Poppe et al. 2000)

conditions within the sediments themselves and at the sediment–water interface (Abril et al. 1999, 2002; Aller 1998). Such conditions have been documented as occurring in WLIS and in parts of CLIS over the past decade (Cuomo and Valente 2003; Cuomo et al. 2004, 2005; Valente and Cuomo 2001, 2005). Once established at the sediment–water interface, such conditions may continue to act as an oxygen-sink and further enhance both the preservation of additional organic carbon in the sediments as well as depressed oxygen levels in the waters directly in contact with those immediately above the sediment–water interface.

4.2.1.2 Bioturbation

The organic carbon content of marine sediments is strongly influenced by the amount, extent, depth, and duration of bioturbation (Rhoads and Boyer 1982; Rhoads and Germano 1982, 1986). Studies conducted in LIS and other estuaries over the past 40 years (Aller 1978, 1980a, b, 1982a, b, 1994b; Aller and Aller 1998; Aller and Yingst 1985; Berner 1978; Cuomo and Valente 2003; Gilbert et al. 2003; Green et al. 1992; House 2003; Ingalls et al. 2000; Kristensen 2000; Kristensen et al. 1991; Michaud et al. 2009, 2010; Reaves 1986; Rhoads and Boyer 1982; Rhoads and Germano 1982, 1986; Sun and Dai 2005; Sun et al. 1999; Waldbusser et al. 2004; Westrich and Berner 1984; Yingst and Rhoads 1978; Zhu et al. 2006) have shown that introduction of oxygenated water into anaerobic bottom sediments by benthic organisms (Fig. 4.4) effectively aerates the sediments and leads to more efficient conversion of sedimentary organic carbon via aerobic decomposition pathways. Aller (1994a, b, 2004) has demonstrated that alternation of aerobic and anaerobic conditions in organic marine sediments actually facilitates the decomposition of organic matter contained within the sediments. The frequency, duration, and orientation of such redox oscillations are tied directly to the activities of bioturbating organisms present in the sediments. Studies examining bioturbation in LIS sediments (e.g., Michaud et al. 2009, 2010; Rhoads and Germano 1982; Waldbusser et al. 2004) have shown that the presence of a deeply bioturbating infauna (>10 cm) is generally correlated with relatively low

Fig. 4.4 Relationship between sediment oxygen demand and bioturbation (taken from Graf 1992). Note the intrusion of oxygen into otherwise anoxic sediments brought about by the burrowing activity of benthic organisms

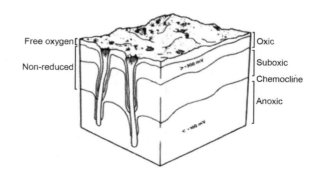

organic carbon preservation. In other words, either by direct consumption by the benthos or by facilitating aerobic/anaerobic microbial decomposition, LIS sediments subjected to significant bioturbation have lower organic carbon content than sediments lacking a vigorously bioturbating infauna. In general, deeper bioturbating infauna are most commonly found in the deeper sediments of the CLIS basin, whereas shallow bioturbation is common at most stations in WLIS—especially during the summer months. The lack of a deeply bioturbating infauna at many stations in WLIS during the late summer months may be tied directly to the presence of hypoxic and anoxic conditions occurring at the sediment–water interface in this area (Cuomo and Valente 2003; Cuomo et al. 2004, 2005).

These studies and others (Cuomo et al. 2011) have revealed a marked seasonality to the efficacy of bioturbation as a regulating factor on sediment oxygen demand in LIS. Studies conducted by Rhoads et al. (1978) and others (Reid et al. 1979; Sanders 1956; Serafy et al. 1977) during the late 1970s through early 1980s revealed an abundant and somewhat diverse benthos present in the sediments of WLIS and CLIS during different times of the year. These studies described seasonal shifts in benthic composition and abundance within the muddy sediment of CLIS and WLIS. Aller (1982a) and Rhoads and Boyer (1982) suggested that the sediments in these areas experienced a shift in sediment geochemistry over the course of a year, in response to the activities of the shifting benthos. Thus, organic-rich sediments underwent a yearly seasonal shift in chemistry, marked by an initial dominantly anaerobic stage colonized primarily by small, surface-deposit feeders and organisms that had survived the winter. This first stage occurred during the late winter-early spring bloom and was followed by increasing oxidation of the sediments linked to increasing bioturbation by relatively abundant deeper-deposit feeders during the late spring-late summer. The sediments rebounded in the fall to a more anaerobic state as organisms died or ceased actively bioturbating. This pattern then repeated the following year. This seasonal pattern was marked by an increase in the depth of the apparent Redox Potential Discontinuity (RPD) in the sediments of the deeper regions of WLIS and CLIS over the course of a year (Fig. 4.5) and resulted in exposure of sedimentary organic carbon to aerobic decomposition processes.

Studies (Bessier and Cuomo 2010; Cuomo 2005; Cuomo and Valente 2003; Cuomo et al. 2004; Valente and Cuomo 2001, 2005) conducted in WLIS and CLIS during the late 1990s–early 2000s reveal the emergence of a different pattern in WLIS than was present during the 1970s–1980s. Organic-rich sediments in the deeper regions of WLIS are initially sparsely colonized by shallow, surface-deposit-feeding benthos in the late winter-early spring of the year. These are, in some areas, followed by slightly deeper bioturbating organisms in the late spring-early summer. However, the apparent RPD—even in the late summer-early fall—does not ever reveal either a deeply bioturbating benthos or an abundant benthos. Rather, bioturbation is limited primarily to depths significantly ≤ 10 cm. Sediment-profile photographs (Valente and Cuomo 2005) reveal that during the fall season, organic-rich sediments in WLIS are often devoid of deeply bioturbating benthos and are instead either barren or colonized by small, surface-deposit feeders.

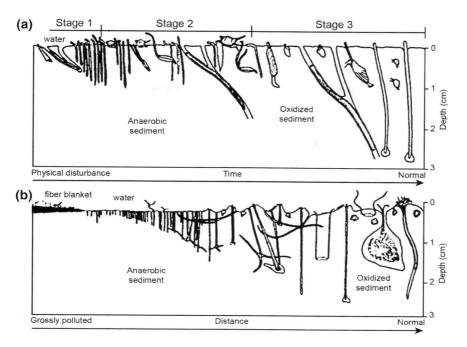

Fig. 4.5 "Ideal" animal-sediment response to (**a**) physical disturbance of the bottom and (**b**) pollutant loading [taken from Rhoads and Germano (1982) and based on models presented in Rhoads et al. (1978) and Pearson and Rosenberg (1978)]. Sediments in the left part of (**a**) and (**b**) display the greatest organic enrichment. Most marine communities are mixtures of these, although the greater the organic enrichment the more likely the movement toward Stage I communities. It should be noted that once a system moves toward the *left* part of the diagram, more carbon is stored in the sediments creating an environment that further favors Stage I type communities. Benthic communities in much of present day WLIS appear to be dominated by Stage I and Stage II communities

This effect is especially pronounced in the far western portions of LIS. Additionally, the density of organisms present at these sites (Cuomo and Valente 2003; Cuomo et al. 2004, 2005, 2011; Valente and Cuomo 2001, 2005) is less than those reported in the earlier studies done in the 1970s–1980s.

The observed change in benthic composition and its associated shallowing of the depth of bioturbation may have implications for SOD since one of the primary effects of a deeper bioturbating benthos is to clear out the carbon inventory of sediments, either by direct consumption or by facilitating microbial consumption (Aller 1978; Aller and Aller 1998; Aller and Yingst 1985; Aller et al. 2001; Arias et al. 2004; Green et al. 1992; Gilbert et al. 2003; Ingalls et al. 2000; Kristensen 2000; Michaud 2010; Michaud et al. 2009; Sun and Dai 2005; Sun et al. 1999; Waldbusser et al. 2004; Yingst 1978; Yingst and Rhoads 1980). "Shallowing-out" the depth of bioturbation that a sediment experiences over the course of the year means that the organic carbon inventory present in the sediment may not get completely utilized. Thus, the organic carbon content of sediments in WLIS may start to accumulate carbon, albeit at a slower rate. Should this pattern continue—and for WLIS this appears to be the case—then carbon will be stored in the sediments each year, slowly raising the TOC content of the sediments. If, in addition to the labile organic carbon derived from phytoplankton, there is a secondary source for the carbon (e.g., terrestrial carbon-containing debris carried to the Sound via run-off), the potential for an increase in the oxygen-demanding capacity of the sediments becomes even more likely. Microbial degradation of such carbon will tend to occur via anaerobic rather than aerobic processes, and the oxygen demand of the sediments will increase slowly over time—even without changing the amount of carbon input into the system. Thus, the SOD becomes, to some extent, decoupled from the overlying water column processes. Anaerobic microbial degradation of an increasing sediment carbon inventory will lead to release of additional nitrogen (N), in the form ammonium to the sediments, which, in turn, may further fuel phytoplankton production in the Sound.

4.2.1.3 Seasonal Variations in Sedimentary Organic Carbon

Organic Matter Pulses

The influx of organic matter to LIS sediments is neither a continuous occurrence nor a uniform one across the entire Sound. The majority of organic matter is carried to the sediments of LIS in pulses—usually, but not only, correlated both with the spring and fall plankton blooms as well as with times of high freshwater runoff (Anderson and Taylor 2001; De Jonge et al. 1994). A significant amount of this pulsed organic matter comes to reside in sediments of the central and western basins, carried there by the currents of LIS.

The historical organic carbon record in LIS sediments reveals an increase in organic carbon storage, possibly related to an increase in primary production, between the 1950s and the 1980s (Thomas et al. 2000; Varekamp et al. 2003a, b). This increase, estimated to be ~200–300 g C/m^2/yr, represents more than a 50 % increase in primary productivity in LIS and has been attributed to riverborne nutrients. De Jonge et al. (1994) reported that riverborne nutrients, related to both increases in spring discharge from the Connecticut and Thames Rivers in Eastern Long Island Sound (ELIS) and decreased discharge during the summer months, are entering the WLIS and CLIS basins in the early spring, fueling phytoplankton productivity.

De Jonge's work (De Jonge et al. 1994) is not the only study of plankton productivity in LIS. Numerous studies have investigated the distribution and behavior of Chlorophyll a (Chl a) in LIS waters and sediments (Anderson and Taylor 2001; Goebel et al. 2004; Ingalls et al. 2000; Sun and Dai 2005; Sun et al. 1991, 1993, 1994, 1998, 1999, 2002; Wu et al. 1997). Overall, these studies have determined that Chl a measurements vary from surface to bottom waters at the same sites in LIS with concentrations at depth generally being 75–100 % of those measured in surface waters. This decline in Chl a concentration reflects both the consumption of organic matter by organisms, as well as its microbial decomposition during its transit through the water column. It also highlights the presence of an inherent lag between production in surface waters and deposition in the sediments of LIS—a lag which Aller and Cochran (1976) have shown to be on the order of a few days. Given both the water column consumption of Chl a-rich particles and their relatively short residence time, it appears that phytoplankton blooms may occur on more rapid time scales in the Sound than are presently accounted for by current monitoring schemes (Anderson and Taylor 2001). It is possible that present monitoring efforts may, in fact, be missing a significant amount of the actual organic carbon being deposited onto the surface of WLIS and CLIS sediments during both the spring and fall plankton blooms. It is quite likely that phytoplankton delivery to bottom waters and sediments in WLIS and CLIS occurs not as a short, continuous raining down during the spring and fall but rather as a rapid series of pulses that continue throughout the spring and fall. Indeed, this was shown clearly by Gerino et al. (1998) in their biweekly sampling of a station in CLIS (station "PULSE"). This study demonstrated that Chl a inventories in the sediment varied in a non-steady state manner, tracking both productivity in the overlying water and decomposition of labile organic matter in the sediment.

Decomposition of Sedimentary Organic Carbon

This section briefly discusses the present state of knowledge regarding the decomposition of known sedimentary organic carbon pools within specific regions of LIS rather than a general discussion of organic carbon decomposition in marine sediments. The diagenesis of organic carbon in marine sediments is a subject that has received considerable investigation over the past 40 years and readers are referred to Bianchi (2007) for a discussion of sedimentary organic carbon diagenesis in estuarine and marine systems under both aerobic and anaerobic environments.

In LIS and other estuaries, the ultimate fate of sedimentary organic matter is determined by the nature and stability of redox conditions within marine sediments, the amount of physical disturbance experienced by the sediments, and the presence/absence of a bioturbating infauna. Measurements of algal chloropigments (e.g., Chl a) in sediments provide a window into the conditions and processes that the organic matter was subjected to during degradation (Bianchi 2007; Bianchi et al. 2000). Common biomarkers include phaeophytin a and phaeophorbide. These algal biomarkers, produced during the microbial breakdown (e.g., phaeophytin) and/or metazoan digestion (e.g., phaoephorbide) of algal products such as Chl a, have been used to track changes in algal accumulation and decomposition in estuarine sediments (Bianchi 2007), including LIS.

The overall distribution and concentrations of chloropigments in the sediments of LIS vary on both a spatial and temporal basis. The sediments of WLIS have significantly higher chloropigment inventories than those of CLIS, indicating both higher inputs of organic carbon into the sediments as well as more anoxic sediment conditions. Overall, shallow water sediments throughout LIS (i.e., <25 m deep) have higher chloropigment inventories than sediments found at greater depths (>25 m), once again pointing to an increase in organic matter deposition at such sites. Aller et al. (1980) (see Sects. 4.1, 4.2), using ^{234}Th inventories, showed that lateral resuspension and redistribution processes occurring in the Sound appear to be the main factors controlling the distribution of organic carbon inputs into these areas. The work also pointed out the temporal lag (on the order of 1–2 months) that exists between the maximum Chl a concentration measured in the LIS water column in early spring and the maximum Chl a sediment inventory, which shows up in the late spring. This lag is primarily controlled by time-dependent zooplankton grazing in the water column and T-dependent metabolic processes of microbial decomposition in the sediments.

Chl a inventories in the sediments of CLIS and WLIS, as in other estuaries (Bianchi 2007), decrease with increasing depth, suggesting the significant role that aerobic and anaerobic sedimentary microbial processes play in chloropigment preservation and distribution in sediments undergoing redox oscillations. Gerino et al. (1998) and Sun et al. (1993) examined the degradation of Chl a in LIS sediments. Both groups observed an exponential decrease in the presence of Chl a with depth under oxic conditions, supporting the idea that Chl a undergoes rapid degradation in oxygenated LIS sediments. The study by Sun et al. (1993), however, also revealed that Chl a in LIS sediments degraded differently, depending on the redox conditions present. Under oxic conditions, both Chl a concentration and activity decreased steadily with depth while under anoxic conditions only Chl a activity decreased with depth. Work conducted by Wu et al. (1997) suggests that the decomposition of Chl a, as measured by acetate oxidation rates, increases with increasing T. Thus, Chl a degradation in LIS undergoes a seasonal increase from March to June. Since acetate represents a key intermediary in the diagenesis of the more labile fractions of organic matter in sediments, the cycling of acetate in sediments can be affected by pulsed inputs of organic carbon (phytoplankton pulses) to the sediments. The acetate pool in LIS sediments was also affected by bioturbation (Wu et al. 1997), which increased acetate transport by a factor of three. This, in turn, further facilitates additional decomposition of the organic carbon delivered to the sediments. Given that T, bioturbation, and sedimentary redox conditions change temporally in LIS sediments, Chl a degradation and the preservation of algal biomarkers in these sediments must also vary on a seasonal basis.

Phaeophytin a, another algal degradation product associated with several microbial decomposition pathways, forms under both oxic and anoxic conditions but is only stable under anoxic conditions (Bianchi 2007). It rapidly degrades under oxic conditions, making its accumulation in estuarine sediments an indicator of anoxic sedimentary conditions. Phaeophorbide, an algal decomposition product associated with metazoan grazing, when present in surface sediments, reflects the ingestion and partial digestion of organic matter as it passes through the water column on its way to the sediment surface. Increasing phaeophorbide concentrations in oxic sediments reflect additional organic matter degradation by meiobenthos and macrobenthos. In CLIS and WLIS, both phaeophorbide and phaeophytin a are present in surface sediments at similar concentrations, indicating both microbial decomposition and metazoan grazing of algae under oxic conditions (Sun et al. 1993). The ratio of phaeophytin a to Chl a increases with depth, indicative of both continued microbial degradation of organic matter under anoxic conditions and preservation of phaeophytin a. Some of this preservation results from the fact that phaeophytin a is more resistant to chemical reactions associated with the lowered pH of anoxic sedimentary conditions.

Bacterial Biomass

The primary consumers of particulate organic carbon (POC) and dissolved organic carbon (DOC) in the water column of LIS are bacterioplankton and particle-attached bacteria (Wu et al. 1997). As such, these two groups represent the largest source of biological oxygen demand (BOD) in the water column. Work examining the distribution and activity of bacteria within the waters of LIS (Aller and Yingst 1980, 1985; Michaud et al. 2009; Murchelano and Brown 1970; Sun et al. 1997; Yingst 1978) has revealed important differences between surface and bottom waters, as well as significant temporal differences. The bacterial biomass present in surface waters in LIS steadily increases from July to August and appears to be T-dependent. This increase parallels the overall decrease measured in DO in the same waters for the same period of time. Bottom water bacterial biomass for the same period does not exhibit the same pattern. Rather, bottom water bacterial biomass reaches peak densities between July and early August and decreases to relatively low levels throughout the fall. This increase in bottom water bacterial biomass in late summer may represent the arrival in the bottom waters of a portion of the phytoplankton produced in the surface waters earlier in the season. It also appears to correlate with the time of maximum stratification of the waters in LIS. Interestingly, the late August-October bottom water hypoxia event described by Cuomo et al. (2004, 2005) occurs after the peak of bottom water bacterial biomass, suggesting that this hypoxic event may be driven by bacterial biomass in the sediments rather than in the water column.

Cuomo (2007) proposed that NH_4^+ and S^{2-} released from the sediments underlying much of WLIS in late August–October, depending on the year, produce a chemical oxygen demand in the bottom waters, which drives down oxygen levels in bottom waters of the Sound, even as the fall overturn occurs. The release of these reductants from the sediments is coupled with enhanced aerobic and anaerobic degradation of labile organic matter carried to the sediments following the late summer–early fall plankton bloom. Thus, bacterial biomass actually increases in the sediments and very bottom waters of WLIS, even as it decreases substantially in the waters ≥ 1 m above the bottom. Oscillating redox conditions within the surficial sediments at this time would also aid in the decomposition of organic matter and increase the flux of metabolites from the sediments into the overlying waters, again contributing substantially to a chemical oxygen demand (COD) in the waters in contact with the sediment–water interface in LIS, especially in the western basin.

The arrival of labile organic carbon to sediments can act as a "bacterial boosting" mechanism and facilitates the rapid growth of microbial biomass and, by their activities, the decomposition not only of the fresh labile organic matter, but also of a portion of the more stable residual carbon already present in the sediments. Aller (2004) and others (Abril et al. 2010) proposed that redox oscillations in the sediment promote losses of bacterial biomass and efficient recycling of carbon by aerobic grazers via the benthic microbial loop. These papers suggest that sedimentary redox oscillations might produce lower eventual carbon storage as these oscillations, in effect, accomplish the same end-point as predation. They produce a periodic death of a significant portion of existing microbial populations. These dead organisms themselves then become the basis for decomposition and, in fact, may "boost" the next round of microbial population growth by providing a nutrient flush. In the long run, this, may promote more efficient overall net remineralization of sedimentary organic carbon, resulting in minimization of organic carbon burial and storage. However, a paper by Dai et al. (2009) examining the effects of redox oscillations and co-metabolism on the overall degradation of organic matter (e.g., diatoms, salt marsh plants) demonstrated that redox conditions actually exerted a limited influence on labile organic matter decomposition. Additionally, this study revealed that degradation of organic matter was retarded when it consisted of a mixture of labile (diatoms) and relatively refractory (saltmarsh grass) materials. Overall, then, it appears that such a mixture of organic material coupled with the frequency and duration of redox oscillations will ultimately determine the adaptations of the microbes involved and whether more organic carbon is synthesized or destroyed as a result of the process.

The interplay between oxic and anoxic pore water states, and all that influences them, becomes an extremely important controller of the fate of organic carbon in the sediments of LIS. Although short-term oxic remineralization processes appear to be most effective in promoting net remineralization and reoxidation of bulk organic carbon sedimentary components, Aller (2004) suggests that even brief exposure to oxygen, on the order of 20% of the time, may approach some of the same results through alternate microbial pathways.

Presently, organic carbon reaching the sediments of ELIS or CLIS has a greater chance of being completely remineralized via processes including, but not limited to, aerobic and anaerobic bacterial decomposition, and consumption by both meiofaunal and macrofaunal organisms—all set against a backdrop of an oxic-to-hypoxic water column and sediments that contain a deeply bioturbating infauna or are coarse-grained and well-drained. Alternatively, organic carbon reaching the sediments in a majority of areas in WLIS has a much greater chance of being preserved due to a depression of redox oscillations resulting from a lack of deeply bioturbating infauna, fine sediment accumulation, multiple types of organic carbon (ranging from highly labile to highly refractory) present in the sediments, and bottom waters that are severely hypoxic-to-anoxic in summer and fall. Work by Aller (2004) has shown that stimulated biomass synthesis and subsequent destruction of organic matter appear to be associated with redox oscillations with a phase lag of ~1 week. Given that for WLIS, the phase lag between oxygenated and anoxic conditions in the sediments appears to be significantly greater than 1 week, a net storage of organic carbon might be expected to occur within at least some of the sediments of WLIS.

4.2.2 Chemical Fluxes Relative to Environmental Conditions

The complete decomposition of organic carbon in estuarine environments involves a suite of bacterial species and transformations resulting, ultimately, in the release of, among other chemicals, volatile fatty acids, alcohols, CO_2, H_2S, NH_4^+, PO_4^{3-}, $SiO_{2(d)}$, and CH_4 into sediment pore waters and into the waters immediately in contact with the sediment–water interface (Aller 2001). Of special interest to scientists seeking to understand the chemical dynamics of LIS are those chemical species associated with nutrient regeneration in LIS (i.e., NO_3^-, NO_2^-, PO_4^{3-}, and $SiO_{2(d)}$) and with anaerobic degradation of organic matter (e.g., NH_4^+, S^{2-}, and CH_4). In this section, the authors will focus on the present state of knowledge relative to sedimentary controls on four of these species: NH_4^+, S^{2-}, PO_4^{3-}, and CH_4 in LIS.

4.2.2.1 Ammonia

Ammonia concentrations in LIS sediments and pore waters result from a dynamic equilibrium among dissolved, exchangeable, and fixed ammonia and ammonium. This equilibrium is largely dependent on the redox state of the sediments. For example, under anoxic conditions, ammonium adsorption by sediments is an important process in the diagenesis of N in nearshore LIS sedimentary environments. Mackin and Aller (1984) have shown that NH_4^+, once released into solution, can be reoxidized to NO_3^-, taken up by organisms, adsorbed onto particles, diffused via pore waters to other portions of the sediments, or released to the overlying water column. In anoxic sediments, such as those found in WLIS, reversible ion exchange on particles is the dominant pathway by which NH_4^+ is removed from solution (Mackin and Aller 1984).

The adsorption of NH_4^+ onto particles depends primarily on chemical variables—the absolute concentrations of major ions present for ion exchange and the specificity of certain ions for adsorption sites (Mackin and Aller 1984). Sediments with the highest adsorption properties are clays; thus, NH_4^+ adsorption onto particles is expected to be highest in the sediments of WLIS and certain sections of CLIS and relatively unimportant in the majority of environments in ELIS. It is further facilitated by the presence of organic matter coatings on the particles. Once adsorbed onto particles, ammonium concentrations appear to be depth independent within any single environment within LIS.

Engstrom et al. (2005) documented that anaerobic ammonium oxidation (anammox), leading to N_2, NO_2^-, and NO_3^-, occurs in LIS sediments. The anammox pathway utilizes anaerobic bacteria to facilitate the movement of N between bottom sediments and the water column. This reaction is an autotrophic process that is thermodynamically favored under anaerobic conditions and has been documented as a major pathway for N loss from marine sediments, estimated to contribute anywhere from 2–6 % of total worldwide benthic N production. The relative importance of anammox for N_2 release from LIS sediments is inversely correlated with Chl a in surface sediments, reduced pore water solute concentrations, and benthic O_2 consumption. This inverse correlation is related to, among other things, competition by reductants for pore water nitrite, and the availability of Mn-oxides during diagenesis. The rates of anammox activity, both absolute and relative in marine coastal sediments, result from these interacting factors. For the majority of WLIS sediments rich in Chl a, the anammox pathway does not likely represent a major denitrification mechanism. However, for sediments of the CLIS that experiences significant bioturbational activities as well as redox oscillations, anammox may be an important denitrification pathway.

Work by Cuomo et al. (2005) has demonstrated that the sediments of WLIS release ammonia into near-bottom waters on a seasonal cycle that is out of phase with the normal occurrence of hypoxia in LIS (Fig. 4.6). These studies point to a spring and fall release of ammonia from surface sediments into bottom waters, following the spring and fall plankton blooms. These same studies indicate that ammonia release from WLIS sediments is somewhat negatively correlated with sulfide release from the same sediments and positively correlated with T. Significant ammonia release from WLIS sediments occurred at temperatures ≥ 14 °C; highest release occurs at temperatures of 14 °C in the presence of an oxic water column.

Ammonia release from CLIS sediments displays a similar seasonality (Cuomo et al. 2011). Bottom water (within 5 cm of the sediment–water interface (SWI) and near bottom water (NBW, 1 m above the SWI) measurements at selected sites within Smithtown Bay over the course of 2 years reveal a seasonal release of ammonia from sediments into bottom waters at the more organic-rich sites within the bay (Fig. 4.7). While the magnitude of the release varied from year to year, the pattern was one of high ammonia release in May–June, with decreasing values in July, and lowest values in August–September—during the times of maximum measured hypoxia. This follows the pattern seen in the previous study of ammonia release from WLIS sediments (Cuomo et al. 2005).

4.2.2.2 Sulfide

Sulfide production in marine systems is tied to anaerobic decomposition of organic matter by sulfate-reducing bacteria. Sulfide release from marine sediments is affected by a variety of factors, including the availability of dissolved metal species, pH, bioturbation, and Aller (1980b) and others (Aller and Yingst 1980; Goldhaber et al. 1977), working on anoxic sediments from the Friend of Anoxic Mud (FOAM) site in CLIS and other areas, observed that total reduced sulfides (TRS) increased

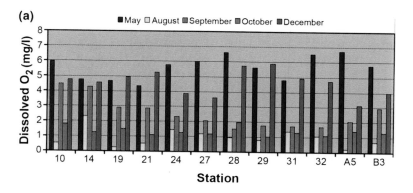

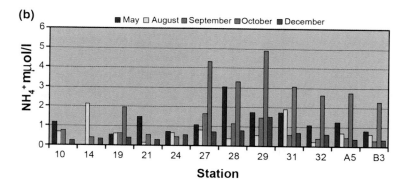

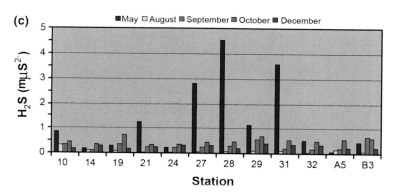

Fig. 4.6 WLIS bottom water (**a**) dissolved oxygen, (**b**) ammonia and (**c**) sulfide during 2002 (Cuomo and Valente 2003). Note that the highest dissolved oxygen measurements in May correspond to some of the highest sulfide an ammonia values observed during the study at certain stations in WLIS, while lowest oxygen values in August correspond, in general, with low ammonia and sulfide values. Highest ammonia values were recorded at certain WLIS sites in October. **a** Dissolved O₂ (*bottom*). **b** Ammonia (*bottom*). **c** Hydrogen sulphide (*bottom*)

(a)

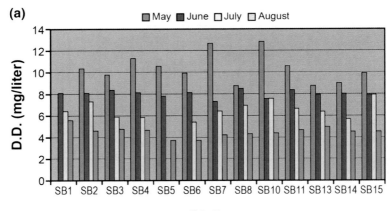

(b)

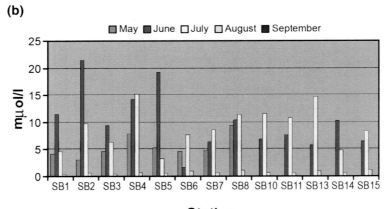

(c)

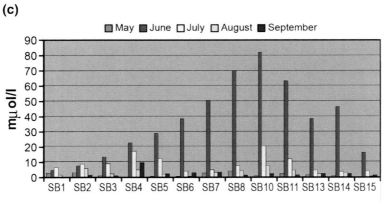

◀ **Fig. 4.7** CLIS bottom water (**a**) dissolved oxygen, (**b**) ammonia and (**c**) Sulfide measured in during 2009 (Cuomo at al. 2011). Note that the highest dissolved oxygen measurements in May are associated with slightly elevated ammonia releases at certain stations within Smithtown Bay. Highest sulfide values and elevated ammonia values were measured during June and correlate with fully oxygenated conditions at all stations. Interestingly, lowest dissolved oxygen measurements in August correlate with no discernible sulfide or ammonia in the bottom waters. **a** Bottom water dissolved oxygen Smithtown Bay 2009. **b** Ebb bottom ammonia 2009. **c** Ebb bottom sulphide 2009

rapidly down to 4 cm and then remained relatively constant throughout the rest of the sediment column while acid-volatile sulfides (AVS), a group composed of metastable iron sulfide minerals and dissolved S^{2-} species, reached a peak value between 4 and 6 cm and then decreased significantly after that. Morse and Rickard (2004) point out that the AVS maximum in Aller's study (1980b) correlates with the base of the TRS formation zone, as well as with the depth of the lowest dissolved iron values, while detectable S^{2-} increases quickly. They also point out that AVS appears to vary greatly from one study site to another, making any generalities about AVS in LIS sediments difficult to support. Comparisons of patterns of AVS distribution in sediments from the Gulf of Mexico and other southern estuaries with Aller's work (1980b) in CLIS revealed a wide variability; few sites resembled AVS patterns found at the CLIS FOAM site. AVS in the sediments studied ranged from comprising the bulk of TRS to hardly any detectable AVS concentrations. At the FOAM site in CLIS, Aller (1980b) also observed seasonal variability in AVS concentrations, with lower concentrations occurring in winter, associated with decreased bacterial activity and decreased bioturbation.

Recent work by Cuomo (2007), Cuomo and Valente (2003), Cuomo et al. (2004, 2005, 2011), and Valente and Cuomo (2001, 2005) on WLIS and CLIS sediments has revealed a seasonal release of sulfides into bottom waters (Figs. 4.6 and 4.7). The spring release of sulfides from organic-rich WLIS sediments occurs at temperatures below 10 °C and is associated both with deposition of the spring plankton bloom and initiation of bioturbation. The fall release of sulfides from these same sediments correlates with high bacterial activity, temperatures in excess of 20 °C, and the deposition of the fall plankton bloom. Sedimentary C_{org} content was also positively correlated with higher sulfide release. Measurements (Cuomo et al. 2011) taken at a variety of sites within Smithtown Bay, located in CLIS, reveal a pattern both consistent with and yet different from that observed in WLIS. Sulfides were released only from the most organic-rich sediments within Smithtown Bay, with the highest releases of sulfide occurring in May and June. Sulfide release decreased at almost all stations during July–September. Sampling did not continue into late fall, leaving a question as to whether sulfide releases from CLIS sediments in the late fall as it does from WLIS sediments.

4.2.2.3 Phosphate

In addition to the production of ammonia and sulfide, the anaerobic decomposition of organic matter in marine sediments results in the release of soluble reactive phosphate to bottom waters (Blackburn and Henriksen 1983; Klump and Martens 1981; Krom and Berner 1980, 1981; Martens et al. 1978; Van Cappellen and Gaillard 1996). The release of nitrogenous compounds and phosphate to bottom waters may provide enough nutrients to stimulate phytoplankton productivity in the overlying water. For LIS, as elsewhere, whether or not pore water phosphate derived from organic matter decomposition on and within the sediments is released from, or retained within, the sediments depends upon several factors, including the nature of the carbon source, the sedimentation rate, the amount of bioturbation and bioirrigation in the sediments, and the type and amount of diagenetic remineralization reactions occurring in the sediment (Ruttenberg and Berner 1993).

Much of the information on phosphate cycling in LIS comes from work conducted in CLIS over the past 3 decades. The majority of these studies have focused on three sites within CLIS—FOAM, NWC (Northwest Control), and DEEP (Deep). FOAM is a relatively shallow station containing a mix of surficial deposit-feeders; NWC, as described earlier, is moderately well-bioturbated sediment located at an intermediate water depth while DEEP, which is further offshore, contains an even more well-developed benthic community consisting of permanent deeply burrowing infauna, including maldanid polychaetes and anemones. Aller (1980b) observed that phosphate fluxes for these stations were in good agreement with values predicted from Fick's first law except for two measurements—those at NWC in the fall and those at DEEP in the summer. Comparison of phosphate fluxes at all three of these CLIS stations revealed a much higher release of HPO_4^{2-} at NWC during the summer and fall (temperatures 22 and 15 °C, respectively) than in winter (4 °C), relative to the other stations. Fluxes from DEEP were higher than predicted during the summer months and lower than predicted in the fall. Aller attributed these to enhanced macrofaunal excretion during the fall while the summer measurements at DEEP were attributed to oxidation and scavenging of phosphate by Fe oxides. Aller also suggested that HPO_4^{2-} fluxes, like NH_4^+, are controlled primarily by production and not by transport processes. Krom and Berner (1980) determined that a significant portion of the dissolved phosphate within the upper 10 cm bioturbated zone of CLIS sediments was derived from the liberation of adsorbed phosphate from depth. At depths below 10 cm, phosphate release to solution occurs via organic matter decomposition. At the time, they postulated that the release of phosphate into bottom waters was controlled by phosphate release from sediments within 1 cm of the sediment–water interface, suggesting that release of phosphate from organic matter must occur either at the time of burial or immediately prior to it.

Ruttenberg and Berner (1993) further examined the major factors influencing phosphate burial and retention at FOAM. They concluded that phosphate diagenesis within CLIS sediments results in the burial of phosphorus (P) associated with different phases (e.g., authigenic carbonate fluorapatite, organic-P) than those with which it was originally deposited. The net result is that P is almost totally retained in sediments in CLIS once it is buried. Ruttenberg and Berner (1993) also observed that P storage in CLIS sediments appears to be similar under both oxic and oscillating redox conditions; both of these conditions also display higher P retention than anoxic sediments. This enhanced retention under oxic and oscillating conditions is attributed to immobilization facilitated by bacteria.

The authigenic formation of carbonate fluorapatite (CFA) has been observed at the FOAM site in CLIS (Ruttenberg and Berner 1993) and is believed to be one of the ways in which P is retained in the sediments. Additionally, solid-phase iron oxyhydroxides in pore waters scavenge dissolved P from oxic sediment and release it back into the under anoxic conditions. In CLIS sediments, the concentration of P in pore waters is determined primarily by the release of phosphate from organic matter via bacterial decomposition processes, the dissolution of fish bones and other biogenic phosphates, the release of phosphate from iron oxyhydroxides under anoxic conditions, formation of CFA and other authigenic phosphate minerals, and diffusion into the overlying water column. At FOAM, Ruttenberg and Berner (1993) determined that the diffusive flux of phosphate from FOAM sediments varied seasonally. (Gobler et al. 2006) measured dissolved inorganic nitrogen (DIN) and dissolved organic phosphorus (DIP) at selected stations across LIS and found values consistent with the observations of Ruttenbery and Berber (1993).

Summertime phosphate concentrations in ELIS and CLIS were ≤ 0.02 mg/l, consistent with Ruttenberg and Berner's (1993) observations. Highest summertime phosphate concentrations were found in WLIS near-bottom waters. However, these numbers do not refer to actual phosphate concentrations in waters at the sediment–water interface. Given that many sites in WLIS are known to contain anoxic sediments, the phosphate concentration in waters immediately in contact with the sediment–water interface is expected to be higher than reported.

4.2.2.4 Methane

Methane emission rates are highly variable in estuaries worldwide although highest rates are associated with marshes and mudflats (Abril and Borges 2004; Abril and Iverson 2002; Middleburg et al. 2002). The topic of methanogenesis is too extensive for this chapter and is beyond the scope of this book. Additionally, the relationships among bacterial nitrate and sulfate reduction, biogenic methane production, and between each of these and anaerobic methane consumption have been examined in depth by a number of investigators (see, for example, Martens and Berner 1974; Whiticar et al. 1986; Zengler et al. 1999) and will also not be reviewed here. The reader is referred to Reeburgh (2007) for an excellent summary of methane dynamics in marine systems. Within the LIS system, however, research on biogenic methane appears limited. It has been found associated primarily with fine-grained sediments in WLIS, CLIS, and ELIS (Martens and Berner 1974). In ELIS, methane-charged sediments have been identified in the Thames River glacial delta as well as in Fishers Island Sound, while in CLIS methane-rich sediments have been identified off the coast of Connecticut, between Branford and Clinton, and extending out from there into the middle of the Sound, toward the north shore of Long Island, NY (Lewis and DiGiacomo-Cohen 2000).

Methane-charged sediments have also been captured in sediment-profile images from WLIS sediments, including some taken within borrow pits (Rhoads and Germano 1982; Valente and Cuomo 2001, 2005). Work by Martens and Berner (1974) suggested that methane in LIS pore waters reaches appreciable concentrations only after 90 % of the sulfate has been removed by sulfate-reducing bacteria. Such conditions are primarily favored in the sediments of WLIS.

Details regarding the formation, consumption, and emission of methane from LIS sediments, especially those present in much of WLIS, and the relationships between any and all of these and factors such as salinity (S), T, dissolved oxygen, and bacterial biomass generally remain unknown at this time. Work on other estuaries suggests that methane concentrations in coastal waters are highly variable, although they generally exceed atmospheric equilibrium (Bianchi 2007).

4.2.3 Manganese and Iron

4.2.3.1 Factors Controlling Manganese and Iron Distribution

The co-precipitation or adsorption of metals by manganese and iron oxyhydroxides and oxides (hereafter referred to as oxyhydr(oxides) (OHO)) exerts a major influence on pore water and sediment concentrations of metals in a majority of marine environments underlain by terrigenous sediments (Goldberg 1954), including LIS. In these systems, for example, Balistrieri and Murray (1986) demonstrated that increases in the solid Mn contents of sediments in the Panama Basin enhanced the binding of Zn, Pb, Co, Cd, and Ba, while other metals (i.e., Cs, Be, Sc, Pu, Sn, and Fe) remained unaffected. The main sources of Mn and Fe OHO to an estuary or coastal environments are the rocks and sediments within the estuary's watershed. Since both Mn and Fe OHO are found in relatively minor quantities (0.072 and 3.6 % by mass, respectively) in the surface rocks and sediments, increased weathering of rocks containing Mn and Fe OHO generally results in a higher portion of Mn and Fe OHO reaching an estuary. Given that OHO of Mn and Fe have low solubilities and are relatively unaffected by weathering processes, the majority of Mn and Fe OHO reach estuarine and coastal waters having undergone little to no chemical alteration. While a substantial amount of the total Mn and Fe content of rivers in the United States is composed of Mn (75 % on average) and Fe (35 % on average) OHO (Canfield 1997), it should be pointed out that the rest (25 % Mn and 65 % Fe) are contained within silicate minerals. Similar ratios of oxides to silicates are found in marine sediments (Canfield 1989; Raiswell and Canfield 1998; Thamdrup et al. 1994), pointing to the fact that the initial control on the amount of Fe and Mn present in marine and estuarine sediments is the composition of the source rock within the watershed. Other factors affecting the Mn and Fe content of marine sediments include the amount of biogenic material present in the sediments, the amount of sorting experienced by the sediments, and the clay and silt fraction of the sediments.

Since both Mn and Fe OHO are involved in the anaerobic bacterial decomposition of organic carbon in marine sediments, it follows that anything that affects bacterial decomposition and/or organic carbon accumulation will ultimately influence the fate and transport of these metals into and out of marine sediments. Aller (1990), working in the Panama Basin, demonstrated that in non-sulfidic sediments, biogenic reworking is the dominant control on Mn cycling. In such sediments, bacterial manganese-reduction is facilitated by the mixing of reactive organic carbon and MnO_2 into suboxic-anoxic deposits. Aller (Aller 1990; Aller and Rude 1988) postulated that, for sulfidic sediments, bioturbation must enhance a second, sulfur-based Mn reduction pathway that brings MnO_2 into contact with iron-sulfides and results in the release of Mn^{+2}. In these sediments, burrows create a series of microenvironments that further increase Mn reduction–oxidation and enhance the depletion of Mn from deeper pore waters.

4.2.3.2 Manganese and Iron Distribution in LIS

In the LIS watershed, the rocks consist of a variety of igneous, metamorphic, and sedimentary rocks—good sources for both Mn and Fe. In general, LIS sediments are dominated by terrigenous materials, despite the significant amount of biogenic carbonate precipitated in the form of shells. Furthermore, the silt–clay fraction of sediments within LIS accumulates primarily within the WLIS basin, portions of the CLIS, and in the harbors and inlets of the WLIS, CLIS, and ELIS regions (Fig. 4.2). Mn and Fe may also become locally enriched through groundwater seepage in certain areas of the Sound.

In general, the redox chemistry of manganese and iron in LIS is closely tied to the organic carbon content, sediment grain size, and oxidation state of the sediments (Figs. 4.1, 4.2, and 4.3). Both metals readily mobilize under anoxic conditions and both precipitate readily out of seawater in the presence of oxygen. Sediment controls on manganese and iron chemistry are directly related to the depth of penetration of oxygen into the sediments and the presence/absence of oxygen in waters in contact with the sediment–water interface (Hunt and Kelly 1985). Overall, depletion of Mn and Fe can be expected in anoxic sediment horizons, while enrichment of Mn and Fe, in the form of oxides and hydroxides, can be expected in oxic regions, especially directly above the anoxic–oxic interface in marine systems (Aller 1980; Thamdrup et al. 1994). Horizontal transport of Mn may also occur due to the lateral transport of released Mn from areas containing reduced sediments and anoxic waters to areas containing oxic waters (Canfield et al. 1993; Thamdrup et al. 1994). Iron, with its lower solubility and faster oxidation kinetics, does not undergo as significant a post-depositional redistribution in estuarine and marine sediments as does Mn.

In LIS, the general trends for sediment metal inventories are higher sediment metal concentrations in the WLIS and CLIS areas, and lower ones in the areas of ELIS (Figs. 4.8 and 4.9) (Mecray and Buchholtz ten-Brink 2000). Overall, highest

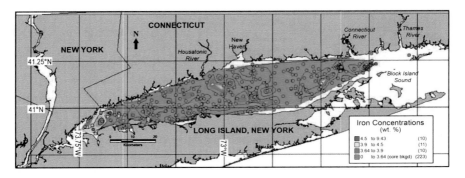

Fig. 4.8 Iron distribution in the surficial sediments of Long Island Sound (taken from Mecray and Buchholtz ten-Brink 2000)

metal concentrations are found in the various harbors of the Sound, with particularly high levels present in the sediments of WLIS. Since it is true that a significant amount of the metals (e.g., Hg, Pb, Cu, and Cd) in the sediments and water column of LIS are the result of human activities within the Sound's watersheds, they are beyond the scope of this chapter. Readers interested in metal pollution within LIS are referred to Chap. 5 for a thorough treatment of this issue.

4.2.3.3 Biogeochemistry of Manganese and Iron in LIS Sediments

Sedimentary Mn acts as a redox intermediate in the sediments of LIS (Aller 1994b). The two primary forms of Mn in the marine environment are the reduced form, Mn(II), and the oxidized state, MnO_2. Aller (1994b) attributes 30–50 % of the benthic oxygen flux in LIS to sedimentary Mn. In nutrient-rich, well-oxygenated surface waters, Mn(II) reacts with oxygen to form MnO_2 particles. These particles sink back to the bottom wherein they are reduced upon exposure to HS^-,

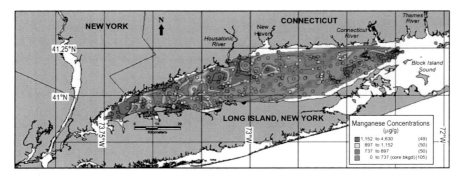

Fig. 4.9 Manganese distribution in Long Island Sound sediments (taken from Mecray and Buchholtz ten-Brink 2000)

regenerating Mn(II) to the water column (Johnson 2006). Under continuous anoxic sedimentary conditions, Mn(II) will be released from the sediments and can accumulate in the tissues of benthic organisms that come in contact with such sediments. Draxler et al. (2005) measured Mn in gill tissues of lobsters taken from LIS during the 1999 lobster die-off and related the concentration of Mn to length of time exposed to hypoxic-anoxic sediments.

Fluxes of Mn from LIS sediments display a seasonal pattern, with spring fluxes higher than fall fluxes, and highest fluxes occurring in late summer (Aller 1994b). The flux of Mn increases in LIS from east to west (Aller 1980) with highest fluxes of 2.2 mmol Mn/m^2/day coming during the late summer in WLIS. This same study also found that spring fluxes of Mn (average 0.43–0.94 mmol/m^2/day) were higher than fall and winter fluxes and attributed this spring release of Mn to the presence of the spring phytoplankton bloom. Aller (1994b) postulated that the increasing organic carbon content associated with increasing planktonic debris in sediments as one moves further into WLIS fueled anaerobic degradation in the sediments of WLIS which increased the net flux of Mn from the sediments into the bottom waters:

$$2MnO_2 + C_{org} + 3CO_2 + 2H_2O \rightarrow Mn^{2+} + 4HCO_3^-$$

In a recent study, Daniel (2011) confirmed the seasonal nature of Mn fluxes observed by Aller (1994b). Dissolved Mn^{2+} was highest in WLIS during the summer (1.5–1.8 μmol/L), and values decreased toward ELIS. Bottom water dissolved Mn^{+2} was also greater than concentrations in surface water in the highly stratified CLIS.

Lyons and Fitzgerald (1980) measured trace metal fluxes into LIS sediments at two nearshore sites in LIS—one in ELIS and one in CLIS. At both sites, a percentage of the metal accumulation was ascribed to local human activities with the highest anthropogenic metal flux occurring near their source. In ELIS, Fe, and Mn sediment concentrations from the Mystic River estuary were low compared with other sediments and actually increased with depth, along with the organic C content. Sedimentary Fe correlated positively with Mn, total P, and organic carbon and negatively with Zn and Pb. Sediment data from Branford Harbor in CLIS demonstrate Fe content correlating with anthropogenically derived metals. Concentrations of both sedimentary Mn and Fe were also higher here than in the sediments of ELIS with Mn reaching a peak of 421 μg/g at a depth of 20 cm, while Fe stayed consistent throughout the core.

4.2.3.4 Sediment Inventories

Greig et al. (1977), in their study of metal distribution in LIS, observed that Mn distribution in LIS sediments was very different from that of other metals (Fig. 4.9). Relatively high Mn concentrations occurred in ELIS sediments as well as in WLIS sediments, with the lowest concentrations occurring in

Gardiners Bay and at the extreme end of Long Island. Mecray and Buchholtz ten-Brink (2000) compared their sediment Mn measurements with those of Greig et al. (1977); sedimentary Mn increased by 53 % over the 19-year time period, although some of this can be attributed to different analytical methods. The distribution of Mn, however, was similar in both studies, with elevated levels measured east of the Connecticut River, in sandy sediments where strong bottom currents dominate, and in WLIS. Compared with the spatial distribution of other metals in LIS, Mn distribution in LIS is much more variable. Sediment Mn concentrations increase from coarse-grained to fine-grained environments, with enrichment values for the whole Sound that are above natural background levels. Such enrichment has been postulated to be the result of the formation of oxide crusts, especially on feldspar and quartz grains in ELIS. These crusts then may serve as a reactive surface for the accumulation of anthropogenic metals.

Sediment inventories of Mn^{+2} vary over the course of the year, especially in WLIS sediments. Aller (1994b) found low to no sediment inventories of Mn^{+2} during the summer months and high sediment inventories during the fall and winter months. Low summer values correspond to a time of year when the oxygen penetration into the sediment is least and Mn^{+2} loss from the sediments is highest. Winter values can be explained by the fact that Mn oxidation increases when oxygen values are high and highest oxygen values in waters overlying the sediment–water interface are reached in the winter months in LIS.

4.2.4 Isotope Geochemistry of LIS Sediments

Isotopes can serve as powerful tools in aquatic systems because they frequently have well-constrained sources and their geochemical behavior permits them to be used as tracers of source or as chronometers for the transport of water or particles. For example, radionuclides of the naturally occurring uranium and thorium decay series have been demonstrated to be useful tracers of particle dynamics and advection of water in LIS. Indeed, the Sound was one of the first estuaries in which many of the applications of such radionuclides were developed (see Turekian et al. 1980 for a summary). The useful radionuclides fall into two categories— those that interact strongly with particles (e.g., ^{234}Th, ^{210}Pb) and those that are less reactive and can be used to trace processes associated with circulation in the Sound (e.g., ^{222}Rn, the Ra isotopes). Isotopes with an anthropogenic source, such as the osmium isotopes, are useful to trace the dispersion of the anthropogenic signal away from the source. Much of the research using these isotopic tracers was carried out at three universities (Yale University, University of Connecticut, and Stony Brook University) and the US Geological Survey in fieldwork from 1970s to 1990s. The highlights of these studies are summarized in the sections that follow.

4.2.4.1 Thorium-234

Thorium-234 (half-life $=$ 24 days) is produced from the decay of ^{238}U. Uranium is present in seawater as the stable anionic uranyl carbonate complex and its concentration is essentially conservative with S. As a consequence, the production of ^{234}Th in the LIS water column (dpm cm^{-2}) can be calculated as:

$$A_U \times \left(\frac{27}{35}\right) \times \hbar$$

where A_U is the activity of ^{238}U in open ocean seawater (2.45 dpm/L at 35 psu), 27 is the average S of LIS, and $\hbar$ is the mean depth of LIS (20 m, but expressed as cm in the equation). Thus, on average, 4 dpm ^{234}Th are produced in the water column per cm^2 of Sound bottom area. The high particle reactivity of ^{234}Th was noticed by Matsumoto (1975) based on its disequilibrium with respect to ^{238}U in surface waters, and shortly thereafter Aller and Cochran (1976) made the first measurements of ^{234}Th in LIS water and bottom sediments. Their measurements showed that this radionuclide is effectively scavenged from the water column and is present in the bottom sediments in excess of its parent ^{238}U activity. There it is distributed in the upper few centimeters as a consequence of mixing by the benthic fauna. Aller and Cochran (1976) also showed that the profiles of excess ^{234}Th in the sediments varied with season, in conjunction with the change in faunal activity. Aaboe et al. (1981) observed that scavenging of the cosmogenic radionuclide, ^7Be, from the LIS water column was similarly rapid and its distribution in the sediments provides another tracer for particle mixing (Krishnaswami et al. 1984).

A more detailed study of the temporal variation of excess ^{234}Th profiles at a single site was made by Gerino et al. (1998) as part of the PULSE project. Samples were collected biweekly for 7 months in 1992–1993 to evaluate the role of the spring bloom in supplying organic matter to the sediments and the mixing and decomposition of the labile organic material following its deposition at the sea floor. In addition to ^{234}Th, Chl a, and luminopheres (fluorescently tagged particles added to the sediment in situ) were measured, permitting bioturbation rates to be compared among the different tracers. The results showed increases in the rate of mixing following the pulsed input of organic matter to the sediment. The sediment inventory of excess ^{234}Th remained relatively constant throughout the sampling (1.6 $\pm$ 0.7 dpm cm^{-2}), but was somewhat less than its production in the overlying water column (station depth 16 m), suggesting that lateral transport of particles influenced ^{234}Th inventories in bottom sediments of LIS.

Indeed, this idea had been developed by Aller et al. (1980), who mapped the distribution of ^{234}Th in LIS on a broader scale and compared inventories in bottom sediments with the values expected based on the depth of water at a given station. These data showed that on average the inventory of ^{234}Th in bottom sediments matched that expected from production in the overlying water column, but there was not a good correlation between the ^{234}Th inventory at a given station and water depth at that station. This pattern also was seen in a larger survey of ^{234}Th by Cochran, Aller, and colleagues conducted as part of the Long Island Sound Study (Cochran and Hirschberg 1991).

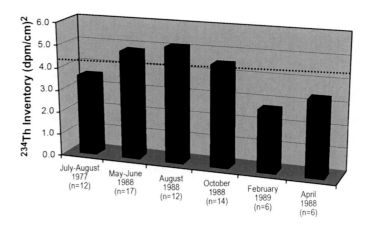

Fig. 4.10 Inventories of excess ^{234}Th in bottom sediments of Long Island Sound sampled at different times. *Dashed line* indicates theoretical production of ^{234}Th from ^{238}U decay in the water column (for a mean depth of LIS of 20 m). 1977 data from Benninger et al. (1979); other sampling times, Cochran (unpublished). Lower average inventories in February and April 1989 likely reflect the small number of stations sampled compared with the other sampling times

Figure 4.10 shows that the average inventory in LIS sediments is comparable to that expected from production in the water column, but Fig. 4.11 reinforces the conclusion of Aller et al. (1980) that the inventory at a given station is not a simple function of the production of ^{234}Th in the overlying water column. These results support the hypothesis that lateral exchanges of particles through the Sound, mediated both by the physical circulation and spatial variations in the rate and depth of particle mixing in the sediments, control the inventory of excess ^{234}Th at a given

Fig. 4.11 Cross plot of ^{234}Th inventories in sediments of Long Island Sound and water depth. Each point represents a single station and the *dashed line* shows the production of ^{234}Th from ^{238}U decay as a function of water depth. The lack of any correlation suggests lateral transfer of particles in LIS on a ^{234}Th time scale (i.e., several months). Sources of data given in caption to Fig. 4.10

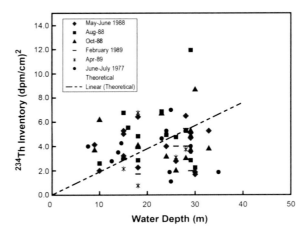

station. Moreover, such exchanges must take place on the time scale of the half-life of ^{234}Th.

4.2.4.2 Lead-210

Lead-210 (half-life $=$ 22.3 years) is added to LIS from the atmosphere, where it is produced from the decay of ^{222}Rn that has emanated from rocks and soils. The atmospheric flux of ^{210}Pb to LIS is essentially constant on an annual basis ($\sim$1 dpm cm^{-2}/year) (Renfro 2010; Turekian et al. 1980), and supports a steady-state standing crop of 32 dpm cm^{-2}. Once supplied to the water column of LIS, ^{210}Pb is rapidly scavenged onto particles and deposited in sediments, but its longer half-life allows it to be distributed deeper in the sediment column. The ^{210}Pb scavenged from the water column is designated as "excess" ^{210}Pb (^{210}Pb$_{xs}$) and is determined by subtracting the supported ^{210}Pb activity (taken to be the activity of its grandparent, ^{226}Ra) from the measured ^{210}Pb. Benninger (1978) calculated a mass balance for ^{210}Pb in LIS and concluded that the mean inventory in sediments was comparable to that expected from the direct atmospheric flux. Additional cores collected as part of the Long Island Sound Study support this argument (mean excess ^{210}Pb inventory $=$ 34 $\pm$ 13 dpm cm^{-2}; Cochran, unpublished data). Thus, other sources of ^{210}Pb to LIS, including exchange with the East River and BIS and in situ production from dissolved ^{226}Ra, are relatively small. Cochran et al. (1998) used this result, coupled with metal and ^{210}Pb inventories in salt marsh deposits bordering LIS, to estimate the fraction of contaminant metal (Pb, Cu, Zn, Cd) inventories in LIS derived from the atmospheric flux. Their results showed that 70–90 % of the contaminant Pb in LIS sediments was supplied from the atmosphere; for Cu and Zn, the fraction was <50 %.

Both sediment accumulation and mixing by infauna act to transport ^{210}Pb to depth in the sediment. As with ^{234}Th, sediment inventories of ^{210}Pb vary spatially in the Sound, correlated with the depth and rate of particle mixing. Indeed both radionuclides support the concept of lateral redistribution of particles in LIS, on time scales ranging from monthly to multi-decadal. Although ^{210}Pb has often been used to determine sediment accumulation rates of coastal sediments, including those of LIS, early studies of ^{210}Pb distributions in LIS sediment cores (Benninger et al. 1979; Krishnaswami et al. 1984; Thomson et al. 1975) showed that particle mixing exerts a strong—perhaps dominant—control on the depth profiles of ^{210}Pb, and sediment accumulation rates can be overestimated if mixing is neglected. For example, Benninger et al. (1979) determined a ^{210}Pb-derived rate of sediment accumulation of 0.11 cm/year for a LIS core, yet Benoit et al. (1979) found that the long-term rate based on radiocarbon in the same core was 0.075 cm/year. Krishnaswami et al. (1984) determined a sediment accumulation rate of 0.094 cm/year on the carbonate fraction of a core from the FOAM site, although in that case, the rate was somewhat larger than that derived from ^{210}Pb in the upper 15 cm. Other cores from LIS (e.g., Moore et al. 2006) show ^{210}Pb-derived rates of sediment accumulation that are approximately a factor of 10 greater than the radiocarbon rates determined by Benoit et al. (1979) and Krishnaswami et al. (1984) or the rate derived by Bokuniewicz and Gordon (1980) from a sediment mass balance (~0.03–0.04 cm/year). Indeed, Moore et al. (2006) determined sediment accumulation rates of 0.4–1.7 cm/year using both ^{210}Pb and ^{137}Cs. The latter radionuclide has an anthropogenic source, having been introduced into the environment through the atmospheric testing of atomic weapons that peaked in the early 1960s prior to the nuclear test ban treaty. These rates are greatly in excess of the long-term rate of sediment input to LIS.

Other applications of ^{210}Pb and ^{137}Cs to determine sediment accumulation rates in LIS have involved the determination of contaminant chronologies (e.g., Hg—Varekamp et al. 2003b; chlordane—Yang et al. 2007). For example, Yang et al. (2007) explicitly modeled both bioturbation and sediment accumulation in their attempts to decipher chronologies of chlordane inputs to LIS. They found that sediment mixing by the benthic fauna was necessary to reproduce the observed profiles and determined accumulation rates of 0.5–0.7 cm/year in three cores collected in harbors and marginal embayments of WLIS. Although these rates are greater than the long-term rate suggested by the studies of Benninger et al. (1979), Benoit et al. (1979), Bokuniewicz and Gordon (1980), and Krishnaswami et al. (1984), the settings of the cores (harbors) may have biased the rates toward values greater than the open LIS.

4.2.4.3 Radium Isotopes and Radon

The radium isotopes (226, 228, 223, and 224) and radon-222 are produced in sediments from decay of Th parents (and ^{226}Ra in the case of ^{222}Rn) and are mobilized to sediment pore waters through processes (e.g., recoil) associated with their production. Ra can be transported (via molecular diffusion or bioirrigation) through the sediment pore water, but also can interact with solid phases in the sediment (e.g., via adsorption onto manganese oxides). Cochran (1979, 1984) made the first measurements of ^{226}Ra (half-life = 1600 years) and ^{228}Ra (half-life = 5.7 years) in the pore water and sediments of LIS and found large enrichments in pore water compared with overlying water. This gradient in Ra concentration drives fluxes into the overlying water. Cochran (1979, 1984) used disequilibrium in solid phase ^{228}Ra/^{232}Th activity ratio to calculate ^{228}Ra fluxes of 1.3–3 dpm cm^{-2}/year from the muddy sediments to the waters of LIS. These values are in general agreement with that obtained by Turekian et al. (1996) in their study of ^{228}Ra in LIS. These authors measured ^{228}Ra distributions in LIS to determine a ^{228}Ra mass balance and estimate both the flux of water from the East River and the residence time of water in the Sound. Estimates of the latter ranged from 63 to 166 days, depending on the flux of ^{228}Ra from the sediments.

A survey by Torgersen et al. (1996) of the short-lived ^{224}Ra (half-life = 3.6 days) in LIS reinforced some of the processes responsible for distributing Ra in the Sound, namely fluxes from bottom sediments and lateral mixing. Torgersen et al. (1996) calculated horizontal eddy diffusion rates of 5–50 m^2 s^{-1} and suggested that ^{224}Ra fluxes into the deep water were enhanced by reduced scavenging of Ra by manganese oxides under conditions of hypoxia. Sun and Torgersen (2001) developed this idea further by modeling dissolved and adsorbed ^{224}Ra in LIS sediments as a function of varying degrees of oxygen and MnO_2 concentrations, and concluded that ^{224}Ra was scavenged into the sediments under conditions of high bottom water oxygen. This hypothesis linking the flux of ^{224}Ra from LIS sediments with bottom water DO was confirmed by the recent work of Cochran and colleagues (Bokuniewicz and Garcia-Orellana; see Daniel 2011), who observed higher concentrations of ^{224}Ra in WLIS and CLIS during summer hypoxia compared with oxygenated conditions in the spring.

The role of bioirrigation of sediments by the benthic infauna in influencing the fluxes of solutes across the sediment–water interface can be characterized by another U-series radionuclide, radon-222. Parallel surveys for ^{222}Rn (half-life $= 3.8$ days) conducted by Torgersen et al. (1997) in WLIS at the same time as the Ra sampling described above (Turekian et al. 1996; Torgersen et al. 1996) demonstrated the importance of bioirrigation in LIS. In contrast to Ra, radon is chemically unreactive and can be transported effectively out of the sediments by diffusion and bioirrigation following its introduction to the sediment pore water by recoil. Bioirrigation is sensitive to conditions of hypoxia because the benthic fauna are less active with low bottom water DO. Direct measurements of the ^{222}Rn flux from sediment cores showed lower fluxes in August, a time of minimum dissolved oxygen in the western Sound and presumably minimum bioirrigation. Torgersen et al. (1997) also observed ^{222}Rn gradients from the margins of LIS into the deep water and used these gradients to calculate horizontal dispersion coefficients of >5–50 m^2 s^{-1}. These values are consistent with those estimated from ^{224}Ra (Torgersen et al. 1996) and also with DO distributions in LIS.

4.2.4.4 Osmium Isotopes

The element osmium has seven stable isotopes one of which, ^{187}Os has a radiogenic contribution from the radioactive beta decay of ^{187}Re (half-life $= 4.6 \times 10^{10}$ years). The ^{187}Os abundance in Os is dependent on the Re/Os ratio and the age of the source of Os. Turekian (1982) and Luck and Turekian (1983) used the ^{187}Os/^{186}Os ratio to infer the origin of platinum group elements (specifically Ir) at the Cretaceous-Tertiary boundary where it was inferred that the Ir concentration indicated a major meteorite impact with its consequences for the history of life at that boundary. In the quest for understanding the cycle of Os on the Earth's surface, a detailed study of LIS was assayed, and we summarize those results here.

Originally, the variation of ^{187}Os relative to all the invariant Os isotopes was measured via ^{187}Os/^{186}Os. It was later discovered that under certain geologic conditions involving mantle-derived rocks, variations in ^{186}Os could be detected due to the radioactive decay of ^{186}Pt as the result of slight variations in the Pt/Os in the mantle-derived rocks. The effect on surface processes is negligible but the method

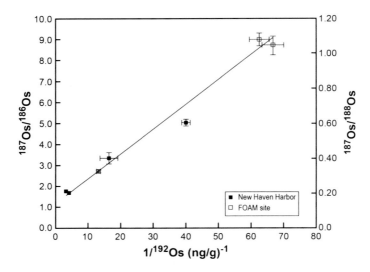

Fig. 4.12 Osmium isotope composition versus $1/^{192}\mathrm{Os}$ concentration in New Haven Harbor surface sediments and in FOAM site surface and deep sediment. The correlation is consistent with two-component mixing between anthropogenic Os ($^{187}\mathrm{Os}/^{186}\mathrm{Os} = 1$) and non-anthropogenic Os derived from crustal weathering and/or seawater ($^{187}\mathrm{Os}/^{186}\mathrm{Os} \geq 9$). Adapted from Esser and Turekian (1993a, b)

of reporting the variation of $^{187}\mathrm{Os}$ now is commonly as the ratio of $^{187}\mathrm{Os}/^{188}\mathrm{Os}$. The conversion between the two ratios is:

$$^{187}\mathrm{Os}/^{188}\mathrm{Os} = 0.120343\ ^{187}\mathrm{Os}/^{186}\mathrm{Os}.$$

Esser and Turekian (1993b) measured the isotopic composition and concentration of Os in New Haven Harbor and in coastal deposits eastward in LIS along the Connecticut shore. They found that much of New Haven Harbor sediments contained Os with an isotopic composition typical of Os produced from ores derived from mantle sources, the most likely source being the Bushfeld Complex of South Africa (Fig. 4.12). They proposed that the source of this Os was mainly from laboratories associated with hospitals using osmium tetroxide to fix tissues for analysis by scanning electron microscopy. This discovery led to a more extensive study of LIS sediments adjacent to New Haven. One can look at two locations along the coast eastward from New Haven to see how the activities in New Haven impacted the environment further afield at a salt marsh at the East Haven-Branford border (Morris Creek marsh) and the FOAM site—an anoxic site offshore studied extensively by geochemists tracking the deposition and reactivity in sediments where the organic flux is sufficiently high to maintain an anoxic mud environment (Goldhaber et al. 1977).

The Os concentration of the top 5 cm of the Morris Creek high salt marsh is 88 ng g^{-1} (dry sample), according to Esser and Turekian (1993b), and the $^{187}\mathrm{Os}/^{186}\mathrm{Os}$ is 5.34. On the assumption that Os in the background terrestrial Connecticut terrain has an $^{187}\mathrm{Os}/^{186}\mathrm{Os}$ of about 9 and the anthropogenic Os has

Table 4.1 Osmium concentrations and $^{187}Os/^{188}Os$ of aerosols from New Haven, CT (Kline Geology Laboratory, Yale University)

Date (mm/dd/yyyy)	$^{187}Os/^{188}Os$	[Os] (pg m^{-3})
5/27/1994	0.347	0.012
6/3/1994	0.463	0.003
11/3/1997	0.139	0.006
11/5/1997	0.378	0.015
11/6/1997	1.049	0.010
11/20/1997	0.920	0.026

a ratio of about 1.2, then half the Os in the top 5 cm is inferred to be of anthropogenic origin. The salt marsh was accreting at ~0.25 cm/year for the past 100 years (McCaffrey and Thomson 1980), which means the top 5 cm represents about 20 years. If one assumes that from the year of collection of the sample in 1989 Os was used progressively more and more in hospital laboratories for tissue fixation for scanning electron microscopic studies since about 1969, then the average anthropogenic Os deposition flux was about 4 pg Os cm^{-2}y^{-1}.

Williams and Turekian (2002) measured the Os concentration and isotopic composition of aerosols collected in New Haven, CT in 1994 and 1997. Most of the air samples showed a mixture of local dust with the characteristic Os isotope value of Connecticut soils and anthropogenic Os (Table 4.1). One sample (collected 11/03/97) was almost totally anthropogenic Os that was inferred to be a direct result of release of a fairly volatile osmium tetroxide during hospital incineration. Because atmospheric concentrations of aerosols are determined not only by supply but also by washout events prior to sampling, a simple relationship involving mixing of two end members is not possible.

If the samples that have been corrected for background Os isotope values (Esser and Turekian 1993a), are used together with the one virtually purely anthropogenic sample (11/03/97; Table 4.1), one can get a rough calculation of the atmospheric flux of anthropogenic Os using its relationship to the atmospheric ^{210}Pb flux. The average air ^{210}Pb composition and its atmospheric flux via precipitation have been extensively studied for New Haven (Graustein and Turekian 1986). Using an anthropogenic Os air concentration of 0.0057 pg Os m^{-3} (Table 4.1) and a ^{210}Pb concentration of 0.034 dpm m^{-3} combined with a ^{210}Pb deposition flux of 1 dpm cm^{-2}/year, we get an anthropogenic Os flux of 0.17 pg Os cm^{-2}y^{-1}. This value is considerably lower than the value of 4 pg Os cm^{-2}y^{-1} calculated for the salt marsh east of New Haven, and the difference implies that most of the anthropogenic Os added to the salt marsh was not delivered via the atmosphere. The transport of anthropogenic Os from sewage sludge and deposited sediments contaminated by this sludge as particles eastward along the Connecticut coast must then have provided the anthropogenic osmium trapped in the salt marsh.

Indeed, further to the east along the Branford coast at the FOAM site, Esser and Turekian (1993b) calculated a flux of anthropogenic Os of about 20 pg Os cm^{-2}/year for the 20 years prior to 1993. Based on the Os isotopic composition of LIS

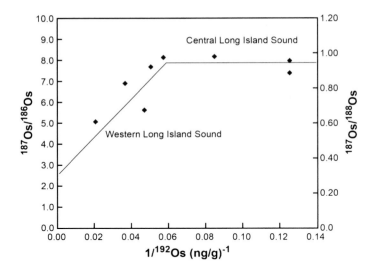

Fig. 4.13 Osmium isotope composition versus $1/^{192}Os$ for bulk sediments collected between August 1974 and November 1988 in Long Island Sound. Stations in the WLIS lie on a mixing line between an anthropogenic end-member with a low $^{187}Os/^{186}Os$ ratio (≈ 1; see Fig. 4.10) and the ratio observed in ambient CLIS sediments. Adapted from Williams et al. (1997)

Table 4.2 Osmium concentrations and isotopic composition of waters in and around Long Island Sound

Location	Salinity	[Os] (fM)	$^{187}Os/^{188}Os$
Hudson river (Newburgh)	0	68	1.265
Vineyard sound	31	46	1.07
Whitestone bridge (km = 0)	20	49	0.94
Western LIS (km = 7)	26	52	0.994
Central LIS (km = 39.8)	28.8	47	1.019

water (Turekian et al. 2007), it is clear that the chemical precipitation of metals from the water cannot explain the high flux of anthropogenic Os at that site. As has been shown above the atmospheric flux is also not sufficient. Therefore, transport from New Haven Harbor remains the most likely source. The higher deposition rate of this Os compared to the salt marsh, which is closer to New Haven, is due to the efficiency with which the anoxic sediments at the FOAM site trap other metals and co-precipitate the flocs bearing the anthropogenic Os carried from New Haven Harbor.

It is evident that if locally derived anthropogenic Os, inferred to be from hospital usage of Os for analytical purposes, is found in New Haven Harbor and transported laterally along the coast to other sites, a similar result should be seen in WLIS. The sources of anthropogenic Os from New York City hospitals along the East River will impact that waterway and then be transported into LIS. The results

Table 4.3 Modification of the Os signature in the East River

Location	Salinity	[Os] (fM)	$^{187}Os/^{188}Os$
Hudson River	0	68	1.265
New York Bight[a]	31	50	1.07
Mixing of New York Bight and Hudson River water in the East River			
No removal	20	56	1.153
With removal	20	40	1.153
Required addition of anthropogenic Os	20	10	0.13
Net (removal + addition of anthropogenic Os)[b]	20	50	0.94

[a]Assumed equal to Vineyard Sound (Table 4.2)
[b]Value observed at Whitestone Bridge station (Table 4.2)

of Williams et al. (1997) demonstrate that indeed there is transport of particles bearing anthropogenic Os from the East River into WLIS (Fig. 4.13).

Aside from the transport of particulate Os through LIS, dissolved Os is also transported. Turekian et al. (2007) made a detailed study of the waters of the East River, the Hudson River, LIS, and the adjacent open ocean (Tables 4.2 and 4.3). Based on the assumption that the dissolved anthropogenic Os is derived from hospital-fed sewers along the East River, as indicated by the sediment studies reported above, the mass balance for Os concentrations and isotopic ratios yields the fraction of stream-borne Os that is removed in the East River estuarine system (about 25 % of the dissolved Os flux carried by the Hudson River). The results also indicate that Os behaves conservatively through LIS over a S range from ~20 to 31 psu (Table 4.2), unlike other reactive species such as ^{210}Pb and the thorium isotopes described above.

4.3 Summary

The sediments of LIS represent a complex, often interconnected geochemical system involving living organisms (e.g., phytoplankton, bioturbating benthos, aerobic, and anaerobic bacteria), chemical species (including redox-sensitive species, metals, and radionuclides), and geological materials (e.g., sands, silts, and clays) all within waters whose physical movements are tightly constrained by the very geography of the Sound itself. The authors of this chapter have endeavored to present an overview of the present state of knowledge of key aspects of the sediment geochemistry of LIS. In particular, attention has been focused on those elements that relate most directly to ongoing management issues within the Sound: the distribution of sediments and organic carbon within the WLIS, CLIS, and ELIS basins, the aerobic and anaerobic decomposition of organic materials in LIS sediments and the factors that affect their accumulation and breakdown, the relationships among microbes, bioturbation, and organic matter accumulation and decomposition within the sediments of LIS, reduced chemical species present in LIS sediments and the factors affecting their release, manganese and iron cycling in LIS sediments and their behavior under anoxic and oxic conditions, and finally, the behavior of certain specific radionuclides within LIS, insofar as they address processes occurring in LIS.

Although it appears that similar chemical processes operate across the Sound's sedimentary environments, different ones dominate depending on the organic carbon content of the sediment, the sediment grain size, the amount and extent of bioturbation present in the sediments, and the proximity of an area to riverine inputs. Overall, the sub-tidal, fine-grained sediments found in WLIS and in the majority of harbors, inlets, and bays in the Sound are dominated by reducing conditions unless a significant population of benthic organisms is present. The coarse sediments that form a large component of the environments in ELIS are dominated by aerobic processes in the absence of large inputs of organic carbon. The sediments of CLIS exist in a dynamic balance between anoxic and oxic conditions, moving continuously between these states.

Organic carbon inventories of LIS sediments, as reflected by TOC and by preservation of chloropigments, also vary from west to east in the Sound, with the highest inventories found in WLIS and in some harbors in CLIS and ELIS. Although a primary source of the carbon input to the sediments appears to be phytoplankton, there is evidence of a terrestrial carbon component along a portion of the Sound's northern shore, near Norwalk Shoals and along its southern shore, within Smithtown Bay. Much of the carbon introduced into the sediments of LIS is either directly consumed by benthic organisms or undergoes microbially mediated decomposition. At the present time it appears that breakdown of a significant amount of the organic carbon present in the sediments of WLIS undergoes anaerobic microbially mediated decomposition rather than aerobic decomposition or direct consumption by benthos. These latter processes occur in WLIS sediments but do not appear to dominate, whereas they do in CLIS and ELIS.

The anaerobic decomposition of organic carbon releases ammonia, sulfide, phosphate, and methane into the pore waters of LIS sediments. Whether or not these species are mobilized out of the sediments into the overlying water column depends on the amount of bioturbation present in the sediments, as well as the T and DO content of the overlying water. Both ammonia and sulfide have been measured in bottom waters of WLIS during the late spring, correlating with deposition of plankton and the onset of bioturbation. They have also been measured in the late summer-fall, after the water column had destratified; this late fall flux has been correlated mainly with plankton deposition and sediment temperatures. The extent to which ammonia and sulfide may or may not be released from the majority of CLIS sediments is not known at the present time, although it is known that, at least for some stations within CLIS, sulfide and ammonia accumulate in sediment pore waters. These species may undergo oxidation while still within the sediments or may be released from the sediments into the overlying water via the activities of organisms. Although phosphate release from CLIS sediments has been shown to be correlated with bioturbation and with the oxidation and scavenging of phosphate by iron oxides, the majority of phosphate produced during diagenesis is retained in the sediments of CLIS, albeit in a different form than it was originally deposited in. Biogenic methane, resulting from the microbially mediated anaerobic decomposition of organic matter, is found in association with fine-grained sediments across LIS.

The distribution of redox-sensitive metals in the sediments of LIS is clearly correlated with TOC content, grain size, and oxidation state of the sediments. In general, higher metal inventories occur in WLIS and CLIS, and lower ones in ELIS. However, manganese exhibits a more variable distribution. The highest inventories of manganese in LIS can be found in WLIS and ELIS; the lowest manganese inventories are found in CLIS. The sediment inventories of both manganese and iron, two redox-sensitive metals, are highest in WLIS and CLIS, and lower throughout the majority of ELIS sediments. Seasonal cycling in these metals is evident, at least for Mn, which shows enhanced fluxes from the sediments of WLIS during summer as hypoxia develops in the bottom water.

Isotope tracers have proved useful in studying sediment-associated processes in LIS. The naturally occurring U-decay series radionuclides ^{234}Th and ^{210}Pb associate strongly with particles, and their distributions in the sediment can be used to reconstruct rates of bioturbation. Their inventories vary spatially, suggesting lateral exchange and transport of particles within the Sound and a strong dependence of the inventories on the rate and depth of particle mixing. The Ra isotopes and radon are produced in and released from the sediments of LIS and are useful tracers of water mixing in the Sound. The element osmium and its stable isotopes, including the radiogenic ^{187}Os, are tracers of the input of anthropogenic Os from hospitals (where it is used to fix tissue for image analysis) via wastewater inputs. This source is particularly evident from the East River and in major harbors such as New Haven Harbor, and sediment data show its imprint on the sediments of WLIS and CLIS.

4.4 Science and Management Recommendations

1. Nitrogen and phosphate remineralization from organic-rich sediments are known to occur and to be related to bacterial processes that are, in part, enhanced by warming temperatures. Yet, N and phosphate remineralization from LIS sediments—especially in WLIS and in the organic-rich sediments of CLIS—is, at present, not well characterized, especially within a temporal framework. Research that has been conducted in the Sound over the past 50 years has demonstrated that such remineralization is, in fact, happening. It appears that sediment N and phosphate remineralization may support the growth of phytoplankton in shallow coastal waters in WLIS as well as benthic microalgae and macroalgae, the growth of which may, in turn, upon degradation, be contributing to the organic carbon content of the sediments and the formation of hypoxic bottom waters.

2. Given the potential for warmer conditions in LIS, it is strongly recommended that attention be paid to determine the extent of shifts in the redox boundary within the sediments of LIS—especially WLIS and CLIS—over the course of a year. Research suggests that WLIS sediments may not be as oxygenated as they once were which may, in turn, be leading to an increasing carbon inventory in the sediments. Anaerobic degradation of organic carbon favors the formation of hypoxic and anoxic waters at the sediment–water interface which, in turn,

favors the increased storage of organic carbon in the sediments. Such conditions, coupled with increasingly warm sediment temperatures, favor the formation of anoxic conditions within the benthic zone resulting in the potential loss of habitat for benthic organisms.

3. There needs to be a better characterization of the organic matter present in the sediments of the bays and inlets in LIS. Additionally, there needs to be a more detailed temporal study of phytoplankton production in the Sound in order for scientists and managers to be better able to gauge the relationship between N sources (e.g., wastewater treatment facilities, riverine inputs, groundwater, and sediment pore waters) and phytoplankton production in LIS, especially in WLIS and CLIS. Such information will also assist researchers to better understand the exchanges, if any that occur between the main stem of the Sound and its bays and inlets.

4. When one examines the studies conducted in LIS, one is struck by the fact that the majority of studies conducted in CLIS occurred between the late 1970s and the early 1990s. Yet, much has changed in the Sound since then. CLIS has the potential to undergo major shifts in bioturbation, organic carbon accumulation, and sediment oxygen demand as temperatures warm. It is strongly suggested that several sites in CLIS be established (including some of the ones that were studied in the 1970s–1990s) and monitored with regard to organic carbon content, inputs, redox oscillations, benthos (including the pattern and rate of biomixing), sediment accumulation, and bacterial biomass in order to detect if any shifts are occurring that favor enhanced carbon burial and increased anaerobic decomposition, especially in the context of climate change.

5. Given that the release and/or precipitation/adsorption of many metals are strongly affected by changes to the redox state of sediments, it is suggested that a clearer understanding of the temporal dynamics of redox-sensitive metals present in the sediments and/or bottom waters of WLIS and CLIS is needed in order to better manage the Sound.

6. Isotopic studies have often proved extremely useful in understanding sediment- and water-associated processes in LIS. It is suggested that these tracers be employed in efforts to further characterize exchanges across the sediment–water interface in WLIS and CLIS sediments. It is also suggested that such tracers can be used to better characterize exchanges occurring between the main stem of LIS and its bays and inlets.

References

Aaboe E, Dion EP, Turekian KK (1981) [7]Be in Sargasso Sea and Long Island Sound waters. J Geophys Res 86:3255–3257

Abril G, Borges AV (2004) Carbon dioxide and methane emissions from estuaries. In: Tremblay A, Varfalvy L, Roehm C, Garneau M (eds) Greenhouse gas emissions: fluxes and processes, hydroelectric reservoirs, and natural environments. Springer, Berlin, pp 187–212

Abril G, Iverson N (2002) Methane dynamics in a shallow, non-tidal, estuary (Randers Fjord, Denmark). Mar Ecol Prog Ser 230:171–181

Abril G, Etcheber H, Le Hir P, Bassoullet P, Boutier B, Frankignoulle M (1999) Oxic–anoxic oscillations and organic carbon mineralization in an estuarine maximum turbidity zone (The Gironde, France). Limnol Oceanogr 44:1304–1315

Abril G, Nogueira E, Hetcheber H, Cabecadas G, Lemaire E, Brogueira MJ (2002) Behaviour of organic carbon in nine contrasting European estuaries. Estuar Coast Shelf Sci 54:241–262

Abril G, Commarieu MV, Etcheber H, Deborde J, Deflandre B, Živadinović MK, Chaillou G, Anschutz P (2010) In vitro simulation of oxic/suboxic diagenesis in an estuarine fluid mud subjected to redox oscillations. Estuar Coast Shelf Sci 88:279–291

Aller RC (1978) Experimental studies of changes produced by deposit-feeders on pore water, sediment and overlying water chemistry. Am J Sci 278:1185–1234

Aller RC (1980a) Quantifying solute distributions in the bioturbated zone of marine sediments by defining an average microenvironment. Geochim Cosmochim Acta 44:1955–1965

Aller RC (1980b) Diagenetic processes near the sediment–water interface of Long Island Sound: II. Fe and Mn. Adv Geophys 22:351–415

Aller RC (1982a) The effects of macrobenthos on chemical properties of marine sediment and overlying water. In: McCall PL, Tevesz MJS (eds) Animal sediment relations: the biogenic alteration of sediments. Plenum Press, New York, pp 53–102

Aller RC (1982b) Carbonate dissolution in nearshore terrigenous muds: the role of physical and biological reworking. J Geol 90:79–95

Aller RC (1990) Bioturbation and manganese cycling in hemipelagic sediments. Trans R Soc Lond A 331:51–68

Aller RC (1994a) Bioturbation and remineralization of sedimentary organic matter: effects of redox oscillation. Chem Geol 114:331–345

Aller RC (1994b) The sedimentary Mn cycle in Long Island Sound: its role as intermediate oxidant and the influence of bioturbation, O_2, and C_{org} flux on diagenetic reaction balances. J Mar Res 52:259–295

Aller RC (1998) Mobile deltaic and continental shelf muds as suboxic, fluidized bed reactors. Mar Chem 61:143–155

Aller RC (2001) Transport and reactions in the bioirrigated zone. In: Boudreau BP, Jørgensen BB (eds) The benthic boundary layer. Oxford University Press, New York, pp 269–301

Aller RC (2004) Conceptual models of early diagenetic processes: the muddy seafloor as an unsteady, batch reactor. J Mar Res 62:815–835

Aller RC, Aller JY (1998) The effect of biogenic irrigation intensity and solute exchange on diagenetic reaction rates in marine sediments. J Mar Res 56:905–936

Aller RC, Cochran JK (1976) $^{234}Th/^{238}U$ in nearshore sediment: particle reworking and diagenetic timescales. Earth Planet Sci Lett 29:37–50

Aller RC, Rude PD (1988) Complete oxidation of solid phase sulfides by manganese and bacteria in anoxic marine sediments. Geochim Cosmochim Acta 52:751–765

Aller RC, Yingst JY (1980) Relationships between microbial distributions and the anaerobic decomposition of organic matter in surface sediments of Long Island Sound, USA. Mar Biol 56:29–42

Aller RC, Yingst JY (1985) Effects of the marine deposit-feeders Heteromastus filiformis (Polychaeta), Macoma balthica (Bivalvia), and Tellina texana (Bivalvia) on averaged sedimentary solute transport, reaction rates, and microbial distributions. J Mar Res 43:615–645

Aller RC, Benninger LK, Cochran JK (1980) Tracking particle associated processes in nearshore environments by use of $^{234}Th/^{238}U$ disequilibrium. Earth Planet Sci Lett 47:161–175

Aller RC, Aller JY, Kemp PF (2001) Effects of particle and solute transport on rates and extent of remineralization in bioturbated sediments. In: Aller JY, Woodin SA, Aller RC (eds) Organism-sediment interactions. University of South Carolina Press, Columbia, pp 315–334

Anderson TH, Taylor GT (2001) Nutrient pulses, plankton blooms, and seasonal hypoxia in Western Long Island Sound. Estuaries 24:228–243

Arias VA, Skoog A, Sanudo S, Beck A (2004) Effect of oxygen concentrations on fluxes of dissolved organic matter nutrients, iron, and manganese over the sediment-water interface. Long Island Sound Res Conf Proc 2004:11–17

Balistrieri LS, Murray JW (1986) The surface chemistry of sediments from the Panama Basin: the influence of Mn oxides on metal adsorption. Geochim Cosmochim Acta 50:2235–2243

Benninger LK (1978) ^{210}Pb balance in Long Island Sound. Geochim Cosmochim Acta 42:1165–1174

Benninger LK, Aller RC, Cochran JK, Turekian KK (1979) Effects of biological sediment mixing on the ^{210}Pb chronology and trace metal distribution in a Long Island Sound sediment core. Earth Planet Sci Letts 43:241–259

Benoit GJ, Turekian KK, Benninger LK (1979) Radiocarbon dating of a core from Long Island Sound. Estuar Coast Mar Sci 9:171–180

Berner RA (1978) Sulfate reduction and the rate of deposition of marine sediments. Earth Planet Sci Letts 37:492–498

Bossier T, Cuomo C (2010) Sediment characteristics in relation to hypoxic events in Smithtown Bay. In: Abstract of Long Island Sound research conference, Stamford, CT, p 53

Bianchi TS (2007) Biogeochemistry of estuaries. Oxford University Press, New York, p 706

Bianchi TS, Johansson B, Elmgren R (2000) Breakdown of phytoplankton pigments in Baltic sediments: effects of anoxia and loss of deposit-feeding macrofauna. J Exp Mar Biol Ecol 251:161–183

Blackburn TH, Henriksen K (1983) Nitrogen cycling in different types of sediments from Danish waters. Limnol Oceanogr 28:477–493

Bokuniewicz HJ, Gordon RB (1980) Sediment transport and deposition in Long Island Sound. In: Saltzman B (ed) Advance in Geophysics, vol 22. Academic, New York, p 424

Canfield DE (1997) The geochemistry of river particulates from the continental United States: major elements. Geochim Cosmochim Acta 61:3349–3365

Canfield DE (1989) Reactive iron in marine sediments. Geochim Cosmochim Acta 53:619–632

Canfield DE, Thamdrup B, Hansen JW (1993) The anaerobic degradation of organic matter in Danish coastal sediments: Fe reduction, Mn reduction and sulfate reduction. Geochim Cosmochim Acta 57:3971–3984

Cochran JK (1979) The geochemistry of ^{226}Ra and ^{228}Ra in marine deposits. PhD Dissertation, Yale University, New Haven, CT, p 260

Cochran JK (1984) The fates of uranium and thorium decay series nuclides in the estuarine environment. In: Kennedy VS (ed) The estuary as a filter. Academic, New York

Cochran JK, Hirschberg D (1991) 234Th as an indicator of biological reworking and particle transport. In: Cochran JK, Aller RC, Aller JY, Hirschberg DJ, Mackin JE (eds) Long Island Sound Study: sediment geochemistry and biology. EPA final report, CE 002870026

Cochran JK, Hirschberg DJ, Wang J, Dere C (1998) Atmospheric deposition of metals to coastal waters (Long Island Sound, New York, USA): evidence from salt marsh deposits. Estuar Coast Shelf Sci 46:503–522

Cuomo C (2007) Assessment of the effects of bottom water temperature and chemical conditions, sediment temparature, and sedmentary organic matter (type and amount) on release of sulfide and ammonia from sediments in WLIS: a laboratory study. Final report, USEPA, Long Island Sound Office, p 11

Cuomo C, Valente R (2003) Monitoring of bottom water and sediment chemical conditions in Western Long Island Sound. Final report, USEPA, Long Island Sound Office, p 31

Cuomo C, Valente R, Dogru D (2004) Monitoring of bottom water and sediment conditions at critical stations in Western Long Island Sound. In: Proceedings of the 6th biennial Long Island Sound conference, October 2002, Connecticut Sea Grant: pp 13–21

Cuomo C, Valente R, Dogru D (2005) Seasonal variations in sediment and bottom water chemistry of Western Long Island Sound: implications for lobster mortality. J Shellfish Res 24:805–814

Cuomo C, Swanson RL, Wilson R (2011) National fish and wildlife federation final report: evaluation of hypoxia in Smithtown Bay (NY) #2005-0333-017

Dai J, Sun M-Y, Randolph AC, Noakes JE (2009) A laboratory study on biochemical degradation and microbial utilization of organic matter comprising a marine diatom, land grass, and salt marsh plant in estuarine ecosystems. Aquat Ecol 43:825–841

Daniel JWR (2011) Coupled Ra and Mn cycling in Long Island Sound. Masters thesis, School of Marine and Atmospheric Sciences, Stony Brook University, Stony Brook, New York, p 73

de Jonge VN, Boynton W, D'Elia CF, Elmgren R, Welsh BL (1994) Responses to developments in eutrophication in different North Atlantic estuarine systems. In: Dyer KR, Orth RJ (eds) Changes in fluxes in estuaries. Olsen and Olsen, Fredensborg, pp 179–196

Draxler AFJ, Sherrell RM, Wieczorek D, LaVigne MG, Paulson AJ (2005) Manganese concentration in lobster (*Homarus Americanus*) gills as an index of exposure to reducing conditions in western Long Island Sound. J Shellfish Res 24:815–820

Engstrom P, Dalsgaard T, Hulth S, Aller RC (2005) Anaerobic ammonium oxidation by nitrite (anammox): implications for N_2 production in coastal marine sediments. Geochim Cosmochim Acta 69:2057–2065

Esser BK, Turekian KK (1993a) The osmium isotopic composition of the continental crust. Geochim Cosmochim Acta 57:3093–3104

Esser BK, Turekian KK (1993b) Anthropogenic osmium in coastal deposits. Environ Sci Technol 27:2719–2724

Gerino M, Aller RC, Lee C, Cochran JK, Aller JY, Green MA, Hirschberg D (1998) Comparison of different tracers and methods used to quantify bioturbation during a spring bloom: 234-thorium, luminophores and Chlorophyll *a*. Estuar Coast Shelf Sci 46:531–547

Gilbert F, Aller RC, Hulth S (2003) The influence of macrofaunal burrow spacing and diffusive scaling on sedimentary nitrification and denitrification: an experimental simulation and model approach. J Mar Res 61:101–125

Gobler CJ, Buck NJ, Sieracki ME, Sanudo-Wilhemy SA (2006) Nitrogen and silicon limitation of phytoplankton communities across an urban estuary: the East River–Long Island Sound system. Estuar Coast Shelf Sci 68:127–138

Goebel NL, Kremer JN, Edwards CA (2004) Temporal and spatial variability in photosynthetic characteristics and primary production, and testing a formulation for primary production in Long Island Sound. In: Abstract of proceedings of the Long Island Sound research conference 2004, p 93

Goldberg ED (1954) Marine geochemistry 1: chemical scavengers of the sea. J Geol 62(3):249–265

Goldhaber MB, Aller RC, Cochran JK, Rosenfeld JK, Martens CS, Berner RA (1977) Sulfate reduction, diffusion and bioturbation in Long Island Sound sediments: report of the FOAM (Friends of Anoxic Mud) group. Am J Sci 277:193–237

Graf G (1992) Benthic-pelagic coupling: A benthic view. Oceanogr. Mar. Biol. Annu. Rev. 30: 149–190

Graustein WC and Turekian KK (1986) ^{210}Pb and ^{137}Cs in air and soils measure the rate and vertical distribution of aerosol scavenging. J Geophys Res 91:14,355–14,366

Green MA, Aller RC (2001) Early diagenesis of calcium carbonate in Long Island Sound sediments: benthic fluxes of Ca^{2+} and minor elements during seasonal periods of net dissolution. J Mar Res 59:769–794

Green MA, Aller RC, Aller JY (1992) Experimental evaluation of the influences of biogenic reworking on carbonate preservation in nearshore sediments. Mar Geol 107:175–181

Green MA, Aller RC, Aller JY (1993) Carbonate dissolution and temporal abundances of foraminifera in Long Island Sound sediments. Limnol Oceanogr 38:331–345

Green MA, Aller RC, Aller JY (1998) Influence of carbonate dissolution on survival of shell-bearing meiobenthos in nearshore sediments. Limnol Oceanogr 43:18–28

Greig RA, Reid RN, Wenzloff DR (1977) Trace metal concentrations in sediments from Long Island Sound. Mar Poll Bull 8:183–188

Hannides AK, Dunn SM, Aller RC (2005) Diffusion of organic and inorganic solutes through macrofaunal mucus secretions and tube linings in marine sediments. J Mar Res 63(5):957–981

House WA (2003) Factors influencing the extent and development of the oxic zone in sediments. Biogeochemistry 63:317–333

Hunt CD, Kelly JR (1985) Manganese cycling in coastal regions: response to eutrophication. Estuar Coast Shelf Sci 26:527–558

Ingalls AE, Aller RC, Lee C, Sun MY (2000) The influence of deposit-feeding on chlorophyll *a* degradation in coastal marine sediments. J Mar Res 58:631–651

Johnson KS (2006) Manganese redox chemistry revisited. Science 313(5795):1896–1897

Klump JV, Martens CS (1981) Biogeochemical cycling in an organic-rich coastal marine basin, 11: nutient-sediment-water exchange processes. Geochim Cosmochim Acta 45:101–121

Knebel HJ, Poppe LJ (2000) Sea-floor environments within Long Island Sound: a regional overview. J Coast Res 16(3):533–550

Krishnaswami S, Monaghan MC, Westrich JT, Bennett JT, Turekian KK (1984) Chronologies of sedimentary processes of the FOAM site, Long Island Sound, Connecticut. Am J Sci 284:706–733

Kristensen E (2000) Organic matter diagenesis at the oxic/anoxic interface in coastal marine sediments, with emphasis on the role of burrowing animals. Hydrobiology 426:1–24

Kristensen E, Aller RC, Aller JY (1991) Oxic and anoxic decomposition of tubes from the burrowing sea anemone *Ceriantheopsis americanus*: implications for bulk sediment carbon and nitrogen balance. J Mar Res 49:589–617

Krom MD, Berner RA (1980) The diffusion coefficients of sulfate, ammonium, and phosphate ions in anoxic marine sediments. Limnol Oceanogr 25:327–337

Krom MD, Berner RA (1981) The diagenesis of P in a nearshore marine sediment. Geochim Cosmochim Acta 45:207–216

Lewis RS, DiGiacomo-Cohen M (2000) A review of the geologic framework of the Long Island Sound basin, with some observations relating to postglacial sedimentation. J Coast Res 16:522–532

Luck J-M, Turekian KK (1983) $^{187}Os/^{186}Os$ in manganese nodules and the Cretaceous-Tertiary boundary. Science 222:613–615

Lugolobi F, Varekamp JC, Thomas E, Buchholtz ten Brink MR (2004) The use of stable carbon isotopes in foraminiferal calcite to trace changes in biological oxygen demand in Long Island Sound. In: Proceedings of the 6th biennial Long Island Sound meeting, Groton, CT, October 2002, pp 47–51

Lyons WB, Fitzgerald WF (1980) Trace metal fluxes to near-shore LIS sediments. Mar Poll Bull 11:157–161

Mackin JE, Aller RC (1984) Ammonium adsorption in marine sediments. Limnol Oceanogr 29:250–257

Martens CS, Berner RA (1974) Methane production in the interstitial waters of sulfate-depleted marine sediments. Science 185(4157):1167–1169

Martens CS, Berner RA, Rosenfeld JK (1978) Interstitial water chemistry of anoxic Long Island Sound sediments, 2: nutrient regeneration and phosphate removal. Limnol Oceanogr 23:605–617

Matsumoto E (1975) Th-234-U-238 radioactive disequilibrium in the surface layer of the oceans. Geochim Cosmochim Acta 39:205–212

McCaffrey RJ, Thomson J (1980) A record of the accumulation of sediment and trace elements in a Connecticut salt marsh. In: Saltzman B (ed) Estuarine physics and chemistry: studies in Long Island Sound. Advances in Geophysics, vol 22. Academic, New York

Mecray EL, Buchholtz ten Brink MR (2000) Contaminant distribution and accumulation in the surface sediments of Long Island Sound. J Coast Res 16(3):575–590

Meyers PA (1994) Preservation of elemental and isotopic source identification of sedimentary organic matter. Chem Geol 144:289–302

Michaud E, Desrosiers G, Aller RC, Mermillod-Blondin F, Sundby B, Stora G (2009) Spatial interactions in the *Macoma balthica* community control biogeochemical fluxes at the sediment-water interface and microbial abundances. J Mar Res 67:43–70

Michaud E, Aller RC, Stora G (2010) Sedimentary organic matter distributions, burrowing activity, and biogeochemical cycling: natural patterns and experimental artifacts. Estuar Coast Shelf Sci 90:211–234

Middleburg JJ, Nieuwenhuize J, Iverson N, Hogh N, DeWilde H, Helder W, Seifert R, Christof O (2002) Methane distribution in European tidal estuaries. Biogeochemistry 59:95–119

Moore J, Galvin Gutierrez E, Mecray EL, Buchholtz ten Brink MR (2006) Physical properties of Long Island Sound sediment cores. USGS Open File Report 02-372, USGS, Washington, DC

Morse JW, Rickard D (2004) Chemical dynamics of sedimentary acid volatile sulfide. Environ Sci Technol 38:131a–136a

Murchelano RA, Brown C (1970) Heterotrophic bacteria in Long Island Sound. Mar Biol 7:1–6

Pantoja S, Lee C (1999) Molecular weight distribution of proteinaceous material in Long Island Sound sediments. Limnol Oceanogr 44:1323–1330

Pearson TH, Rosenberg R (1978) Macrobenthic succession in relation to organic enrichment and pollution of the marine environment. Oceanogr. Mar. Biol. Ann. Rev. 16: 229–311

Poppe LJ, Knebel HJ, Mlodzinska ZJ, Hastings ME, Seekins BA (2000) Distribution of surficial sediment in Long Island Sound and adjacent waters: texture and total organic carbon. J Coast Res 16:567–574

Premuzic ET, Benkovitz CM, Gaffney JS, Walsh JJ (1982) The nature and distribution of organic matter in the surface sediments of world oceans and seas. Organic Geochem 4:63–77

Raiswell R, Canfield DE (1998) Sources of iron for pyrite formation in marine sediments. Am J Sci 298:219–245

Reaves CM (1986) Organic matter metabolizability and calcium carbonate dissolution in nearshore marine muds. J Sed Petrol 56:486–494

Reeburgh WS (2007) Oceanic methane biogeochemistry. Chem Rev 107(2):486–513

Reid RN, Frame AB, Draxler AF (1979) Environmental baselines in Long Island Sound, 1972–1973. NOAA National Marine Fisheries Service technical report SSRF-738US, Department of Commerce

Renfro A (2010) Particle-reactive radionuclides (^{234}Th, ^{7}Be and ^{210}Pb) as tracers of sediment dynamics in an urban coastal lagoon (Jamaica Bay, NY). PhD Dissertation, Stony Brook University, Stony Brook, New York, p 308

Rhoads DC, Boyer LF (1982) The effects of marine benthos on physical properties of sediments: a successional perspective. In: McCall PL, Tevesz MJS (eds) Animal sediment relations: the biogenic alteration of sediments. Plenum Press, New York, pp 3–52

Rhoads DC, Germano JD (1982) Characterization of organism-sediment relations using sediment profile imaging: an efficient method of remote ecological monitoring of the seafloor (Remots™ system). Mar Ecol Prog Series 8:115–128

Rhoads DC, Germano JD (1986) Interpreting long-term changes in benthic community structure: a new protocol. Hydrobiol 142:291–308

Rhoads DC, McCall P, Yingst J (1978) The effect of disturbance on the ecology of the estuarine seafloor. Amer Scientist 66(5):577–587

Ruttenberg KC, Berner RA (1993) Authigenic apatite formation and burial in sediments from non-upwelling, continental margin environments. Geochim Cosmochim Acta 57:991–1007

Sanders HL (1956) Oceanography of Long Island Sound, 1952–1954:X—the biology of marine bottom communities. Bingham Oceanogr Collect 15:346–414

Serafy DK, Hartzband DJ, Bowen M (1977) Army engineer waterways experiment station. Technical report D-77-6, Vicksburg, Mississippi, November 1977 (11 Tables, 4 Appendices, 11 Figs, 96 Refs), p 238

Skoog A, Hall POJ, Julth S, Paxeus N, Van Der Loeff R, Westerlund S (1996) Early diagenetic production and sediment-water exchange of fluorescent dissolved organic matter in the coastal environment. Geochim Cosmochim Acta 60:3619–3629

Sun MY, Dai J (2005) Relative influence of bioturbation and physical mixing on degradation of bloom-derived particulate organic matter: clue from microcosm experiments. Mar Chem 96:201–218

Sun MY, Torgersen T (2001) Adsorption-desorption reactions and bioturbation transport of ^{224}Ra in marine sediments: a one-dimensional model with applications. Mar Chem 74:227–243

Sun MY, Aller RC, Lee C (1991) Early diagenesis of chlorophyll a in Long Island Sound sediments: A measure of carbon flux and particle reworking. J Mar Res 49:379–401

Sun MY, Lee C, Aller RC (1993) Anoxic and oxic degradation of 14C-labeled chloropigments and a 14C-labeled diatom in Long Island Sound sediments. Limnol Oceanogr 7:1438–1451

Sun MY, Aller RC, Lee C (1994) Spatial and temporal distributions of sedimentary chloropigments as indicators of benthic processes in Long Island Sound. J Mar Res 52:149–176

Sun MY, Wakeham SG, Lee C (1997) Rates and mechanisms of fatty acid degradation in oxic and anoxic coastal marine sediments. Geochim Cosmochim Acta 61:341–355

Sun MY, Wakeham SG, Aller RC, Lee C (1998) Impact of seasonal hypoxia on diagenesis of phytol and its derivatives in Long Island Sound. Mar Chem 62:157–173

Sun MY, Aller RC, Lee C, Wakeham SG (1999) Enhanced degradation of algal lipids by benthic macrofaunal activity: effect of *Yoldia limatula*. J Mar Res 57:775–804

Sun MY, Aller RC, Lee C, Wakeham SG (2002) Effects of oxygen and redox oscillation on degradation of cell-associated lipids in surficial marine sediments. Geochim Cosmochim Acta 66:2003–2012

Thamdrup B, Fossing H, Jorgensen BB (1994) Manganese, iron, and sulfur cycling in a coastal marine sediment, Aarhus Bay, Denmark. Geochim Cosmochim Acta 58(23):5115–5129

Thomas E, Gapotchenko T, Varekamp JC, McCray E, Buchholtz ten Brink MR (2000) Benthic foraminifera and environmental changes in Long Island Sound. J Coast Res 16(3):641–655

Thomson J, Turekian KK, McCaffrey RJ (1975) The accumulation of metals in and release from sediments of Long Island Sound. In: Cronin LE (ed) Estuarine research, vol 1. Academic, New York

Torgersen T, Turekian KK, Turekian VC, Tanaka N, DeAngelo E, O'Donnell J (1996) [224]Ra distribution in surface and deep water of Long Island Sound: sources and horizontal transport rates. Cont Shelf Res 16:1545–1559

Torgersen T, DeAngelo E, O'Donnell J (1997) Calculations of horizontal mixing rates using [222]Rn and the controls on hypoxia in western Long Island Sound, 1991. Estuaries 20:328–345

Turekian KK (1982) Potential of [187]Os/[186]Os as a cosmic versus terrestrial indicator in high iridium layers of sedimentary strata. In: Silver LT, Schultz PH (eds) Geological implications of impacts of large asteroids and comets on the Earth. Geological Society of America Special Paper, vol 190, pp 243–249

Turekian KK, Cochran JK, Benninger LK, Aller RC (1980) The sources and sinks of nuclides in Long Island Sound. In: Saltzman B (ed) Estuarine physics and chemistry: studies in Long Island Sound: advance in geophysics, vol 22. Academic Press, New York

Turekian KK, Tanaka N, Turekian VC, Torgersen T, DeAngelo EC (1996) Transfer rates of dissolved tracers through estuaries based on [228]Ra: a study of Long Island Sound. Cont Shelf Res 16:863–873

Turekian KK, Sharma M, Williams Gordon G (2007) The behavior of natural and anthropogenic osmium in the Hudson River-Long Island Sound estuarine system. Geochim Cosmochim Acta 71:4135–4140

Valente R, Cuomo C (2001) Investigations into the cause of American lobster mortality in Long Island Sound: REMOTS[TM] sediment-profile imaging and water quality monitoring from August to November 2000. Final report USEPA, Long Island Sound office, vol 1 and 2, p 55

Valente RM, Cuomo C (2005) Did multiple sediment-associated stressors contribute to the 1999 lobster mass mortality event in western Long Island Sound, USA? Estuaries 28(4):529–540

Van Capellan P, Gaillard JF (1996) Biogeochemical dynamics in aquatic sediments. In: Lichtner PC, Steefel CI, Oelkers EH (eds) Reactive transport in porous media: reviews in mineralogy, vol 34. Mineralogical Society of America, Washington

Varekamp JC, Thomas E, Buchholtz ten Brink M, Altabet MA, Cooper S (2003a) Environmental change in Long Island Sound in the recent past: eutrophication and climate change. In: Proceedings of the 3rd Long Island Sound lobster health symposium. CT sea grant publication 03–02, 12–14

Varekamp JC, Kreulen B, Buchholtz ten Brink MR, Mecray EL (2003b) Mercury contamination chronologies from Connecticut wetlands and Long Island Sound sediments. Environ Geol 43:268–282

Waldbusser GG, Marinelli RL, Whitlatch RB, Visscher PR (2004) The effects of infaunal biodiversity on biogeochemistry of coastal marine sediments. Limnol Oceanogr 49:1482–1492

Westrich JT, Berner RA (1984) The role of sedimentary organic matter in bacterial sulfate reduction: the *G* model tested. Limnol Oceanogr 29:236–249

Whiticar MJ, Faber E, Schoell M (1986) Biogenic methane formation in marine and freshwater environments: CO_2 reduction vs. acetate fermentation—isotope evidence. Geochim Cosmochim Acta 50(5):693–709

Williams G, Marcantonio F, Turekian KK (1997) The behavior of natural and anthropogenic osmium in Long Island Sound, an urban estuary in the eastern U.S. Earth Planet Sci Letts 148: 341–347

Williams G, Turekian KK (2002) Atmospheric supply of osmium to the oceans. Geochim Cosmochim Acta 66:3789–3791

Wu H, Green M, Scranton MI (1997) Acetate cycling in the water column and surface sediment of Long Island Sound following a bloom. Limnol Oceanogr 42:705–713

Yang L, Li X, Crsius J, Jans U, Melcer ME, Zhang P (2007) Persistent chlordane concentrations in Long Island Sound sediment: implications from chlordane, ^{210}Pb and ^{137}Cs profiles. Environ Sci Technol 41:7723–7729

Yingst JY (1978) Patterns of micro- and meiofaunal abundance in marine sediments, measured with the adenosine triphosphate assay. Mar Biol 47:41–54

Yingst JY, Rhoads DC (1978) Sea floor stability in central Long Island Sound: part II—biological interactions and their potential importance for seafloor erodibility. In: Wiley MA (ed) Estuarine interactions. Academic, New York

Yingst JY, Rhoads DC (1980) The role of bioturbation in the enhancement of bacterial growth rates in marine sediments. In: Tenore KR, Coull BC (eds) Marine benthic dynamics. University of South Carolina Press, Columbia

Zengler K, Richnow HH, Rosselló-Mora R, Michaelis W, Widdel F (1999) Methaneformation from long-chain alkanes by anaerobic micororganisms. Nature 401:266–269

Zhu Q, Aller RC, Fan Y (2006) A new ratiometric, planar fluorosensor for measuring high resolution, two-dimensional pCO_2 distributions in marine sediments. Mar Chem 101:40–53

Chapter 5
Metals, Organic Compounds, and Nutrients in Long Island Sound: Sources, Magnitudes, Trends, and Impacts

Johan C. Varekamp, Anne E. McElroy, John R. Mullaney
and Vincent T. Breslin

5.1 Introduction

Long Island Sound (LIS) is a relatively shallow estuary with a mean depth of 20 m (maximum depth 49 m) and a unique hydrology and history of pollutant loading. These factors have contributed to a wide variety of contamination problems in its muddy sediments, aquatic life, and water column. The LIS sediments are contaminated with toxic compounds and elements related to past and present wastewater discharges and runoff. These include nonpoint and stormwater runoff and groundwater discharges, whose character has changed over the years along with the evolution of its watershed and industrial history. Major impacts have resulted from the copious amounts of nutrients discharged into LIS through atmospheric deposition, domestic and industrial waste water flows, fertilizer releases, and urban runoff. All these sources and their effects are in essence the result of human presence and activities in the watershed, and the severity of pollutant loading and their impacts generally scales with total population in the watersheds surrounding LIS. Environmental legislation passed since the mid-to-late 1900s (e.g., Clean Air Act, Clean Water Act) has had a beneficial effect, however, and contaminant loadings for many toxic organic and inorganic chemicals and nutrients have diminished over

J. C. Varekamp (✉)
Earth and Environmental Sciences, Wesleyan University, Middletown, CT 06459, USA
e-mail: jvarekamp@wesleyan.edu

A. E. McElroy
School of Marine and Atmospheric Sciences, Stony Brook University,
Stony Brook, NY 11794, USA

J. R. Mullaney
US Geological Survey, East Hartford, CT 06108, USA

V. T. Breslin
Science Education and Environmental Studies, Southern Connecticut State University,
New Haven, CT 06515, USA

J. S. Latimer et al. (eds.), *Long Island Sound*, Springer Series on
Environmental Management, DOI: 10.1007/978-1-4614-6126-5_5,
© Springer Science+Business Media New York 2014

the last few decades (O'Shea and Brosnan 2000; Trench et al. 2012; O'Connor and Lauenstein 2006; USEPA 2007). Major strides have been made in reducing the inflow of nutrients into LIS, but cultural eutrophication is still an ongoing problem and nutrient control efforts will need to continue. Nonetheless, LIS is still a heavily human impacted estuary (an "Urban Estuary," as described for San Francisco Bay by Conomos 1979), and severe changes in water quality and sediment toxicity as well as ecosystem shifts have occurred since the European colonization in the early 1600s (Koppelman et al., 1976). The Sound has seen the most severe environmental changes over the last 400 years during its 10,000 year history (Lewis, this volume), suggesting that human impacts have overwhelmed the natural forces at play.

The main rivers that discharge into LIS are the Housatonic and Connecticut Rivers on the north, and the Thames River at the northeastern end of LIS, with the Quinnipiac and several other smaller rivers also coming in from Connecticut. The East River, a tidal strait connecting with New York Harbor through the heart of the New York City metropolitan region, is at the head of LIS at its western boundary. The Housatonic, Quinnipiac, Connecticut, and Thames river basins drain agricultural, urban, and industrial lands in a watershed that extends from Connecticut north to Canada. The Sound receives contaminants from many sources within and outside its contributing watershed, including direct discharges from coastal industries, wastewater treatment facilities (WWTF), urban runoff, and atmospheric deposition. New England has a long history of industrial activity, with factories that once crowded its riverbanks and shores now having succumbed to economic forces that drove manufacturing overseas. Relict deposits with legacy pollutants in upland sediments persist and combined with modern runoff sources from an increasingly densely populated watershed, continue to be a source of contaminants for LIS. While toxic exposure from legacy and active sources has diminished over the years as wastewater treatment has improved and industries are closed or moved away, pockets of contamination still have consequences for many embayments and coves, particularly near urbanized areas of western LIS.

The loading of nutrients and carbon has been of recent concern in LIS because of the extensive impacts observed since the mid-1980s. Excess nutrients not only create inhospitable conditions for higher forms of aquatic life through reduced oxygen levels and disrupting trophic dynamics, but also by altering the local biogeochemistry. As a result, the release of toxic substances into the water column may be enhanced in hypoxic waters, thus exerting a toxic effect or enhancing incorporation of toxic pollutants into the food web, and exacerbating bioaccumulation and biomagnification in tissues of various species, including those consumed by humans. This combined human impact is not unique to LIS, but common to many urban estuaries worldwide.

Urban estuaries worldwide are beset by a host of environmental problems. Environmental problems of other US urban estuaries were reviewed in detail, e.g., for the Chesapeake Bay (e.g., Smith 2003; Boesch et al. 2001; Cooper and Brush 1993; Kemp et al. 2005; Jackson et al. 2002); Puget Sound (e.g., Puget Sound Partnership 2011, http://www.psp.wa.gov/scienceupdate.php); San Francisco Bay (e.g., Kuivila and Foe 1995; Flegal et al. 1996); as well as from LIS (Mitch and Anisfeld 2010).

This chapter reviews and synthesizes the sources and distribution of contaminants and pollutants in LIS, and some of their impacts on the biosphere. We use the terms "contaminants" (substances or elements that occur above their local natural background) and "pollutants" (contaminants whose elevated concentrations have an impact on the local ecosystem) as originally accepted by GESAMP (http://www.gesamp.org/) interchangeably throughout the text. This chapter starts with a discussion of the sources of the various pollutants, followed by a section on metal concentrations in the sediment of LIS basin-wide, and a short section on metal concentrations in embayments and harbors. The next section reviews and discusses organic pollutants, sediment toxicity, and pollutant concentrations in biota, followed by a section that reviews the nutrient fluxes into LIS and their variations over time, and nutrient concentrations in the water column. The last section reviews core data for carbon and nitrogen, and puts the current data into a historical context. The chapter ends with recommendations for further research and data needs.

5.2 Data Sources

5.2.1 Metals and Organic Compounds

A limited number of peer-reviewed papers have been published on contaminants in LIS, the most recent by Mitch and Anisfeld (2010). Contaminants data are available in many governmental and technical reports (e.g., EPA, NOAA, and Brownawell et al. 1991). Mitch and Anisfeld (2010) reviewed the literature and available databases on metal and organic contaminant levels in sediment and biota in LIS. They discussed discrepancies among the data sources, examined trends over the last few decades, compared sediment contaminant levels with benchmark levels for toxicity, and provided an overview of contaminant issues within the region.

Metal pollutants in LIS sediment were discussed by Mecray and Bucholtz ten Brink (2000) and Varekamp et al. (2000, 2003, 2004, 2005), and studies on metal contamination in the fringing saltmarsh sediments were provided by Cochran et al. (1998) and Varekamp (1991). Buchholtz ten Brink et al. (2000) discussed the presence of WWTF effluents in LIS sediment using spores of *Clostridium perfringens*, and Turekian and co-workers traced WWTF effluents using Osmium and its isotopic composition (Cuomo et al. Chap. 4, in this volume).

The primary data sources on contaminants in LIS come from federally funded monitoring programs. These programs generally focus on determining concentrations of organic contaminants in sediments, and resident mussels, and/or measures of sediment toxicity. The programs initially reported levels of a suite of metals, chlorinated pesticides, polycyclic aromatic hydrocarbons (PAHs), polychlorinated biphenyls (PCBs), and in some cases polychlorinated dibenzodioxins and furans and polybrominated diethyl ethers (PBDEs). NOAA's National Status and Trends (NS&T) Program monitors levels of organic and inorganic contaminants in surface sediments and blue mussels (*Mytilus edulis*) collected from coastal embayments during the winter, in efforts commonly referred to as the Benthic Surveillance and Mussel Watch Programs, respectively. The NS&T program provides the longest dataset collected on a regular basis throughout US coastal areas since 1986, with the most recent available data online from 2008. There are nine NS&T sites within LIS, although samples were not consistently collected from all sites each year. NS&T program data were reviewed by O'Connor and Lauenstein (2006) and most recently by Kimbrough et al. (2009).

The USEPA's National Coastal Assessment (NCA) collected surface sediments from a large number of sites during each summer between 2000 and 2006. There were 488 discrete sediment samples collected from the Connecticut and New York waters covered in the NCA. About 50 of the Connecticut stations were in LIS or in tidal portions of rivers feeding into LIS; about half of the 50 New York NCA sites were located in LIS, mostly located in coastal embayments on the New York side of the western sections of LIS. The NCA measured contaminant concentrations in sediments and performed amphipod (*Ampelisca abdita*) sediment toxicity tests on the collected sediments. The NCA data collected between 2000 and 2002 were reviewed in a report on the entire national program (USEPA 2008) and in a more focused report on LIS (USEPA 2007). Data up through 2006 are available online at http://www.epa.gov/emap/nca/html/regions/ne0006/index.html.

An earlier, large survey of organic contamination and effects was conducted within USEPA's Environmental Monitoring and Assessment Program (EMAP) during the early 1990s (Paul et al. 1999). Monitoring programs for the New York–New Jersey Harbor Estuary Program conducted by NOAA in 1991 and 1993 as part of the NS&T program were reviewed by Long et al. (1995). Other surveys were conducted in 1993, 1998, 2003, and 2008, as part of USEPA, Regional EMAP or REMAP programs, of which the first two were discussed by Adams et al. (1998) and Adams and Benyi (2003). During those surveys, surface sediments were collected for analysis of metals and organic pollutants, and amphipod sediment toxicity tests were conducted. A special issue of the journal *Estuaries* (*Estuaries* 14(3), September 1991) reviewed many aspects of contaminant loadings, distributions, and effects on selected biota in LIS.

The USGS conducted an extensive sampling campaign to evaluate the distribution of trace metals and elements in surface sediments and selected sediment cores in 1996. Their reports and analyses of sedimentary properties were published in a special issue of the *Journal of Coastal Research* (Knebel et al. 2000). The USGS compiled a database of LIS sediment properties and contaminant levels, which is available online (Mecray et al. 2003). A map was compiled by Mitch and Anisfeld (2010; Fig. 5.1) of sampling locations from these national programs as well as for sites sampled for dredged material management purposes housed in the Sediment Quality Information database (SQUID) maintained by the Connecticut Department of Energy and Environmental Protection (CTDEEP). Also indicated in Fig. 5.1 are the boundaries among the three basins of LIS, Western LIS (WLIS) extending from the East River to a line between the mouth of the Housatonic River in Connecticut and Port Jefferson Harbor in New York, central LIS (CLIS) extending from WLIS to

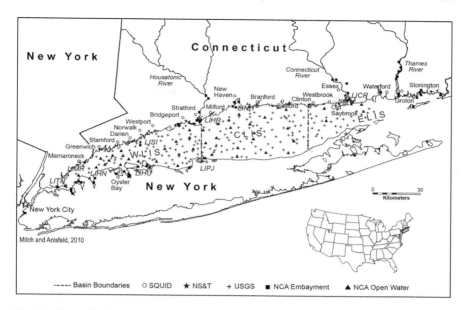

Fig. 5.1 Map of LIS showing the three major basins: western LIS (WLIS), central LIS (CLIS), and eastern LIS (ELIS), and the sampling locations within LIS sampled by several large national and local monitoring programs including the State of Connecticut Sediment Quality Information Database (SQUID), the National Status and Trends Program (NS&T) of NOAA, the National Coastal Assessment (NCA) of the USEPA. Two types of sites, embayment, and open water, are shown for the NCA study. Abbreviations for the NS&T sites from west to east are as follows: Throgs Neck (LITN), Mamaroneck (LIMR), Hempstead Harbor (LIHH), Huntington Harbor (LIHU), Sheffield Island (LISI), Port Jefferson (LIPJ), Housatonic River (LIHR), New Haven (LINH), and Connecticut River (LICR). Map from Mitch and Anisfeld 2010)

a line between Clinton, CT and Mattituck, NY, and eastern LIS (ELIS), the area east of CLIS out to The Race between Plum and Fishers Islands.

We reviewed these primary data sources on metals and organic contaminants in LIS surface sediment, including a treatment of the historical record of contamination based on sediment core data from LIS, as well as from surface sediment samples taken over the last 25 years. Body burdens of contaminants in LIS biota were reviewed within the context of the limited data available on toxicity benchmarks, data on sediment toxicity tests discussed, as well as the very limited data available on sublethal effects observed in resident biota. Data from original research reports were also included in our analysis, e.g., an analysis of recent data on the metal distribution in several Connecticut embayments (Conklin 2008; Church 2009; Lee 2010; Titus 2003). The record of mercury contamination is discussed in some detail because of the abundance of studies and data on this element in the LIS region. It should be noted that chemical contaminant data are reported in various ways, usually normalized to the dry weight of sample (either tissue or sediment) as a mass unit per unit dry weight (dw). Although particularly for tissue samples, values are often reported normalized to wet weight (ww) or sometimes lipid weight (lw), as body burdens of organic contaminants can be strongly influenced by lipid content of an organism (McElroy et al. 2011). When not otherwise indicated, the reader can assume concentrations are expressed on a basis of dry weight.

5.2.2 Nutrients and Related Pollutants

Macronutrient (N, P, Si) sources into LIS have been the topic of many state and government reports, and are a major focus of the National Estuary Program's Long Island Sound Study (LISS). These data have been used to establish the Total Maximum Daily Load (TDML) as a means to improve LIS water quality, especially the hypoxia in WLIS (USEPA 2000; NYSDEC and CTDEP 2000). Both N and P have received increasing attention and control as a national priority for management under the Clean Water Act. Despite meaningful reductions in WWTF N fluxes into the Sound (36 % overall reduction by 2011 compared to 1990) each summer, portions of CLIS and most of WLIS are hypoxic from as early as June to as late as September (Fig. 5.2a).

Anthropogenic sources of nutrients and organic carbon-based biological oxygen demand (BOD) are primary drivers of cultural eutrophication that has significantly disrupted the LIS ecosystem. After the massive fish kills of the 1970s and 1980s, more subtle ecosystem changes have been documented (Lopez et al. Chap. 6, in this volume). A short overview on sediment burdens of organic carbon, carbon sources, and carbon isotope data complements the more in-depth analysis of nutrient loading. Concentrations of N species and isotopes in LIS sediment and water-column samples are considered as well. Water quality monitoring data from the USGS and CTDEEP for macronutrients and organic carbon at stream gaging stations throughout the Connecticut watershed were discussed by Sprague et al. (2009). Here, we provide an in-depth analysis of the 40-year history of these element fluxes from the main rivers that discharge into LIS as well as from the WWTFs that discharge directly into LIS from data compiled by CTDEEP for their Nitrogen Credit Exchange Program, and discuss the potential implications for local primary productivity and carbon storage in the LIS sedimentary system. We also summarize some data from recent theses related to nutrient dynamics and impacts in LIS (Boon 2008; Andersen 2005; Lugolobi 2003).

5.3 Contaminant Sources to LIS Sediments

5.3.1 Metal Loading to Sediments

Many point and nonpoint source discharges contribute metals to LIS sediment. A review of sources of toxic contaminants to LIS by the Long Island Sound Study (1994) indicated that upstream riverine sources contributed most to delivery of the

Fig. 5.2 A: Frequency of hypoxia in LIS from 1991 to 2011 (From CTDEEP) 2B: Geography of the LIS watershed, with the main river basins feeding into LIS, locations of major permitted discharges, and WWTFs that discharge directly or indirectly into LIS and bars of population growth over the period 1750–2000 (in 50 year increments – yellow bar is 1750–1800, pale blue bar 1800–1850, and so on). From: The United States Geological Survey http://wo odshole.er.usgs.gov/project-pages/longislandsound/ Research_Topics/ Contaminants.htm

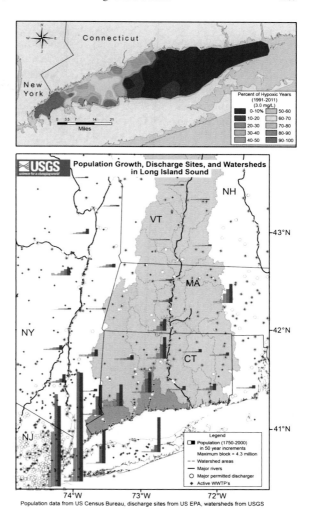

majority of toxic metals to LIS, except for lead (Pb) which has a major source from urban runoff. Significant sources of copper (Cu) and zinc (Zn) once existed along the Housatonic River and along some embayments (e.g., Bridgeport and New Haven). Connecticut was the center for brass production in the 1700–1800s, with Waterbury and Naugatuck as the two leading centers (Weigold and Pillsbury, Chap. 1, in this volume; Varekamp et al. 2005). Brass sheet metal and wire as well as clocks, buttons, and weaponry were all made in the Brass Valley of the Naugatuck River watershed. According to Andersen (2004), by the early 1900s, nearly 4,000 manufacturers were active in Connecticut, with 381 factories lining the rivers of the Brass Valley.

The Remington Company in Bridgeport has made ammunition and weapons in extensive factories since 1867, and may have been an important source of Cu, Zn, and Pb to the local watershed. Eli Whitney's development of inter-changeable parts in the manufacturing of arms was a boon to the arms industry in Connecticut, along with the success of his Whitney Armory, which became the Whitney Arms Company in 1863 and provided arms for the American Civil War. Parker's Meriden Machine Company was also under Union contract to produce 10,000 repeating rifles and 15,000 Springfield rifles. Whitney Arms was eventu-ally sold to the Winchester Repeating Arms Company in New Haven, which was active in the 1800s and closed only recently. Other plating and metal industries were active in Hartford and Springfield (MA), and discharged part of their waste into the Connecticut River. Meriden and Wallingford (CT) were known as "silver cities" due to the large number of sterling flatware, hollowware, and related silver (Ag) plated and other products that were manufactured there by companies such as International Silver, Wallace Silversmith, and Meriden Cutlery. Connecticut has been and continues to be a center of the metal industry, and waste fluids rich in met-als have been discharged over the centuries into the main rivers draining into LIS.

Danbury and surrounding towns (Bethel) were the center of the hat-making industry, with the town of Norwalk as a second focus. The manufacturing of felt, initially from beaver fur and later from rabbit fur, involves sprinkling a solution of mercury (Hg) nitrate on the fur before it is hot pressed into felt and then formed into hats. The Hg-nitrate apparently makes the larger hairs crinkly and they wrap around the finer hairs to make a dense mat of felt. The subsequent heating process decomposes the Hg-nitrate again, leading both to the Hg exposure of hat-makers as well as constituting a significant source of environmental elemental Hg con-tamination. The resulting felt hat is relatively mercury-free and presumably almost all Hg used in the process is thus lost to the environment. The elemental Hg was possibly oxidized and then became attached to particles, providing a reservoir of strongly contaminated sediment in the uplands in Danbury around the former hat-making factories and in the Still River and Housatonic River. Similarly, in Norwalk several hat-making factories existed in the old center around Water Street and provided contaminated upland soils that were eroded over time and trans-ported into the Norwalk River. A second atypical source of concentrated mercury pollution possibly existed in the HELCO power plant in Hartford, CT that used mercury vapor (instead of steam) as the working fluid in the plant. Spills during recharge and accidents may have provided mercury to Connecticut River sediment (Varekamp 2011). Other, lesser Hg sources relate to the use of Hg bearings in some WWTFs and discharges from companies making switches and thermostats.

Most WWTFs (locations shown in Fig. 5.2b) process a variety of waste water streams, including sewage that contains human fecal matter, industrial waste flu-ids, and liquid household waste. Studies of suspended solids in WWTF fluids and in sludges report high levels of Ag, Hg, Zn (Scancar et al. 2000; Sañudo-Wilhelmy and Flegal 1992) and a variety of other elements and compounds (e.g., Osmium; see Cuomo et al., Chap. 4, in this volume), reflecting the omnipresence of many metal-lic elements in common household and hospital products. Particularly, concen-trations of Cu, cadmium (Cd), Zn, Pb, chromium (Cr), and Hg can be high. After the 1988 federal law that banned ocean dumping of sewage sludge, currently most sludge from around LIS is disposed of through incineration (CT) or transformed into biopellets as a fertilizer (NY) or exported for landfill storage to other states. Some fine-grained matter and dissolved fractions escape with the fluids, providing a flux of metals to LIS. In addition, runoff from urban areas may contain metals, which wash off paved areas and from contaminated soils. In some urban areas, including New York City, stormwater passes through combined sewer overflow systems (CSOs) and may contribute to relatively high metal fluxes into rivers and estuaries. Sediments in some LIS embayments are strongly contaminated with metals, usually related to local sources (e.g., Pb and Cr contamination in the sediments of the Mill River in the Southport section of Fairfield, CT, from the former Exide Battery plant) and one cove also showed low-level contamination with radioisotopes from the nuclear power industry (Benoit et al. 1999). Small patches of highly contaminated sediment removed from urban harbors around LIS occur at the dredge material disposal sites in LIS, although most of these have been capped with clean sand to minimize environmental threats (Fredette et al. 1992; Fredette and French 2004).

5.3.2 *Organic Contaminant Loadings to Sediments*

Primary sources of organic contaminants to LIS are direct inputs from WWTFs, urban runoff, CSOs, and atmospheric deposition. Riverine inputs, particularly in Connecticut and the East River, extend the range of contaminant sources, as these rivers, in addition to serving inland population centers, also historically have functioned as corridors for industrial activity. Generally only organic contaminants that are sufficiently hydrophobic to become associated with sediment particles are analyzed in monitoring programs, so very little is known about the sources or distributions of dissolved organic contaminants in LIS, and these will not be discussed in this chapter. A map of population density over the period 1750–2000, location of WWTFs, major discharges, and watersheds for the region (Fig. 5.2b) clearly shows the sources of anthropogenic contaminants in WLIS. Wolfe et al. (1991) cited LIS riverine discharges and WWTFs as primary sources of nutrients and metals, with urban runoff as the primary source for most organic contaminants and lead. They recognized atmospheric deposition as another important source, both for distribution of new organic contaminants and redistribution of more volatile compounds. In contrast, the WWTF inputs of organic contaminants were considered to be secondary to river inputs in the CCMP, but are still significant. Sources for polycyclic aromatic hydrocarbons (PAH) in the New York/New Jersey estuary were mainly urban runoff, while atmospheric deposition was a significant source for PAHs with molecular weights of <200 g/mol (Rodenburg et al. 2010). Reviews by the National Academy of Sciences (NAS 1985, 2003) showed that urban runoff was also a major source of petroleum hydrocarbons to the sea. The PCBs that enter LIS largely follow the same pathways as the PAHs; significant sources of PCBs in the upper reaches of the Housatonic River to the north and from the Hudson River to the west do not seem to reach LIS

(Pringle et al. 2011). Chlorinated pesticides such as DDT and its metabolites, DDE, dieldrin, and chlordane, although banned in the 1970s and 1980s, are still found at appreciable concentrations in upland soils and LIS sediments in some locations; these legacy sources continue to enter LIS through pathways similar to those described for PAHs and PCBs. Once in the system, particle reactive and/or hydrophobic contaminants will tend to be associated with fine, organic-rich sediments. Although all but the largest sediment-sorbed contaminants (those with log octanol water coefficients >6.5) are generally available for bioaccumulation by resident biota, only those that cannot be broken down via metabolism biomagnify to top predators through trophic interactions.

5.3.3 Macronutrients

The nutrients N and P are essential for primary productivity and Si for diatoms as well because they use silica in their skeletons. The flux of N and P into LIS has increased dramatically over the last two centuries, leading to eutrophication (e.g., Lugolobi et al. 2004) though recent management efforts have significantly reduced point source loadings of N and P from WWTFs. A direct record of current and past nutrient fluxes is provided by monitoring data of rivers (USGS) and WWTFs (CTDEEP), whereas longer term, more indirect data can be extracted from sediment core analyses for particle-reactive species (Varekamp et al. 2004). The N and P fluxes today have a strong anthropogenic component, whereas the Si fluxes are only modulated by human activity (e.g., dam building; Triplett et al. 2008) but are not a human-caused effluent (derived from rock weathering processes).

The sources of N and P include atmospheric deposition (Luo et al. 2002), nonpoint and stormwater discharge from the human-altered landscape (Mullaney et al. 2002), excess fertilizer flows, excess particulate organic carbon from the land that becomes oxidized in LIS to release its N and P, but most of all, nutrient releases from WWTFs, especially those in the highly populated environs of New York City (Fig. 5.2b). Presently, the WWTF's in New York and Connecticut with discharge into LIS and its tributaries contribute an average of 61,200 kg of N per day, down from a baseline value of 95,250 kg N/day in 1990, but still quite significant (CTDEEP 2011). The effluents of the WWTFs are monitored and a TMDL has been established to meet dissolved oxygen water quality standards in Connecticut and New York by limiting the N inputs into the Sound (CTDEP and NYSDEC 2000). The TMDL target is to reduce the input of N into the Sound by 58.5 % (relative to a ca. 1990 "average" baseline load) by 2014. Reductions are accomplished primarily by upgrading WWTF's to tertiary or advanced wastewater treatment. This process presently reduces rates of ammonium discharge at some plants by an order of magnitude and DIN concentrations by as much as 70 % (CTDEEP 2011). Additional upgrades at many plants are planned which may further increase the efficiency of the removal process.

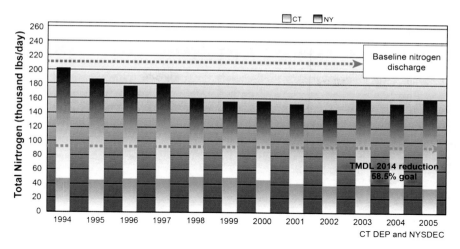

Fig. 5.3 Total N loadings of LIS from CT and NY from 1994 to 1005, showing a steady decrease over the first 9 years. Taken from CTDEP and NYSDEC data

Progress toward that total N target of 39,900 kg N/day is measured against a baseline flux of 95,250 kg N/day (Fig. 5.3). To reach the target, the phased (5 year increments) management of WWTF effluents in Connecticut and New York is emphasized and accomplished using a variety of regulatory tools including a "bubble" permit approach in New York City and N "trading" in Connecticut (CTDEEP, 2011). As of 2011, NY and CT had accomplished approximately 70 % of this goal; additional upgrades are scheduled, including several New York City plants coming online between 2012 and 2014. In terms of concentration, reductions are required from >8 mg/L total N in WWTF fluids to a final goal of 5.6 mg/L total N.

5.4 Metal Contamination

5.4.1 Spatial Patterns of Metal Contamination in LIS Surface Sediment

The absolute concentrations of contaminants in sediment are strongly influenced by grain size and organic carbon contents, and these tend to be correlated to each other in many environments as well (Windom et al. 1989). The association of metal contaminants with fine-grained organic-rich sediment thus impacts the spatial distribution of contaminants as well as the distribution of their maximum and minimum concentration values. Trace metal concentrations in LIS sediment show a wide range, caused by these sedimentary factors and variations in local anthropogenic source inputs. The mean and maximum values for a suite of trace metals, together with their preindustrial (background) values (after Mecray and Buchholtz

Table 5.1 Mean and maximum concentrations (µg/g) for metallic contaminants in LIS surface sediment (upper 2 cm).

	Ag	Cu	Cd	Hg	Pb	As	Zn	Cr	V	Ni	Ba
Mean	1.5	117	2	0.7	83	6	160	78	101	26	230
Maximum	10.1	7720	35	17	3284	61	4800	2000	160	665	445
Background	0.05	8	0.2	0.1	23	2.5	68	59	90	25	377
Mean EF	29.8	14.6	9.9	6.5	3.6	2.5	2.4	1.3	1.1	1.0	0.6

The natural background values are after Mecray and Buchholtz ten Brink 2000 and the Enrichment Factor (EF) is the mean value divided by the natural background concentration (for Arsenic few core data exist, background value estimated from the USGS surface sediment data base for relatively unpolluted sediment)

ten Brink 2000) and mean enrichment factors (EF), are shown in Table 5.1. Silver is the most strongly enriched metal (30x), followed by Cu, and then Cd and Hg. Barium, vanadium (V), and nickel (Ni) have close to natural background values. The high EF for Ag probably reflects the large inputs from WWTFs (Sañudo-Wilhelmy and Flegal 1992) and the high EF value for Cu reflects the inputs from the Housatonic River, which drains the historically industrialized Naugatuck River, and several smaller sources located along local embayments (Breslin and Sañudo-Wilhelmy 1999). Metal concentrations in LIS sediment increase with proximity to New York City, and higher concentrations are associated with fine-grained deposits that expectedly have higher surface-to-volume ratios relative to larger size fractions (Mitch and Anisfeld 2010; Mecray and Buchholtz ten Brink 2000; Greig et al. 1977; Long Island Sound Study 1993). The concentration maps for Cu and Pb exemplify this trend (Fig. 5.4), with some patchiness, but clearly higher values are found in the central and western parts of the Sound. Sediment grain size generally decreases from east to west (Knebel et al. 2000), with depositional sedimentary environments in CLIS and WLIS. To correct metal abundances for these grain size influences, we normalized the metal data on their sample Fe concentrations, which has been shown to correlate closely with grain size (Fig. 5.5). We then evaluated the resulting Fe-normalized east–west patterns for different metals.

Nickel has only modest contributions from anthropogenic sources, but non-normalized Ni concentrations show a clear increase from east to west (Fig. 5.6a). The Fe-normalized Ni concentrations show a much flatter east–west trend (Fig. 5.6b), however, indicating that Ni concentrations in LIS sediment are probably largely controlled by natural processes. In contrast, Fe-normalized trends for Cu, Hg, Ag, and Cr (Fig. 5.7a–d) show increases from east to west, indicating that metal sources increase to the west. Sources near the central and western part of the Sound are the industrial sources in the Housatonic River basin for Cu, Zn, and Hg, and urban sources along the East River (Ag, Hg). In particular, many large WWTFs that discharge directly into central and western LIS or a short distance up tributary rivers may be responsible for the Ag enrichment in sediment.

Element ratios that are characteristic of specific sources can be used to trace the metal origins. These source signals were determined from metals analyses

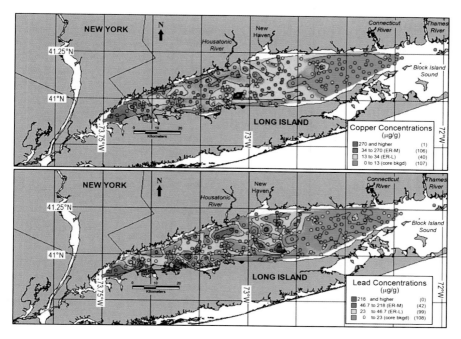

Fig. 5.4 Lead and Copper concentrations in LIS surface sediments, showing the highest concentrations in the western section of LIS, with a more patchy pattern in central LIS. Data source: http://pubs.usgs. gov/of/2000/of00-304/htmldocs/chap06/index.htm

Fig. 5.5 Relationship between sediment iron content and mean grain-size for LIS embayments (Lee 2010; Church 2009 and unpublished data). Similar results were obtained for 'open LIS' surface sediment data (Varekamp et al. 2000)

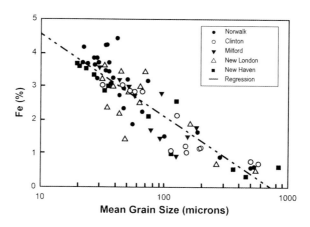

in the sediment of the Housatonic River estuary, which carries the characteristic source signals of the metal industry in the Brass Valley (Cu, Cr) and the Hg from the hat-making industry in Danbury and surrounding areas (mainly the Still River effluents). Connecticut River metals were characterized from sediment cores

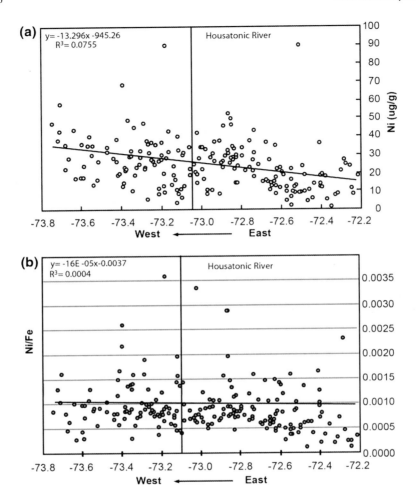

Fig. 5.6 Spatial distribution of Ni in LIS surface sediments. **a** Ni concentrations based on dry weight sediment, **b** Ni concentrations normalized to Fe concentrations. Much of the E–W trend in Ni concentrations is an artifact of the grain size differences between the east and west sections of LIS. Data source: http://pubs.usgs.gov/of/2000/of00-304/htmldocs/chap06/index.htm

taken from Chapman Pond, a small cove south of East Haddam, and Great Island in the mouth of the Connecticut River estuary. Metal concentrations in sewage sludge were estimated from sediment samples taken in the New York Bight where a large amount of sewage sludge had historically been disposed until passage of the Ocean Dumping Ban Act of 1988, which ended sludge disposal in 1991 (e.g., http://pubs.usgs.gov/fs/fs114-99/fig3.html), using correlations between sewage indicators and metal concentrations.

In addition to Cu and Zn, sediment upstream of the Housatonic River estuary has up to 7,000 mg Cr/kg (Varekamp et al. 2005), providing a Cr-rich sediment

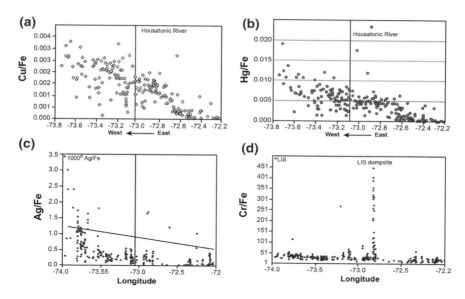

Fig. 5.7 Spatial distribution of Metal/Fe ratios in LIS surface sediments for Cu, Hg, Ag, and Cr. These elements show a significant E to W trend after normalization on Fe, (in contrast to Ni, Fig. 5.6), indicating the presence of strong metal sources along central and western LIS

source for LIS. The Cr/Fe in LIS sediment indeed increases from east to west (Fig. 5.7d). The area around the CLIS dredge material disposal site also shows high Cr/Fe and these sediments are presumably partially derived from dredged Housatonic River sediment. Many LIS sediments have Cu/Ag similar to sewage, but a group of LIS sediments with high Cu/Ag must have additional contributions from Cu-rich Housatonic River sediment. A similar pattern to Cr emerges for Cu and Zn concentrations (Fig. 5.8). Mixing calculations based on element ratios (Fig. 5.9) indicate that up to 20 % of Housatonic River sediment can explain the high Cr and Cu concentrations in many LIS sediment samples, with extremes carrying up to 40 % of the highly metal-polluted Housatonic River sediment.

5.4.2 Metal-Laden Sediments in Coastal Embayments

Harbor or embayment sediments may have much higher metal concentrations than open LIS sediment due to the restricted water circulation and the proximity to multiple sources of industrial and municipal wastewater (Breslin and Sañudo-Wilhelmy 1999; Rozan and Benoit 2001; Luoma and Phillips 1988). Most embayments are characterized by a variety of sedimentary environments

Fig. 5.8 Element ratios in LIS surface sediment and some of the sources of these metals: Cu–Ag, Cu–Zn, Cu/Cr–Cu/Zn

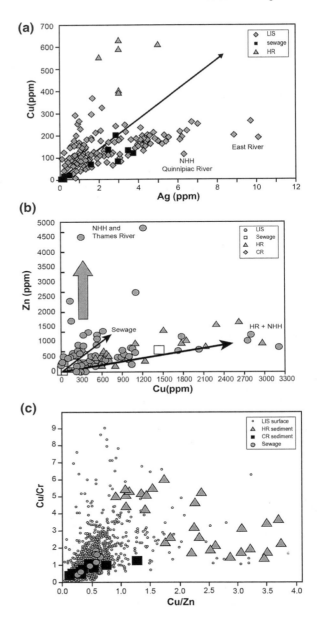

with high concentrations of Zn, Cu, Pb, and Cd (Breslin and Sañudo-Wilhelmy 1999; NOAA 1994; Rozan and Benoit 2001). The high metal loadings are a concern, particularly in embayments that support commercial and recreational shellfish industries (O'Connor 1996). The mean and median sediment Zn and

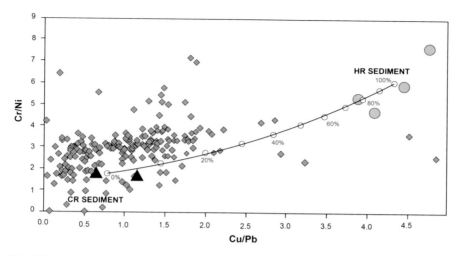

Fig. 5.9 Calculated potential contributions of Housatonic river (HR) and Connecticut River (CR) sediment to LIS metal contaminant budgets

Cu concentrations for several Connecticut embayments (Table 5.2 and Fig. 5.10) all exceed natural background values and display a much greater variability than open LIS sites. Sediment metal contents in Port Jefferson Harbor, NY for instance vary by over an order of magnitude over distances of only 50–500 m (Breslin and Sañudo-Wilhelmy 1999), suggesting highly variable sedimentary environments or the influence of local discharges. With the exception of Clinton Harbor (Cu and Zn) and the Housatonic River (median Zn), all mean and median Cu and Zn values are similar to or exceeded NOAA Effects Range Low (ERL) thresholds. For most Connecticut embayments, both mean and median sediment Zn concentrations are similar to the ERL Zn threshold. Only Clinton Harbor (below the ERL) and the Bridgeport and Norwalk Harbors (above the ERL) differ significantly from the ERL Zn threshold. Both Cu and Zn concentrations were similar to or exceed NOAA Effects Range Median (ERM) thresholds for one or more locations in four of the eight harbors examined in ELIS (New London), CLIS (New Haven), and WLIS (Bridgeport and Housatonic River; Fig. 5.10). The ERL and ERM values were set based on empirical data gathered across the country by NOAA linking sediment contaminants levels with observed toxicity, where the ERL corresponds to the 10th and the ERM the 50th percentiles, respectively, of toxicity effects (Long and Morgan 1990). Hot spots in embayments are defined as locations where sediment Cu and Zn concentrations exceed the 90th percentile values. The number of hot spots identified within Connecticut embayments range from one (Clinton, Branford, and Milford) to eight (New Haven).

Linear regression analyses of the sediment Cu and Zn concentrations within each harbor show that Cu and Zn co-vary, with regression coefficients from

Table 5.2 Comparison of copper and zinc concentrations in Connecticut harbors with previously published data and reference levels

Harbor	Stations	Copper (mg/kg)			Zinc (mg/kg)		
		Range	Mean	Median	Range	Mean	Median
New London[a]	n = 35	3.2–252	62	61	10.0–642	156	165
Clinton[b]	n = 14	0.4–49	22	13	10.1–247	108	70
Branford[c]	n = 18	17.2–148	67	67	35.7–274	159	156
New Haven[d]	n = 128	6.3–405	82	74	7.5–463	151	146
Milford[b]	n = 11	19.9–104	67	70	37.0–236	147	161
Housatonic River[e]	n = 31	6.3–685	146	93	11.3–517	159	104
Bridgeport[d]	n = 26	18.5–491	182	146	5.6–677	234	196
Norwalk[f]	n = 30	6.9–218	99	98	33.0–387	174	170
ELIS[g]	n = 302	2.8–96	53		23.5–186	110	
CLIS[g]	n = 323	8.6–185	84		43.3–221	137	
WLIS[g]	n = 453	14.8–216	116		39.2–315	183	
Effects Range Low[h]			34			150	
Effects Range Median[h]			270			410	
Natural Background[i]			8			68	

[a]Lee (2010)
[b]Church (2009)
[c]Breslin et al. (2005)
[d]Titus (2003) and unpublished data
[e]Conklin (2008)
[f]Unpublished data
[g]Data from Mitch and Anisfeld (2010) compiled from multiple sources. Range represents 10th and 90th percentile values
[h]Effects Range data from NOAA (1999)
[i]Natural background from Mecray and Bucholtz ten Brink (2000)

0.73 for Norwalk Harbor to 0.96 for Clinton Harbor. For most Connecticut harbors, the slope of the regression line defining the relationship between Cu and Zn was similar for ELIS and CLIS harbors in this study (0.22–0.49). In contrast, the slope of the regression line increased (0.58–0.81) for WLIS harbors (Bridgeport, Housatonic, and Norwalk) indicating a disproportionately higher sediment Cu content relative to Zn content for these harbors (Table 5.3). Linear regression analysis shows that Cu is strongly correlated with both sediment Loss on Ignition (LOI) (a proxy for organic matter) and sediment Fe (= Iron %) for all Connecticut harbors/rivers studied (Fig. 5.11a and b; Table 5.3). Steeper slopes of the Cu versus LOI and Cu versus Fe relationships indicate disproportionately higher source strength for Cu within WLIS river estuaries and embayments, such as the Housatonic River.

Mecray and Buchholtz ten Brink (2000) showed a significant correlation between metal concentrations and % fines in open LIS sediment, whereas the Mitch and Anisfeld (2010) analysis of the National Coastal Assessment 2000–2002 and State of Connecticut Sediment Quality Information Database (SQUID) showed no strong relationship between sediment metal concentration and grain size. The coastal embayments generally show a greater range of sediment Cu and Zn variability compared to their respective open water sediment location in LIS

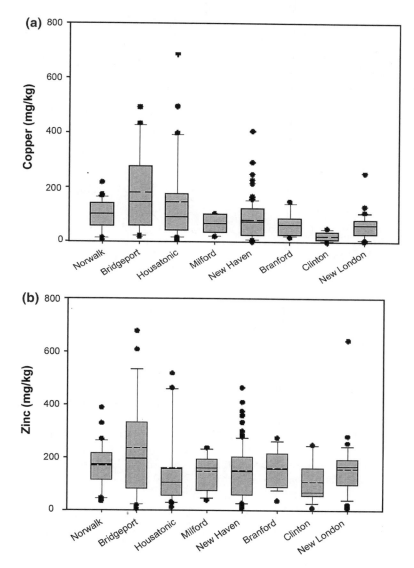

Fig. 5.10 Comparison of copper (**a**) and zinc (**b**) sediment metal concentrations in Connecticut harbors. Embayments are arranged along the X-axis from left to right, west to east. The whiskers indicate the 10th and 90th percentile values. The bottom and top of each *box* indicate the 25th and 75th percentile values, respectively. The *solid line* within the *shaded box* indicates the median concentration and the *dashed line* indicates the mean concentration. The *solid circles* represent outliers; sediment metal values higher or lower than the 90th and 10th percentiles

(Mitch and Anisfeld 2010), which is here largely explained by variations in sediment physical properties and local depositional environment. The "hot spots" are characterized by the presence of fine-grained, high-LOI sediment, and are most

Table 5.3 Coefficients of determination[a] (r^2) and p-values for testing the population slope ($\beta 1$) = 0 between sediment copper and zinc, copper and iron, and copper and loss on ignition for Connecticut embayment datasets

Harbor	Cu vs Zn			Cu vs Fe[b]			Cu vs LOI[c]		
	r^2	Slope	p	r^2	Slope	p	r^2	Slope	p
New London	0.92	0.41	<0.001	0.68	31.0	<0.001	0.32	5.1	<0.001
Clinton	0.96	0.22	<0.001	0.93	16.5	<0.001	0.95	5.7	<0.001
Branford	0.88	0.53	<0.001	0.49	29.9	0.001	0.51	5.1	<0.001
New Haven	0.74	0.49	<0.001	0.62	28.7	<0.001	0.52	12	<0.001
Milford	0.83	0.42	<0.001	0.93	30.9	<0.001	0.79	7.1	0.001
Housatonic River	0.94	0.81	<0.001	0.79	121	<0.001	0.84	39	<0.001
Bridgeport	0.88	0.71	<0.001	0.72	71.6	<0.001	0.66	25.7	<0.001
Norwalk	0.73	0.58	< 0.001	0.51	35.2	<0.001	0.68	14.9	<0.001
Port Jefferson, NY[d]	0.96	0.51	<0.001	0.74	26.6	<0.001			

[a]Coefficient of Determination (r^2) and slope estimates derived from a linear regression analysis of the dataset for each harbor in Table 5.2. The number of stations (n) analyzed for each regression analysis for each harbor is given in Table 5.2
[b] Regression analysis for sediment copper versus iron shown for each harbor in Fig. 5.11b
[c]Regression analysis for sediment copper versus loss on ignition shown for each harbor in Fig. 5.11a
[d]Coefficient of Determination (r^2) and slope estimates derived from a linear regression analysis of data available in Breslin and Sañudo-Wilhelmy (1999)

frequently located in the inner (northern) reaches of the harbors (dredged channels, river mouths, river coves) proximate to contaminant sources. Copper and Zn sediment concentrations at these "hot spot" locations in the harbors throughout LIS exceed ERL and ERM thresholds. The two main WLIS harbors (Housatonic River and Bridgeport) have disproportionately high sediment Cu concentrations compared to other LIS harbors, showing the influence of the Housatonic River inputs from the "Brass Valley" industries of the past. The correlation of high Cu and Zn with LOI may reflect metal association with particulate organic carbon, and if so, microbially mediated repartitioning and mobilization of these elements to more bioavailable forms in the solution phase, may then be a management concern.

5.4.3 Core Records of Historic Metal Contamination

Many sediment cores in LIS have been dated with radioisotopes and thus provide a time record of metal contamination in LIS (Buchholtz TenBrink, personal communication; Varekamp and Thomas 2010). Examples for several metals are shown for two cores, positioned on the A transect of the USGS coring cruise of 1996 (Fig. 5.12a). Core A4C1 is positioned in shallow water close to the Connecticut coast, whereas core A1C1 is located in the middle part of CLIS. Sediment accretion

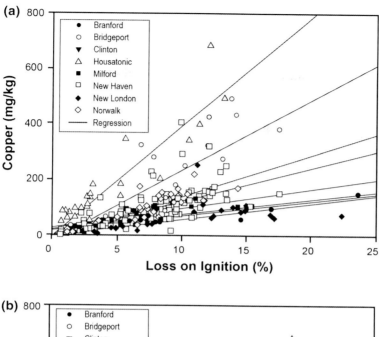

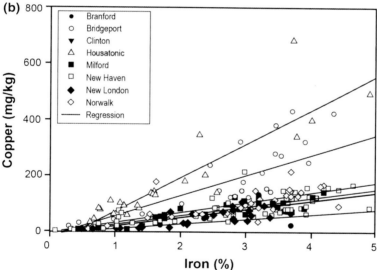

Fig. 5.11 Co-variance between sediment Cu concentrations and **a** sediment loss on ignition (LOI) and **b** sediment Fe concentrations for Connecticut harbors and embayments

rates in core A1C1 (0.6 mm/year mean sedimentation rate) are much lower than those in A4C1 (2 mm/year), providing a higher time resolution in core A4C1.

The metal profiles versus age (Fig. 5.12b) show background concentrations in precolonial times and gentle increases in concentration in the early to mid-1800s

and stronger enrichments in the 1900s. In most cores, metal concentrations
decreased between 1980 and 2010, the period of enhanced environmental control required by the Clean Water Act. The metals Ag, Cu, Zn, Cr, Hg, and Pb
show EFs over natural background values ranging from ~3 to 15. The element
Ni shows no enrichment in sediment deposited since 1850 and appears to have
no major anthropogenic contributions. A similar conclusion for Ni was obtained
earlier in the Ni/Fe values along the longitudinal transect of LIS. Core A4C1
shows a strong concentration spike in Cu, Zn, Cr, and Hg around 1955, a period
when two hurricanes hit central Connecticut. This distribution of elements is
characteristic for sediment from the Housatonic River, and this thin metal-
enriched layer was caused by a 100 year hurricane flood deposit in the central
LIS area (Varekamp et al. 2005).

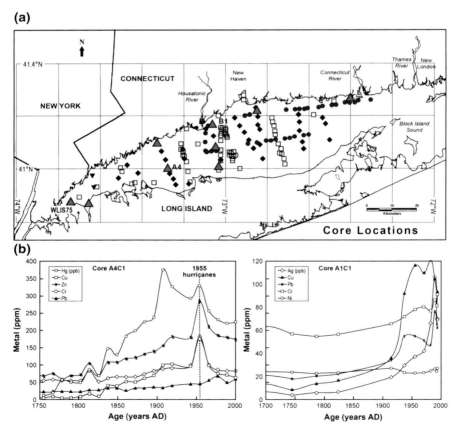

Fig. 5.12 Core profiles of metal concentration versus age in LIS cores A1C1 and A4C1 (see
index map for locations) **a** Cu, Zn, Cr, Pb, and Hg in core A4C1 (location A4, core 1) **b** Ag, Cu,
Cr, Pb, and Ni in core A1C1 (location A4, core 1)

5.4.4 Mercury Contamination in LIS

The geochemistry of mercury (Hg) as a contaminant in LIS has been studied extensively over the last 20 years. Sediment concentrations range from low ppb (ng Hg/gr dry sediment) Hg values in the east up to 800 ppb Hg in WLIS (Figs. 5.13a, 5.14a). The Hg/Fe values also trend higher going west (Fig. 5.7b), like many other metals with sources in WLIS. Most LIS core profiles for Hg show a large increase in the mid-1800s (in dated cores based on ^{210}Pb and ^{14}C ages), which correlates with the concentration increase in spores of *Clostridium perfringens* (Fig. 5.13b), a sewage indicator (Buchholtz ten Brink et al. 2000; Varekamp et al. 2000, 2003). This synchronicity between sewage input and Hg enrichment is not necessarily causal: Both relate to industrialization and the increase in population density over the last 150 years.

Besides the common far-field Hg sources such as coal and solid waste incineration, western Connecticut has a large Hg point source in the historic hat-making industry (Varekamp et al. 2005). Upland sediments around the old hat-making towns of Danbury and Norwalk have mercury concentrations in the thousands to hundred thousands ppb Hg. These sediments are remobilized in a steady state fashion at low level, but more intensely during major rain storms, hurricanes, and extended wet periods. The marshes and mudflats of the Housatonic River estuary are strongly enriched in Hg (Varekamp et al. 2005; Table 5.4), and the offshore delta deposits also show a strong spike at about the 1955 level, which is the time of major flooding in Connecticut (two hurricanes in 2 weeks; Fig. 5.14b, c). In western LIS, a core near Execution Rock (core WLIS75GGC1) shows a strongly Hg-enriched layer that is directly underlain by a coarser deposit with small coal and debris fragments (Fig. 5.14d). Most likely, this is also part of a hurricane deposit, possibly from the 1955 hurricanes, although the direct source of this Hg

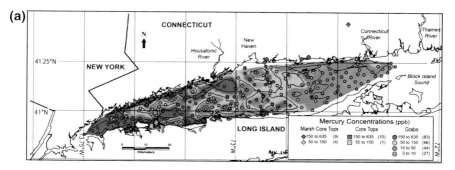

Fig. 5.13 a Mercury contamination pattern in LIS surface sediment and **b** with two core records for Hg and **c** perfringens versus depth (Core A7C1 is the most southerly coring site on the A transect of Fig. 5.12). The core concentration profiles show the strong increase in Hg and a retreat in the top of the core, whereas the C. perfringens concentrations remain high also in the core top. Data source: http://pubs.usgs.gov/of/2000/of00-304/htmldocs/chap06/index.htm

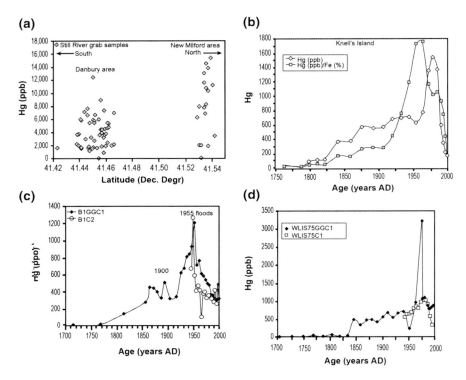

Fig. 5.14 Mercury data from the Housatonic River watershed and western LIS. **a** High mercury concentrations in the Still river that drains the main hat-making area in CT (Danbury) **b** Mercury profile of core KI from Knells Island in the Housatonic River estuary **c** Mercury profile from cores at the B1 site on the Housatonic River delta **d** Mercury profile from the WLIS75 coring site near Execution Rock in WLIS (Fig. 5.13b). All cores show evidence for a steep rise in Hg in the mid-1800s, and spikes around 1955 from hurricane activity

Table 5.4 Mercury concentrations in cores from LIS, salt marshes, and coves around LIS; the Housatonic River (HR) shows strong contamination from the uplands with the hat-making industry near Danbury, CT (after Varekamp et al. 2000, 2003)

	Hg peak range (ng Hg/gr)	Inventory excess Hg (ng/cm^2)
LIS	Up to ~800	2120–9200
Salt marshes (excluding HR)	160–470	490–3770
HR marshes	1500–7000	>5000

in the far western Sound is not known. The Connecticut River has carried Hg-rich fine-grained sediment to the Sound over the last 60–80 years, presumably from various industrial sources and possibly from an experimental power plant that used Hg as its working fluid (Varekamp 2011). Clay deposits in small coves and inlets on the river flood plain contain up to 3,000 ppb Hg (Varekamp 2011). Presumably, this fine-grained sediment was also carried into the Sound over the years, constituting a heretofore unrecognized Hg source for LIS.

Core profiles in LIS and its surrounding marshes show that sediments deposited in the 1960–1970s periods display the strongest Hg contamination. These cores show a decrease in Hg concentrations in the core tops, whereas concentrations of *C. perfringens* increased also over these last 50 years (Fig. 5.13b), because WWTF effluent discharges kept increasing in volume over this time. Besides the direct hurricane layers, many LIS and coastal salt marsh cores show a double peak in Hg concentrations, one at ~1900 and the second peak in the 1960–1970s. The late 1800s and early 1900s were a very wet period and presumably more Hg was exported from the heavily contaminated watersheds of the Housatonic and Norwalk Rivers The wet climate also possibly enhanced the "flush out" of Hg from the atmospheric reservoir (Varekamp et al. 2003).

Records of Hg deposition from ponds on Block Island, RI, east of LIS (i.e., in Block Island Sound), where the primary source of Hg is atmospheric, show a different pattern than the open LIS core records (Fig. 5.15). These Block Island records were obtained from a freshwater pond and a freshwater marsh on an island with no local Hg contamination sources (Neurath 2009) and no influx from LIS sediment. These records show only a very gentle increase in Hg concentration over the last part of the nineteenth century (from 30–60 ppb Hg). The first large increase in Hg concentrations (from 70 to 200–300 ppb Hg) and Hg accumulation rates only starts in ~1935–1940 in these cores dated with ^{210}Pb, ^{137}Cs, and ^{14}C. This relatively recent and rapid increase in Hg contamination is also found in some ice core records (Schuster et al. 2002) and other remote lake records (Perry et al. 2005), and reflects far-field atmospheric deposition of Hg. These results contrast strongly with almost all records from the open LIS basin and its coastal marshes with their steep rise in Hg in the mid-1800s. The substantial local Hg sources from the hat-making industry started at the beginning of the nineteenth century, and have influenced the Hg distribution in LIS for more than 200 years, overwhelming the variations in the strength of the far-field atmospheric input. The latter is deposited both in situ on the Sound and in the watershed, the latter focused through the watersheds and then transported through riverine sediment into LIS. If the hat-making Hg has such a large influence on the Hg distribution patterns and concentrations in most of CLIS, we have to assume that fine-grained sediment is circulated throughout the Sound with the strong tidal currents. Satellite images of the sediment plume of the Connecticut River during Hurricane Irene (September 2011) clearly show that the river sediment plume is dispersed dominantly to the west but to a lesser degree also to the east-southeast.

Wastewater treatment facilities provide a flux of Hg into the Sound and correlations with the abundance of the sewage tracer *C. perfringens* spores (Fig. 5.16) suggest that up to 25 % of total Hg in Sound sediment was derived from WWTF effluents (Varekamp et al. 2003). The hat-making Hg may have been responsible for 20–30 %, based on Cu-based mass flux constraints of Housatonic River sediment, whereas the remainder may have come in with the sediment from the Connecticut River and smaller watersheds.

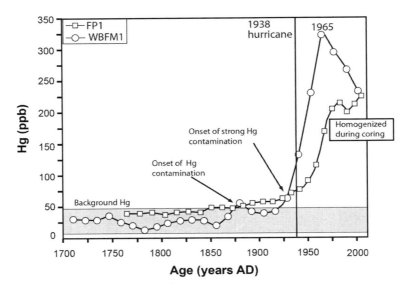

Fig. 5.15 Mercury concentrations in two bogs (FP1 and WBFM1) on Block Island (Rhode Island) to the east of the LIS map in Fig. 5.12), showing a very different pattern compared to those from LIS (see text for discussion; Fig. 5. after Neurath 2009)

A modern Hg budget for LIS was presented by Balcom et al. (2004), who considered the Hg inputs and outputs to LIS and cycling within LIS (Fig. 5.17). The inorganic pool of dissolved Hg and Hg adsorbed onto fine particulate matter from the main rivers feeding LIS (East River, Connecticut River, Thames River and Housatonic River) provides the dominant input flux (close to a 1,000 mol/year), whereas the WWTFs along the LIS coastline supply a much smaller amount (~60 mol/year). Atmospheric deposition is a sizeable source term at 130 mol/year, but rather surprising is the magnitude of the estimate of the volatile elemental Hg(0) escape from LIS into the atmosphere (~400 mol/year). The inorganic Hg that is brought in by the various sources may become reduced during aqueous bacterial reactions; this leads to a buildup of Hg^o (aq) (Rolfhus and Fitzgerald 2001; Tseng et al. 2003) that ultimately escapes to the atmosphere. The estimated flux of Hg from LIS waters back into the atmosphere is about three times as large as the estimated direct atmospheric Hg deposition rate (Balcom et al. 2004). Thus, the evasion of elemental Hg from the surface waters of LIS is a return pathway of Hg that was brought into LIS associated with particles through rivers from a variety of point and nonpoint sources. Mercury may also escape from coastal salt marsh areas through reduction and vapor evasion, but the magnitude of that flux is not well-known (Lee et al. 2000). The LIS coastal zone thus is an area with active processing of Hg, where the input consists of focused atmospheric deposition from the watershed, in situ deposition on LIS, and point source Hg in dissolved and particulate form. Bacterial reduction in the water column with subsequent evasion to the atmosphere forms an important Hg return flux.

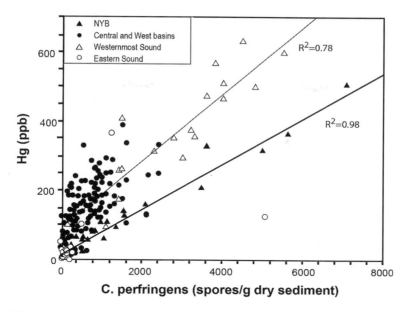

Fig. 5.16 Correlation between Hg and C. perfringens for LIS surface samples. Most data points plot above the sewage correlation line (heavy black line associated with NYB points, data from NY bight sewage dumpsite), indicating that sources other than sewage contribute to the Hg loadings in LIS

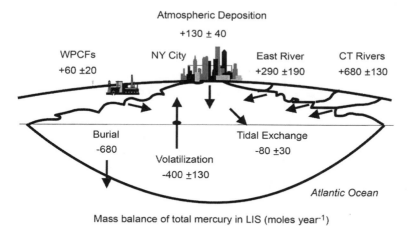

Mass balance of total mercury in LIS (moles year^{-1})

Fig. 5.17 Mercury budget for LIS based on measured fluxes from rivers, WWTFs and atmospheric deposition (after Balcom et al. 2004). The Hg revolatilization flux from LIS surface water into the atmosphere is substantial (three times the atmospheric Hg deposition flux)

Methylation and ecosystem uptake occur in LIS and marsh fringes, and methylation rates in marsh environments can be substantial (Langer et al. 2001). Fish in LIS have Hg enrichments up to 1 mg/kg wet weight (Skinner 2009), suggesting that a fraction of the inorganic mercury has been transformed into methylmercury through bacterial reactions (Whalin et al. 2007). The methylation of Hg in LIS depends on the substrate that carries the Hg and presence of bacteria capable of sulfate or Fe reduction and methylation. Sediment in WLIS with high Hg concentrations has the lowest methylation rates (Hammerschmidt et al. 2008; Fitzgerald et al. 2000), believed to be the result of the presence of abundant organic carbon that tightly binds the Hg to the substrate. With potential decreases in eutrophication in the future, the deposition rate of organic carbon may decrease, which may lead to higher methylation rates and hence increased trophic transfer of Hg in that part of the Sound (i.e., the concept of "growth dilution" for Hg in reverse; Fitzgerald et al. 2007; Ward et al. 2010), although sulfide abundances in waters also play a role in the kinetics of these Hg conversions.

5.5 Organic Contaminants (PAHs, PCBs, and Chlorinated Pesticides)

5.5.1 Spatial Patterns of Organic Contaminants in LIS Surface Sediment

The USGS, as part of their comprehensive evaluation of sediment contamination in LIS sediments begun in 1995, gathered all available data on sediment contaminant levels for priority pollutants from federal and state monitoring programs as well as dredge material permits and published papers. Their dataset is available online at http://pubs.usgs.gov/of/2003/of03-241/. An overview of spatial patterns in the toxic fraction of petroleum hydrocarbon contaminant levels in sediments was obtained by calculating total PAHs by summing all 23 PAHs analyzed in the dataset after removing data reported as being below the limit of detection, and data from nonpeer-reviewed studies conducted as part of dredge permitting applications (e.g., SQUID). The PAH values were grouped into four categories, showing low levels approximately corresponding to the 85th percentile ranking of all values reported by NOAA's NS&T program (<2,000 ng/g dry weight) in blue, slightly higher levels (from 2,000 to 4,000 ng/gdw) in green, levels exceeding the ERL (4000–45,000) in yellow, and levels exceeding the ERM (<45,000) in red (Fig. 5.18). Particularly in ELIS and CLIS, most sediment sampled showed PAHs levels below 2,000 ng/gdw (blue symbols), indicating that they rank among the least contaminated sites throughout the United States. In ELIS and CLIS, values exceeding the NS&T 85 % percentile (green symbols) are only seen near the CLIS dredge material disposal site in central LIS and at coastal locations, particularly in Connecticut, although it should be noted that there is a paucity of data along the eastern shore of Long Island. In WLIS, there are many more data points and clear evidence of organic contaminant pollution. In this region, most sediment PAH levels exceed the NS&T 85 % percentile, and many exceed the ERL (yellow),

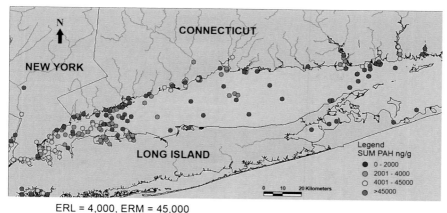

ERL = 4,000, ERM = 45,000

Fig. 5.18 Total PAH concentrations in LIS surface sediments (USGS database, years 1975–2000) ERL (effects range low), and ERM (effects range medium) specify the 10th and 50th percentile values of a given pollutant that have a noticeable biological effect, Long and Morgan 1990)

not only along the coast, but also in samples taken along the main stem of LIS. Sediment samples exceeding the ERL are also found in some coastal embayments along the Connecticut shore, particularly New Haven Harbor, and in some river samples in mid- and central LIS. Total PAH levels exceeding the ERM are found only in the extreme western Sound and outlet of the upper East River.

The USGS dataset contains data for 18 individual PCB congeners, but none of the more toxic co-planar PCBs. Toxic reference values for total PCBs are much lower than for total PAHs, with the ERL at 23 and the ERM at 180 ng/gdw, so data were plotted showing levels below the ERL (blue), between the ERL and ERM (green), values exceeding the ERM but <500 ng/g (yellow), and values exceeding 500 ng/g (red) (Fig. 5.19). Generally, a similar distribution pattern is observed for PCBs as for PAHs, with the lowest sediment concentrations (blue) observed in open waters of ELIS and CLIS with the exception of the former mud dump site, and elevated levels exceeding the ERL common at coastal Connecticut locations. In contrast, sediment PCB levels exceeding the ERL are routine both in coastal areas and open waters of WLIS. Sediment PCB levels exceeding the ERM are fairly common in WLIS and in a number of coastal areas further east in Connecticut. Extremely high levels (>500 ng/g) are only found closer to the Hudson River and in one site on the Mystic River.

These overall patterns were also reported in more recent data compilations such as Mitch and Anisfeld (2010), who reviewed contaminant data derived from the period 1994–2006, pooling all data for WLIS, CLIS, and ELIS (Table 5.5). Values within each sub-basin were reported as percentile values, and as overall means for each area. Mean levels of sediment PAHs, PCBs, and DDT generally decreased from west to east, although some very high levels of PAHs were observed in CLIS, leading to a very high

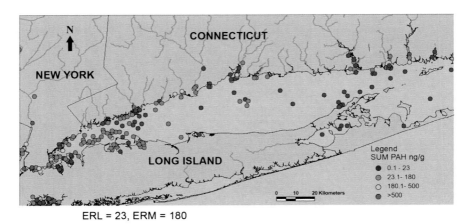

ERL = 23, ERM = 180

Fig. 5.19 Total PCB levels in LIS surface sediments (from USGS data base, 1975–2000 all years combined)

90th percentile value of 10,900 ng/g, and a mean value of 2,860 ng/g, higher than that observed in WLIS (2,470 ng/g). Concentrations at the 90th percentile exceeded ERL values for all three groups of contaminants in all three LIS areas. In WLIS, values in the 50th percentile exceeded ERL values for DDT and PCBs, but not PAHs. The dataset for metals is much more extensive than those for organic contaminants, but generally, the two datasets show the same trends, with more WLIS stations having sediments that exceed the ERL values. Metal ERM values were only exceeded for Hg in WLIS.

Mitch and Anisfeld (2010) also compared sediment data from embayments with those from open-water regions of LIS for PAHs, PCBs, and DDT and found no significant differences between these two regions. This conclusion contradicts the indications from earlier datasets, and the much larger datasets compiled by the USGS. This apparent lack of coastal enrichment in organic contaminants could have been driven by the preponderance of open water NCA sampling sites in WLIS, or the high variability in values observed for most contaminants. In a comparison of sediment contaminant levels reported by NS&T from 1994 to 2004, only the Housatonic River, Mamaroneck and Throgs Neck showed levels of PAHs or PCBs exceeding the NS&T 85th percentile value (1870 and 14.5 μg/g, respectively). Mitch and Anisfeld, comparing EMAP data from 1990 to 1992 to NCA data from 2000 to 2002, reported no significant changes in sediment PCB concentrations, while sediment PAH concentrations fell by a factor of approximately three during this period. These temporal patterns are consistent with the more persistent nature of PCBs compared to PAHs, which are more easily metabolized. Mitch and Anisfeld also evaluated data on chlorinated pesticides, including total chlordanes, chlorpyrifos, total DDT and total dieldrin, total endosulfan, and total hexachlorocyclohexane (Lindane). They reported levels exceeding the NS&T 85th percentile value for almost all pesticides in the Housatonic River, the Mamaroneck, and the Throgs Neck sites. These same three sites also had high levels of the contaminant metals Cr, Cu, Pb, Hg, Ag, and Sn. A

Table 5.5 Summary of sediment concentrations of organic and inorganic contaminants 1994–2006. Microgram per gram dry weight for metals; nanogram per gram dry weight for organics

Contaminant	Area of Long Island Sound														
	WLIS					CLIS					ELIS				
	N	10th	50th	90th	Mean	N	10th	50th	90th	Mean	N	10th	50th	90th	Mean
Arsenic	366	ND	6.85	12.2	6.68	193	ND	4.98	10.56	5.69	268	ND	3.75	11.2	4.92
Cadmium	452	ND	0.63	2.6	1.18	311	ND	0.21	2.16	0.92	304	ND	0.2	2.37	0.82
Chromium	453	13.4	64	110	64.9	288	20.7	51.4	108	62	310	9.2	30.1	69.3	36.1
Copper453	14.8	89	216	116	323	8.6	51.7	185	83.8	302	2.8	33.2	96.3	52.5	
Lead	453	12.6	57	162	87	322	9.0	37	85.9	45.6	306	4.8	24.8	82.2	43.9
Mercury	454	0.02	0.3	1.00	0.49	300	0.02	0.15	0.47	0.21	302	ND	0.12	0.56	0.24
Nickel	451	7.4	23	37.6	23.9	306	6.8	19.4	37	22.5	303	5.6	15.2	30	18.1
Selenium	56	ND	ND	5.3	1.3	34	ND	0.2	3.91	2.22	26	ND	0.03	2.12	0.74
Silver	142	0.05	0.54	2.05	0.97	164	ND	0.31	1.65	0.71	60	ND	0.06	0.6	0.25
Tin	36	1.65	5.54	9.8	5.88	23	1.11	2.94	5.34	6.77	18	0.45	1.51	7.95	2.99
Zinc	450	39.2	164	315	183	305	43.3	113	221	137	299	23.5	82.2	186	110
DDTs	72	ND	3.71	15.3	6.37	39	ND	ND	3.68	2.22	30	ND	ND	3.95	1.29
PAHs	72	61.1	880	4350	2370	36	69.1	561	10900	2860	30	ND	463	4610	1810
PCBs	72	3.21	36.5	174	162	36	ND	2.75	35.3	32.6	30	ND	1.37	31	15.2

>ERL = Bold, > ERM = Bold, > NS&T national 85th percentile value = *Italics*
Adapted from Mitch and Anisfeld (2010)

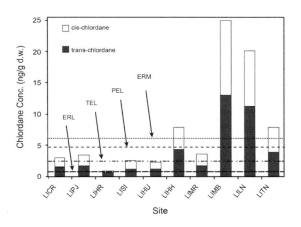

LITN:Throgs Neck
LILN:Little Neck Bay
LIMB:Manhasset Bay
LIMR: Mamaroneck River
LIHH: Hempstead Harbor
LIHU:Huntington Harbor
LISI: Sheffield Island
LIHR: Housatonic River
LIPJ: Port Jefferson
LICR: Connecticut River

Levels of total chlordane in surface
sediments collected in 2005 and
2006. ERL= Effects Range Low;
ERM=Effects Range Medium;
TEL=Threshold Effect Level;
PEL=Probable Effects Level.

After Yang et al.,2007

Fig. 5.20 Chlordane in LIS surface sediments (data from LIS NS&T sites)

recent evaluation of the chlordane data by Yang et al. (2007) indicates that levels of this persistent but banned hydrophobic insecticide have not consistently diminished over the past 2 decades. Most concentrations are higher in WLIS, exceeding those thought to cause toxicity (Fig. 5.20).

Butyltins resulting from use in antifouling paints are another organic contaminant of concern, particularly in port and harbor areas where boats are moored and serviced. Unfortunately, data on butyltins in sediment samples from LIS are scarce, and the NS&T data portal only contains values collected in 1995 and 1996. Despite their ban, periodic sampling for this important group of contaminants in coastal sediments is needed, particularly given recent increases in butyltin concentrations in biota from some areas discussed below.

5.5.2 Sediment-Associated Emerging Contaminants

All the studies discussed previously report on priority pollutants that have been the focus of monitoring programs over the last 20–30 years. Until recently, emerging contaminants have received relatively little attention. The only relatively new contaminants that have been added to national monitoring programs are the polybrominated diethyl ethers (PBDEs), a family of recently recognized persistent bioaccumulative toxic compounds widely used as fire retardants until they were banned by international treaty during the last 10 years. Due in part to their propensity to bioaccumulate, most of the monitoring associated with tracking PBDEs in the environment has been conducted on tissue samples. Recently released NOAA data from the Mussel Watch program reported concentrations of mono to hepta PBDEs in mussel tissue and sediment from around the country (Kimbrough et al. 2009). Unfortunately, PBDEs were not measured in any samples from LIS, but those collected in New York Harbor ranged from 19–41 ng/gdw, some of the highest concentrations measured nationwide. Concentrations of PBDEs

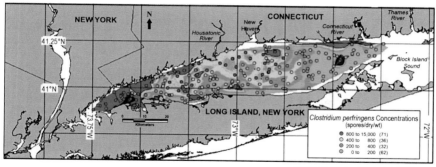

After Buckholtz ten Brink et al. 2000

Fig. 5.21   Distribution of C. perfringens in LIS surface sediment, with highest values in western LIS and some patchy zones in eastern and central LIS

measured in mussels collected from LIS are discussed below. Use of these lower molecular weight PDBEs has now been banned in the United States and throughout much of the world. However, use of the deca PBDEs is still allowed.

Data collected as part of the USGS survey characterizing sediment properties of LIS in 1996 mapped the distribution of *C. perfringens* spores in LIS sediments (Buchholtz ten Brink et al. 2000) as a marker for sewage inputs to the system (Fig. 5.21). The *C. perfringens* spores are abundant in WLIS as well as in pockets of elevated spore counts at locations near the Connecticut and New York coasts and smaller areas in the main stem of LIS. Due to their persistence and density, spores are likely to be distributed with fine sediments throughout LIS.

Brownawell's research group has recently measured a new group of sewage tracer compounds, a class of disinfectants called quaternary ammonium compounds or QACs. In a transect of surface sediments along the main stem of LIS from near the Throgs Neck Bridge to Mt. Sinai, NY, concentrations of total QACs decreased from a high of almost 14 μg/gdw to <1 μg/gdw, clearly indicating the relative significance of sewage input to WLIS (Brownawell, SBU unpublished data). The WWTFs are also a source of natural and synthetic hormones as well as hormone mimics such as the detergent breakdown products, alkylphenol polyethoxylates. No comprehensive survey of these compounds has yet been conducted in LIS. However, a recent examination of sex ratios in local Atlantic silversides (*Menidia menidia*) from coastal areas around Long Island (Duffy et al. 2009) reported a significant correlation between longitude and sex ratio with fish from the western, more urban portions of Long Island showing female biased sex ratios (Fig. 5.22).

A reconnaissance study could be a useful option for evaluating other emerging organic contaminants of potential concern. Chief among these are perfluorinated compounds (PFCs), including perfluorooctanoic acid (PFOA), and perfluorooctane sulfonate (PFOS), which are components of many nonstick and stain resistant coatings with known Immunotoxicant and endocrine disrupting potential (DeWitt et al. 2009). Although production of some of these materials ended in the United States and most of Europe, a review of recent data indicates that levels in biota continue to increase in some

Fig. 5.22 Sex ratio in *Menidia menidia* as a function of distance from the East River entrance into LIS (after Duffy et al. 2009). Sex ratio regressed onto longitude ($P < 0.001$, adj. $R^2 = 0.478$ of arcsin transformed data). Dotted line represents a 1:1 sex ratio

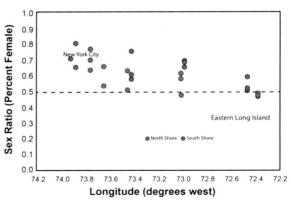

Adapted from Duffy et al. MEPS 2009

areas (Houde et al. 2011). Evenmore troubling is the almost exponential increase in PFC levels in the United States population identified by the National Health and Nutrition Examination Survey (NHANES) (Kao et al. 2011). Given the potential for PFCs to bioaccumulate and biomagnify, analysis of levels in LIS biota is clearly warranted.

Also of interest are pharmaceuticals and personal care products (PPCPs). The QACs discussed above belong to this group. Newer analytical techniques using mass spectroscopy (MS) with or without prior separation using liquid chromatography (LC), principally MS/MS and LC/MS–MS, facilitate measurement of these compounds with highly diverse structures and physical chemical properties. As evident from the USGS's landmark national reconnaissance study (Kolpin et al. 2002), with better detection, these compounds are now being found in surface waters in many parts of the United States. A corresponding study of coastal waters and coastal sediments has not been attempted. Considering the proximity of LIS to the New York metropolitan area, and the relatively high density of people and urban areas throughout the LIS coastline, the potential for inputs of human-derived PPCPs is high. Given, the very low levels detected in surface waters by the USGS, it is unlikely that the individual compounds are present at concentrations high enough to have a biological impact in coastal water. The potential for interactive effects with other PPCPs or other types of legacy contaminants exists, so there is a need to at least determine their ambient levels.

5.6 Contaminants in Biota and Their Effects

5.6.1 Organic and Inorganic Contaminants in Biota

Data on levels of contaminants in tissues of LIS aquatic species are limited. The largest dataset consists of tissue contaminant levels in blue mussels collected at nine sites (Connecticut River, Hempstead Harbor, Housatonic River, Huntington

Table 5.6 Comparison of Contaminants in LIS with other Urban Estuaries in the USA

Location	Sediments number of sites	Mussels Number of values in top 20	Number of sites	Number of values in top 10
Hudson/Raritan Estuary	6	81	6	32
Boston/Salem Harbors	6	56	5	20
Los Angeles coast	9	33	5	16
Long Island Sound	7	22	9	17
Sand Diego Bay and coast	6	18	5	14
Tampa Bay	7	13	ND	ND
San Francisco Bay	9	13	5	9
Puget Sound	10	12	3	5
Delaware Bay	9	5	ND	ND
Galveston Bay	7	3	ND	ND

Adapted from Robertson et al. (1991)
ND No data

Harbor, Mamaroneck, New Haven, Port Jefferson, Sheffield Island, and Throgs Neck) as part of the NS&T program, annually from 1986 to 1994, and every other year since then. Data for the Mussel Watch Program from NS&T collected since 1989 are available on an easily searched data portal (http://egisws02.nos.noaa. gov/nsandt/index.html#).

Robertson et al. (1991) conducted a detailed assessment of contaminant status in LIS as it related to national trends for both inorganic and organic contaminants in mussels and sediments from the 1986–1988 NS&T data. Data for the nine Mussel Watch stations and the two Benthic Surveillance stations in WLIS and ELIS were plotted on cumulative distributions for the entire country for both sediment and mussel tissue contaminant data. Several patterns are evident from these plots. LIS sites tend to rank low for inorganic contaminants compared to other sediment sites on the national distribution, with no site standing out as being highly contaminated on a national basis. In contrast, organic contaminants generally show higher rankings, with most LIS sites ranking above national means. A summary of national rankings for all contaminants is provided in Table 5.6, which shows that with regard to sediment or mussel contaminants, LIS ranked fourth highest in the nation, behind only the Hudson/Raritan estuary, Boston/Salem Harbors and the Los Angeles coast, but ranking above San Diego, Tampa Bay, San Francisco Bay, Puget Sound, Delaware Bay, and Galveston Bay. The Throgs Neck station in extreme WLIS ranked particularly high in this assessment for tPAHs, tClordane, tPCBs, tDDT, and Pb, being more than one standard deviation above the national mean, thus ranking in the top 16 % of values nationwide.

Turgeon and O'Connor (1991) reviewed Mussel Watch data for LIS from 1986 to 1988, focusing on comparisons among sites within LIS, and reported contaminant

Table 5.7 Time trends in LIS mussel watch data; analysis by spearman rank correlation

Location	N	Lind	tCld	tDDT	tDld	tPCB	tTBT	tPAH	As	Cd	Cu	Hg	Pb
Connecticut river	11	NT	**	**	NT	*	NT	*	NT	NT	NT	NT	NT
New Haven	13	**	**	**	**	**	*	NT	NT	*	NT	NT	NT
Housatonic river	12	NT	**	NT	**	NT	**	NT	NT	NT	NT	NT	NT
Sheffield island	10	NT	**	**	NT	**	NT	NT	NT	NT	NT	NT	NT
Mamaroneck	12	**	**	**	NT	**	**	**	NT	NT	*	NT	NT
Throgs neck	13	**	**	**	NT	**	**	**	NT	NT	**	N	**
Hempstead harbor	13	**	**	**	NT	**	**	NT	NT	**	**	NT	**
Huntington harbor	12	**	**	**	**	**	**	NT	NT	NT	*	NT	NT
Port jefferson	13	**	**	**	**	**	**	**	#	NT	NT	#	NT

Lind Lindane, *tCld* total chlordanes, *tDDT* total DDts, *tPCB* total PCBs, *tTBT* total organotins, *tPAH* total PAHs

NT no significant time trend

** Decreasing trend with 95 % of confidence

* Decreasing with 90 % confidence

Increasing with 90 % confidence

Adapted from O'Connor and Lauenstein (2006)

levels in mussels from WLIS (particularly the Throgs Neck site) to be high relative to national standards. Median concentrations at the Throgs Neck site were 4,900, 1,300, and 210 µg/gdw for total PAH, PCB, and DDT, respectively, for the 3-year period studied, higher than any other LIS site. O'Connor and Lauenstein (2006) published a detailed analysis of the National Mussel Watch dataset analyzing time trends in data from 1986 to 2003. For the organic contaminants Lindane, tClordane, tDDT, tPCB, tPAH, and tButyltin, significant decreases were observed at most LIS sites for most contaminants with the exception of tDieldrin and tPAHs, for which no significant time trends were observed for five out of the nine sites (Table 5.7). In contrast, very few of the metals showed time trends, with only Cd, Cu, and Pb showing significant decreases at a couple of sites. Interestingly, As and Hg showed statistically significant increasing trends at Port Jefferson over time. Coal combustion can be a source of both As and Hg to the environment, but there are no data to support why this would be a factor in Port Jefferson Harbor as coal has not been used at the power plant there for decades. If these trends continue, attempts should be made to identify the source. For the organic contaminants analyzed, national median values during these periods (1988–2003 for tPAHs, and 1989–2003 for tButyltin) decreased on average by a factor of four, with tButyltin showing the largest decrease (8x) while tDDT, tPCB, and tPAH only decreased by a factor of two to three. The enhanced reduction of tButyltin is likely due to legislation banning its use on vessels <25 m in length beginning in 1988, and the voluntary "sunsetting" of the United States production in 2001. These trends were significant at the 95 % level for each of these groups of contaminants. Of contaminant metals, only Cd showed significantly decreasing trends over this period, by a factor of only 1.5.

A closer inspection of all available data for PAHs and butyltins in LIS, including the 2004–2008 data from the NS&T data portal, suggests that although levels are now lower than observed in the 1980s, decreasing trends are no longer evident.

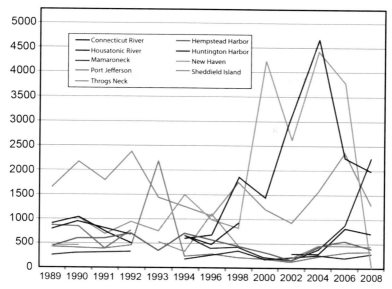

* data reported only for PAHs included in all years. (1,6,7-trimethylnahpthalene, 1-methylnapthalene,
1-methylphenanthrene, 2,6-dimehtylnaphthalene, 2-methylnaphthalene, acenaphthene, acenaphthylene,
antrhacene, benz[a]antrhacene, benzo[a]pyrene, benzo[b]fluoranthene, benzo[e]pyrene, benzo[g,h]perylene,
benzo[k]fluoranthene, chrysene, dibano[a,b]anthracene, fluoranthene, fluorene, indeno
[1,2,3-c,d]pyrene,naphthalene, perylene, phenanthrene, and pyrene)

Fig. 5.23   Time trends in median total PAH concentrations in mussel tissue from NS&T sites in
LIS over the period 1989–2008. Concentrations in ng/gdw

Furthermore, there appear to be differences emerging at some sites where levels
could be increasing, and overall levels from the WLIS sites are not necessarily the
highest observed Sound-wide. Some contaminants at some of the LIS sites are
no longer decreasing nor are values from the WLIS necessarily still the highest
for individual contaminants in all cases. These points out the need for continued
monitoring, and more process-oriented work at some sites to determine potential
sources of these contaminants to particular environments.

Data for tPAHs in mussel tissue from NS&T over the period 1989–2008
(Fig. 5.23) show median values over this entire period from a low level of
314 ng/gdw in Port Jefferson to a high level of 1,580 ng/gdw at Throgs Neck.
Median levels for the period exceed 1,000 μg/g for both the Housatonic River and
New Haven Harbor, including some of the highest values recently measured. Median
values for all LIS sites exceeded the national median of 220 ng/gdw reported by
O'Connor and Lauenstein (2006) for 2002–2003. These accumulated body burdens
are still well below values predicted to cause acute toxicity (around 100 μg/gdw),
although some PAHs can act via receptor-mediated mechanisms, leading to developmental effects at much lower levels (McElroy et al. 2011). The observed trend of
increasing tissue concentration values for tPAH is a cause for concern. Since mussels filter particulates from the water column, current values in these organisms
should be reflective of recent trends in water column bulk concentrations of these

contaminants. The presence of increasing body burdens could reflect increasing inputs to the watershed or at least enhanced resuspension of relict polluted sediments.

Data for butyltins in the NS&T Mussel Watch program illustrate spatial trends unlike those observed for most other organic contaminants in that the Throgs Neck site is not the most contaminated. In an early review looking at data from 1986 to 1989, Turgeon and O'Connor (1991) found the highest levels at Mamaroneck (500 ng/gdw), Port Jefferson (300 ng/gdw), and then Huntington Harbor (250 ng/gdw with lower levels at the six other sites. Throgs Neck ranked fifth among the sites sampled. These levels were considered low from a national perspective, approximately 10 times less than concentrations observed in oysters or mussels at sites near marinas in other parts of the country. Recent NS&T data for butyltins (mono-, di-, tri-, and tetra-butyltin, 1989–2008) were only available from five of the Mussel Watch sites (Hempstead Harbor, Mamaroneck, Huntington Harbor, Port Jefferson, and Throgs Neck). Tetra-butyltin was not detected in any sample. But of the three organotin compounds measured, tributyltin (TBT) usually accounted for the majority of the residues with dibutyltin, a metabolic breakdown product of tributyltin, sometimes found at levels approaching tributyltin. Data for the sum of mono-, di-, and tributyltin for 1989 through 2010 are shown in Fig. 5.24. Maximum levels were generally observed in 1989 or 1990, with the highest levels (exceeding 500 ng/gdw) observed in Mamaroneck and Port Jefferson. Peak levels (about 400 ng/g) were observed later (1993) at Throgs Neck. Port Jefferson experienced a secondary peak of almost 100 in 1994, with periodic increases observed throughout the sampling period, including

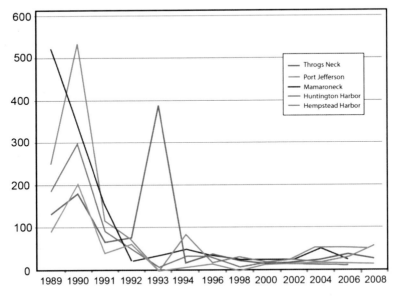

Fig. 5.24 Time trends in Butyltin (mono + di + tri) concentrations in mussel tissue from five locations in western LIS (from NS&T Mussel Watch Stations) over the period 1989–2008. Units: ng/gdw

the most recent data available in 2008, where levels of 56 ng/g were reported. Throgs Neck and Mamaroneck mussels still contain butyltins at levels exceeding 25 ng/g, whereas most other sites have dropped to <15 ng/g. Patterns observed in butyltin levels in mussel tissues likely reflect decreasing environmental concentrations resulting from the 1988 partial ban of TBT antifouling paints in the United States. More recent spikes at Throgs Neck and Port Jefferson indicate either new sources or remobilization of high TBT sediment reservoirs, continuing well past the partial ban. Although body burdens of butyltins in mussels are still well below levels known to cause mortality (48,000 ng/gdw), growth impairment (3200 ng/gdw), or even imposex in sensitive neogastropod species (320 ng/gdw) (Meador 2006), increased levels despite the ban are a cause for concern. Sources could include resuspension of older more heavily contaminated harbor sediments due to dredging projects or potentially more recent inputs from either illegal use of TBT paints or from vessels exempt from the 1988 ban.

Apart from the NS&T data, there are only limited and sporadic additional data available on levels of organic or inorganic contaminants in LIS organisms. Most of the additional data have resulted from periodic surveys conducted by the NYSDEC, often in partnership with CTDEEP, NYSDOH, CTDOH, and the USEPA, which has monitored PCB levels primarily not only in striped bass (*Morone saxitilis*), but also in bluefish (*Pomatomus salatrix*). More recently, as part of a large survey done in 2006 and 2007 funded by the LISS (Long Island Sound Study), PCBs and mercury were analyzed in striped bass, bluefish, American eel (*Anguilla rostrata*), weakfish (*Cynoscion regalis*), and American lobster (*Homarus americanus*) hepatopancreas, as well as Cd, dioxins, and furans in lobster hepatopancreas from LIS (Skinner et al. 2009). The majority of data on PCB levels in fish has been collected as part of the NYSDEC striped bass monitoring program, which has been focused on samples from the Hudson River. However, periodic sampling of striped bass from LIS has occurred since 1984, with the most recent comprehensive survey completed in 2006 and 2007, introduced above (Skinner et al. 2009). In evaluating these recent samples, Skinner et al. also reviewed the earlier data and trends in PCB levels in striped bass and bluefish. For PCBs, no significant differences were observed between sampling areas roughly corresponding to WLIS, north and south regions of CLIS, and ELIS in any of the fish species sampled, although collections of eel and weakfish were insufficient to test for differences among all four of the areas sampled. The PCB body burdens were not correlated with wet weight of striped bass, although large bluefish (>508 mm) had significantly higher body burdens. Average concentrations of PCBs were 0.333, 0.110, 0.565, and 0.512 μg/g wet weight (gww) for striped bass, small (305–508 mm) bluefish, large (>506 mm) bluefish,, American eel, and weakfish, respectively. Levels of PCBs reported for striped bass declined by >80 % from levels observed in 1985 on a wet weight basis (from >2 μg/gww to <0.5 μg/gww); during this period the lipid content also dropped significantly (Fig. 5.25). On a lipid-normalized basis, PCB levels dropped by 50 % on average between the mid-1980s and 2006–2007. Combining all the data for the entire record, PCB body burdens were significantly correlated with lipid content. Similar time trends were observed with bluefish, where declines of 70 % in wet weight normalized in tissue levels were observed, while no significant declines in

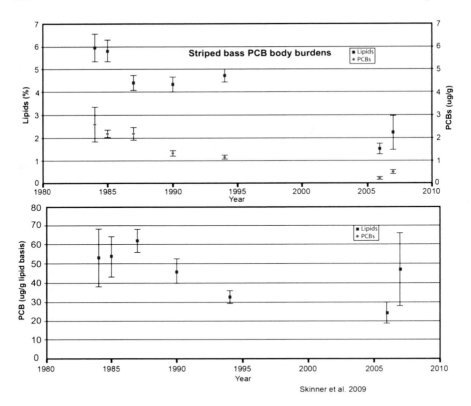

Fig. 5.25 The PCB body burdens in striped bass over the period 1984–2007 (after Skinner et al. 2009)

lipid-normalized body burdens were found for the same period. Skinner concluded from these analyses that ambient levels of PCBs in the LIS environment have not significantly changed since the mid-1980s. Insufficient data on American eel and weakfish prevent any conclusions regarding temporal differences. In the NYSDEC 2006–2007 study, the mean concentration of PCBs in lobster hepatopancreas for LIS was 1.31 μg/gww. Interestingly, significantly higher concentrations were observed in male lobsters, possibly due to depuration by female lobsters through egg production. The differences between concentrations in males and females created more complex special patterns. On a wet weight basis, lobsters from ELIS appeared to have lower PCB concentrations, but this pattern was not apparent when body burdens were normalized to lipid content. Skinner did not comment on temporal trends in PCB concentrations over time due to insufficient data.

Although mean body burdens of PCBs in all species analyzed were below the FDA action limit of 2 mg/kgww (2 ppm), it is important to note that significant individual variation in PCB content was observed in all species analyzed. Maximum PCB levels in some species approached or exceeded the FDA limit. Maximum PCB concentrations observed were 1.45, 3.17, 1.85, and 2.83 μg/gww for striped bass,

Table 5.8 Total PCB concentrations (ng/gr wet weight tissue) in LIS marine organisms (From NCA data 2000–2006)

	Mean	SD	Maximum	N
Blue crab	31	27.3	87.3	27
Lobster	9.4	7.1	20.9	5
Summer flounder	56.1	55.3	421	92
Winter flounder	38.4	43.3	216	41
Scup	28.6	27.4	137	64
White perch	660	513	1680	8

large (>508 mm) bluefish,, weakfish, and lobster hepatopancreas, respectively. This implies some health risk to both wildlife and humans in consuming these species.

Tissues of blue crab (*Calinectes sapidus*), American lobster, scup (*Stenotomus chrysops*), summer flounder (*Paralichthys dentatus* also commonly called fluke or Northern fluke), winter flounder (*Pseudopleuronectes americanus*), bluefish, brown bullhead (*Ameiurus nebulosus*), channel catfish (*Ictalurus punctatus*), and white perch (*Morone americana*) were also analyzed as part of the NCA assessment, although analysis of these data has not been published to our knowledge. In most cases, filets were analyzed, but sometimes the viscera, also known as "offal" or whole organisms, were analyzed. Data on PCB body burdens in fish and shellfish collected were examined as part of this review. Twenty-one individual PCB congeners were analyzed, including three of the more toxic, co-planer PCBs (IACUC# 77, 118, and 126). Polychlorinated biphenyls were detected in most but not all samples analyzed. Sums of all PCBs in filets with five or more samples (Table 5.8) indicate average concentrations in filets were generally low relative to the FDA action limit of 2000 mg/kgww, with average body burdens below 100 mg/kgww for all species except white perch. Although only eight white perch specimens were analyzed, the maximum value recorded (1,680 mg/kgww) was within 20 % of the FDA action limit. The NCA data are for filet or muscle tissue only, so they cannot be directly compared to the hepatopancreas data reported in the NYDEC study for lobsters. The NCA dataset on PCB body burdens is not extensive, but it does provide data on species that are prominent members of the local sport fishery for which we have no other recent data. Individual variability in this dataset is also large with coefficients of variation approaching or exceeding 100 %. Despite this, the very low levels reported indicate minimal human exposure to PCBs as a result of the consumption of these species, with the exception of white perch for which maximum values are close to the FDA limit.

The PBDEs are similar in structure and chemical properties to PCBs, and are a more recently recognized bioaccumulative organic toxicant, particularly in urban areas. Kimbrough et al. (2009) recently reviewed the limited data available as part of the NS&T program on PBDE levels in bivalves. Elevated levels were found in mussel tissue collected near urban areas in many parts of the country, including Seattle, Los Angeles, Boston, and Baton Rouge, but the largest concentrations of hot spots on the national map were associated with the New York/New Jersey Harbor Estuary and LIS. Elevated PBDE levels were observed in 57 locations nationwide in samples collected in 1996 or between 2004 and 2007, and five were in LIS. Concentrations (expressed in ng/g lipid) were 697 at Throgs Neck, 495 at New Haven, 493 at the Housatonic River, 350 at the Connecticut River, and 334 at Hempstead Harbor. Mussels from sites in the Hudson River and New York/New Jersey Harbor estuary generally showed higher levels, ranging from 2190 ng/g lipid at Governor's Island in the Hudson River to 594 ng/g lipid in Raritan Bay. Six of the 10 highest levels nationwide were reported from mussels collected in this area. Using the national dataset, Kimbrough determined there was a statistically signif- icant relationship between population density and PBDE levels in local bivalves. Given the relatively high levels observed in LIS biota and the high population den- sity in the LIS watershed, continued sampling for PBDEs in biota is warranted.

The NYSDEC also measured total Hg in striped bass, bluefish, weakfish, American eel, and lobster tissues as part of their 2006–2007 survey in LIS (Skinner et al. 2009). In both striped bass and bluefish, total Hg levels in muscle tissue were significantly correlated with size, with the highest concentrations being observed in the north portion of CLIS due to the greater numbers of larger fish caught in this area. Mean tissue Hg concentrations in Area 2 were 0.528 μg/gww in striped bass, and 0.400 μg/gww in bluefish. Mercury levels were lower in weakfish (0.141 μg/gww) and American eel (0.110 μg/gww). The smaller sample size for these two species prevented spatial comparisons to be made. Mercury was also measured in the hepatopancreas of American lobster. Regardless of location, all values were found to be <200 ng/gww. Lobsters from ELIS showed a small but statistically significant elevation in Hg concentrations (111 vs. 61 ng/gww) as com- pared to lobsters from the WLIS and ELIS combined. This was not considered to be extremely meaningful, given the overall low levels. Mercury, as a class B metal, strongly binds to sulfhydryl functional groups of proteins (Kuwabara et al. 2007), and evaluation of Hg levels in lobster muscle rather than hepatopancreas may have yielded higher concentrations. Comparison of striped bass Hg data collected in 1985 by the NYSDEC to values obtained in 2006–2007 indicated no significant differences in mean Hg concentrations over this 20-year period.

Hammerschmidt and Fitzgerald (2006) quantified both total Hg and methylmer- cury (MeHg) levels in four species of fish (alewife (*Alosa pseudoharengus*), winter flounder, bluefish,, and tautog (*Tautoga onitis* also known as blackfish)) as well as American lobster collected from CLIS in 2002. Virtually all (98 %) Hg recovered from animal tissue was in the form of MeHg, allowing relatively straightforward comparison between their data on MeHg with the data on total Hg reported by NYSDEC. Mean MeHg levels in muscle were similar in bluefish and tautog (137 and 191 ng/gww), but much lower in winter flounder muscle (21 ng/gww), and in whole bodies of alewife (27 ng/gww). Levels in lobster tail muscle (140 ng/gww) were similar to those found in bluefish and tautog. Collectively, these data indi- cate no significant spatial trends for Hg body burdens in fish or shellfish from LIS. However, given the relatively high levels observed in some of the species evaluated, unrestricted human consumption is ill-advised with a consumption advisory set by the USEPA of 300 ng/gww (USEPA 2001). Research to develop an improved quan- titative understanding of Hg trophic transfer in LIS seems advisable.

5.6.2 Toxicity of LIS Sediments and Health Impacts on Resident Biota

Assessment of toxicity can be inferred by comparison to body burdens associated with toxicity in other studies. However, potential or realized effects can also be directly measured through sediment toxicity tests or analysis of local fauna. In support of the LISS NOAA conducted contaminant toxicity assessment in LIS during the period 1988–1991. Wolfe et al. (1994) published the first comprehensive analysis of these data. Sediment samples were collected at 20 coastal stations along the entire coast of Connecticut, and from the WLIS coast out to Cold Spring Harbor, and 11 stations along the main stem of LIS from the west to a point northeast of Montauk. At each sampling location, three sediment samples were collected and three independent toxicity assays were performed: the 10-day whole sediment toxicity test with *A. abdita*, a 48 h sediment elutriate test embryo test with *Mulinia lateralis,* and a microbial bioassay (Microtox™[1]) performed on organic sediment extracts.

The *A. abdita* toxicity test, which exposes organisms to bedded sediments, is probably the most realistic exposure scenario for sediment-sorbed contaminants. About 80 % of samples collected from the coastal embayments showed significant toxicity in this assay, while tests of samples from the main stem of LIS with the exception of a site in CLIS showed little to no toxicity. The *M. lateralis* test was less sensitive, yielding fewer toxic sites. The Microtox™ assay identified a similar number of toxic sites as the amphipod test, yet the results were not correlated, indicating these tests are responding to different sediment properties. Sites that show toxicity in 1, 2, or 3 toxicity tests are given in the data summary (Fig. 5.26). Coastal embayments in the WLIS, including Little Neck Bay, Oyster Bay, and Manhasset, showed the most significant toxicity in this survey, although sites in extreme WLIS as well as sites in most coastal embayments along Connecticut showed some evidence of sediment toxicity as well (Fig. 5.26).

The *A. abdita* test is also being used as part of the NCA program. Results obtained from 2000 to 2006 showed significant mortality (60–80 % of control survival) in 19 out of 310 tested samples (Fig. 5.27; data taken from http://www.epa.gov/emap/nca/html/regions/ne0006/index.html). Also shown in this map are samples with even greater mortality (<60 % of control survival). Despite the general reduction in toxic sites identified, the NCA data indicate that toxic sediments are still present throughout coastal embayments along both the New York and Connecticut shores of LIS with approximately equal numbers of toxic sites in each state. The presence of toxic sediments in coastal sites throughout LIS is also indicated by data from McElroy's group who analyzed sediment toxicity using the Mictorox test on extracts of LIS sediments collected in 2002. This test measures reduction in light emission from luminescent bacteria in response to test solutions.

---

[1] Any use of trade, firm, or product names is for descriptive purposes only and does not imply endorsement by the U.S. Government.

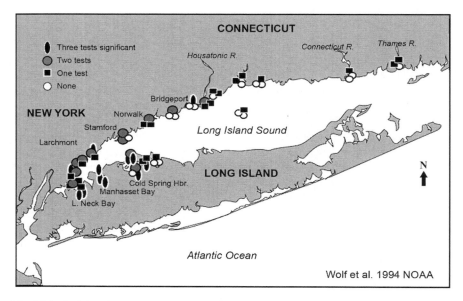

Fig. 5.26 Toxicity of surface sediment in LIS with one, two or three tests done (Wolfe et al. 1991)

It is considered a nonspecific general indicator of relative toxicity among samples. Evidence of sediment toxicity is fairly widespread in LIS (Fig. 5.28). The Mutatox™ assay uses a dark mutant of these bacteria that no longer is capable of producing light. After a suitable incubation period of the dark mutant with a test substance, mutations leading to restoration of the light emitting phenotype indicate that the test substance is mutagenic. Results from the Mutatox™ assay indicate that some solvent extractable materials in these sediments are also capable of causing mutations in bacterial test species, and thus represent a potential mutagenic risk to higher organisms.

Toxicity studies on indigenous species were performed as part of the NS&T assessment in the late 1980s, and were reported in the 1991 special editions of *Estuaries*. Few related studies have been conducted since that time. Nelson et al. (1991) evaluated reproductive success of winter flounder from Shoreham, Madison, New Haven, Milford, Norwalk, and Hempstead in comparison to flounder from two sites near WWTF outfalls in Boston Harbor in 1986 through 1988 by examining egg and larval properties. Evaluation of % fertilization, embryo lipid content, and larval size indicated New Haven Harbor to be the most impacted of the study sites in LIS, although fish from both Shoreham and Milford scored poorly in some indices. The embryos from all LIS sites were in better condition than those sampled from Boston Harbor. Within LIS there were no differences in sediment, adult liver, or embryo PCB concentrations, nor was there any apparent correlation between embryo PCB concentrations and embryo health measures. No significant site specific differences in PCB concentrations in flounder livers were

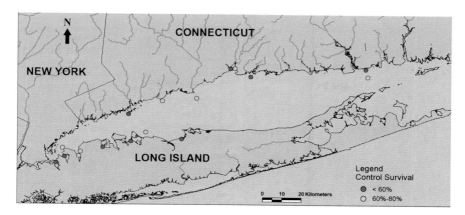

Fig. 5.27 Toxicology of various coastal LIS sites using *Ampelisca abdita* tests, with <60 % survival rate indicated as the most toxic sites. Data taken from USEPA 2010, USEPA, National Coastal Assessment Northeast 2000–2006 Summary Data at http://www.epa.gov/emap/nca/html/regions/ne0006/index.html

observed in LIS, and all were lower than those from fish collected near WWTF outfalls in Boston Harbor. No correlation was found between embryo or sediment metal concentrations and toxicity. Their data indicate that some other factors, such as a contaminant not analyzed, or other environmental or biological variables likely contributed to the toxicity observed. Nelson et al. (1991) concluded that winter flounder reproduction was impaired at some LIS locations, but that neither PCBs nor metals appeared to be linked to the health status of the fish. A related study examined sublethal abnormalities in winter flounder embryos and found developmental deficits and elevated incidence of cytotoxicity and DNA damage in fish embryos (Perry et al. 1991). Fish embryos from New Haven were most significantly affected, although abnormalities were also observed in embryos from Hempstead, Shoreham, and the Boston Harbor sites.

Gronlund et al. (1991) reported on a more detailed study assessing overall health of winter flounder collected at three of the NS&T sites: New Haven, Niantic and Norwalk. Histological examination of liver from adults indicated fish from the New Haven site to be most severely impacted with almost 90 % showing some measure of histological lesions. Flounder from Norwalk ranked next, with 70 % showing lesions, while only about 20 % of fish from Niantic showed liver lesions. Levels of DNA adducts were highest in fish from New Haven, which had prevalences around 35 %, similar to that observed in Boston Harbor. Fish from both Norwalk and Niantic had levels of hepatic DNA adducts of <10 %. Of the three sites examined, New Haven had the highest levels of chemical contaminants including PAHs, PCBs, Ag, Ni, Pb, Cu, Cr, and Zn. Although no cause and effect relationships could be determined from this study, levels of PAHs at the New Haven site of 4000 ng/gdw were similar to those associated with liver lesions in other flounder species at other sites.

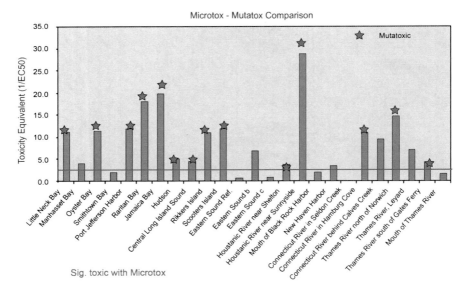

Fig. 5.28 Toxicity equivalent comparison for various locations using the Microtox™ and Mutatox™ luminescent bacterial assays (McElroy, unpub. data)

Cytochrome P450 monooxyganase (Cyp1a) has been used as a biomarker of exposure to organic contaminants, particularly PCBs and PAHs in fish for almost 30 years. Monosson and Stegeman (1994) evaluated Cyp1a activity in livers of winter flounder from a variety of locations throughout the Northeast, including LIS, in the early 1990s. Their analysis indicated widespread exposure to chemicals that induce Cyp1a throughout the region. As measured by Cyp1a catalytic activity, only fish from Georges Bank appeared to be free of contamination. Fish from Hempstead Harbor and Niantic in LIS were evaluated as part of this study. Hempstead Harbor fish expressed the highest levels of any examined in this study, and activity at both sites exceeded that observed in Boston Harbor fish. Cyp1a activity correlated with PCB concentrations in liver from the same fish.

A more comprehensive study, part of NOAA's Benthic Surveillance Program (Collier et al. 1998), reported Cyp1a data on winter flounder from samples collected from 1988 to 1994 at 22 sites around the Northeast, and demonstrated that with the exception of Maine, Cyp1a induction was widespread throughout the region. Furthermore, there appears to be no evidence of diminished responses in the more recently collected fish samples. The results of this study indicate that exposure to organic contaminants is widespread in fish collected from urban areas throughout the northeast. More recent work in McElroy's lab examining Cyp1a gene expression in livers from young-of-the-year winter flounder collected in 2008 and 2009 from Port Jefferson, Oyster Bay, Manhasset Bay, and Little Neck Bay in LIS, and Shinnecock and Jamaica Bay on Long Island's south shore also indicates widespread evidence of exposure to organic contaminants in LIS fish (Romany 2010).

5.7 Nutrient and Carbon Patterns and Trends

Like so many other urban estuaries, LIS suffers from nutrient and carbon pollution, creating directly (carbon) or indirectly (nutrients) a large biogenic oxygen demand (e.g., Diaz 2001; Boesch et al. 2001; Bricker et al. 2003). The direct effect of the enhanced nutrient fluxes is cultural eutrophication with the recurring effect of seasonal hypoxia–anoxia in WLIS and CLIS (O'Shea and Brosnan 2000; Welsh and Eller 1991; Parker and O'Reilly 1991; Lee and Lwiza 2008).

5.7.1 Currently Measured and Estimated Nutrient Fluxes

The magnitude of the various nutrient fluxes and their relationship to cultural eutrophication impacts today can be estimated and measured directly from river data. River loads quantify the aggregated contributions of nutrients from all sources within the watershed, and are a starting point for their allocation among specific sources such as WWTFs, urban, and agricultural runoff, and atmospheric deposition that comprise the major contributors of nutrient enrichment (NYSDEC and CTDEP 2000). Groundwater nutrient fluxes are less well-known but may be significant as well.

We estimated atmospheric wet deposition of inorganic N to LIS watersheds, an area of approximately 4.3 Mha, for calendar years 1994–2009 using data from the National Atmospheric Deposition Program (NADP 2009). The data from the NADP have been synthesized into a gridded format based on interpolation at each of the monitoring locations. The mean deposition rate was calculated for each year by taking the average of the gridded data for the LIS watershed, including the Pawcatuck River basin. The watershed area used did not include the surface of LIS, because the grids were clipped to the coastlines.

Wet deposition of inorganic N on the LIS watershed, generally referred to as "indirect deposition," as opposed to direct deposition on the waterbody's surface, generally decreased since the mid-1990s. This decrease is parallel to decreases in emissions of N oxides from vehicles and electrical power generation. The declines in atmospheric deposition of reactive N are primarily in the form of nitrate, whereas emissions of NH_3 do not appear to have significant trends, indicating that reduced forms of N are becoming a larger fraction of the atmospheric N budget (Pinder et al. 2011). Estimated wet deposition of inorganic N ranged from 28 Mkg/year (6.5 kg/ha/year) in 1996 to 15 Mkg/year in 2009 (3.5 kg/ha/year) (Fig. 5.29). The direct atmospheric input of N onto LIS is estimated at ~2.5 Mkg/year (Castro and Driscoll 2002).

The loading of the other nutrients such as Si (an essential element for diatoms, the main primary producer in the Sound) and P is largely determined by the river fluxes, where the element originates through weathering of rocks on land. Secondary Si fluxes may stem from phytoliths (small silica grains in plants) in

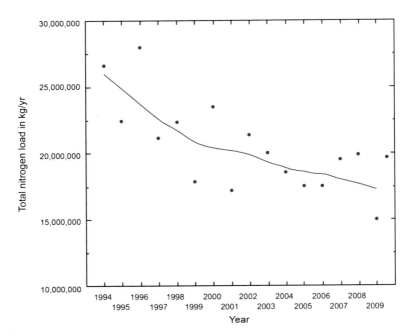

Fig. 5.29 Atmospheric wet deposition of inorganic N, averaged for the LIS Watershed, 1994–2009 (data from National Atmospheric Deposition Program 2009)

association with particulate organic matter discharge. The combined annual flux of Si from the Connecticut and Housatonic Rivers is estimated at 1,500 Mkg Si/year, while the WWTF silica flux is estimated at ~1 % of the riverine flux (Boon 2008). The total P load derived from rock weathering and WWTFs is estimated at 0.55 Mkg P/year.

5.7.2 Trends in Nutrient Fluxes over the Last 40 Years

The United States Geological Survey has maintained a network of stream gages and water quality monitoring sites in the LIS watershed, such as NASQAN (Ficke and Hawkinson 1975) and USGS National Water Quality Assessment or NAWQA (Gilliom et al. 1995), summarized in Fig. 5.30. These long-term monitoring sites are largely stations above the tidal influence ("fall line" stations), and nutrient loads below these stations are not measured. The monitoring sites on the major rivers (Connecticut, Farmington, Housatonic, Naugatuck, Quinnipiac, Shetucket, and Quinebaug) represent 82 % of the drainage area of the LIS watershed; however, the majority of the point sources of nutrients (N and P) are in coastal areas, downstream of these sites. Data on the loads and trends of N and P have previously been summarized by Mullaney et al. (2002), Colombo and Trench (2002), Trench and Vecchia (2002), Trench (1996, 2000), and Sprague et al. (2009). Data for selected water-quality monitoring stations and associated flow data from USGS stream gaging sites were analyzed to estimate the N, P, Si, and total particulate organic carbon loads from 1974 to 2008. Loads were calculated using the linear regression method LOADEST developed by Runkel et al. (2004). Given a time series of streamflow and constituent concentration, LOADEST assists the user in developing a regression model for the estimation of constituent load. Explanatory variables in the regression model include various functions of streamflow, decimal time, and additional user-specified data. The formulated regression model then is used to estimate loads over a user-specified time interval. Several trend types were analyzed, including flow-adjusted trends in concentrations and loads, and nonflow-adjusted trends in concentrations and loads. These trend analysis methods were presented in detail by Sprague et al. (2009).

Our analysis shows that the concentrations and loads of total N have declined during the 1974–2008 study period, probably related to the implementation of the Federal Clean Water Act (CWA) of 1972 and the related benefits of management actions taken under its authority. Trends in flow-adjusted concentrations and loads and nonflow-adjusted concentrations and loads are shown in Table 5.9. Significant flow-adjusted trends in concentrations and nonflow-adjusted concentrations of total N were observed at most selected sites, with the exception of the Quinnipiac River at Wallingford and the Saugatuck River at Redding (Table 5.9). The Quinnipiac River is dominated by point source N discharges from three wastewater treatment facilities totaling 26,500 m^3/day (Mullaney et al. 2002). Mullaney et al. (2009) estimated that about 360 ha of impervious area were added to this watershed from 1985 to 2002 as a result of urbanization. The Saugatuck River watershed at Redding (Fig. 5.30) has remained mostly forested and rich in wetlands (80.7 %) (Fig. 5.30). This station is an example of concentrations and loads of N that can be expected in similar, mostly forested watersheds of the LIS basin.

Downward trends in N loads were significant at seven of the 11 sites selected for this summary (Table 5.9). The Housatonic, Norwalk, Quinnipiac, and Saugatuck River sites showed no significant trends. Total N loads (nonflow-adjusted) declined significantly from the other seven stations, averaging 1.08 %/year over the period of record. Although over this long period of analysis (water years 1974–2008) there are significant downward trends in total N, concentrations generally showed no significant trend in later years, as described by Sprague et al. (2009).

Total N loads from the seven major fall line stations described above declined dramatically during the period of study, from about 21 Mkg/year to 15 Mkg/year. The most notable 5 decreases occurred at the Connecticut, Farmington, Naugatuck, Quinebaug, and Shetucket Rivers (Fig. 5.31; Tables 5.9, 5.10). Yields (loads/watershed area/time) of N during the study period ranged from 3.0 kg/ha/year for watersheds with minimal development to 21.4 kg/ha/year in watersheds with dense urban development and major wastewater discharges. Nitrogen yields are strongly correlated with percent of impervious surface in each watershed, which serves as a measure of the overall urban development and magnitude of wastewater discharge (Fig. 5.32).

Table 5.9 Trends in nutrient constituents at USGS fall-line monitoring stations on rivers in Connecticut 1974–2008 [p-values in bold indicate significance at the = 0.05 level]

Rivers[a]	Trend in flow-adjusted concentration or load, in percent/year	Flow-adjusted trend in load, in kg/day/year	p value for flow-adjusted trend in load	Trend in concentration, in percent/year	p-value for non-flow-adjusted concentration	Trend in load, in percent/year	Trend in load, in kg/d/year	p-value for trend in load
Total nitrogen								
Shetucket	−0.92	−10.21	<0.0001	−0.90	<0.0001	−0.97	−10.83	0.0032
Quinebaug	−0.84	−23.97	<0.0001	−0.78	<0.0001	−1.20	−34.23	<0.0001
Connecticut	−1.04	−325.63	<0.0001	−1.06	<0.0001	−0.94	−295.47	<0.0001
Farmington	−0.46	−11.82	0.0004	−0.42	0.0173	−0.53	−13.54	0.0361
Hockanum	−1.21	−11.49	0.0001	−1.37	<0.0001	−0.96	−9.09	0.0118
Salmon	−0.65	−1.12	0.0030	−0.68	0.0019	−0.96	−1.64	0.0417
Quinnipiac	−0.16	−2.07	0.2337	−0.16	0.4141	−0.15	−2.00	0.5184
Housatonic	−0.67	−21.89	<0.0001	−0.67	<0.0001	−0.22	−7.11	0.6938
Naugatuck	−1.96	−79.65	<0.0001	−1.88	<0.0001	−2.03	−82.38	<0.0001
Saugatuck	0.02	0.00	0.9492	0.01	0.9567	−0.05	−0.01	0.9471
Norwalk	−1.02	−0.84	<0.0001	−1.01	<0.0001	−0.98	−0.81	0.0662
Total phosphorus								
Shetucket	−1.97	−1.75	<0.0001	−1.97	<0.0001	−2.00	−1.77	<0.0001
Quinebaug	−2.02	−5.69	<0.0001	−2.01	<0.0001	−2.18	−6.14	<0.0001
Connecticut	−1.61	−36.78	<0.0001	−1.62	<0.0001	−1.54	−35.17	<0.0001
Farmington	−1.39	−4.41	<0.0001	−1.36	<0.0001	−1.42	−4.50	<0.0001
Hockanum	−1.90	−1.94	<0.0001	−2.06	<0.0001	−1.71	−1.75	0.0005
Salmon	−1.51	−0.07	0.0005	−1.56	0.0003	−1.72	−0.09	0.0022
Quinnipiac	−1.46	−2.90	<0.0001	−1.46	<0.0001	−1.46	−2.89	<0.0001
Housatonic	−1.66	−2.66	<0.0001	−1.64	<0.0001	−1.38	−2.22	0.0058
Naugatuck	0.27	0.86	0.2934	0.62	0.0949	0.10	0.32	0.7187
Saugatuck	−1.03	−0.01	0.0097	−1.03	0.0095	−1.07	−0.01	0.1204
Norwalk	−1.28	−0.05	0.0003	−1.27	0.0005	−1.24	−0.05	0.0562
Total organic carbon								

(continued)

Table 5.9 (continued)

Rivers[a]	Trend in flow-adjusted concentration or load, in percent/year	Flow-adjusted trend in load, in kg/day/year	p value for flow-adjusted trend in load	Trend in concentration, in percent/year	p-value for non-flow-adjusted concentration	Trend in load, in percent/year	Trend in load, in kg/d/year	p-value for trend in load
Shetucket	−0.45	−25.77	0.0008	−0.47	0.0026	−0.55	−31.38	0.2675
Quinebaug	−0.59	−91.85	0.0001	−0.62	<0.0001	−1.07	−164.79	0.0037
Connecticut	−0.65	−916.20	<0.0001	−0.63	<0.0001	−0.49	−692.30	0.1274
Farmington	−0.64	−64.11	<0.0001	−0.65	<0.0001	−0.74	−75.02	0.0560
Hockanum	−0.60	−7.85	0.0941	−0.59	0.0982	−0.10	−1.35	0.8506
Salmon	−0.60	−7.34	0.0016	−0.66	0.0013	−0.95	−11.66	0.0641
Quinnipiac	−1.01	−23.75	<0.0001	−1.01	<0.0001	−1.00	−23.61	0.0027
Housatonic	−0.65	−94.05	0.0010	−0.62	0.0018	−0.16	−22.83	0.7956
Naugatuck	−1.34	−91.58	<0.0001	−1.32	<0.0001	−1.55	−106.20	<0.0001
Saugatuck	−0.85	−2.23	<0.0001	−0.86	<0.0001	−0.90	−2.36	0.1608
Norwalk	0.55	1.65	0.0188	0.56	0.0320	0.62	1.86	0.4056
Silica								
Shetucket	0.05	5.14	0.5726	0.05	0.6026	−0.05	−4.55	0.9172
Quinebaug	−0.32	−43.29	0.0290	−0.37	0.0127	−0.87	−117.99	0.0282
Connecticut	0.45	740.40	0.0004	0.47	0.0003	0.69	1126.18	0.0591
Farmington	0.16	21.06	0.2582	0.15	0.2682	0.03	3.51	0.9480
Hockanum	−0.26	−6.28	0.2774	−0.30	0.2090	0.21	5.08	0.6349
Salmon	0.16	3.70	0.0581	0.19	0.0376	−0.20	−4.46	0.6497
Quinnipiac	0.30	12.92	0.0438	0.29	0.0552	0.30	13.29	0.3893
Housatonic	0.70	64.15	0.0731	0.86	0.0448	1.63	149.09	0.0906
Naugatuck	−0.22	−15.68	0.0219	−0.21	0.0277	−0.61	−43.31	0.0572
Saugatuck	0.42	1.59	0.0246	0.42	0.0240	0.35	1.33	0.6248
Norwalk	−0.43	−2.56	0.1898	−0.42	0.2011	−0.38	−2.27	0.5627

[a] Shetucket R. at S. Windham, CT, 01122610; Quinebaug R. at Jewett City, CT, 01127000; Connecticut R. at Thompsonville, CT, 01184000; Farmington R. at Tariffville, CT, 01189995; Hockanum R. near East Hartford, CT, 01192500; Salmon R. near E. Hampton, CT, 01193500; Quinnipiac R. at Wallingford, 0119650O; Housatonic R. at Stevenson, CT, 01205500; Naugatuck R. at Beacon Falls, CT, 01208500; Saugatuck R. near Redding, CT, 01208990; Norwalk R. at Winnipauk, CT, 01209710

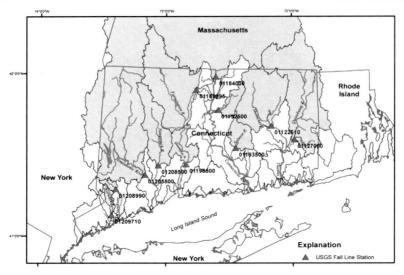

USGS station number	USGS station name	Drainage area (km²)	Period of record used (water years)	Urban land use, in percent	Agricultural land use	Forested land cover, in percent	Wetlands, in percent	Impervious area, in percent	2000 population density, in people/km2
01122610	Shetucket River at South Windham, Conn.	1,060	1974-2008	10.0	7.6	68.7	10.0	1.8	99
01127000	Quinebaug River at Jewett City, Conn.	1,850	1974-2008	11.4	10.1	60.9	13.6	2.6	98
01184000	Connecticut River at Thompsonville, Conn.	25,000	1974-2008	6.9	6.8	78.0	3.5	1.6	45
01189995	Farmington River at Tariffville, Conn.	1,490	1974-2008	13.7	5.8	71.3	4.9	3.1	126
01192500	Hockanum River near East Hartford, Conn.	190	1990-2008	41.4	9.8	40.9	5.2	13.2	515
01193500	Salmon River near East Hampton, Conn.	259	1974-2008	11.3	8.3	62.5	14.6	1.8	101
01196500	Quinnipiac River at Wallingford, Conn.	298	1974-2008	53.2	2.8	36.9	4.1	17.5	534
01205500	Housatonic River at Stevenson, Conn.	4,000	1974-2008	10.4	13.4	66.6	5.0	2.3	88
01208500	Naugatuck River at Beacon Falls, Conn.	673	1974-2008	25.5	8.9	59.4	2.9	8.6	358
01208990	Saugatuck River near Redding, Conn.	54	1974-2008	13.2	3.7	75.2	5.5	1.3	129
01209710	Norwalk River at Winnipauk, Conn.	86	1981-2008	27.6	2.7	63.2	4.6	6.8	279

Fig. 5.30 Water-quality stations in Connecticut used for flux monitoring by USGS. (Land use/ land cover data interpreted from The United States Environmental Protection Agency (2001a, b)

Nitrogen loads from Connecticut's WWTFs were generally lower in 2010 than in 1995. Nitrogen loads from these point sources have been declining, particularly since 2002, when the Nitrogen Credit Exchange (NCE) program began (CTDEEP 2011). Nitrogen loads from wastewater treatment facilities in Connecticut and New York have been managed in watershed-based zones (Fig. 5.33). End-of-pipe discharge of N is an order of magnitude greater in management zones 8 and 9 in the New York City area, compared to other management zones in Connecticut and New York (graphs of these zones are highlighted in the Fig. 5.5 because they are at a different scale). End-of-pipe N loads averaged 21,000 kg/day in Connecticut and 64,000 kg/day in New York in 1995. In 2010, they totaled 11,300 kg/day, and

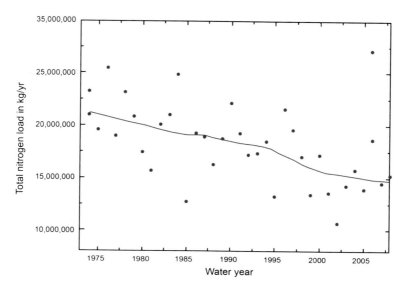

Fig. 5.31 Sum of estimated total N loads for fall-line stations in Connecticut, water years 1974–2008 (Stations include Connecticut River (01184000), Farmington River (01189995), Quinnipiac River (01196500), Housatonic River (01205500), Naugatuck River (01208500), Shetucket River (01122610), and Quinebaug River (01127000), fitted line is a Lowess smooth curve)

50,000 kg/day, respectively (Joseph Salata, USEPA Long Island Sound Office, written communication, March 3, 2011).

Total P concentrations and loads (Fig. 5.34) have also declined in the majority of stations analyzed for this summary (Tables 5.9, 5.10), probably due to improvements in wastewater treatment practices that are required by state and federal law, and state bans on phosphate in most detergents (Litke 1999). Total P loads (nonflow-adjusted) declined significantly at eight stations, averaging 1.6 %/ year over the period of record. Significant downward trends in total P occurred over this long period of analysis (water years 1974–2008), but there are indications of increases in P concentrations and loads toward the end of the record. Analysis of data water years 1993–2003 (Sprague et al. 2009) indicates significant increasing flow-adjusted and nonflow-adjusted trends in total P at five of these stations. Phosphorus yields ranged from 0.1 kg/ha/year in watersheds with minimal development to 3.3 kg/ha/year in the past at sites with high-density urban development and major wastewater discharge (Table 5.10). Total P loads from the seven major fall line stations described above declined during the period of study, from about 2.5 Mkg/year to 1.3 Mkg/year. The trends identified by these analyses indicate large declines in N and P loads during 1974–2008 with associated changes in the ratios of these nutrients. Although changes in Si/N ratios were initially implicated as a cause for shifts away from diatom dominance in coastal waters, changes in TN/TP may also have implications for the dominant phytoplankton in LIS (diatoms and dinoflagellates; Sommer 1994; Hodgkiss and Ho 1998). The TN/TP

Table 5.10 Loads and yields of nutrient constituents at USGS fall-line monitoring stations on rivers in Connecticut, water years 1974, 1990, 2000, and 2008 [Loads in kg/year, yields in kg/ha/year]

| Water Year | 1974[b] | | 1990[c] | | 2000 | | 2008 | |
River[a]	Load[d]	Yield[e]	Load	Yield	Load	Yield	Load	Yield
Total nitrogen								
Shetucket	1,420,000	7.7	1,300,000	7.0	988,000	5.4	897,000	4.9
Quinebaug	633,000	6.0	612,000	5.8	438,000	4.1	377,000	3.6
Connecticut	14,400,000	5.8	14,600,000	5.8	11,500,000	4.6	10,600,000	4.2
Farmington	973,000	6.5	1,270,000	8.5	1,030,000	6.9	816,000	5.5
Hockanum			341,000	17.9	364,000	19.1	354,000	18.6
Salmon	120,000	4.6	146,000	5.6	92,300	3.6	80,600	3.1
Quinnipiac	589,000	19.8	638,000	21.4	542,000	18.2	529,000	17.8
Housatonic	2,660,000	6.7	2,870,000	7.2	2,270,000	5.7	2,020,000	5.1
Naugatuck	1,420,000	21.1	1,410,000	20.9	802,000	11.9	431,000	6.4
Saugatuck	19,900	3.7	21,000	3.9	16,000	2.9	16,600	3.1
Norwalk	34,100	4.0	50,400	5.9	37,200	4.4	51,400	6.0
Total phosphorus								
Shetucket	166,000	0.9	87,100	0.5	52,600	0.3	45,300	0.2
Quinebaug	71,400	0.7	29,000	0.3	19,100	0.2	20,100	0.2
Connecticut	1,780,000	0.7	1,080,000	0.4	789,000	0.3	922,000	0.4
Farmington	162,000	1.1	110,000	0.7	87,500	0.6	86,600	0.6
Hockanum			43,100	2.3	32,400	1.7	30,800	1.6
Salmon	5,830	0.2	4,840	0.2	2,670	0.1	2,750	0.1
Quinnipiac	98,900	3.3	67,600	2.3	51,300	1.7	48,300	1.6
Housatonic	214,000	0.5	118,000	0.3	87,200	0.2	104,000	0.3
Naugatuck	185,000	2.7	154,000	2.3	151,000	2.2	190,000	2.8
Saugatuck	1,020	0.2	982	0.2	599	0.1	536	0.1
Norwalk	1,950	0.2	3,230	0.4	2,140	0.3	2,770	0.3
Total organic carbon								
Shetucket	8,790,000	47.6	8,040,000	43.5	5,940,000	32.2	6,190,000	33.5
Quinebaug	4,740,000	44.9	3,840,000	36.3	2,870,000	27.2	3,510,000	33.2

(continued)

Table 5.10 (continued)

Water Year	1974[b]	1974	1990[c]	1990	2000	2000	2008	2008
River[a]	Load[d]	Yield[e]	Load	Yield	Load	Yield	Load	Yield
Connecticut	95,100,000	38.0	86,300,000	34.5	72,400,000	28.9	94,100,000	37.6
Farmington	5,840,000	39.1	5,460,000	36.5	4,520,000	30.2	4,810,000	32.2
Hockanum			506,000	26.6	637,000	33.5	547,000	28.8
Salmon	974,000	37.6	966,000	37.3	668,000	25.8	736,000	28.4
Quinnipiac	1,760,000	59.1	1,360,000	45.7	950,000	31.9	1,160,000	39.0
Housatonic	14,100,000	35.3	12,300,000	30.8	11,400,000	28.5	13,000,000	32.5
Naugatuck	4,320,000	64.2	3,010,000	44.7	2,090,000	31.0	2,160,000	32.1
Saugatuck	247,000	45.4	207,000	38.1	145,000	26.7	154,000	28.3
Norwalk	98,600	11.5	271,000	31.7	216,000	25.3	257,000	30.1
Silica								
Shetucket	9,150,000	49.5	8,670,000	46.9	6,450,000	34.9	6,770,000	36.7
Quinebaug	6,080,000	57.5	6,080,000	57.5	4,910,000	46.5	5,070,000	48.0
Connecticut	90,800,000	36.3	109,000,000	43.6	104,000,000	41.6	123,000,000	49.2
Farmington	6,920,000	46.3	7,530,000	50.4	6,710,000	44.9	6,920,000	46.3
Hockanum			944,000	49.7	1,060,000	55.8	1,030,000	54.2
Salmon	1,420,000	54.8	1,560,000	60.2	1,250,000	48.3	1,290,000	49.8
Quinnipiac	1,920,000	64.5	2,370,000	79.6	2,090,000	70.2	1,970,000	66.2
Housatonic	12,400,000	31.0	14,800,000	37.0	13,400,000	33.5	17,400,000	43.5
Naugatuck	3,780,000	56.1	3,770,000	56.0	3,070,000	45.6	3,240,000	48.1
Saugatuck	282,000	51.8	325,000	59.7	274,000	50.4	273,000	50.2
Norwalk	181,000	21.2	431,000	50.4	350,000	40.9	397,000	46.4

[a] Shetucket R. at S. Windham, CT, 01122610; Quinebaug R. at Jewett City, CT, 01127000; Connecticut R. at Thompsonville, CT, 01184000; Farmington R. at Tariffville, CT, 01189995; Hockanum R. near East Hartford, CT, 01192500; Salmon R. near E. Hampton, CT, 01193500 Quinnipiac R. at Wallingford, 01196500; Housatonic R. at Stevenson, CT, 01205500; Naugatuck R. at Beacon Falls, CT, 01208500 Saugatuck R. near Redding, CT, 01208990; Norwalk R. at Winnipauk, CT, 01209710

[b] Norwalk River, beginning water year is 1981

[c] Hockanum River, beginning water year is 1992

[d] Load is the amount delivered from the basin upstream from the monitoring station over the course of the water year

[e] Yield is the annual load divided by the drainage area, and allows for comparisons among stations

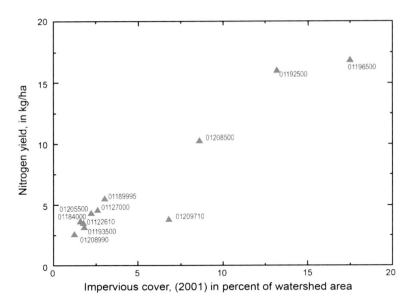

Fig. 5.32 Relationship between impervious cover and N yield in selected watersheds that drain into LIS (Impervious cover data interpreted from The United States Environmental Protection Agency 2001)

ratios were lowest (11–26) among sites with the most developed watersheds (Hockanum, Naugatuck, and Quinnipiac Rivers), and the highest ratios (43–76) were found for the least developed watersheds (Salmon and Saugatuck Rivers). The TN/TP generally increased from water years 1974 to the mid-1990s, and then began to decline until present. This is generally due to declines in TP loads relative to TN loads in the early part of the record followed by slightly increasing TP loads from the late 1990s through 2008 (Fig. 5.35).

Flow-adjusted concentrations and loads of total organic carbon (Fig. 5.36) declined significantly at most of the stations during the study period. Only three stations had significant declines in nonflow-adjusted loads of total organic carbon (Table 5.9).

The trends in Si loads (Fig. 5.37) and concentrations were not consistent over the time period of study (Table 5.9). The Quinebaug and Naugatuck Rivers had a significant decline in flow-adjusted concentrations and loads of Si, while the Connecticut, Quinnipiac, and Saugatuck Rivers had a significant increase in flow-adjusted Si concentrations and loads. Only the Quinebaug River site had a significant decline in nonflow-adjusted loads. The Connecticut River had a marginally significant ($p = 0.059$) increase in Si load that amounted to 0.69 %/year, or 24 % during the total studied period of record. Silica concentrations measured at the fall-line stations were similar to those commonly expected for natural waters (SiO_2) (Hem 1985), and ranged from about 3–12 mg/L as SiO_2. The median yield of silica was 48 kg/ha/year, with an interquartile range of 37–57 kg/ha/year. The lowest yields occurred in the Connecticut and Housatonic Rivers, and the highest

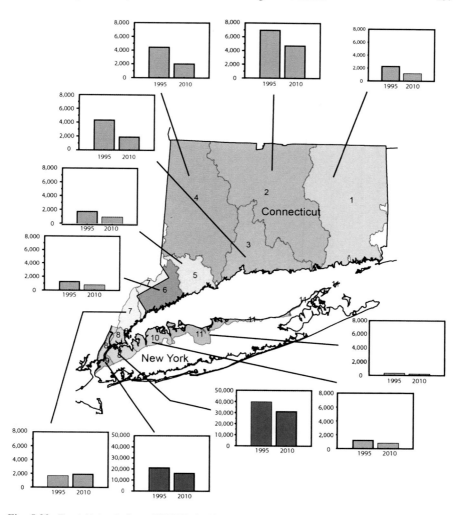

Fig. 5.33 Total N loads from WWTFs in N management zones for LIS, 1995 and 2010. The units along the Y axis are in mean kg/day (Data from Joseph Salata, USEPA Long Island Sound Office 2010, written communication)

yield was in the Quinnipiac River (Table 5.10). Lower yields of dissolved silica have been associated with the accumulation of biogenic silica from diatom productivity in the impoundments (Humborg et al. 2000; Triplett et al. 2008). The higher yields associated with the Quinnipiac River may be due to higher silica concentrations in groundwater in sediments of the Mesozoic Valley of Connecticut relative to sediments derived from other local bedrock types (Grady and Mullaney 1998). Increasing development of watersheds and climate change may affect the baseflow index and therefore the Si loads, due to the introduction of additional

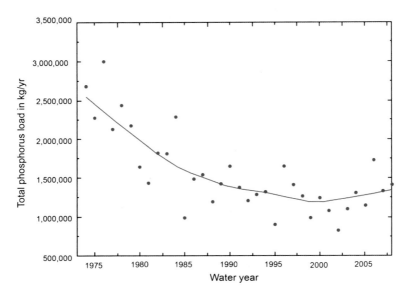

Fig. 5.34 Sum of estimated total P loads for fall-line stations in Connecticut, water years 1974–2008 (Stations include Connecticut River (01184000), Farmington River (01189995), Quinnipiac River (01196500), Housatonic River (01205500), Naugatuck River (01208500), Shetucket River (01122610), and Quinebaug River (01127000), fitted line is a Lowess smooth curve)

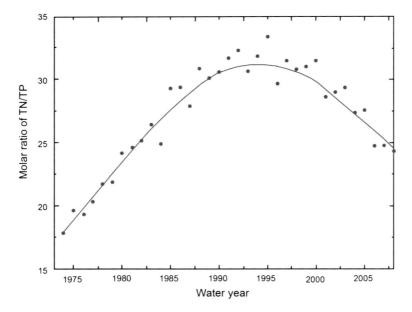

Fig. 5.35 The TN:TP ratios for combined fall-line stations in Connecticut, water years 1974–2008 (Stations include Connecticut River (01184000), Farmington River (01189995), Quinnipiac River (01196500), Housatonic River (01205500), Naugatuck River (01208500), Shetucket River (01122610), and Quinebaug River (01127000), fitted line is a Lowess smooth curve)

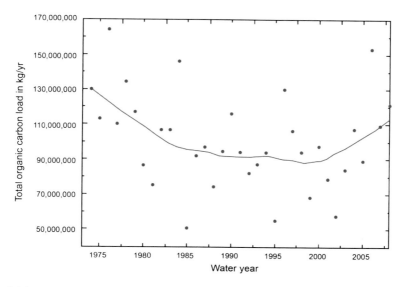

Fig. 5.36 Sum of estimated total organic C loads for fall-line stations in Connecticut, water years 1974–2008 (Stations include Connecticut River (01184000), Farmington River (01189995), Quinnipiac River (01196500), Housatonic River (01205500), Naugatuck River (01208500), Shetucket River (01122610), and Quinebaug River (01127000), fitted line is a Lowess smooth curve)

impervious areas (Loucaides et al. 2007), and changes in recharge rates and the time distribution of streamflow. Carey and Fulweiler (2011) report relations among land use/land cover and silica loads for streams in southern New England, with forested watersheds delivering the smallest loads and developed land delivering the largest loads. Further information on silica loading from the watershed is necessary to understand the mechanisms that control the loading of this nutrient.

Groundwater loading is an important transport mechanism of N entering LIS, more so than for P, which tends to be low in groundwater underlying urban development in Connecticut (Grady and Mullaney 1998). Phosphorus from groundwater can still constitute a substantial part of the watershed load of P (Mullaney 2007) and can be mobile under some geochemical conditions (Denver et al. 2010). Phosphorus loads from groundwater have not been studied extensively in the LIS watershed; most P is likely delivered by point sources and stormwater runoff. More information is needed to understand the loading of Si from direct and indirect (tributary) discharge of groundwater to LIS. Concentrations of Si in groundwater are similar to those in surface waters of the LIS watershed, and concentrations can be related to aquifer source material and land use (Grady and Mullaney 1998).

The loading of N from groundwater can be indirect, as a component of tributary loads (Mullaney 2007) or can be direct discharge to coastal embayments or LIS. Loading of N from groundwater may require different management mechanisms than the management of stormwater. The groundwater travel time from the

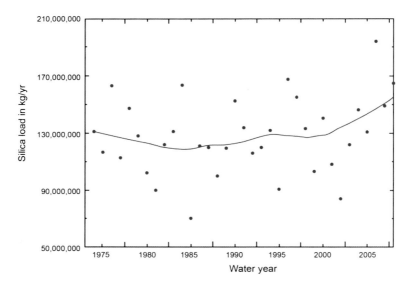

Fig. 5.37 Sum of estimated Si loads for fall-line stations in Connecticut, water years 1974–2008 (Stations include Connecticut River (01184000), Farmington River (01189995), Quinnipiac River (01196500), Housatonic River (01205500), Naugatuck River (01208500), Shetucket River (01122610), and Quinebaug River (01127000), fitted line is a Lowess smooth curve)

source of recharge to the tributary or embayment of LIS can be long (years to decades), delaying the effects of land-use changes or management actions on receiving waters. Loading of N from groundwater from Long Island to LIS was estimated at 0.75 Mkg/year (Scorca and Monti 2001). In Connecticut, the percent base flow of streams relatively unaffected by human activities is determined by the percentage of each watershed underlain by coarse-grained glacial stratified deposits (Mazzaferro et al. 1979). Connecticut is underlain by about 18 % coarse-grained glacial stratified deposits, which equates to a base-flow index of 46 % of total runoff, indicating that groundwater discharge may be a significant part of the tributary flow in Connecticut watersheds. Further research is needed to understand the importance of this groundwater source on both the Connecticut and New York parts of LIS, and the mechanisms of transport from the groundwater system to coastal waters. Information is lacking on the importance of residential and municipal on-site wastewater management in contributing to this subsurface load of nutrients.

In summary, total N loads from tributaries in Connecticut have declined relative to 1974, but appear to have stabilized or slightly increased in the more recent part of the record. Total P loads have generally declined, but may have increased slightly toward the end of the record for this study (2008). Silica loads appear to vary with discharge, but are possibly affected by impoundments in some of the tributaries. Changes to the time distribution of stream flow caused by withdrawals for public water supply, and increased impervious surfaces, and other changes to land cover may have an effect on the yields of silica as well.

Atmospheric wet deposition of inorganic N appears to have declined since 1994. Nitrogen yields from forested watersheds generally were smaller than the amount of wet deposition of inorganic N they received in 1994, indicating retention of N in these areas. In 2008, wet deposition of N was only slightly higher than the total N loss from forested watersheds. Questions remain about whether these areas will become saturated with N, increasing the losses of N from the watershed, similar to those described by Aber et al. (2003).

The data presented here are useful in understanding the loads of TN from the major tributaries. The TN load from the Connecticut River represents a large part of the overall tributary load. Other sources of N with limited information include the net flux of N from the East River in the western end of LIS, the importance of the loading of N from direct and indirect groundwater sources, and the processing of N in tidally affected rivers.

5.7.3 LIS Water Column Nutrient Data

The CTDEEP carries out biweekly surveys of water quality in LIS along a series of fixed stations. Analytical data from water samples from these and several other cruises provide a snapshot of the distribution of nutrient concentrations and $\delta^{15}N$ values in LIS in 2007 (Boon 2008). The nutrient concentrations and isotopic ratios in LIS waters are determined by the various inputs and by the sinks, mainly primary productivity, but also exchange with the open ocean through tidal exchange (Bowman 1977). The isotopic ratio of nitrogen ($\delta^{15}N$) shows the pathways that the N has gone through before arrival in LIS. Terrestrial soil fixed nitrogen (NO_3) usually has $\delta^{15}N$ values of 0–5 ‰ whereas atmospheric deposition nitrate has $\delta^{15}N$ of 0–3 ‰. Consumption of land plants by higher organisms leads to higher $\delta^{15}N$ values in their tissues and waste products. During denitrification (turning NO_3^- into N_2 gas), the bacteria prefer the light isotope and what remains behind is isotopically heavy. As a result, manure, animal remnants, and WWTF effluents tend to have relatively high $\delta^{15}N$ values (with values >10 ‰).

The concentrations of the various N species display differences for surface and bottom waters and a strong E–W trend with higher concentrations in WLIS (Table 5.11). Records from CTDEEP from 1991 to 2004 of TDIN (in mg N/L) versus time show the characteristic cyclic pattern of build-up during the winter and

Table 5.11 Ranges of nutrient concentrations (μmol/L) in LIS waters (after Gobler et al. 2006)

	East river	WLIS	CLIS	ELIS
TDIN	25–47	2.6–2.7	1.4–9.5	1.0–9.5
TDIP	1.5–3.8	0.6–2.8	0.4–2.1	0.5–1.5
DSi	26–78	10–85	13–78	15–41

TDIN total dissolved inorganic nitrogen, *TDIP* total dissolved inorganic Phosphorus, *DSi* dissolved silica

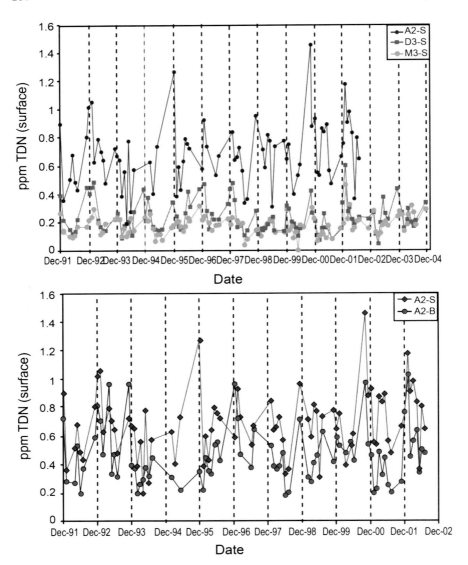

Fig. 5.38 **a** Temporal variations in total dissolved N in LIS surface waters in western LIS (A2), central LIS (D3), and eastern LIS (M3) over the period 1991 to 2002 (CTDEEP data). **b** Comparison of time trends in total N in surface and bottom waters at site A2 (western LIS)

withdrawal during the early spring bloom (Fig. 5.38). Phytoplankton (e.g., diatoms) blooms occur in late winter/early spring (sunlight induced algal blooms), and late summer and early fall from nutrient build-up, warm waters, and strong sun exposure. Nitrogen concentrations drop markedly during these blooms. Experiments with nutrient addition (Gobler et al. 2006) demonstrate that LIS primary production is largely N-limited. The NO_3/NH_3 ratio is determined by the various inputs as well as by the extraction by photosynthesizing organisms, some preferring NH_3, others NO_3. The input of WWTF ammonia is an important factor as is the ammonia flux from the decomposition of organic material in the bottom waters (Cuomo et al. 2005). The molar ratios of nitrate to ammonium in the two main input rivers vary from 2.5 to 20 (with seasonal fluctuations), with a mean of ~10, whereas WWTFs have much lower ratios (~1.5–2). The molar NO_3/NH_3 in LIS varies between 0.15 and 0.75, which is very different from the nitrate-dominated riverine environment. The variations over time for surface and bottom waters in station A2 indicate that in WLIS the surface waters tend to be enriched in TDIN with respect to the bottom waters, whereas the reverse may be true in waters from CLIS and ELIS.

The Si concentrations reach very low values during the blooms, and locally, Si-limitation may curb diatom productivity (Gobler et al. 2006; Boon 2008). A detailed analysis of the extensive dissolved nutrient data from LIS collected over the last few decades in conjunction with the measured inputs summarized above is a necessity for a better understanding of nutrient cycling and primary productivity in LIS over time, but is outside the scope of this chapter.

The spatial distribution of dissolved nitrogen, P, and silica in LIS varies with the season, but a snapshot is provided for the year 2007 (Figs. 5.39, 5.40). The WLIS surface and bottom waters are rich in P and ammonium, whereas the dissolved Si shows a simple variation with salinity (S) (indicating that most of the Si is brought in by rivers). The dissolved NO_x pattern is more complex, with depletions in CLIS (probably related to local productivity) and enrichments in WLIS.

The isotopic composition of nitrogen ($\delta^{15}N$) is a function of its ultimate source, with, in general, isotopically heavier N stemming from material higher up in the food chain (Cravotta 2002). In addition, the removal of N through photosynthetic activity may fractionate the N isotopes, and thus the resulting N isotope ratios are a complex function of N supply, speciation, and withdrawal. Denitrification is extensively used in WWTFs to remove N, and WWTF effluents may show $\delta^{15}N(NH_3) = +6$ ‰ to $+25$ ‰, and $\delta^{15}N(NO_3) = -9$ ‰ to $+16$ ‰ (Boon 2008). Riverine TDIN is a mixture of all N sources which include effluents of local WWTFs and has $\delta^{15}N(NH_3)$ from $+2$ to $+4$ ‰ and $\delta^{15}N(NO_3)$ from 0 to $+10$ ‰ (data from samples collected between February and August 2007 in the Connecticut River in various locations above the estuarine mixing zone; Boon 2008)). In the Housatonic River, $\delta^{15}N(NH_3)$ ranges from $+2$ to $+10$ ‰ and $\delta^{15}N(NO_3)$ from $+4$ to $+12$ ‰ for the same period. The modern mean $\delta^{15}N$ of total riverine N is ~$+5$ ‰, whereas that for WWTFs is heavier at ~$+10$ to $+12$ ‰, albeit with large seasonal and local variations. The N isotope data thus can be used to broadly trace the N sources and possibly the fate of the relatively heavy WWTF N within the LIS foodweb, and ultimately in the buried remains of photosynthate in sedimentary marine organic matter. Observed $\delta^{15}N$ in nitrate in LIS waters ranges from -8.6 ‰ to $+5.7$ ‰ in surface waters, and in bottom waters from -2.7 ‰ to $+11.4$ ‰ (Fig. 5.39). The $\delta^{15}N$ values for dissolved NH_3 are not yet available, but presumably on the heavy end of the range, up to $+20$ %.

A study by Anisfeld et al. (2007) showed the origin of N compounds in the Quinnipiac and Naugatuck Rivers in Connecticut and apportioned the sources of

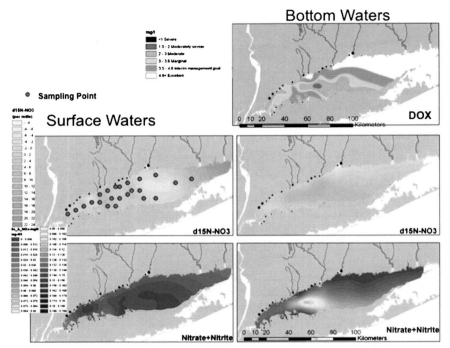

Fig. 5.39 Spatial trends in LIS during a period of water column stratification and hypoxia in bottom waters in August 2007. Samples collected during CTDEEP sampling cruise in central and western LIS (red dots are sampling points). Spatial patterns of dissolved oxygen (DOX) in bottom waters, $\delta^{15}N$ in nitrate, and nitrate+nitrite in surface and bottom waters are shown. The black dots along the coastlines are WWTPs, with the dot size reflecting their relative N-output. Colors at the eastern end of LIS are an artifact of the spatial interpolation model, most easterly data point is located just west of the mouth of the Connecticut River. Note the correlation between low dissolved oxygen and high dissolved nitrate+nitrite and high $\delta^{15}N$ in nitrate in the bottom waters of CLIS. Figure adapted from Boon, 2008; measurements made in the laboratory of Dr. Mark Altabet (SMAST, Univ. of Massachusetts, Dartmouth, MA)

N in these rivers based on the $\delta^{15}N$ and the $\delta^{18}O$ in nitrate. They found for these rivers that at base flow, the atmospheric deposition of N was a small contributor, whereas during storm flow a significant part of the N came from the atmospheric reservoir. The contributions from sewage were hard to determine from that study and the authors showed a very wide range in estimates of sewage contributions.

5.7.4 The Historic Record of Organic Productivity from Sediment Core Data

The modern LIS nutrient concentration data discussed above do not aid directly in the identification of trends in nutrient fluxes beyond the last 4 decades. The

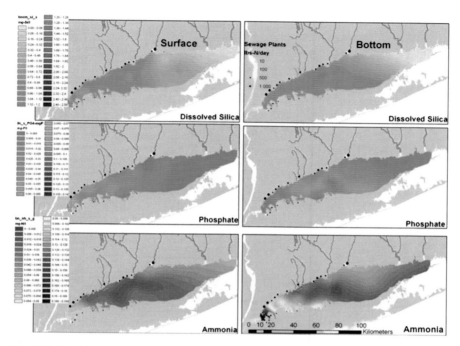

Fig. 5.40 Spatial trends in LIS for dissolved Si, P and NH$_3$ in surface and bottom waters of LIS in August 2007 (see caption of Figure 5.39 for sampling details). Note the high NH$_3$ concentrations in extreme western LIS

increased nutrient fluxes of the last 100–200 years probably enhanced the primary productivity in LIS, which in conjunction with global climate change (e.g., enhanced stratification of LIS; see O'Donnell et al. Chap. 3, in this volume), led to the seasonal hypoxia in CLIS and WLIS. The hypoxic conditions of the last few decades are probably also impacted by the organic carbon that is now stored in the sediment bed, which provides a long-term oxygen sink. Many processes play a role in the creation of hypoxia, and one approach to unravel this multitude of drivers is to consider the historic record of eutrophication in sediment cores to assess when eutrophication started in conjunction with other factors that may have a causal role.

The organic carbon and N concentrations in sediment cores indicate the amount of buried organic matter, which is commonly recast as organic carbon accumulation rates (mg Corg/cm^2 year) if the sedimentation mass accumulation rates are adequately known. These carbon accumulation rates are an indirect proxy for historic nutrient fluxes into the Sound because higher nutrient fluxes usually translate into higher rates of primary productivity (the system is N-limited; e.g., Gobler et al. 2006) and higher burial rates of organic matter. Integration of core data over the whole bottom of the Sound to obtain LIS-wide burial rates of organic carbon and N is fraught with many uncertainties, but the trends in representative cores provide proxies for changes in the overall fluxes of carbon and N over time. The organic fractions of sediment are mixtures of organic carbon imported from the watersheds (terrestrial carbon) and in situ production of marine algal matter. Mixing calculations based on $\delta^{13}C$ and N/C values (Perdue and Koprivnjak 2007) of the organic fraction of the sediment are used to deconvolute these mixtures, so pure marine organic matter accretion rates can be calculated (Varekamp et al. 2009). The burial rate of marine organic carbon is an indirect proxy for primary productivity and for the nutrient fluxes that sustained that productivity.

The isotopic composition of N in LIS sediment is, among others things, an indicator of the N sources, although the N isotopes are also fractionated during the photosynthetic process. The modern increase in N fluxes stems partly from enhanced WWTF flows, which have more positive $\delta^{15}N$ values than most other sources. Mapping the $\delta^{15}N$ in LIS water and sediment cores thus is another mode of tracing the magnitude of increased N fluxes. A caveat is that changes in the rates of denitrification and in the food chain may also cause changes in $\delta^{15}N$ in the buried organic matter. We present data on these aspects for two sediment cores: core WLIS75GGC1, taken near Execution Rock in the far western part of WLIS and core B1GGC2, taken on the delta of the Housatonic River in CLIS (Fig. 5.12).

Evidence is found in the sediment of hypoxia beginning in the early to mid-1800s, as indicated by lighter $\delta^{13}C$ values in carbonate (Lugolobi 2003; Lugolobi et al. 2004). The oxidation of photosynthate in the bottom waters and sediment porewaters leads to consumption of dissolved oxygen, but also to an isotopically lighter carbon pool of DIC in the bottom waters (the photosynthate is isotopically light compared to marine DIC). The presence of this isotopically light DIC is reflected in the lower values of $\delta^{13}C$ in carbonate tests of benthic foraminifera (Thomas et al. 2000, 2004). The records of $\delta^{13}C$ in carbonate of foraminifera from the two cores show more or less constant values during precolonial times, but then drop by 1–3 ‰ since the early 1800s. Part of this drop in $\delta^{13}C$ is a reflection of the isotopically lighter CO_2 in the atmosphere, which has decreased by about 1 ‰ over the last 150 years (Keeling et al. 2005), That decreasing trend is also shown in marine carbonates in various other studies (e.g., Bohm et al. 2002). After correction for this atmospheric isotopic change, it appears that most cores in WLIS show an additional decrease in $\delta^{13}C$ in foraminiferal carbonate of about 1 %, indicating potentially the onset of eutrophication and associated hypoxia in WLIS about 150 years ago. This time scale is coincident with the increase in population density and growth in industrialization around LIS and, presumably, increased nutrient fluxes.

Particulate organic carbon (POC) is a contaminant in LIS because it creates biological oxygen demand that ultimately is one of the causes of the hypoxia. The POC in LIS stems from the local marine primary productivity (marine organic carbon or CORGM) and from import from land (terrestrial carbon or CORGT) as well as a particulate CORG flux from WWTFs (CORGWW). The three forms of POC can be distinguished based on their $\delta^{13}C$, C/N, and $\delta^{15}N$ values (Meyers 1997; Varekamp et al. 2009), although we do not distinguish explicitly here the CORGWW component (similar in C/N to marine organic matter with C/N~8). The CORGM accumulation rates in sediment cores are a proxy for marine primary productivity although the cores only represent the buried (and preserved) fraction of the total marine organic

Fig. 5.41 Total organic C accumulation rates in cores from central and western LIS (see Fig. 5.12 for locations). Data sources: Varekamp et al. 2009; Varekamp and Thomas 2010; Varekamp unpublished data)

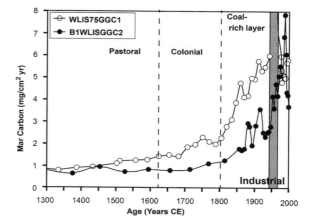

Fig. 5.42 Terrestrial and marine organic C accumulation rates for core WLIS75 in the extreme western end of LIS. Data sources: same as Fig. 5.41

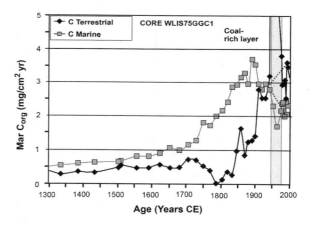

productivity. Other fractions of the marine productivity are consumed higher up in the food chain and possibly exported from LIS (fish catch, movement of fish to other areas). Most LIS cores show higher CORG concentrations in the more recent sediment, and bulk CORG accumulation rates increased five to eight times from precolonial background levels to recent times (Fig. 5.41).

The CORGM accumulation record in the WLIS core WLIS75GGC1 shows a strong increase in late colonial times, but decreased strongly over the last 100 years (Fig. 5.42). The CORGT dropped to almost zero in the mid- to late 1700s, most likely the result of the extensive deforestation at that time. After 1800, the CORGT flux increases again and by 1900, the CORGT and CORGM accumulation rates are roughly equal, and CORGT is the dominant form of organic carbon in core sediments deposited over the last 50 years. The Housatonic River delta core B1GGC2 record has a similar pattern, be it less extreme, and the CORGT in the sediment dominates over the last 100 years (Varekamp et al. 2000).

Fig. 5.43 Biogenic silica accumulation rates in two cores from western and central LIS, showing strong increases between late 1700s and mid 1900s, followed by a strong decline (after Andersen 2005)

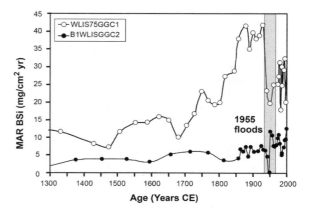

Obviously, the contributions of both marine and terrestrial organic matter to LIS have fluctuated strongly over the last few hundred years, a reflection of changes in landscape and nutrient fluxes. The common hypothesis that increased nutrient fluxes drove up the marine primary productivity of the Sound and thus caused hypoxia is not the complete story; the impact of the abundant CORGT on the occurrence of hypoxia should probably be considered as well (see e.g., Latimer et al. 2003). The accumulation rates of biogenic silica (BSi) were measured in many LIS cores (Andersen 2005) and have increased strongly since 1800, but decreased over the last 50 years (Fig. 5.43). The Execution Rock core had higher background (4–5x) and much higher peak (10x) BSi accumulation rates compared to those in the Housatonic delta core. Higher nutrient concentrations may have been common in the western Sound over time, which seems to translate into higher sedimentary BSi values. Apparently, the productivity of diatoms decreased, presumably as a result of silica limitation, and possibly other species took over (e.g., dinoflagellates; see Lopez et al. Chap. 6, in this volume).

The sedimentary N content and its isotopic composition provide a direct insight into sources of N over time, and in the following section we present preliminary, unpublished data from two cores. The sedimentary N increased simultaneously with CORG concentrations over time as has the $\delta^{15}N$ of the bulk sediment (Fig. 5.44). The precolonial $\delta^{15}N$ background values differ by ~1 ‰ in between the two studied cores, but both cores show an increase of 1.0–1.5 ‰ over the last 150–200 years. This strongly suggests the arrival of heavier N in LIS waters, most likely sewage-derived N. This may have created, possibly with enhanced fluxes of carbon as CORGWW, higher $\delta^{15}N$ values in the Sound waters over the last 150 years. The bulk $\delta^{15}N$ values of the CORG in sediment represent a mixed signal, however, influenced by the proportions of CORGT and CORGM in the sediment. The $\delta^{15}N$ signature of the bulk CORG was once more deconvoluted into that of CORGM, using a $\delta^{15}N$ of +2.5 ‰ for CORGT, using the proportions of the two types of CORG calculated before. These values provide a detailed record of the $\delta^{15}N$ of locally produced marine organic matter and presumably CORGWW over the last few hundred years (Fig. 5.45).

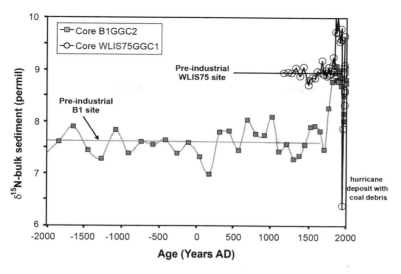

Fig. 5.44 Nitrogen isotope composition of bulk organic matter in the same two cores from LIS. Both cores show isotopically heavier N in sediments deposited since ~1800 AD. Data sources: same as Fig. 5.41

Fig. 5.45 Nitrogen isotope trends in marine organic carbon in sediment samples from the same two cores. Note the increase in the early 1800s, and the decline in N isotope ratio since the mid twentieth century in both cores. Data sources: same as Fig. 5.41

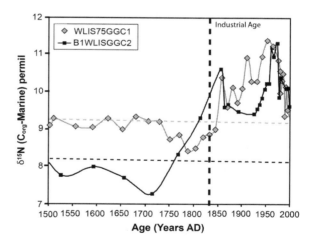

The two cores show very similar records and the precolonial $\delta^{15}N$ in CORGM from the Housatonic River delta core is ~+8.2 ‰ (based on a 4000 year record in this core; Varekamp et al. 2000) and rose to ~+10 % by 1850. The $\delta^{15}N$ in CORGM of the Execution Rock core rose from ~+9.2 ‰ in the precolonial background to ~+10.2 ‰ in industrial times. Both core records show a decrease in $\delta^{15}N$ in marine organic matter of ~1–1.5 ‰ over the last 50 years. The latter is a puzzling observation that seems to contradict the simple scenario of strong

Fig. 5.46 Marine organic C
accumulation records, driven
by two separate N sources:
WWTFs and nonpoint source
N similar to precolonial
N. Data sources: same as
Fig. 5.41

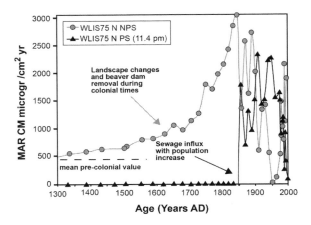

primary productivity stimulation by the abundant N from WWTFs with high $\delta^{15}N$
for the last half century. This aspect needs more study.

On a more speculative basis, we estimated the burial rate of marine organic
matter partitioned according to N derived from WWTFs and N from nonpoint sources. These calculations do not consider the N isotope fractionations
(or changes in those over time) during photosynthesis, and these estimates are
also sensitive to the assumed mean value of $\delta^{15}N$ in bulk WWTF effluents. The
observed mean value for $\delta^{15}N$ in WWTF effluent in 2007 was $+10$ ‰, but the final
value had to be taken equal to or larger than the heaviest $\delta^{15}N$ value in the core
records. We used $+11.4$ ‰ as the mean for $\delta^{15}N$ in WWTF effluents and assumed
a zero WWTF N flux prior to 1850. We use the precolonial $\delta^{15}N$ values in each
core as representative of the $\delta^{15}N$ from nonpoint source N, although we realize
that those values may have changed over time as well.

The mass accumulation rates of marine organic carbon productivity driven by
nonpoint source N inputs in the Execution Rock core sediment show an increase
in the 1600–1700 periods, probably as a result of landscape changes and beaver
dam removal (Varekamp 2006), and continued to rise until 1850. This is followed
by a decrease between 1850 and 1950 till almost zero contributions, but rising
strongly again since the mid-twentieth century. The WWTF N-driven productivity increased since 1850 and became dominant between 1900 and 1950, but then
again decreased to low values as found today.

The CORGM record of the Housatonic River delta core partitioned according to
N source is similar to that of WLIS75GGC1, with recent productivity equally divided
between WWTF and nonpoint source N (Fig. 5.46). A shift in primary productivity
from diatoms to dinoflagellates over the last 50 years may have both impacted the
rate of marine carbon burial in western LIS (Dortch et al. 2001) as well as changed
the $\delta^{15}N$ of the buried organic matter (Lopez et al. Chap. 6, in this volume).

In conclusion, nutrient pollution of the Sound is ongoing, although both riverine
and direct WWTF inputs of N and P have been diminishing over the last few decades.
The environmental effects of these processes are obvious in the seasonal hypoxia that

occurs with variable intensity in mid- to late summer in CLIS and WLIS. The trends in cores on stored organic carbon and its isotopic signatures indicate that the flux of terrestrial carbon into LIS may also be an important driver for hypoxia, although terrestrial carbon may be more refractory than marine carbon (Meyers 1997). The records also show that the increased marine productivity over the last 150 years is related to both enhanced N fluxes from land as well as from WWTFs.

5.8 Summary of Data and Research Needs and Recommendations

Long Island Sound represents an estuarine system significantly impacted by anthropogenic activities. Data from sediment cores show without any doubt the profound human impacts on the water and sediment quality of LIS since precolonial times. Inputs of both organic and inorganic contaminants increased steadily during the first three quarters of the twentieth century and peaked during the 1960–1970s. Fortunately, with implementation of upgraded sewage treatment associated with the Clean Water Act, and laws regulating use and releases of persistent toxic contaminants such as PCBs and chlorinated pesticides, general water quality has improved in LIS since the mid-1980s. However, despite these improvements, there remains a legacy of persistent contaminants in LIS sediments, and measurable contaminants are found in many LIS biota. Contaminant levels remain elevated in WLIS and CLIS relative to ELIS, and hot spots for various contaminants remain in coastal embayments, particularly along the Connecticut shores of LIS receiving inputs from major cities and/or rivers. A summary with broad sediment quality indicators (Fig. 5.47) show that in WLIS, almost 50 % of the habitat areas can still be rated as "poor." Recent trends in contaminant levels in indigenous bivalves in some cases show that levels are no longer decreasing, and in some cases appear to be on the rise again, albeit much below than those found in the 1980s. Of particular concern are methylmercury levels in biota. In addition to continuing inputs from atmospheric deposition and "legacy Hg" from the watersheds, decreased degrees of eutrophication in some areas may lead to diminished organic carbon burial (an important stable host for Hg) and thus could accelerate the release of methylmercury from LIS sediments. Despite reductions in contaminant concentrations in surface sediments, low level sediment toxicity appears to still be common in many areas of WLIS, but also in a number of coastal bays. Data from the NCA, as well as several smaller independent research projects, indicate that the potential for negative impacts on organisms throughout LIS still exists. These trends are illustrated in a recent survey comparing sediment quality indices estimated from levels of contaminants in sediment, sediment toxicity as measured by the *A. abdita* test, and organic carbon content for the three subareas of LIS using data collected from 2000 to 2004 as part of the NCA (USEPA 2008). This analysis indicated from 9 to 46 % of the area within each sub-basin was in poor condition (mainly WLIS) while between 18 and 69 % of the area was in good condition (Fig. 5.17a; USEPA 2008).

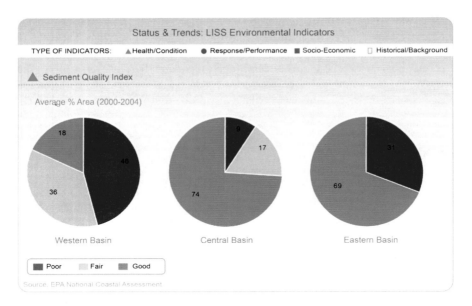

Fig. 5.47 Broad sediment quality indicators in the three LIS zones over the period 2000–2004. From: USEPA National Coastal Condition report (2008)

Of all human impacts, one of the most pervasive and disrupting has been the increase in nutrient fluxes into LIS. The loading of nutrients to LIS has been dynamic over time, but especially over the last 40 years. Actions to control the effects of wastewater discharges have been initiated, affecting the loads and ratios of N and P in the major tributaries. Remarkable success has been observed in the decrease in N loads from many sources, yet the occurrence and extent of hypoxia continue. The onset and persistence of hypoxia are influenced by meteorology (supply of oxygen; O'Donnell et al. Chap. 3, in this volume) as well as by respiration and decay processes of organic materials (consumption of oxygen; Cuomo et al. Chap. 4, in this volume), and decreases in N may not be the only factor influencing the extent and severity of hypoxia. Climate change and a hysteresis effect of the stack of sediment rich in organic matter may all influence the location and severity of LIS hypoxia.

Population and land development in the watershed have increased during the last few decades, adding nonpoint source loads of nutrients, as well as increasing wastewater volume. Much of the focus of research has been on N loading to LIS. Phosphorus loads from the major tributaries in Connecticut have generally declined from 1974 to 2008, but may be increasing in recent years. A coherent dataset to understand the trends in P loads from point sources in the LIS watershed is not available. The changing N/P and Si/N values have implications for changes in the composition of dominant phytoplankton in LIS. Further research is needed to understand the links among external and internal (from the sediment bed) sources of N, P, and Si, and hypoxia.

It is uncertain how climate change may affect nutrient loads from forested watersheds. Predictions of the water budget under a range of global emissions scenarios indicate likely increases in groundwater recharge and decrease in snowfall over the next century, especially over southern parts of the LIS watershed (Bjerklie et al. 2011). These changes will have implications for tree species shifts, deposition, and processing of N in forests, as well as possible changes in the timing and delivery of N to LIS.

The largest N sources for LIS are associated with the East River (Fig. 5.33) due to the sewage inputs of N from the New York City area. Currently, a delivery factor of 21 % has been used in N management from this zone, based on modeling done for the LIS TMDL (CTDEP and NYSDEC 2000). Studies to determine the net flux across this boundary would be essential to constrain the magnitude of this large source of N. In addition, limited information exists on the N losses in the tidal reaches of the major rivers. For example, the Connecticut, Housatonic, and Quinnipiac Rivers have extensive tidal marshes and brackish zones that may use substantial amounts of N. Knowledge of such estimates has implications for accurate equivalency factors used in N trading among municipalities in the LIS watershed.

Long Island Sound is fortunate to have been the focus of several large-scale environmental assessments by NOAA, USGS, and the USEPA, and state agencies, some of which such as NCA and the NS&T continue to provide data today. The extensive CTDEEP dataset on nutrient levels in LIS should be evaluated in detail, and variations in fluxes and riverine inputs should be assessed within the context of measured nutrient levels in LIS. Potential trends in pH of LIS waters are just beginning to be monitored in recent years by CTDEEP, changes in DIC over time are currently completely unknown, and studies are needed that focus on existing datasets and/or sediment proxies. Periodic detailed sampling and analyses of both metals and organic contaminants in sediments should be continued at multiple year (possibly five year) intervals, particularly in coastal embayments with high population density. The fate of legacy contaminants such as Hg, Cr, and Cu from the Housatonic River and several harbor areas should be monitored, especially after major hurricanes or other wet events. The main data available today on contaminant levels in biota come from NS&T's Mussel Watch dataset, a few studies on PCBs by the NYSDEC, and some recent work on metals in bivalves in Connecticut harbors. A useful option for expanding this work is to include species harvested for human consumption as well as key resource species of importance to aquatic food chains. Assessment of contaminant-related effects in indigenous biota is particularly lacking as is assessment of how global climate change may influence sources of contaminants to LIS, their cycling within LIS, particularly from sediment reservoirs into biota. Also lacking is an assessment of the combined impact of environmental factors such as hypoxia, lowered pH and nutrient, and toxic chemical additions on local organisms. Monitoring plans also need to include advective transport and internal cycling of emerging contaminants such as PBDEs, PFCs, pharmaceuticals, hormones, and personal care and cleaning products, particularly those that have a potential to act synergistically with other contaminants or influence disease susceptibility in indigenous organisms.

In conclusion, extensive water quality monitoring, studies on biota, and sediment samples have documented the plethora of environmental problems that LIS has faced and to some degree is still facing today. Legislation to limit these fluxes of a variety of contaminants has improved or stabilized the water quality in LIS, but lingering legacy pollutants and toxic sediment beds remain. Future research would benefit from a focus on the linkages among the different processes that impact the LIS system. Resource management should use an ecosystem-based approach that takes direct anthropogenic inputs, climate change, and changes in the watershed into account.

References

Aber JD, Goodale CL, Ollinger SV, Smith M-L, Magill AH, Martin ME, Stoddard JL (2003) Is nitrogen deposition altering the nitrogen status of northeastern forests? Bioscience 53:375–390

Adams D, Benyi S (2003) Final report: sediment quality of the NY/NJ harbor system: A 5-year revisit 1993/4–1998. EPA/902-R-002. USEPA-Region 2, Division of Environmental Science and Assessment. Edison, NJ

Adams DA, O'Connor JS, Weisberg SB (1998) Final report: sediment quality of the NY/NJ Harbor system—an investigation under the regional environmental monitoring and assessment program (REMAP). EPA/902-R-98-001. USEPA-Region 2, Division of Environmental Science and Assessment. Edison, NJ

Andersen T (2004) This fine piece of water. An environmental history of Long Island Sound. Yale University Press, New Haven 274

Andersen N (2005) The history of eutrophication and hypoxia in Long Island Sound. Undergraduate thesis, Wesleyan University, Middletown CT, 159

Anisfeld SC, Barnes RT, Altabet MA, Wu T (2007) Isotopic apportionment of atmospheric and sewage nitrogen sources in two Connecticut rivers. Env Sci Technol 41:6363–6369

Balcom PH, Fitzgerald WF, Vandal GM, Lamborg CH, Rolfhus KR, Langer CS, Hammerschmidt CH (2004) Mercury sources and cycling in the Connecticut River and Long Island Sound. Mar Chem 90:53–74

Benoit G, Rozan TF, Patton P, Arnold CL (1999) Trace metals and radionuclides reveal sediment sources and accumulation rates in Jordan Cove Connecticut. Estuaries Coasts 22(1):65–80

Bjerklie DM, Trombley TJ, Viger RJ (2011) Simulations of historical and future trends in snowfall and groundwater recharge for basins draining to Long Island Sound. Earth Interact 15:1–35

Boesch DF, Brinsfield RB, Magnien RE (2001) Chesapeake Bay eutrophication: Scientific understanding, ecosystem restoration, and challenges for agriculture. J Environ Qual 30(2):303–320

Böhm F, Haase-Schramm A, Eisenhauer A, Dullo W-Chr, Joachimski MM, Lehnert H, Reitner J (2002) Evidence for preindustrial variations in the marine surface water carbonate system from coralline sponges. Geochem Geophys Geosys 3(3):10.1029/2001GC000264

Boon N (2008) Nutrient dynamics of Long Island Sound. Master's thesis, Wesleyan University, Middletown CT, p 280

Bowman MJ (1977) Nutrient distributions and transport in Long Island Sound. Est Coast Mar Sci 5:531–548

Breslin VT , Sañudo-Wilhelmy SA (1999) High spatial resolution sampling of metals in the sediment and water column in Port Jefferson Harbor. Estuaries 22(3A):669–680

Breslin VT, Dart P, Clough D, Guancial S, Navarro J (2005) Sediment metal contamination in the Branford River and Harbor. Final Report CCMS-3-2005. Center for Coastal and Marine Studies, Southern Connecticut State University, 39 pp

Bricker SB, Ferreira JG, Simas T (2003) An integrated methodology for assessment of estuarine trophic status. Ecol Modelling 169(2003):39–60

Brownawell BJ, Fisher NS, Naeher L (1991) Characterization of data based on toxic chemical contaminations in Long Island Sound. Final Report to LISS, 112 pp (plus appendices)

Buchholtz ten Brink MR, Mecray EL, Galvin EL (2000) *Clostridium perfringens* in Long Island Sound sediments: an urban sedimentary record. J Coastal Res 16:591–612

Carey JC, Fulweiler RW (2011) Human activities directly alter watershed dissolved silica fluxes. Biogeochem. doi:10.1007/s10533-011-9671-2

Castro MS, Driscoll CT (2002) Atmospheric nitrogen deposition to estuaries in the mid-Atlantic and northeastern United States. Environ Sci Technol 36:3242–3249

Church VME (2009) A comparative study of sediment metal Clinton and Milford Harbors, CT. Dissertation, Southern Connecticut State University, New Haven, 55 pp

Cochran JK, Hirschberg DJ, Wang J, Dere C (1998) Atmospheric deposition of metals to coastal waters (Long Island Sound, New York U.S.A.): Evidence from salt marsh deposits. Estuarine Coastal Shelf Sci 46:503–522

Collier TK, Anulaction BF, Bill BD (1998) Hepatic CYP1A in winter flounder (Pleuronectes americanus) along the Northeast Coast: Results from the National Benthic Surveillance Project. Mar Poll Bull 37:86–91

Colombo MJ, Trench ECT (2002) Trends in surface-water quality in Connecticut: US Geological Survey Water-Resources Investigations Report 02-4012, 39

Conklin J (2008) Metal concentrations in the sediment of the Housatonic River. Undergraduate thesis, Southern Connecticut State University, New Haven CT, 58 pp

CTDEEP (2011) Report of the nitrogen credit advisory board for calendar year 2010 to the Joint Standing Environment Committee of the General Assembly. CTDEEP, Hartford, 47 pp http://www.ct.gov/dep/lib/dep/water/municipal_wastewater/nitrogen_report_2010.pdf

Conomos TJ (1979) San Francisco Bay: the urbanized estuary, Pacific Division. American Association for the Advancement of Science, San Francisco, 493 pp

Cooper S, Brush GS (1993) A 2,500-year history of anoxia and eutrophication in Chesapeake Bay. Estuaries 38:617–626

Cravotta III CA (2002) Use of isotope of carbon, nitrogen, and sulfur to identify sources of nitrogen in surface waters in the lower Susquehanna River Basin, Pennsylvania. USGS Water-Supply Paper, 2497, p. 99

Cuomo C, Valente R, Dogru D (2005) Seasonal variations in sediment and bottom water chemistry of western Long Island Sound: implications for lobster mortality. J Shellfish Res 24(3):805–814

Diaz RJ (2001) Overview of hypoxia around the world. J Environ Qual 30:275–281

Denver JM, Cravotta III CA, Ator SW, Lindsey BD (2010) Contributions of phosphorus from groundwater to streams in the Piedmont, Blue Ridge, and Valley and Ridge Physiographic Provinces, Eastern United States. U.S. Geological Survey Scientific Investigations Report 2010–5176, New Cumberland

DeWitt JC, Shnyra A, Badr MZ, Loveless SE, Hoban D, Frame SR, Cunard R, Anderson SE, Meade BJ, Peden-Adams MM, Luebke RW, Luster MI (2009) Immunotoxicity of perfluorooctanoic acid and perfluorooctane sulfonate and the role of peroxisome proliferator-activated receptor alpha. Crit Rev Toxicol 39:76–94

Dortch Q, Rabalais NN, Turner RE, Qureshi NA (2001) Impacts of changing Si/N ratios and phytoplankton species composition, in Coastal Hypoxia-consequences for living resources and ecosystems. Am Geophys Union, Washington, pp 37–48

Duffy TA, McElroy AE, Conover DO (2009) Variable susceptibility and response to estrogenic chemicals in *Menidia menidia*. Mar Ecol Prog Ser 380:245–254

Ficke JF, Hawkinson RO (1975) The national stream quality accounting network (NASQAN)-some questions and answers: U.S. Geol Surv Circ 719:23

Fitzgerald WF, Vandal GM, Rolfhus KR, Lamborg CH, Langer CS (2000) Mercury emissions and cycling in the coastal zone. J Environ Sci 12:92–101

Fitzgerald WF, Lamborg CH, Hammerschmidt CR (2007) Marine biogeochemical cycling of mercury. Chem Revs 107(2):641–662

Flegal AR, River-Duarte I, Ritson PI, Scelfo GM, Smith GJ, Gordon MR, Sañudo-Wilhelmy SA (1996) Metal contamination in San Francisco Bay waters: Historic perturbations, contemporary concentrations, and future considerations, in San Francisco Bay: the ecosystem, Pacific Division, American Association for the Advancement of Science, San Francisco, pp 173–188

Fredette TJ, Germano JD, Kullberg PG, Carey DA, Murray P (1992) Chemical stability of capped dredged material disposal mounds in Long Island Sound USA. Chem Ecol 7:173–194

Fredette TF, French GT (2004) Understanding the physical and environmental consequences of dredged material disposal: History in New England and current perspectives. Mar Poll Bull 49(1–2):93–102

Gilliom RJ, Alley WM, Gurtz ME (1995) Design of the National Water-Quality Assessment Program—Occurrence and distribution of water-quality conditions: U.S. Geol Surv Circ 1112:33

Gobler CJ, Nathan J, Buck NJ, Sieracki ME, Sañudo-Wilhelmy SA (2006) Nitrogen and silicon limitation of phytoplankton communities across an urban estuary: The East River-Long Island Sound system. Estuar Coast Shelf Sci 68:127–138

Grady SJ, Mullaney JR (1998) Natural and human factors affecting shallow water quality in surficial aquifers in the Connecticut, Housatonic, and Thames River basins: U.S. Geological Survey Water-Resources investigations Report 98-4042, p 81

Greig RA, Reid RN, Wenzloff DR (1977) Trace metal concentrations in sediments from Long Island Sound. Mar Poll Bull 8:183–188

Gronlund WD, Chan S-L, McCain BB, Clark Jr. RC, Myers, MS, Stein JE, Brown DW, Landahl JT, Krahn MM, Varanasi U (1991) Multidisciplinary assessment of pollution at three sites in Long Island Sound. Estuaries 14:299–305

Hammerschmidt CR, Fitzgerald WF (2006) Bioaccumulation and trophic transfer of methylmercury in Long Island Sound. Arch Environ Contam Toxicol 51:416–424

Hammerschmidt CR, Fitzgerald WF, Balcom PH, Visscher PT (2008) Organic matter and sulfide inhibit methylmercury production in sediments of New York/New Jersey Harbor. Mar Chem 109:165–182

Hem JD (1985) Study and interpretation of the chemical characteristics of natural water (third edition): U.S. Geological Survey Water-Supply Paper 2254, 263 pp

Hodgkiss IJ, Ho KC (1998) Are changes in N:P ratios in coastal waters the key to increased red tide blooms? In: Asia-Pacific Conference on Science and Management of Coastal Environment Developments in Hydrobiology, vol 123:141–147

Houde M, De Silva AO, Muir DCG, Letcher RJ (2011) Monitoring of perfluorinated compound sin aquatic biota: an updated review. Environ Sci Technol 45:7962–7973

Humborg C, Conley DJ, Rahm L, Wulff F, Cociasu A, Ittekkot V (2000) Silicon retention in river basins: Far-reaching effects on biogeochemistry and aquatic food webs in coastal marine environments. Ambio 29(1):45–50

Jackson JBC, Kirby MX, Berger WH, Bjorndal KA, Botsford LW, Bourque BJ, Bradbury RH, Cooke R, Erlandson J, Estes JA, Hughes TP, Kidwell S, Lange CB, Lenihan HS, Pandolfi JM, Peterson CH, Steneck RS, Tegner MJ, Warner RW (2002) Historical overfishing and the recent collapse of coastal ecosystems. Science 293:629–638

Keeling CD, Piper SC, Bacastow RB, Wahlen M, Whorf TP, Heimann M, HA Meijer (2005) Atmospheric CO_2 and $^{13}CO_2$ exchange with the terrestrial biosphere and oceans from 1978 to 2000: Observations and carbon cycle implications. In: Ehleringer JR, Cerling TE, Dearing MD (eds) A history of atmospheric CO_2 and its effects on plants, animals, and ecosystems. Springer, New York

Kemp WM, Boynton WR, Adolf JE, Boesch DF, Boicourt WC, Brush G, Cornwell JC, Fisher TR, Glibert PM, Hagy JD, Harding LW, Houde ED, Kimmel DG, Miller WD, Newell RIE, Roman MR, Smith EM, Stevenson JC (2005) Eutrophication of Chesapeake Bay: Historical trends and ecological interactions. Mar Ecol Prog Ser 303:1–29

Kimbrough KL, Johnson WE, Lauenstein GG, Christensen JD, Apeti DA (2009) An assessment of polybrominated diphenyl ethers (PBDEs) in sediments and bivalves of the U.S. coastal zone. Silver Spring, MD. NOAA Technical Memorandum NOS NCCOS, vol 94, 87 pp

Knebel HJ, Lewis RS, Varekamp JC (eds) (2000) Regional processes, conditions and characteristics of the Long Island Sound sea floor. J Coast Res 16(3):519–662

Kolpin DW, Furong ET, Meyer MT, Thurman EM, Zaugg SD, Barber LB, Buxton HT (2002) Pharmaceuticals, hormones, and other organic wastewater contaminants in U.S. streams, 1999–2000: a national reconnaissance. Environ Sci Technol 36:1202–1211

Koppelman LE, Weyl PK, Grant GM, Davies DS (1976) The urban sea: Long Island Sound. Praeger Publishers, New York 223 pp

Kuivila KM, Foe CG (1995) Concentrations, transport, and biological effects of dormant spray pesticides in the San Francisco Estuary California. Environ Toxicol Chem 14(7):1141–1150

Kuwabara JS, Arai Y, Topping BR, Pickering IJ, George GN (2007) Mercury speciation in piscivorous fish from mining-impacted reservoirs. Environ Sci Tech 41(8):2745–2749

Langer CS, Fitzgerald WF, Visscher PT, Vandal GM (2001) Biogeochemical cycling of methylmercury at Barn Island Salt Marsh, Stonington, CT, USA. Wetlands Ecol Manag 9:295–310

Latimer JS, Boothman WS, Pesch CE, Chmura GL, Pospelova V, Jayaramen S (2003) Environmental stress and recovery: the geochemical record of human disturbance in New Bedford Harbor and Apponagansett Bay, Massachusetts. Sci Total Environ 313:153–176

Lee A (2010) Metal concentrations in the sediment of the Thames River and New London Harbor. Dissertation, Southern Connecticut State University, New Haven, CT, 64 pp

Lee X, Benoit G, Hu X (2000) Total gaseous mercury concentration and flux at a coastal salt marsh in Connecticut, USA. Atmos Environ 34:4205–4213

Lee YJ, Lwiza KMM (2008) Characteristics of bottom dissolved oxygen in Long Island Sound, New York. Est Coast Shelf Sci 76:187–200

Litke DW (1999) Review of phosphorus control measures in the United States and their effects on water quality: U.S. Geological Survey Water-Resources Investigations Report 99-4007, 38 pp

Long Island Sound Study (LISS) (1993) Toxic substance contamination-assessment of conditions and management recommendations: The Long Island Sound Study comprehensive conservation and management plan support document, Draft, 68 pp

Long Island Sound Study (LISS) (1994) The comprehensive conservation and management plan, 205 pp, http://longislandsoundstudy.net/wp-content/uploads/2011/10/management_plan.pdf

Long ER, Wolfe DA, Carr RS, Scott KJ, Thursby GB (1995) Magnitude and extent of sediment toxicity in the Hudson-Raritan Estuary. NOAA Technical Memorandum NOS ORCA XX Silver Spring, New York

Long ER, Morgan LG (1990) The potential for biological effects of sediment-sorbed contaminants tested in the National Status and Trends Program. NOAA Technical Memorandum NOS OMA 52. National Oceanic and Atmospheric Administration, Seattle

Loucaides S, Cahoon LB, Henry EJ (2007) Effects of watershed impervious cover on dissolved silica loading in storm flow. J Am Water Resources Assn 43(4):841–849

Lugolobi F (2003) Environmental issues in Long Island Sound. MA thesis, Wesleyan University, Middletown CT 210 pp

Lugolobi F, Varekamp JC, Thomas E, Buchholtz ten Brink MR (2004) The use of Carbon isotopes in foraminiferal calcite to trace changes in biological oxygen demand in Long Island Sound. In: Proceeding of the sixth Biennual LIS research conference, pp 47–51

Luo Y, Yang X, Carley RJ, Perkins C (2002) Atmospheric deposition of nitrogen along the Connecticut coastline of Long Island Sound: a decade of measurements. Atmos Environ 36(28):4517–4528

Luoma SN, Phillips DJH (1988) Distribution, variability, and impacts of trace elements in San Francisco Bay. Mar Poll Bull 19(9):413–425

Mazzaferro DL, Handman EH, Thomas, MP (1979) Water resources inventory of Connecticut, part 8, Quinnipiac River basin. Connecticut Water Resources Bulletin 27, 88 p

McElroy AE, Barron G, Beckvar N, Kane Driscoll SB, Meador JP, Parkerton TF, Preuss TG, Steevens JA (2011) A review of the tissue residue approach for organic and organometallic compounds in aquatic organisms. Int Environ Assess Manag 7:50–74

Meador J (2006) Rationale and procedures for using the tissue-residue approach for toxicity assessment and determination of tissue, water and sediment quality guidelines for aquatic organisms. Hum Ecol Risk Assess 12:1018–1073

Mecray EL, Buchholtz ten Brink MR (2000) Contaminant distribution and accumulation in the surface sediments of Long Island Sound. J Coastal Res 16(3):575–590

Mecray EL, Reid JM, Hastings ME, Buchholtz ten Brink MR (2003) Contaminated sediments database for Long Island Sound and the New York Bight. U.S. Geological Survey open-file report no. 03-241, Available at http://pubs.usgs.gov/of/2003/of03-241

Meyers PA (1997) Organic geochemical proxies of paleoceanographic, paleolimnologic, and paleoclimatic processes. Org Geochem 27:213–250

Mitch AA, Anisfeld SC (2010) Contaminants in Long Island Sound: data synthesis and analysis. Est Coasts 33:609–628

Monosson E, Stegeman JJ (1994) Induced cytochrome P45011A in winter flounder, Pleuronectes americanus, from offshore and coastal sites. Can J Fish Aquat Sci 51:933–941

Mullaney JR (2007) Nutrient loads and ground-water residence times in an agricultural basin in north-central Connecticut. U.S. Geological Survey Scientific Investigations Report 2006–5278, 45 pp

Mullaney JR, Schwarz GE, Trench ECT (2002) Estimation of nitrogen yields and loads from basins draining to Long Island Sound, 1988-98. U.S. Geological Survey Water-Resources Investigations Report 02–4044, 84 pp

Mullaney JR, Lorenz DL, Arntson AD (2009) Chloride in groundwater and surface water in areas underlain by the glacial aquifer system, northern United States. U.S. Geological Survey Scientific Investigations Report 2009–5086, 41 pp

National Research Council (1985) Oil in the sea: inputs, fates, and effects. National Academy Press, Washington

National Research Council (2003) Oil in the sea III: Inputs, fates, and effects. National Academy Press, Washington

National Atmospheric Deposition Program (NRSP-3) (2009) NADP Program Office, Illinois State Water Survey, 2204 Griffith Dr, Champaign, p 61820

National Oceanic and Atmospheric Administration (NOAA) (1999) Sediment quality guidelines developed for the National Status and Trends Program http://response.restoration.noaa.gov/book_shelf/121_sedi_qual_guide.pdf

National Oceanic and Atmospheric Administration (1994) Biological effects of toxic contaminants in sediments from Long Island Sound and environs. NOAA Technical Memorandum NOS ORCA 80. National Ocean Survey, Office of Ocean Resources Conservation and Assessment, Silver Spring, New York

Nelson DA, Miller JE, Rusanowksy D, Greig RA, Sennefelder GR, Mercaldo-Allen R, Kuropat C, Gould E, Thurberg FP, Calabrese A (1991) Comparative reproductive success of winter flounder in Long Island Sound: A three-year study (biology, biochemistry, and chemistry). Estuaries 14:318–331

Neurath R (2009) Atmospheric mercury deposition in an isolated environment: A 150-year record at Block Island, RI. Senior thesis, Smith College, Northampton, 103 pp

NYSDEC and CTDEP (2000) A total maximum daily load analysis to achieve water quality standards for dissolved oxygen in Long Island Sound. Report for LISS; http://www.dec.ny.gov/docs/water_pdf/tmdllis.pdf

O'Connor TP (1996) Trends in chemical concentrations in mussels and oysters collected from the US coast from 1986 to 1993. Mar Environ Res 41(2):183–200

O'Connor TP, Lauenstein GG (2006) Trends in chemical concentrations in mussels and oysters collected along the US coast: update to 2003. Mar Environ Res 62:261–285

O'Shea ML, Brosnan TM (2000) Trends in indicators of eutrophication in Western Long Island Sound and the Hudson-Raritan Estuary. Estuaries 23:877–901

Parker CA, O'Reilly JE (1991) Oxygen depletion in Long Island Sound: a historical perspective. Estuaries 14:248–264

Paul JF, Gentile JH, Scott KJ, Schimmel SC, Campbell DE, Latimer RW (1999) EMAP-virginian province four-year Assessment (1990-93). US EPA NHEERL, Narragansett, EPA/6320/R-99/004, 199 pp

Perdue EM, Koprivnjak J-F (2007) Using the C/N ratio to estimate terrigenous inputs of organic matter to aquatic environments. Est Coast Shelf Sci 73:65–72

Perry ER, Norton SA, Kamman NC, Lorey PM, Haines T, Driscoll CT (2005) Mercury accumulation in lake sediments in the northeastern United States during the last 150 years. Ecotox 14:113–124

Perry DM, Hughes JB, Hebert AT (1991) Sublethal abnormalities in embryos of winter flounder, Pseudopleuronectes Americanus. Estuaries 14:306–317

Pinder RW, Appel KW, Dennis RL (2011) Trends in atmospheric reactive nitrogen for the Eastern United States. Environ Poll 159:3138–3141

Robertson A, Gottholm BW, Turgeon DD, Wolfe DA (1991) A comparative study of contaminant levels in Long Island Sound. Estuaries 14:290–298

Rodenburg, LA, Valle SN, Panero MA, Munoz GR, Shore LM (2010) Mass balances on selected polycyclic aromatic hydrocarbons in the New York-New Jersey Harbor. J Environ Qual 39(2):642

Rolfhus KR, Fitzgerald WF (2001) The evasion and spatial/temporal distributions of mercury species in Long Island Sound, CT-NY. Geochim Cosmochim Acta 65:407–418

Romany JS (2010) Measures of immune system status in young-of-the-year winter flounder (Pseudopleuronectes americanus) from Long Island coastal bays. Dissertation, Stony Brook University, Stony Brook, 47 pp

Rozan TF, Benoit G (2001) Mass balance of heavy metals in New Haven Harbor, Connecticut Preponderance of nonpoint sources. Limnol Oceanogr 46(8):2032–2049

Runkel RL, Crawford CG, Cohn TA (2004) Load estimator (LOADEST)—A FORTRAN program for estimating constituent loads in streams and rivers: U.S. Geol Sur Tech Methods 4(A5): 69 pp

Sañudo-Wilhelmy SA, Flegal AR (1992) Anthropogenic silver in the Southern California Bight: a new tracer of sewage in coastal waters. Environ Sci Tech 26:2147–2151

Scancar J, Milacic R, Strazar M, Burica O (2000) Total metal concentrations and partitioning of Cd, Cr, Cu, Fe, Ni and Zn in sewage sludge. Sci Total Environ 250(1–3):9–19

Schuster PF, Krabbenhoft DP, Naftz DI, Dewayne Cecil I, Olson ML, Dewild JF, Susong DD, Green JR, Abbott MI (2002) Atmospheric mercury deposition during the last 270 years: a glacial ice core record of natural and anthropogenic sources. Environ Sci Technol 36:2303–2310

Scorca MP, Monti Jr J (2001) Estimates of nitrogen loads entering Long Island Sound from ground water and streams on Long Island, New York, 1985-96. U.S. Geological Survey Water-Resources Investigations Report 00-4196, 29 pp

Skinner LC, Kane MW, Gottschall K, Simpson DA (2009) Chemical residue concentrations in four species of fish and the American lobsters from Long Island Sound, Connecticut and New York 2006 and 2007. Report to the Environmental Protection Agency

Smith VH (2003) Eutrophication of freshwater and coastal marine ecosystems: a global problem. Environ Sci Pollut Res Int 10(2):126–139

Sommer U (1994) Are marine diatoms favoured by high Si:N ratios? Mar Ecol-Prog Ser 115:309–315

Sprague LA, Mueller DK, Schwarz GE, Lorenz DL (2009) Nutrient trends in streams and rivers of the United States, 1993–2003. U.S. Geological Survey Scientific Investigations Report 2008–5202, 196 pp

Thomas E, Abramson I, Varekamp JC, Buchholtz ten Brink MR (2004) Eutrophication of Long Island Sound as traced by benthic foraminifera. In: Proceedings of the sixth Biennual LIS research conference pp 87–91

Thomas E, Gapotchenko T, Varekamp JC, Mecray EL, Buchholtz ten Brink MR (2000) Benthic foraminifera and environmental changes in Long Island Sound. J Coast Res 6:641–655

Titus T (2003) Trace metal contents and physical characteristics of sediment in Bridgeport Harbor. Undergraduate thesis, Southern Connecticut State University, New Haven, CT, 74 pp

Trench ECT, Moore RB, Ahearn EA, Mullaney JR, Hickman RE, Schwarz GE (2012) Nutrient concentrations and loads in the northeastern United States—Status and trends, 1975–2003. U.S. Geological Survey Scientific Investigations Report, East Hartford

Trench ECT, Vecchia AV (2002) Water-quality trend analysis and sampling design for streams in Connecticut, 1968–98. U.S. Geological Survey Water-Resources Investigations Report 02–4011, East Hartford

Trench ECT (1996) Trends in surface-water quality in Connecticut, 1969-88: U.S. Geological Survey Water-Resources Investigations Report 96-4161, 176 pp

Trench ECT (2000) Nutrient sources and loads in the Connecticut, Housatonic, and Thames River Basins: U.S. Geological Survey Water-Resources Investigations Report 99-4236, 66 pp

Triplett LD, Engstrom DR, Conley DJ, Schellhaass SM (2008) Silica fluxes and trapping in two contrasting natural impoundments of the upper Mississippi River. Biogeochem 87:217–230

Tseng CM, Balcom PH, Lamborg CH, Fitzgerald WF (2003) Dissolved elemental mercury investigations in Long Island Sound using online Au amalgamation-flow injection analysis. Environ Sci Tech 37:1183–1188

Turgeon DD, O'Connor TP (1991) Long Island Sound: distribution, trends, and effects of chemical contamination. Estuaries 14:279–289

USEPA (2001) 2001 National land cover data: Multi-resolution land characteristics consortium, accessed 1 February 2010 at http://www.epa.gov/mrlc/nlcd-2001.html

USEPA (2000) TMDL report for nitrogen, http://longislandsoundstudy.net/wp-content/uploads/2010/03/Tmdl.pdf

USEPA (2001) Water quality criterion for the protection of human health: Methylmercury. EPA-823-R-01-001, Washington

USEPA (2007) National Estuary program coastal condition report. Chapter 3: Northeast National Estuary program coastal condition, Long Island Sound Study http://water.epa.gov/type/oceb/nep/upload/2007_05_09_oceans_nepccr_pdf_nepccr_nepccr_ne_partg.pdf

USEPA (2008) National coastal condition report III. Chapter 3. Northeast coast coastal condition. http://water.epa.gov/type/oceb/assessmonitor/upload/2008_12_09_oceans_nccr3_chapter3_northeast-a.pdf

Varekamp JC (1991) Trace element geochemistry and pollution history of mudflat and marsh sediments from the Connecticut coastline. J Coastal Res 11:105–123

Varekamp JC (2011) Wethersfield cove, a 300 year urban pollution record. GSA annual meeting, Minneapolis, (Abstract 188–10)

Varekamp JC, Buchholtz ten Brink MR, Mecray EL, Kreulen B (2000) Mercury in Long Island Sound sediments. J Coastal Res 16:613–626

Varekamp JC, Kreulen B, Buchholtz ten Brink MF, Mecray E (2003) Mercury contamination chronologies from Connecticut wetlands and Long Island Sound Sediments. Environ Geol 43:268–282

Varekamp JC, Thomas E, Lugolobi F, Buchh[o]ltz ten Brink MR (2004) The paleo-environmental history of Long Island Sound as traced by organic carbon, biogenic silica and isotope/trace element studies in sediment cores. In: Proceedings of sixth Biennual LIS research conference, pp. 109–113

Varekamp JC, Mecray EL, Zierzow T (2005) Once spilled, still found: Metal contamination in Connecticut wetlands and Long Island Sound sediment from historic industries. In: Whitelaw, DM and Visiglione, GR (eds), America's changing coasts-private rights and public trust. E. Elgar Publishing, Northampton, vol 9, pp. 122–147

Varekamp JC (2006) The historic fur trade and climate change. EOS Trans Am Geophys Union 87:596–597

Varekamp JC, Altabet M, Thomas E, Buchholtz Ten Brink M, Andersen N, Mecray E (2009) Carbon cycling in LIS: nutrient fluxes and landscape development over the last 1000 years. In: Proceedings of the ninth Biennual LIS research conference

Varekamp JC, Thomas E (2010) Report to CTDEP on paleoenvironmental research in Long Island Sound. http://www.wesleyan.edu/ees/JCV/varekamp.html (LIS Report)

Ward DM, Nislow KH, Chen CY, Folt CC (2010) Rapid, efficient growth reduces mercury concentrations in stream-dwelling Atlantic salmon. Trans Am Fisheries Soc 139:1–10

Welsh BL, Eller FC (1991) Mechanisms controlling summertime oxygen depletion in Western Long Island Sound. Estuaries 14:265–278

Whalin L, Kim EH, Mason R (2007) Factors influencing the oxidation, reduction, methylation and demethylation of mercury species in coastal waters. Mar Chem 107(3):278–294

Windom HLS, Schroop SJ, Calder FD, Ryan JD, Smith Jr., RG, Burney LC, Lewis FG, Rawlinson CH (1989) Natural trace metal concentrations in estuarine and coastal marine sediments of the southwestern Unites States. Environ Sci Tech 23:314–320

Wolfe DA, Monahan R, Stacey PE, Farrow DRG, Robertson A (1991) Environmental quality of Long Island Sound: assessment and management issues. Estuaries 14:224–236

Wolfe DA, Bricker SB, Long ER, Scott KJ, Thursby GB (1994) Biological effects of toxic contaminants in sediments from Long Island Sound and environs. NOAA Technical Memorandum, NOS ORCA 80

Yang L, Li X, Crusius J, Jans U, Melcer ME, Zhang P (2007) Persistent chlordane concentrations in Long Island South sediment; implications from chlordane, ^{210}Pb, and ^{137}Cs profiles. Environ Sci Technol 41:7723–7729

Chapter 6
Biology and Ecology of Long Island Sound

Glenn Lopez, Drew Carey, James T. Carlton, Robert Cerrato, Hans Dam,
Rob DiGiovanni, Chris Elphick, Michael Frisk, Christopher Gobler,
Lyndie Hice, Penny Howell, Adrian Jordaan, Senjie Lin, Sheng Liu,
Darcy Lonsdale, Maryann McEnroe, Kim McKown, George McManus,
Rick Orson, Bradley Peterson, Chris Pickerell, Ron Rozsa,
Sandra E. Shumway, Amy Siuda, Kelly Streich, Stephanie Talmage,
Gordon Taylor, Ellen Thomas, Margaret Van Patten, Jamie Vaudrey,
Charles Yarish, Gary Wikfors, and Roman Zajac

6.1 Introduction

Many compelling management issues in Long Island Sound (LIS) focus on how
organisms respond to stresses such as commercial and recreational harvesting,
eutrophication, hypoxia, habitat degradation, invasion of non-native species, ocean
acidification, and climate change. In order to address these complex problems,

G. Lopez (✉) · R. Cerrato · M. Frisk · C. Gobler · L. Hice
D. Lonsdale · B. Peterson · S. Talmage · G. Taylor
School of Marine and Atmospheric Sciences, Stony Brook University,
Stony Brook, NY 11794, USA
e-mail: Glenn.Lopez@stonybrook.edu

M. Van Patten
Connecticut Sea Grant, University of Connecticut, Groton, CT 06340, USA

C. Elphick · C. Yarish
Department of Ecology and Evolutionary Biology, University of Connecticut,
Groton, CT 06340, USA

H. Dam · S. Lin · S. Liu · G. McManus · S. E. Shumway · J. Vaudrey
Department of Marine Sciences, University of Connecticut, Groton, CT 06340, USA

C. Pickerell
Cornell Cooperative Extension of Suffolk County, Riverhead, NY 11901, USA

A. Siuda
Sea Education Association, Falmouth, MA 02540, USA

E. Thomas
Department of Geology and Geophysics, Yale University, New Haven, CT 06511, USA

R. Zajac
Department of Biology and Environmental Science, University of New Haven,
New Haven, CT 06516, USA

J. S. Latimer et al. (eds.), *Long Island Sound*, Springer Series on
Environmental Management, DOI: 10.1007/978-1-4614-6126-5_6,
© Springer Science+Business Media New York 2014

we must first understand the factors controlling biological processes and how organisms interact ecologically. This chapter provides an overview of the major groups of organisms occupying the dominant habitats of LIS.

We begin by describing the biota inhabiting the intertidal and shallow subtidal regions of LIS, including tidal marshes and submerged aquatic vegetation such as seagrasses, seaweeds, and benthic fauna. Next the subtidal and more open water reaches of LIS are discussed; separate sections on plankton, benthos, nekton, and other wildlife (birds, reptiles, and mammals) are included. We end this chapter by focusing on several ecological challenges to the biota of LIS, including hypoxia, introduction of non-native species, climate change, and collapse of commercially important populations.

6.2 Littoral Zone: Habitats and Benthic Ecology

6.2.1 Intertidal Zone

Tidal amplitude ranges in LIS ranges from ~0.7 to 2.2 m. LIS has a variety of littoral or intertidal habitats including rocky areas, cobble and sand beaches, sand

R. DiGiovanni
Riverhead Foundation for Marine Research and Preservation, Riverhead, NY 11901, USA

M. McEnroe
School of Natural and Social Sciences, Purchase College, Purchase, NY 10577, USA

D. Carey
Coastal Vision, Newport, RI 02840, USA

J. T. Carlton
Williams-Mystic Maritime Studies Program, Williams College, Stonington, CT 06355, USA

P. Howell
Marine Fisheries Division, Connecticut Department of Energy and Environmental Protection, Hartford, CT 06106, USA

G. Wikfors
Northeast Fisheries Sciences Center, NOAA, Milford, CT 06460, USA

K. McKown
Bureau of Marine Resources, New York Department of Environmental Conservation, East Setauket, NY 11733, USA

R. Rozsa
Coastal Management Program, Connecticut Department of Energy and Environment Protection Retired, Hartford, CT 06106, USA

R. Orson
Orson Ecological Consulting, Branford, CT 06405, USA

A. Jordaan
University of Massachusetts-Amherst, Amherst, MA 01003, USA

K. Streich
Bureau of Water Management Planning and Standards, Connecticut Department of Energy and Environmental Protection, Harford, CT 06106, USA

and mud flats, and tidal marshes. Because of the interplay between the physiological stresses imposed by periodic aerial exposure and biological interactions related to competition and predation, many of these habitats are characterized by zones or bands in species distributions.

The most common intertidal habitats bordering LIS (including bays, harbors, and other sheltered areas) are salt marshes and shallow sloping intertidal flats (Fig. 6.1). Each represents about 31 % of the LIS coastline. These are followed in importance by cobble, gravel, or riprap areas (19 %), rocky intertidal zone (14 %),

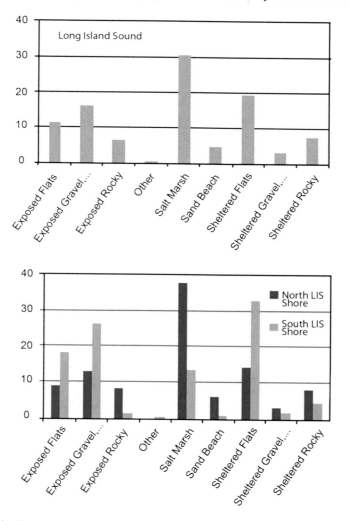

Fig. 6.1 Distribution of intertidal habitats bordering Long Island Sound. Data were derived from Environmental Sensitivity Index (ESI) GIS maps obtained from the NOAA Office of Response and Restoration (http://response.rerstoration.noaa.gov/esi). For each segment of coastline, the most shoreward ESI class was retained, and multiple ESI classes were grouped to simplify the classification

and sand beaches (5 %). The distribution of these habitats varies with region. Salt marshes (37 %) are the most common habitat on the north shore of LIS while intertidal flats (51 %) dominate the south shore of LIS. Both coasts have sizable amounts of exposed gravel, cobble, and riprap, but this habitat represents a large fraction (52 %) of the eastern half of the Long Island shoreline. Overall, the south shore of LIS has about twice the amount of this habitat type compared to the north shore. Although both shores have rocky areas, the natural areas tend to be exposed bedrock on the north shore and glacial erratics on the south shore.

6.2.2 Rocky Intertidal Zone

The rocky intertidal zone (Fig. 6.2) has the most noticeable biotic zonation of all the intertidal habitats (Peterson 1991). This region is dominated by epiflora and epifauna living on the rock surfaces. Rock surfaces provide attachment sites, shelter in the form of crevices, and an attached source of microbial food. Physical factors such as desiccation, exposure to extreme winter and summer air temperatures, wave stress, and S stress are the strongest determinants of the upper limits of the distribution of organisms in this habitat, while biological factors tend to control the seaward limits of their distributions (Berrill and Berrill 1981; Bertness et al. 2001).

Zonation in the rocky intertidal of LIS follows a pattern typical of the north Atlantic coast (Berrill and Berrill 1981; Bertness 1999; Nybakken 2001). At the highest level is the spray or splash zone; it lies above the spring high tide

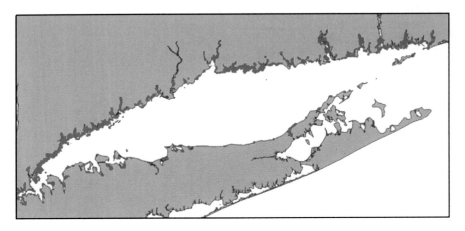

Fig. 6.2 Distribution of rocky intertidal habitats. Data were derived from Environmental Sensitivity Index GIS maps obtained from the NOAA Office of Response and Restoration (http://response.rerstoration.noaa.gov/esi)

line and is dominated by lichens. Below that, in the region where only spring tides reach, is a zone where the rocks are black in color because of the presence of cyanobacteria. The most abundant animal in this zone is the rough periwinkle (*Littorina saxatilis*). Within the high intertidal zone, which is inundated even during neap tides, the dominant animals are the northern rock barnacle (*Semibalanus balanoides*) and the common periwinkle (*Littorina littorea*). Dominant species in the next intertidal zone are determined by the degree of wave action. On wave-exposed shores, the blue mussel (*Mytilus edulis*) is the most successful competitor in the mid-intertidal zone. On wave-protected coastlines, knotted wrack (*Ascophyllum nodosum*), and rockweeds (*Fucus* spp.), become the competitive dominants. Irish moss (*Chondrus crispus*) is the most abundant species in the low intertidal zone, generally below mean low water; this red alga is restricted to this zone because it can tolerate only short exposure times. Other common species include green algae such as sea lettuce (*Ulva spp.*) and *Ulva* tubular *spp.*, along with common periwinkles. In the shallow subtidal, kelp, such as *Saccharina latissima* (formerly classified as *Saccharina latissima*) and a variety of red seaweeds become abundant. On some surfaces, it is common to find the chink shell (*Lacuna vincta* also known as the Northern lacuna). Throughout the intertidal zone, mobile predators such as the long-clawed hermit crab (*Pagurus longicarpus*), the rock crab (*Cancer irroratus*), the green crab (*Carcinus maenas*), the mud crab (*Dyspanopeus sayi*), and the Asian shore crab (*Hemigrapsus sanguineus*) play critical roles in regulating community structure.

The character of the rocky intertidal has changed over time as many key species have been introduced. These include, along with their approximate year of introduction, Asian shore crab (1992), green crab (1820s), common periwinkle (1840s), and green fleece, (*Codium fragile*) (1950s) (Gerard et al. 1999; Steneck and Carlton 2001). It is likely that the rocky intertidal community during the eighteenth century bore little resemblance to that found in LIS today.

6.2.3 Gravel, Cobble, and Riprap

Organisms associated with unconsolidated, coarse-grained intertidal areas are regulated by the mobility of the substrate (Fig. 6.3) (Osman 1977; Sousa 1979). Large cobbles and riprap are resistant to movement and areas dominated by this substrate tend to resemble the rocky intertidal, with mid-tidal surfaces covered by Northern rock barnacles, common periwinkles, rockweeds, and algal crusts (e.g., *Ralfsia* spp., *Hildenbrandtia prototypus*). As in the rocky intertidal, the Asian shore crab has become a dominant predator in this habitat (Lohrer and Whitlatch 1997). Below this zone Irish moss and green fleece are common (Hammerson 2004). Small cobbles and even smaller gravel particles are frequently overturned and shifted; consequently, beaches with this substrate tend to resemble sandy beaches and have few species present.

6.2.4 Sand Beaches

Sand beaches are common on the north shore of LIS (Fig. 6.4). They often appear barren but are populated by low numbers of large animals. Infauna predominate since the sand layer creates a buffer from temperature (T) extremes and desiccation, and the substrate is too mobile to support attached plants and animals (Peterson 1991). Beach characteristics such as slope and particle size result from

Fig. 6.3 Distribution of gravel, cobble, and riprap. Data were derived from Environmental Sensitivity Index GIS maps obtained from the NOAA Office of Response and Restoration (http://response.rerstoration.noaa.gov/esi)

Fig. 6.4 Distribution of sand beaches. Data were derived from Environmental Sensitivity Index GIS maps obtained from the NOAA Office of Response and Restoration (http://response.rerstoration.noaa.gov/esi)

an interaction between wave action and source material (Nybakken 2001). In general, beaches that are coarser-grained and more steeply sloped tend to drain once the tide recedes, while finer-grained, gently sloping beaches retain water for longer periods of time after the tides withdraw. On a beach, sand grain size tends to be coarsest where waves break and becomes finer both seaward and with increased elevation shoreward (Komar 1998). These grain-size variations contribute to biotic zonation (Peterson 1991), and generally more organisms are found on fine-grained beaches (Nybakken 2001).

The only primary producers on sand beaches are microflora such as benthic diatoms, since the sediments are too mobile for attachment of macrophytes (Nybakken 2001). The dominant fauna include suspension feeders, detritus feeders, and scavengers. Semi-terrestrial talitrid amphipods (e.g., *Orchestia grillus* and *Talorchestia* spp.) are found at or above the high tide level (Steinback 1999). Further down, the most common species include the isopods *Politolana* spp. and *Chiridotea* spp. at mid-tidal level, and haustoriid amphipods such as *Haustorius virginiana* and *Amphiporeia virginiana* and the spionid polychaete *Scolelepis squamata* in the swash or surf zone (Croker 1970; Weiss 1995). Some of the fauna, such as the talitrid amphipods, undergo fortnightly tidal migrations in order to maintain their positions relative to the wrack (eelgrass (*Zostera marina*), cordgrass (*Spartina alterniflora*), and macroalgae detritus) as it progressively moves up the beach (Steinback 1999). Wrack provides food, refuge, and insulation against T extremes and desiccation, as well as breeding grounds for a variety of organisms, especially insects (Steinback 1999). Competition for space is not as intense as on rocky shores. In contrast to the slow predators in the rocky intertidal, the main predators on sand flats tend to be highly mobile fishes, crabs, and shorebirds (Peterson 1991). Characteristic birds include piping plovers (*Charadrius melodus*), least terns (*Sternula antillarum*), sandpipers (*Calidris* spp.), ring-billed gulls (*Larus delawarensis*), and herring gulls (*Larus argentatus*) (Hammerson 2004; Weiss 1995). These birds feed on amphipods, polychaetes, and other invertebrates (Hammerson 2004).

6.2.5 Sand and Mud Flats

Sand and mud flats are found in depositional environments, in sheltered areas such as embayments and behind spits (Fig. 6.5) (Whitlatch 1982), and are often bordered by beaches and marshes landward and by eelgrass seaward. The principal producers are microalgae present as films or mats composed of diatoms, euglenoids, dinoflagellates, and cyanobacteria (Whitlatch 1982). Microalgae are often found in the upper intertidal but excluded from lower tidal levels by grazers like the mud snail (*Ilyanassa obsoleta*) (Whitlatch 1982). Macroalgae are generally rare because of the unstable nature of the sediments, but short-lived, highly productive macroalgae such as sea lettuce (including both sheet and tubular forms) sometimes become established for short periods of time (Whitlatch 1982).

Fig. 6.5 Distribution of sand and mud flats. Data were derived from Environmental Sensitivity Index GIS maps obtained from the NOAA Offssssice of Response and Restoration (http://response.rerstoration.noaa.gov/esi)

Mud flats have similar dominant processes and tend to share fauna with fine-grained subtidal areas. The fine-grained sediments on mud flats retain water, have poor exchange with the overlying water, accumulate organic matter, and have high bacterial decomposition rates (Nybakken 2001). As a result, dissolved oxygen is usually only present in the first few millimeters of the sediment. Below that, pore waters are anoxic and rich in sulfides. Because mud flats are low energy areas, permanent burrows are possible (Nybakken 2001), so tube building invertebrates are common at mid-tidal levels. These include the surface deposit-feeding poly-chaetes *Polydora ligni* and *Streblospio benedicti* and the amphipod *Corophium* spp. (Whitlatch 1982). Other common surface deposit feeders include the mud snail, the Baltic clam (*Macoma balthica*), and the snail *Hydrobia truncata* (Whitlatch 1982). The principal suspension feeder at mid-tidal level is the soft shell clam (*Mya arenaria*) which supports an important recreational fishery in LIS. Burrowing fauna include the polychaetes *Lumbrineris tenuis* and *Heteromastus filiformis*. An important carnivore associated with mud flats is the polychaete *Nereis virens* (Whitlatch 1982).

Species zonation patterns are not very distinct but do exist on tidal sand flats, mainly due to physiological stresses associated with limited feeding, respiration times, and biotic interactions such as predation and competition (Peterson 1991; Whitlatch 1982). Common molluscs include the predatory moon snail *Neverita duplicata* (formerly *Polinices duplicatus*) and the small, suspension-feeding bivalve *Gemma gemma*. These tend to occur from mid- to low-tide level (Whitlatch 1982). In the low intertidal zone, larger, suspension-feeding bivalves such as the common razor clam (*Ensis directus*) and the hard clam (*Mercenaria mercenaria*, also commonly called quahog) are found. *Spiophanes bombyx* and *Arenicola marina* are representative deposit-feeding polychaetes on sand flats. Both tend to be found at or above the mid-tide level (Anderson 1972; Whitlatch 1982). The former is a surface deposit feeder and the latter builds a U-shaped burrow, funneling sediments to the mouth at one end and defecating at the other. Horseshoe crabs (*Limulus polyphemus*) are seasonal predators that dig pits on sand flats during high tide, searching for invertebrate prey. They also spawn here in the high intertidal. Other predators utilizing sand flats include fishes such as scup (*Stenotomus chrysops*, also commonly called porgy), sand lance (*Ammodytes americanus*), and several species of flounder (e.g., summer flounder (*Paralichthys dentatus* also commonly called fluke or Northern fluke), fourspot flounder (*P. oblongatus*), and winter flounder (*Pseudopleuronectes americanus*)) and a variety of birds, including sandpipers, least terns, and gulls (Whitlatch 1982).

Reid et al. (1979) described a sand community distributed in shallow areas along the eastern two-thirds of Long Island, including the nearshore region in Smithtown Bay and from Port Jefferson to Mattituck Inlet. The molluscan fauna in this community were dominated by three suspension-feeding bivalves; razor clams, surf clams (*Spisula solidissima*), and *Pandora gouldiana*, a surface deposit feeder; *Tellina agilis*, and two gastropods; the suspension-feeding slipper shell (*Crepidula fornicata*) and the scavenging dog whelk (*Ilyanassa trivittata*). Also abundant were two carnivores/omnivores, the painted worm (*Nephthys picta*) and long-clawed hermit crab, and a tubiculous, deposit-feeding amphipod *Ampelisca vadorum*. All stations with this faunal assemblage occurred at water depths less than 10 m. This geographic region consists of a series of shoals whose major source of sand is from erosion of adjacent bluffs (Bokuniewicz and Tanski 1983). An accumulation of mollusc shells is generally present at the base of the shoals, and further seaward, the sediments become finer-grained. Commercial harvesting of surf clams occurs on the shoals east of Mount Sinai Harbor.

In harbor areas, sand and mud faunal assemblages are affected by many of the same physical and biotic processes and are similar to those found in the deeper subtidal areas of LIS. As a result, no separate discussion of these assemblages is necessary here. It is notable, however, that studies examining the structure of benthic communities in harbor areas tend to clearly identify community characteristics that suggest frequent natural and/or anthropogenic disturbances. For example, Ocean Surveys, Inc. (2010) found that a number of opportunistic species were abundant in Sheffield Harbor, Norwalk, CT. Cuomo and Zinn (1997) characterized the benthic community at most sites in the lower West River in New Haven and West Haven, CT as early to mid-successional (Stage I-Stage II; sensu Rhoads and Germano 1982) and suggested they were being maintained at that level by frequent disturbances. Cerrato and Holt (2008), in an investigation of Oyster Bay/Cold Spring Harbor, Huntington/Northport Bay, and Port Jefferson Harbor, found that portions of each area had benthic community characteristics that suggested the presence of stress. These included lower than expected species richness, anomalously low abundances, and dominance by low successional, opportunistic species. For example, in Oyster Bay/Cold Spring Harbor and Huntington/Northport Bay, more that 60 % of all individuals collected belonged to one of five early successional species. Similarly stressed assemblages were found in Bowery Bay and Flushing Bay in the East River (Cerrato and Bokuniewicz 1986).

6.2.6 Science Gaps and Management Implications

Surprisingly, little quantitative data are available on the spatial and temporal features of the intertidal flora and fauna, and the only shallow water data available are a small part of larger Sound-wide studies (e.g., Reid et al. 1979). As a result, most descriptions in this section relied on generic and/or decades-old sources. Data and knowledge gaps on intertidal and shallow water biotic resources are so large that an accurate, reliable characterization cannot be made at this time. Even so, anthropogenic impacts are still clearly evident using simple measures such as the number of invasive species present, the fraction of marsh area lost, the amount of hardened shoreline present, or the obvious dominance of species that are advantaged by anthropogenic disturbances. The management implications of our current state of knowledge are that while the habitats can be identified and measured in extent, and while we have seen human-mediated change, what we have lost, whether we can recover any of it, and how we should proceed, are largely unknown.

6.3 Tidal Marshes

6.3.1 General Characteristics

Three broad classes of tidal marshes are present in LIS: salt marsh (polyhaline), brackish marsh (mesohaline and oligohaline), and fresh-tidal marsh (fresh); salt marsh is the most common class near the shores of LIS. Each of these marsh types exhibits characteristic plant and animal communities (Nichols 1920). The geology of LIS does not provide extensive shallow waters sheltered from wave action, so tidal marshes in the region are small and confined to shelter embayments and tidal rivers; the largest tidal marsh is only 800 acres. Tidal marsh acreage is estimated at 20,895 acres with 84 % located in Connecticut (see Fig. 6.6).

Prior to the passage of the Connecticut Tidal Wetlands Act in 1969, 30 % of all tidal marshes in the state were lost to unregulated activities such as filling, dredging, and hydrological modifications changing tidal exchange, such as millpond dams and tide gates. Losses were greatest in the urban and least in rural towns. The 1969 regulatory programs in Connecticut, followed by New York's in 1973, have largely arrested the loss of tidal marshes. Dreyer and Niering (1995) provide a more complete description of the history, ecology, and restoration of LIS marshes.

The most familiar tidal marshes around LIS are the salt marshes with their characteristic short grassy meadows; the Wequetequock-Pawcatuck marshes of Stonington, CT have served as the paradigm for southern New England salt marshes (Miller and Egler 1950). Less common are the brackish meadows (typically dominated by salt hay (*Spartina patens*) and blackgrass (*Juncus gerardii*) and brackish reed and fresh-tidal tall reed marshes (dominated by cattail (*Typha*

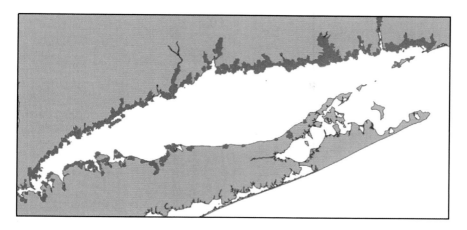

Fig. 6.6 Distribution of salt marshes. Data were derived from Environmental Sensitivity Index GIS maps obtained from the NOAA Office of Response and Restoration (http://response.rerstoration.noaa.gov/esi)

augustifolia) and sedges (e.g., *Scirpus* spp.)). Nearly 30 % of the Sound's tidal wetlands are the fresh-tidal and brackish marshes of the Connecticut River. They constitute the best examples in the northeast and were designated as Wetlands of International Importance under the Ramsar Convention in 1994. Details about these marshes are described by Rozsa et al. (2001). Other small but noteworthy brackish marshes are the Quinnipiac, Housatonic, Stratford (CT), and Nissequogue Rivers (NY). Plant diversity is lowest in salt marshes and highest in fresh-tidal marshes.

Salt marshes (Fig. 6.6) are the most productive intertidal habitats in LIS. Two major trophic pathways occur in marshes: One pathway is supported by benthic algae and phytoplankton; the other is based on marsh grass detritus. The dominant salt marsh grasses (cordgrass and salt hay) are perennial. During the fall, they withdraw nutrients from the leaves and store them in roots and rhizomes. Leaves dieback and over the winter stalks are dislodged by ice and transported to the high intertidal zone. Mats of this decaying wrack can smother existing vegetation and eventually bare patches are opened up to new colonization (Bertness 2007). Once in the high intertidal zone, the plant detritus begins the long process of decay aided by physical fragmentation and microbial (especially fungal) decomposition. This material represents the major food source for a few species of specialist detritivores and a supplemental food for many marsh and coastal species (see Lopez and Levinton 2011). Niering and Warren (1980) estimated that the production of above-ground vegetation in the marshes of Long Island and Connecticut was about 93,000 metric tons per year.

Zonation in intertidal salt marshes is controlled by a combination of tidal flooding, salt stress, physical disturbances (ice damage, burial by wrack), competition, and predation (Bertness 2007). The low marsh, located between mean low and mean high water, is dominated by cordgrass. This salt tolerant plant forms a belt along the seaward edges of the marsh, with the tallest forms nearest the banks. The seaward colonization of cordgrass is limited by the degree of tidal flooding and its landward extent by competition with other plants. The high marsh extends from about mean high water to the highest spring tide level; this zone is flooded during spring tides. The dominant plants in the high marsh are salt hay, spike grass (*Distichlis spicata*), blackgrass, and glasswort (*Salicornia* spp.). These high marsh plants generally outcompete cordgrass, but they are unable to successfully invade low marsh areas because they cannot tolerate the water-logged soil there (Bertness 2007). Above the high marsh is an upland area that is exposed to seawater only during major storms. Major species in this zone include the marsh elder (*Iva frutescens*), red cedar (*Juniperus virginiana*), and eastern Baccharis (*Baccharis halimifolia*). Non-native haplotypes of the common reed (*Phragmites australis*) have been displacing cordgrass in salt marshes. Reasons for this are not completely understood but are probably connected to anthropogenic changes in the hydrologic cycle, nutrient inputs, and physical disturbance (Chambers et al. 1999; Minchinton and Bertness 2003). The native haplotype of the common reed also occurs on the Connecticut River (R. Rozsa, personal communication).

6.3.2 Associated Fauna and Flora

Prominent animals in salt marshes include ribbed mussels (*Geukensia demissa*), mud snails, and fiddler crabs (*Uca pugilator* and *U. pugnax*). Ribbed mussels are suspension feeders found in the low marsh at the roots of cordgrass. Mud snails are deposit feeders that tend to occur in the low marsh and in mud flats adjacent to the marsh. Fiddler crabs are deposit feeders and usually occur in the low marsh; *U. pugilator* is usually associated with sandy sediments and *U. pugnax* with muds. In addition, marshes are important areas for a variety of fishes (e.g., the mummichog (*Fundulus heteroclitus*) and provide food and shelter for birds such as herons and egrets (e.g., black-crowned night-heron (*Nycticorax nycticorax*) and the snowy egret (*Egretta thula*) (Bertness 2007)). Given the muddy nature of most salt marsh sediments, marshes also harbor many benthic species found in mudflats.

Tidal marshes provide extensive ecological value. Killifish such as the mummic-hog, striped killifish (*Fundulus majalis*), and sheepshead minnow (*Cyprinodon variegatus*) consume marsh invertebrates such as amphipods, isopods, and snails while foraging in the marshes during high tide. Retreating to tidal creeks on the ebbing current where they become prey for estuarine species such as bluefish (*Pomatomus salatrix*), striped bass (*Morone saxatilis*), and summer flounder. Marshes are habitat for birds, including saltmarsh sparrow (*Ammodramus caudacutus*), considered globally vulnerable, and the seaside sparrow (*A. maritimus*), also a nationally ranked conservation priority. Both species inhabit the low grassy meadows of coastal tidal marshes. The secretive least and American bitterns (*Ixobrychus exilis* and *Botaurus lentiginosus*) are found in tall reed marshes. On the Connecticut River, the spring freshet floods the upstream freshwater tidal wetlands, which become natural impoundments at a time of waterfowl migration. The submerged aquatic vegetation of the tidal creeks and coves of brackish and fresh-tidal marshes provides a refuge for migrating juvenile alewife (*Alosa pseudoharengus*) and shad (*Alosa sapidissima*).

Decaying salt marsh wrack provides habitat and food for specialist detritivores, in particular the amphipod *Orchestia grillus* and the salt marsh snail *Melampus bedentatus*. Feeding activities of these species stimulate detrital breakdown and convert the low food value detritus into preferred prey for many predators (e.g., Kneib 1986).

6.3.3 Trends in Marsh Management and Restoration

Most tidal marshes have been ditched for mosquito control. Coastal municipalities funded ditching in the early part of the last century. Small, unditched marshes exist but the two largest marshes are the Wheeler Wildlife Management Area in Milford, CT and Great Meadows in Stratford, CT. The Connecticut Department of Health Services abandoned maintenance ditching in 1984 in favor of selective implementation of the more environmentally sound technique known as open marsh water management. This began a coast-wide experiment to determine how to restore pool habitat and meandering tidal creeks to marshes. The unditched tidal marsh had extensive areas of shallow pools, an important habitat for shorebirds, wading birds, and waterfowl.

The Connecticut Coastal Management Act of 1980 included a policy encouraging the restoration of degraded tidal wetlands and the Coastal Area Management Program began the systematic restoration of degraded tidal marshes. The primary restoration technique is to restore tidal flow by removing undersized culverts, tide gates, and fill. More than 1,100 acres have been restored in Connecticut at more than 70 locations. Tidal marsh restoration in New York was delayed for lack of a restoration policy. Fletchers Creek at Silver Sands State Park, Milford, CT, is an exemplary restoration site, where tidal flow of an old municipal landfill was restored and several elevated areas were excavated.

Long-term studies of tidal marsh restoration have shown that tidal flow restoration causes a return of the pre-disturbance plant and animal communities. So salt marsh vegetation returns to former salt marshes but the elevation of degraded tidal marshes is typically lower than natural marshes. The dominant vegetation is cordgrass and is thus a low marsh habitat. Hurricane Carol in 1954 destroyed the tide gates (which promote drainage and peat decomposition) at the Great Harbor Marsh (Guilford, CT), drowned the subsided marsh, and created extensive peat flats. Over the next nearly 60 years, cordgrass gradually colonized the peat flat in a downstream to upstream direction. The marsh above Great Harbor, known as Lost Lake, is still devoid of vegetation, but the culvert under State Route 146 is undersized and impounds water in the tidal lake. High marsh has yet to come back. Warren et al. (2002) showed that marsh invertebrates return at various rates with the slowest being *M. bidentatus*. It took 20 years for this snail to return to densities present at nearby natural reference marshes.

There are several non-native plant species in tidal marshes, but the most disruptive invader is the common reed. This species does not invade salt marshes where the soil S is too high. However, common reed frequently invades upland salt marsh borders in contact with groundwater (i.e., brackish marsh communities). These habitats experience disturbance when the highest spring tides deposit flotsam, including the seeds and rhizome fragments of common reed. Common reed colonizes the seaward edges of brackish reed marshes, because the flotsam cannot penetrate the tall vegetation. Metzler and Rozsa (1987) surmised that common reed might be a non-native variety and that the diffuse form found at Great Meadows in Essex, CT, was the native form. Saltonstall (2002) used DNA analysis to demonstrate that the invasive form was from Europe and historic herbarium collections contained several native varieties to the US. The variety present at the Great Meadows in Essex has been determined to be native and may be the largest population in the northeast.

Tidal marshes are subject to loss due to several mechanisms. In the late 1980s, the Connecticut Coastal Management Program investigated a complaint of a dying marsh on the Five Mile River in Darien. The low marshes west of New Haven Harbor appeared to be gradually subsiding and drowning. Over the course of 2–3 decades, cordgrass has undergone a gradual but progressive stunting with the resulting habitat becoming an intertidal flat. Aerial photo and historic map analyses by Dr. Scott Warren at Connecticut College concluded that submergence began before the beginning of the last century. The reason is unknown. Anisfeld and Hill (2011) dismissed nitrogen (N) enrichment as a cause. Subsidence may have been due to the combination of tidal range and the acceleration of sea level rise in the mid-1800s. The most extensive submergence marsh in the LIS region is the brackish low marsh habitat of the Quinnipiac River dominated by cattail. Several hundred hectares have been lost since the early 1970s.

A rapid loss of cordgrass along creeks and mosquito ditch banks occurred in Connecticut marshes beginning in 1999. This phenomenon has been termed sudden vegetation dieback and is found along the seaboard of the East and Gulf Coast states (Alber et al. 2008). The dieback is rapid, occurring in a single growing season and simultaneously at multiple sites. The characteristics of this dieback suggest a pathogen, and abundance of the fungi *Fusarium* spp. has correlated with dieback (Elmer and Marra 2011). The strain present in southern US areas most closely matches a species from Africa and the northern species appears to be undescribed. Studies in the Caribbean suggest that various pathogens have been transported to the US in African dust. In Connecticut, dieback is only associated with banks that are subject to daily tidal flooding and drainage at low tide. The stunted cordgrass on ditch plugs is an indicator of the lack of drainage and these areas are unaffected by dieback.

6.3.4 Science Gaps and Management Implications

In 1984, maintenance of mosquito ditches for mosquito control ceased in CT. It was anticipated that this would bring the marshes to some semblance of the preditching marsh complex especially with regard to returning pannes/pools to the marsh surface. Nearly 30 years later, pannes and pools are returning to tidal wetlands, and the marshes are beginning to resemble the preditching condition (as documented in a photograph of the Quinnipiac River marshes taken in 1917) (LISS 2002). As pannes continue to develop, breeding of salt marsh mosquitoes appears to have increased, so managers are examining the best way to treat this emerging issue. There has been no analysis to date on the mechanisms of panne formation and the ecological services they provide. At Barn Island, some of the pannes appear to deepen over time and may be leading to pond formation. As ditches fill and surface waters increase, it is likely that drainage will become an important issue; it is not known whether natural incising of creeks will occur in peats that are dense with high marsh vegetation roots. Designated control marshes would provide managers an important comparison to altered marshes to examine the efficacy of approaches such as maintenance ditching or open marsh water management.

In 2007, Connecticut scientists and managers discovered a recurring marsh cycle that appears to be responding to changes in tidal ranges associated with the metonic or lunar cycle. Miller and Egler (1950) described the eroded edge where the upland border of blackgrass and switchgrass (*Panicum virgatum*) vegetation belts eroded and become barren peat. The peat is colonized by herbaceous plants, which in turn are replaced by blackgrass, and the erosion returns. The eroded edge was described again in 1965 (Gross 1966) and then witnessed by scientists and managers in 1983. In 2007, it was apparent that the edge was returning. Miller and Egler (1950) suggested the underlying cause was grazing by cattle and mowing, practices that have long ceased. The current interpretation is that the lunar nodal cycle causes a predictable increase and decrease of the tidal range on the order of 6 cm over the 18.6 year period in southeastern Connecticut. As the tidal range increases, the edge forms new peat, but at the end of the period, when the tidal range is decreasing, the peat becomes aerobic and decomposes. Groundwater is the likely cause of peat removal that had been previously described as erosion. During the 1990s, sea level rise averaged 14 mm year^{-1} as the tidal range was increasing. While others only consider the average rates during the 18.6 year period, this analysis is revealing in that salt marshes survived a sea level rise seven times higher than the average, yet scientists suggest that at values greater than 10 mm year^{-1}, marshes will drown. Managers and scientists are trying to use various models to focus the effect of accelerated sea level rise in the future on marine transgression, yet we clearly know little about how marshes respond to sea level rise during the Metonic cycle. Furthermore, LIS is within the east coast hotspot for sea level rise that has seen a doubling since 1990. Examples of the types of research and monitoring that should help managers make informative decisions include remapping vegetation at places like Barn Island, Ragged Rock Creek, and Great Meadows, resurveying microrelief plots, and continuing long-term transects with an eye toward examination vegetation across the entire marsh, not just edge, for responses to the changes in the Metonic cycle.

The Long Island Sound Study (LISS) funded a project to examine wetland change in six western Long Island Sound (WLIS) embayments using the national wetlands inventory classification and summer aerial photography acquired every 5 years since 1974–2000. These sites are within the portion of central and WLIS that have been experiencing marsh subsidence since the late 1800s. This analysis confirms low marsh gradually converting to tidal flats but did not reveal any patterns such as rate changes that might results from varying sea level rise changes during the Metonic cycle. Dr. Scott Warren (personal communication) has found this dataset to be unique and valuable, and would advise that we continue to update this dataset.

Connecticut DEEP forecasted that the Connecticut River wetlands, designated of international importance under the Ramsar Convention (convention on wetlands of international importance), were likely to be impacted by climate change in several ways. Changes in the snow pack to the north would likely reduce the height and timing (arrive earlier) of the spring freshet. This would change ecological function and affect the upstream limits of the salt wedge, which in turn would affect the distribution of fresh and brackish tidal marshes. In addition, sea level rise will also cause the limit of the salt wedge to move upstream. The CTDEEP funded the installation of two real-time S gages on the river in order to begin to collect baseline data on the salt wedge to advise managers about future changes to wetlands and other resources on the Connecticut River. To this end, it would be valuable to establish reference wetlands with permanent transects and include instruments to measure sea level rise.

6.4 Seagrasses and Seaweeds

6.4.1 Seagrasses

The two fully submerged estuarine vascular plants found in LIS are eelgrass and widgeon grass (*Ruppia maritima*). Eelgrass, the dominant seagrass in LIS, was once prevalent throughout the shallow coastal areas of LIS. The Atlantic-wide die-off in the 1930s resulted in the loss of eelgrass from much of the local area, but healthy populations were reestablished in ELIS by the 1950s. The recovery of eelgrass in WLIS was less successful and today those populations have all but vanished. Since the 1950s, eelgrass populations along the Connecticut coast have suffered additional losses thought to be linked to the effect of N loading on the coastal ecosystem. Additional losses are predicted to occur in response to rising sea levels, increases in storm activity, and increasing temperatures at temperate latitudes, all of which are predicted responses to climate change. Management of watershed activities and use of the coastal waters have the potential to mitigate the factors contributing to eelgrass decline. In addition, seagrass restoration efforts in the ELIS have been successful, indicating improving water quality in these areas.

6.4.1.1 Overview of Habitat Requirements

Light, nutrients, and T are the primary factors controlling these plants (Dennison et al. 1993; Lee et al. 2007). The distributions of eelgrass and widgeon grass are also influenced by S, with eelgrass occurring in a more characteristically marine environment and widgeon grass found in estuarine and freshwater areas (McRoy and McMillan 1977; Short et al. 2002).

Light is generally recognized as the limiting factor to which eelgrass is most sensitive, having a high light requirement relative to macroalgae and phytoplankton (Dennison et al. 1993; Duarte 1995; Longstaff and Dennison 1999; Moore and Wetzel 2000; Hauxwell et al. 2003; Lee et al. 2007). Minimum light requirements of *eelgrass* range from 4 to 44 % of surface irradiance, according to review articles (Duarte 1991; Dennison et al. 1993; Batiuk et al. 2000) and experimental results (Dennison and Alberte 1985; Orth and Moore 1988; Olesen and Sand-Jensen 1993; Koch and Beer 1996). Minimum requirements determined for US east coast populations range between 15 and 35 % of surface irradiance, with studies specific to' LIS and Massachusetts also falling within this range (Dennison and Alberte 1985; Moore 1991; Koch and Beer 1996). The minimum light requirement is influenced by local conditions, including T, sediment and water column oxygen concentration, and sedimentary conditions (organic matter content and sulfide concentration). Under adverse conditions, the plants require more light. For example, higher temperatures increase respiratory rates relative to concomitant increases in photosynthetic rates (Dennison 1987; Lee et al. 2007). In turn, the minimum light requirement is greater under higher temperatures to support productivity sufficient to meet the respiratory demands (Staehr and Borum 2011).

While the minimum light requirement varies with habitat conditions, a conservative estimate can be used to define a maximum depth limit for seagrass beds, based on the light attenuation coefficient (K_d) of the water. The maximum depth limit under relatively ideal light conditions was calculated for LIS using the Lambert–Beer equation, following the method of Batiuk et al. (2000) for Chesapeake Bay. The maximum depth (z) was calculated using a minimum light requirement of 22 % of surface irradiance (I_z/I_0) and a K_d of 0.7 m^{-1}:

$$z = \frac{\ln\left(\frac{i_z}{i_o}\right)}{-k_d} = \frac{\ln(0.22)}{-0.7} = 2.16 \text{ m}$$

Under current water quality conditions, the maximum depth limit for eelgrass in WLIS will be considerably shallower due to higher light attenuation coefficients. Values for K_d in ELIS are typically ~0.5 m^{-1}, yielding a predicted maximum depth of 3.1 m. In WLIS, K_d is closer to 1 m^{-1}, yielding a predicted maximum depth of 1.5 m, with an extreme of K_d reaching 4.4 m^{-1} (Koch and Beer 1996). The maximum distribution depth is further influenced by the tidal range (Koch and Beer 1996). In areas with higher tidal ranges, the light available at the bottom will be reduced at high tide relative to areas with lower tidal ranges. Tidal ranges in LIS increase to the west, ranging from ~0.7 m in the east to ~2.2 m in the west. So the maximum depth will be shallower in WLIS, given the greater tidal range and increased K_d.

Geographic distribution of eelgrass is also constrained by a minimum depth requirement and an apparent requirement for a minimum depth span between the shallow and deep edge of the bed. While some eelgrass is found in intertidal areas, eelgrass in LIS seems to require a minimum depth equivalent to half of the spring tidal range (Koch and Beer 1996), ranging from 0.5 m in ELIS to 1.25 m in WLIS. In addition to the minimum and maximum depth restrictions, modeling of depth distributions in LIS indicates that these populations require a 1 m depth difference between the shallow and deep edge of the bed (Koch and Beer 1996). This vertical depth distribution requirement protects a bed from storm scour. Theoretically, the deep edge of the bed will be more resistant to storm scour, as the wave energy is attenuated with depth; thus, providing a source population for recolonization of damaged areas. Under current K_d conditions, which serve to restrict the deep edge of the distribution, eelgrass is unlikely to survive in WLIS and is marginal in CLIS. These predictions match the current distribution of eelgrass (Fig. 6.7).

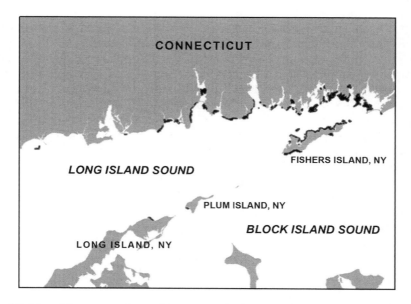

Fig. 6.7 Map of the current eelgrass (*Zostera marina*) distribution in Long Island Sound adapted from Tiner et al. (2010). Meadow polygons were enlarged with a 400′ buffer to increase visibility in this format. Map provided by Lorne Brousseau, CCE

Nutrient limitation of eelgrass in natural environments is unlikely to occur as the plants have relatively low N and P requirements and have access to nutrients in sediments via their root and rhizome network (Zimmerman et al. 1987; Lee et al. 2007). While sediment porewater has higher nutrient concentrations, assimilation of nutrients by leaves and roots can be nearly equal (Lee et al. 2007). Excess nutrients stimulate growth of phytoplankton and macroalgae, both of which require considerably more nutrients than seagrass (Bintz et al. 2003; Hauxwell et al. 2003). Both phytoplankton and macroalgae have faster uptake rates of dissolved inorganic N and faster growth rates than seagrass (Duarte 1995). Thus, the requirement for high light by seagrass also necessitates low nutrient levels or phytoplankton; macroalgae would flourish and shade the seagrass.

Eelgrass exhibits an optimal T range for growth with a worldwide average of 15.3 ± 1.6 °C and an optimal range for photosynthesis of 23.3 ± 2.5 °C (Lee et al. 2007). In the shallow embayments of LIS, summer temperatures often exceed 25 °C and may be detrimental to its success, inhibiting growth, and possibly photosynthesis. Inhibition may result from thermal disruption of metabolic processes or from an increase in the minimum light requirement necessary to compensate for increasing respiratory demands (Staehr and Borum 2011). The higher mid-summer temperatures in LIS and its embayments lead to a bimodal growth pattern in most years, with best growth in the spring and fall and inhibition in the warm summer months (Olesen and Sand-Jensen 1993; Moore et al. 1996; Bintz et al. 2003; Keser et al. 2003; Yarish et al. 2006). During the winter, production is almost nonexistent for eelgrass in LIS (Yarish et al. 2006).

6.4.1.2 Historical and Current Distribution

Eelgrass was common throughout LIS prior to 1930s. By the summer of 1931, most of the eelgrass ranging from North Carolina to New England and in much of the Atlantic had been wiped out by a wasting disease, attributed to the heterokont labyrinthulomycete *Labyrinthula zosterae* (Short et al. 1987). Only an estimated 1 % of the North Atlantic population remained, occurring primarily in the low S waters of upper estuarine areas (Cottam 1933). The eastern portion of LIS experienced a recovery by the 1950s, while recovery in WLIS was spotty and eventually failed (see references cited in: Rozsa 1994; Keser et al. 2003; Yarish et al. 2006; Johnson et al. 2007). In the 1990s, Clinton Harbor was the westernmost location supporting eelgrass (Yarish et al. 2006).

As part of the US Fish and Wildlife Service's National Wetland Inventory Program, aerial surveys of eelgrass extent in LIS were conducted in 2002, 2006, and 2009 (Tiner et al. 2007, 2010). These surveys were initiated with support from the LISS and The State of Connecticut's Office of Long Island Sound Programs within the Connecticut Department of Energy and the Environmental Protection (CTDEEP) (2002 and 2006 surveys). Continued support comes primarily from the LISS (2009 survey) (Tiner et al. 2010). These surveys indicate that the majority of eelgrass acreage is found east of Rocky Neck State Park in Connecticut, and around Fisher's Island and Orient Point in New York (Fig. 6.7; Tiner et al. 2010; Pickerell unpublished data). A few beds still exist west of these locales. As of 2011, two small beds (2.6 ha total) of eelgrass were extant in the bight between Clinton Harbor and Westbrook Harbor associated with the Duck Island breakwater and a very small patch near the mouth of Clinton Harbor (Tiner et al. 2010; Vaudrey and Yarish unpublished data). Three beds (10 ha total) were identified along the north shore of Long Island, but both the naturally occurring small patches and the restored areas at two locations were not identified as part of the aerial survey (Tiner et al. 2010; Pickerell unpublished data).

Tiner et al. (2010) analyzed change in eelgrass coverage among the three aerial surveys. At this point, changes in bed area among the three surveys likely reflect interannual variability; the time series is not yet sufficient to detect long-term trends. For example, the loss noted by the aerial surveys of 4.45 ha in Mumford Cove (Groton, CT) between 2002 and 2006 reflects the shifting of a sand bar in 2006, just prior to the survey. This shift resulted in the smothering of eelgrass in the southeastern portion of Mumford Cove. By the following year, eelgrass was once again growing there (Vaudrey, unpublished data) and was reflected in the 2.8 ha increase between the 2006 and 2009 surveys (Tiner et al. 2010). These small variations contrast with the longer history of eelgrass in Mumford Cove. Beginning in 1946, a wastewater treatment facility (WWTF) discharged effluent into the head of the cove from Fort Hill Brook (Vaudrey et al. 2010). Nutrient inputs increased until the outflow was relocated to the Thames River in October 1987. Water column nutrient concentrations decreased rapidly and by the following year, the recurring massive blooms of sea lettuce had disappeared. Recolonization of the cove by eelgrass occurred slowly over the subsequent 15 years and has been relatively stable at ~20 ha since 2002 (Tiner et al. 2010; Vaudrey et al. 2010).

6.4.1.3 Linking Water Quality Parameters and Watershed Activities to Seagrass Distribution

Research examining habitat criteria for eelgrass in Chesapeake Bay identified light as the primary factor along with several factors and habitat constraints controlling its distribution (Batiuk et al. 1992; Dennison et al. 1993). Yarish et al. (2006) examined habitat criteria for eelgrass in LIS, and their suggested guidelines were applied to three case study sites to further evaluate suitability of the defined criteria in seagrass management along the Connecticut coastline (Vaudrey 2008a, b). Results of these analyses suggest that LIS seagrass requires more conservative standards, erring toward clearer water with less nutrient input relative to Chesapeake Bay (Table 6.1). Even more conservative standards are required to establish a new eelgrass bed.

These studies verified the choice of habitat criteria by comparing field data for the parameters to the extant populations. For example, light attenuation coefficients in three eelgrass sites were compared to suggested criteria (Fig. 6.8).

Table 6.1 Recommended habitat requirements from Chesapeake Bay submerged aquatic vegetation water quality standards compared to recommendations for Long Island Sound (Batiuk et al. 2000; Koch 2001; Yarish et al. 2006; Vaudrey 2008a, b) and to eelgrass restoration site selection parameters for Long Island Sound

	Chesapeake Bay guidelines	Suggested LIS guidelines
Primary requirements		
Minimum light requirement of surface irradiance)	>22	>22
Secondary requirements		
Chlorophyll *a* (μg L^{-1})	<15	<5.5
Dissolved inorganic nitrogen (mg L^{-1})	<0.15	<0.03
Dissolved inorganic phosphorus (mg L^{-1})	<0.02	<0.02
Total suspended solids (mg L^{-1})	<15	<30
Habitat constraints		
Minimum depth limit (m)	$Z_{min}^{a} = (MHHW - MLLW)/2$	
Maximum depth limit (m)	$Z_{max}^{b} = -\ln (0.22)/K_d$	
Minimum for $(Z_{max} - Z_{min})^{c}$ (m)	0.5	1
Sediment organic matter (%)	0.4–12	0.4–10
Sediment grain size	0.4–30 % fines	<20 % silt and clay
Sediment sulfide (μM)	<1,000	<400
Current velocity (cm s^{-1})	5 < X < 180	5 < X < 100

[a]MHHW refers to mean higher high water, the high tide level during spring tides. MLLW refers to mean lower low water, the low tide level during spring tides
[b]The value of 0.22 refers to the minimum light requirement of 22 % of surface irradiance. K_d refers to the light attenuation coefficient for the waters of concern
[c]The minimum required difference between the depth of the shallow and deep edge of the bed

Eelgrass populations in Niantic River and at Bushy Point are still viable, while the bed surveyed in Clinton Harbor between 1993 and 1995 has now disappeared (Yarish et al. 2006; Kremer et al. 2008). While the Clinton Harbor bed surveyed in the 1990s is gone, a small bed of eelgrass has been located in an area <0.5 km away (Vaudrey and Yarish, unpublished data). Higher light attenuation coefficients, as seen in the Clinton site, may indicate eelgrass populations existing under stressful conditions (Fig. 6.8).

6.4.1.4 Seagrass Ecosystem Services

Seagrass communities provide a range of ecosystem services (Hemminga and Duarte 2000; Orth et al. 2006). These communities provide foraging ground and shelter, and act as a nursery to certain species of crustaceans, fishes, and mollusks (Beck et al. 2001; Heck et al. 2003; Orth et al. 2006). Seagrasses are considered ecosystem engineers; they change their environment, often to their own benefit (Bouma et al. 2005; Boer 2007). These changes include trapping of sediment resulting in

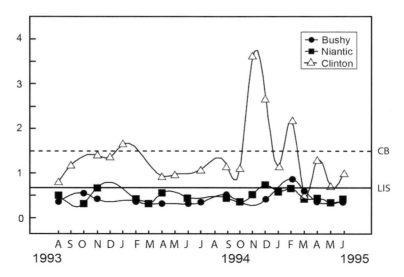

Fig. 6.8 Temporal variation in K_d at three *Zostera marina* sites (Bushy Point, Niantic Bay and Clinton Harbor) along the Connecticut coast. The *solid* and *dotted lines* indicate values suggested as standard habitat requirements for eelgrass in Long Island Sound (LIS) and Chesapeake Bay (CB). From Yarish et al. (2006)

increased water clarity and growing season sediment accretion (Bos et al. 2007). Additionally, these communities are effective at nutrient transformation, including substantial carbon sequestration in the sediment (Duarte et al. 2005; Romero et al. 2006; McGlathery et al. 2007). The major threat to seagrass communities is cultural eutrophication of coastal waters due to increased anthropogenic nutrient inputs (Short and Burdick 1996; Hauxwell et al. 2003; Orth et al. 2006; McGlathery et al. 2007; Waycott et al. 2009). Additional stressors to seagrasses include physical disturbance (boating, commercial fishing), disease, and global climate change (Rasmussen 1977; Short et al. 1987; Walker et al. 1989; Short and Neckles 1999; Neckles et al. 2005). Seagrass communities are considered by some to be one of the most threatened habitats on the planet, but are often overlooked by the general public (Orth et al. 2006; Waycott et al. 2009).

6.4.1.5 Current Trends in Seagrass Management

Management strategies have been shifting toward establishing criteria that are protective of key indicator species such as seagrasses (Lewis et al. 1998; Orth et al. 2002; Greening and Janicki 2006; Steward and Green 2007; Wazniak et al. 2007). Seagrass ecosystems have been described as "coastal canaries," with the implication that loss of seagrass is a sign of degradation with associated loss of important ecosystem services (Orth et al. 2006). The benefit of managing an indicator species (or habitat) versus setting a goal for a single parameter is the possibility of the inclusion of many factors that influence coastal waters; thus, reflecting the diverse and complex interactions of these coastal communities. While many factors affect seagrass, anthropogenic N delivered from watersheds has been identified as the most detrimental. Latimer and Rego (2010) evaluated N load in relation to eelgrass extent in 62 Southern New England embayments, six in LIS. Results matched similar approaches applied to various seagrass species in embayments in Connecticut, Massachusetts, Florida, Australia, and England (Short and Burdick 1996; Valiela and Cole 2002; Hauxwell et al. 2003; Steward and Green 2007; Vaudrey 2008b). Nitrogen loads below 50 kg N ha^{-1} year^{-1} were considered protective of eelgrass habitats, with variability in eelgrass extent attributable to other influencing factors. Above that input, eelgrass coverage was reduced; a complete loss of seagrass typically occurs above N loadings of 100 kg N ha^{-1} year^{-1} (Latimer and Rego 2010).

Connecticut's Coastal Management Act, promulgated in 1980, allows for the protection of eelgrass. The act specifically recognizes eelgrass as a coastal resource "...to protect, enhance, and allow for the natural restoration of eelgrass flats..." (CGS 22a-92(c)(2)(A)). The CTDEEP, administrators of the act, considers eelgrass protection when reviewing and authorizing activities such as dredging and construction of piers and docks. While such activities do present a threat to eelgrass survival, the potential adverse effect of land-derived nutrients on eelgrass communities has been recognized as an important contributor. Connecticut's management of nutrients in estuaries focuses on N reductions through the LIS Total Maximum Daily Load Analysis for dissolved oxygen (CTDEP and NYSDEC 2000). The CTDEEP manages a number of nonpoint and point source programs designed to reduce N loadings to LIS. In addition to these programs and in order to more cost effectively achieve N reductions at wastewater treatment facilities (WWTFs), Connecticut implemented a N credit exchange (NCE) in 2002 (www.ct.gov/dep/nitrogencontrol). This approach has resulted in attainment of an aggregate N reduction limit from 79 WWTFs located throughout the state.

Connecticut has also evaluated the effect of commercial and recreational fishing activities on eelgrass beds as the result of Public Act 01-115. The report workgroup concluded that although fishing activities do negatively impact eelgrass, N enrichment of coastal systems presents a major threat to the remaining eelgrass beds (Johnson et al. 2007).

Further site-specific N reductions may be warranted to protect eelgrass via the restoration and maintenance of eelgrass habitat. The CTDEEP continues to support the collection of information to further scientific understanding regarding eelgrass habitat requirements and the impact of watershed loading on such requirements. It will consider guidelines or standards as appropriate information becomes available.

Seagrass protection in New York is somewhat limited in that there are no explicit protections afforded to eelgrass or widgeon grass. In order to address this deficiency, the New York State Legislature passed a law supporting creation of a Seagrass, Research, Monitoring, and Restoration Task Force in 2006. The task force was charged with developing recommendations on elements of a seagrass management plan with the goal of preserving, restoring, and mapping the native seagrass populations on Long Island. Additionally, the group was charged with recommending means of action that would bring about a lasting restoration of fin-fishes, shellfish, crustaceans, and waterfowl compatible with an improved quality of life and economic growth for the region (NYS Seagrass Task Force 2009).

In response to the findings of the Seagrass Task Force (NYS Seagrass Task Force 2009), members of the New York State Legislature drafted a bill to protect seagrasses in New York State waters in 2010. In June of 2010, the New York State Senate and Assembly passed the Seagrass Protection Act. Although there was no specific language as to how to achieve the proposed goals set forth in this bill, in principle it called for: restrictions in the types of mechanically powered fishing gear used in seagrass meadows identifying pesticides and chemicals harmful to seagrasses and restricting their use near seagrasses; restrictions on applying fertilizer containing phosphorus (P) after 1st November and before April; and development and adoption of seagrass management protective of both selected seagrass beds and traditional recreational activities. Some constituent communities expressed concerns about language within the bill and for this reason, the governor did not sign the original bill and the law was never enacted. In 2011, a revised version of the bill was submitted to the House and Senate for a vote. On July 31, 2012, Governor Andrew Cuomo signed the bill into law; it will take effect at the end of December 2012.

6.4.1.6 Restoration Efforts

The first documented eelgrass restoration attempts in the region were conducted in response to the catastrophic losses from the wasting disease outbreak of 1931 and 1932 (Cottam 1933). In what may have been the first attempt to restore seagrass in the U.S., eelgrass was transplanted at sites along the north mid-Atlantic coast from Virginia to Massachusetts by the US Department of Agriculture, Bureau of Biological Survey, and others during 1935 and 1936 (Lynch and Cottam 1937). As part of this effort eelgrass collected in Friday Harbor, WA and from Pungoteague Creek, VA and Mecox Bay, NY were planted near Jones Beach and in Great South Bay, NY in approximately 1 m^2 plots (Lynch and Cottam 1937). Additional work, although not well described in the available references, involved the first documented use of seeds as a planting method for eelgrass. It appears that most of this work was unsuccessful, but it points to the fact that, as early as the mid-1930s, restoration was considered a worthwhile management tool.

Addy and Johnson (1947) reported on the results of several transplant attempts in Connecticut using eelgrass collected from a meadow in Niantic Harbor. Of the sites planted, Black Hall River (Old Lyme) and Hotchkiss Grove Beach (Branford) were reported as being successful while the Patagausett Cove (East Lyme) and Norwalk River (Norwalk) were not checked and failed, respectively. An attempt in Huntington Harbor, Long Island, also failed. Wasting disease might have played a part in the demise of the Huntington Harbor transplants. By 1953, only the Hotchkiss Grove site still supported eelgrass; this planting may have survived at least until ~1955. In 1955 and 1956, an additional planting was undertaken at the same site to supplement what was already there and replace what had been lost. Anecdotal information indicates that the planting of eelgrass in the Hotchkiss Grove site may have continued annually for 30 years (Rozsa 1994). The presence of this bed was confirmed in 1982 by staff of the National Marine Fisheries Service (Rozsa 1994). A subsequent survey by Yarish et al. (2006) from 1993 through 1995 did not locate this bed. However, the existence of live eelgrass at this site during 2010, and the lack of any proof of recent plantings, leads the authors to believe the plantings from the late 1940s, or at least the mid-1950s, may have persisted to this day (Vaudrey, personal observation).

Although there were several attempts to plant and restore eelgrass to New York waters in the Peconic Estuary (PE) and the South Shore Estuary Reserve (SSER) beginning in the late 1970s and continuing in the 1990s, work in LIS did not begin until after 2000 (Churchill et al. 1978; Pickerell et al. 2007). In 2003, Cornell Cooperative Extension of Suffolk County (CCE) began test plantings in LIS near East Marion, NY. This work followed soon after the discovery of healthy eelgrass meadows along the high-energy open coast on the east end of Long Island near Petty's Bight (East Marion, NY) and Orient Point (Orient, NY). Detailed observations of these meadows lead to a better understanding of the types of habitats that were suited for the growth of eelgrass in LIS. Unexpectedly, these meadows appear to thrive in wave-swept and often high current areas in very coarse sediment composed of sand and gravel with large numbers of rocks and boulders. This extreme physical environment differs substantially from the meadows known to exist historically throughout the PE and SSER, most of which occurred in fine-sediment-dominated bottoms of protected bays, harbors, and creeks. With this information, CCE staff set out to locate and screen suitable sites for eelgrass restoration potential along Long Island's North Fork from Mattituck Inlet east to East Marion.

Using the reference meadows as a cue, the first pilot planting sites were chosen based on the conditions described above as well as their proximity to the eastern side of north facing points along the North Fork: Terry's Point (Orient) and St. Thomas Point (Southold). This leeward location provided partial protection from northwest winds, the direction from which the most consistent and extreme fall and winter winds typically originate. This was important given that plantings were undertaken in the fall and shoots would not have time to root before the windy season began. Pilot-scale (i.e., several hundred shoots) plantings at both sites in 2003 resulted in healthy and expanding patches. Following this success, additional plantings were added during the second and third years at St. Thomas Point to create a very large and stable meadow (Fig. 6.9).

Additional projects were initiated to continue plantings further into LIS as far west as Lloyd's Neck (Huntington) and as far east as Plum Island (Southold) and Great Gull Island (Southold). The western plantings were intended to test the interaction of planting depth, water clarity, and tidal amplitude described by Koch and Beer (1996). Plantings at Lloyd's Neck were only partially successful, and

Fig. 6.9 Restored meadows during early spring when the leaves are still short. **a** St. Thomas Point, East Marion, NY. This site was first planted in 2003 and involved using transplants from several donor meadows in the region. Depth at this site ranges from 2.5 to 3.5 m at low tide. **b** Terry's Point, Orient, NY. This meadow was planted using plants from two natural donor meadows; Orient Point in the Peconic Estuary and Mulford Point in LIS. Depth at this site is 1.5 m at low tide (Photo credits: Chris Pickerell, CCE)

survival was limited to a very narrow band ~1 m below mean lower low water MLLW. During the second growing season, these otherwise healthy shoots were lost to erosion and it is not clear whether eelgrass can survive this far west in LIS based on water quality and tidal range.

At the time of this writing, the oldest restoration site on the north shore of Long Island has persisted for 8 years. In total, eight successful planting sites run from Hortons Point (Southold) in the west to Great Gull Island (Southold) in the east and more work is underway to identify new planting sites and expand on the patches that have been created. Shoot densities at the oldest restoration sites compare favorably with reference sites and, in fact, those in portions of the restored meadows exceed the mean in the densest parts of the reference sites (Pickerell, personal observation). In July 2011, within patch shoot densities at the Petty's Bight (East Marion) reference meadow were 513 ± 22 m^{-2} (mean $\pm$ standard error) while those at St. Thomas and Terry's Point were 401 ± 20 m^{-2} and 811 ± 60 m^{-2}, respectively. In an attempt to apply the experience gained in New York to the Connecticut coast, the latest round of restoration work applies methods used in New York waters to similar areas along the Connecticut coast (Pickerell and Vaudrey 2010).

6.4.1.7   Science Gaps and Management Implications

In order to address the lack of estuary specific habitat criteria data for eelgrass in LIS, the New England Interstate Water Pollution Control Commission (NEIWPCC), with funding provided by USEPA, is sponsoring a research effort to develop a GIS-based model that will include those factors generally considered to be limiting to the growth of eelgrass in LIS (Pickerell et al. 2011). This project will: identify and map areas where eelgrass natural recruitment and/or restoration is possible; identify areas where restoration may be possible if water quality or other parameters improve; identify parameters limiting to eelgrass natural recruitment and/or restoration in areas determined to be unsuitable by the GIS model; and identity areas where eelgrass may colonize in response to sea level rise. This work is a collaborative effort between CCE, University of Connecticut, and NOAA scientists and involves several authors of this chapter (Pickerell et al. 2011).

Seagrass beds constitute a vibrant and important community in LIS, providing habitat and food to a host of commercially and ecologically important species, and serving as sinks for nutrients and suspended solids. Their presence indicates a desirable state of water quality, as the plants are adversely affected by high nutrient concentrations in the water column and the resulting issues often associated with high N inputs (e.g., algae blooms, hypoxia). The greatest threats to seagrasses in LIS have anthropogenic sources: cultural eutrophication and climate change. Continued monitoring of seagrass areal extent, density, and metabolic parameters is a key for understanding and predicting the response of seagrasses to a changing environment. Critical gaps in knowledge include an assessment of the current responses of LIS seagrass communities to locally occurring stresses (e.g., nutrients, T) and widespread data from LIS on the suitability of habitats for the support of eelgrass. An identification of seagrass communities exhibiting signs of stress would aid in focusing attention on locales most in need of management actions (e.g., nutrient and sediment load reductions, etc.). Habitat assessments could identify areas of LIS suitable for eelgrass restoration efforts. A combination of further research into metabolic responses of seagrass to stressors, appropriate management actions in coastal waters and watersheds, and continued restoration efforts will serve to preserve and perhaps expand the extent of eelgrass in LIS.

6.4.2 Seaweeds

Long Island Sound supports a very rich and diverse algal flora, with an estimated 250 species (Van Patten 2006; Schneider et al. 1979; J. Foertch, personal communication 2005). With a few exceptions, they are benthic, sessile species during the adult life stages that inhabit the supralittoral, upper littoral, eulittoral, sublittoral, and subtidal zones on rocky shores of LIS. These algae have many varied shapes and sizes, and while some are perennial year-round residents, others may be annual or ephemeral. As is the case with most algae, species tend to have a highly plastic morphology, readily changing shape and color in response to environmental conditions such as light, T, photoperiod, substrate, wave action, and other parameters (Lobban and Harrison 1994; Lüning 1990).

6.4.2.1 Seasonal Cycles: Abundance and Biomass

The geographic location and orientation of LIS are such that it hosts an interesting and possibly unique assemblage of both cold and warm-T tolerant species. Some of the Arctic-cold-to-temperate Atlantic assemblage includes species that originated in the Pacific and traveled to the Atlantic during Paleo-migration (Lüning 1990). There are also cold-water species that migrated south from Labrador and Greenland via the North Atlantic Boreal Current. These species tend to thrive in late fall, winter, and early spring. Using genetic analysis, Hu et al. (2010) postulated a trans-Atlantic repopulation Irish moss from localized refugia in Europe to North America following the last Pleistocene glacial maximum, noting that its southern range seems to coincide with the 17 °C isotherm (Lüning 1990).

Kelp (principally *Saccharina latissima* and *Saccharina longicruris*, both formerly classified as "*Latissima saccharina*") is one among the largest macroalgae species in the North Atlantic and key species supporting benthic communities. They are good examples of cold-water species at or near the southern limit of their biogeographic range. While these species are perennial in more northern locales, they are biennial in LIS. Late fall, winter, and spring are the best growth seasons; spring and fall are optimal for reproduction. In the summer, abundance decreases as the water heats up, with blades degenerating in August (Egan and Yarish 1990, 1988; Van Patten et al. 1993; Yarish et al. 1980).

During the summer, many warm temperate Atlantic species and sub-tropical species appear, some of which may be endemic and others that travel north via the Gulf Stream. Subtidal examples include graceful red weed (*Gracilaria tikvahiae*), barrel weed (*Champia parvula*), and several *Ceramium* spp., as well as a few invasive species (e.g., *Grateloupia turuturu*, *Gracilaria vermiculophylla*, *Porphyra yezoensis*). The intertidal components of the Sound's marine flora are generally hardy perennial organisms that have adapted to extremes of T and light, particularly rockweeds and knotted wrack, which can be found year-round. Sea lettuce can be found for most of the year in the upper littoral zone, but may bleach in very sunny weather (e.g., Pedersen et al. 2008).

There are also several perennating species that use two distinct morphologies as a way to cope with the extreme seasonal T ranges in LIS; for example, sausage weed (*Scytosiphon lomentaria*) puts out erect branching when T conditions are favorable, and becomes a small discoid crust when they are not (Lobban and Harrison 1994).

6.4.2.2 Distribution and Abundance

In general, the presence of macroalgae in LIS, like all photosynthetic organisms, is determined largely by the quantity and quality of light (e.g., Dring et al. 2001; Dring 1992) and thus are to be found in the photic zone; naturally requirements vary for individual species. In most estuaries, the successful species tend to be both euryhaline and eurythermic (Lee 1989). Wave action is also a significant factor in determining what grows where, and individual tolerances vary.

The historical as well as the current challenges have been to establish what species are in LIS in order to know how they may be interacting and changing. Initially, most work on the LIS macroalgae was taxonomic, with the intent of establishing a baseline of what species inhabited the estuary. Schneider et al. (1979) developed an annotated checklist of Connecticut seaweeds in 1979 for the Connecticut Geological and Natural History Survey. This is probably the last complete survey performed for LIS, although Dominion Nuclear's Millstone Environmental Laboratory has done extensive continuous monitoring and surveying for the eastern Connecticut shore of the Sound for the past 3 decades (e.g., Keser et al. 2010; Foertch et al. 2009). A University of Connecticut digital algal herbarium with many LIS specimens is available online at http://www.algae.uconn.edu.

Many macroalgae begin life as microscopic propagules in the water column that requires hard benthic surfaces to settle on. Therefore, the majority of species is found on rocky shores or where glacial cobble and till have accumulated. There are exceptions, such as some species that colonize mud flats and marshes (e.g., *Polysiphonia subtilissima*) or float (e.g., broadleaf gulfweed (*Sargassum fluitans*)). Algae tend to form distinctive bands or zones and develop specific morphological adaptations to cope with life on rocks, battered by wind and waves. These may include gripping structures, such as holdfasts (a crustose form), alternate life forms depending on seasons or conditions, and phycocolloids that act as both anti-desiccants and glue.

Temperature, light, and day length are primary factors controlling both distribution and productivity of seaweeds in LIS (Lee and Brinkhuis 1988; Brinkhuis et al. 1983; Yarish et al. 1984, 1986, 1987). Light intensity and wavelength are important for growth and reproduction. While individual species tolerances vary, the macroalgae can be grouped by the dominant photosynthetic pigments in their tissues, which accounts for the typical zonation of blue-green, green, brown, and red algae as water depth increases. Light intensity decreases exponentially with depth, so macroalgae are constrained within the photic zone. Within this zone, T is generally the controlling factor for both intertidal and sublittoral distributions of algae in LIS (e.g., Yarish et al. 1984; Lüning 1990; Pedersen et al. 2008); other factors include exposure, nutrient availability, competition for substrate, and predation (Pedersen et al. 2008).

Few, if any, comprehensive studies of algal abundance and biomass for the entire LIS have been performed. Brinkhuis and colleagues examined five of the most common species in LIS—kelp, graceful red weed, green fleece, rockweed, and *Agardhiella subulata*—for their potential as biofuels (Brinkhuis et al. 1982, 1983). Egan and Yarish (1988) described the geographic distribution of kelp on a large geographic scale, noting that except for an outlier deep-water population off the coast of New Jersey, LIS is the southern limit for the species. Several species of kelp exist in LIS. Two members of genus *Saccharina*, one with a flat blade and long, hollow stipe (*S. longricruris*) and the other with a short, solid stipe, and ruffled blade margins (*S. latissima*). Populations were found at Black Ledge (Groton, CT) along the eastern Connecticut shore, Cove Island, off Stamford, CT, and at several other locations, such as Crane Neck and Eatons Neck, along the north shore of Long Island, and Montauk Point (Brinkhuis 1983; Egan and Yarish 1988). Horsetail kelp (*Laminaria digitata*), was found at Montauk Point, NY and along the north shore of LIS from Groton, CT and west to the Thimble Islands, Branford, CT (Egan and Yarish 1988).

Most researchers (e.g., Pedersen et al. 2008) agree that for most of the year, the dominant algal species in LIS are: *Ulva* spp. (both tubular and sheet forming species) and *Blidingia minima* species in the upper littoral zone, rockweeds, particularly *F. vesiculosus*, in the mid-littoral, and Irish moss in the infra-littoral and sublittoral. In a Cove Island study (Pedersen et al. 2008), the invasive green fleece periodically increased in abundance above the infra-littoral zone. Kelp was rarely recorded there. The authors also noted several species of red algae in the genus *Porphyra* in the intertidal zone, but they were difficult to distinguish genetically and documenting the abundance was problematic for this genus because of its morphology and epiphytic habit. Kim et al. (2009) noted that while most *Porphyra* spp. occurs seasonally, one, *Porphyra umbilicalis*, occurs year-round in the eulittoral zone. This species is the most abundant of the many species in LIS (Yarish et al. 1998; Broom et al. 2002; Klein et al. 2003; He and Yarish 2006; Neefus et al. 2008a, b). In the summer, these key species are still present, and sausage weed, red hornweed (*Ceramium virgatum*, formerly *C. rubrum*), and barrel weed emerge. A few kelps are found below 5 m depth (Egan and Yarish 1988; Yarish et al. 1990).

Eastern LIS monitoring by Dominion's Millstone Environmental Laboratory has been invaluable for examining species diversity and abundance in areas both impacted and not impacted by the nuclear power plant's thermal plume. Recently, use of multivariate analyses on the data has shed light on complex interactions at these monitored areas. The decline of *Chondrus crispus* (Irish moss) and *Fucus* spp. (rockweeds) in 1983, for example, with simultaneous increase in green fleece, have been correlated to the second cut for plume out-flow made at Millstone. Subsequent recovery of rockweed has also been doc-umented, and the post-Irish moss dominance of multiple genera of red algae such as *Gelidium*, *Polysiphonia*, *Corallina*, and *Hypnea* has been correlated in Waterford to the addition of Dominion's Unit 3 (Foertch et al. 2009). This find-ing may provide insights into the potential effects of warming waters expected in the next decade and beyond (Keser et al. 2010).

Gerard (1995) observed shallow, hard-bottom algal communities at Crane Neck Point in CLIS and reported continuous dominance by ecologically and economically important species from 1983 to 1991. Irish moss, called the "major groundcover," dominated the infra-littoral to subtidal rock, with kelp forming a canopy above it. She reported a dramatic change in community structure at this location in 1991, when kelp all but disappeared and Irish moss became sparse. Several species of finely branched red algae, i.e., *Cystoclonium purpurea* and *Phyllophora pseudocera-noides*, replaced Irish moss. *Punctaria latifolia* (a brown alga that resembles juve-nile kelp blades) increased in abundance. This phenomenon lasted for about 2 years, after which Irish moss returned but kelp continued to be absent. The cause of the shift in species composition was arbitrarily attributed to a warm summer in 1991 and storm damage. This change was observed at other locations along the north shore (Gerard 1995), while the Connecticut shore kelp seemed to be unaffected. Ecotypic differentiation in T tolerance in these populations might account for this phenom-enon (Egan et al. 1990). Interestingly, Yarish et al. (1990) noted that sporophytes of kelp in WLIS, which routinely experienced warmer temperatures than ELIS kelp, did not exhibit greater survival at warm temperatures (20 °C). However, there is evidence of genotypic differentiation of populations from the southern range of the distribution and those from mid-range populations (Neefus et al. 1993). *S. latissima* from LIS is exposed to higher summer temperatures and high nutrient regimes, and can tolerate more heat stress than the mid-range type (Gerard and Du Bois 1988; Gerard et al. 1987). Meiospore germination and gametophyte growth at the south-ern boundary in LIS have been successful in July at temperatures as high as 25 °C. Optimal growth temperatures for gametophytes shift from 10–15 °C in March to 15–20 °C in July, exhibiting patterns of seasonal T acclimation. Optimal growth of young sporophytes for the species is 10–15 °C all year, but sporophytes from LIS have been found to survive up to 20 °C for all months except January (Egan et al. 1989, 1990; Lee and Brinkhuis 1988). The survival of plants in LIS at higher tem-peratures is attributed to the ability to accumulate and store higher levels of N in their tissues. The additional N reserve bolsters the photosynthetic apparatus and pos-sibly contributes to production of protective heat shock proteins (Gerard 1997).

Of course, T is not always the primary factor affecting distribution. Kim et al. (2008, 2009), in studies of *Porphya* spp. at different tidal elevations, found a correlation between nitrate uptake (which may be enhanced by desiccation) and vertical distribution patterns.

So far, this discussion has focused on hard-bottom algae, but there are estuarine soft-bottom species such as *Polysiphonia subtilissima, Bostrichyia radicans, Caloglossa leprieurii,* and *Neosiphonia harveyi* that commonly grow at the base of cordgrass in shallow embayments, salt marshes, and muddy areas of LIS. These epiphytic, finely branched rhodophytes use the marsh vegetation for support and shade (Yarish and Edwards 1982; Yarish and Baillie 1989).

6.4.2.3 Productivity

No comprehensive quantitative assessment has been made to our knowledge of the overall productivity of LIS macroalgae, but some studies estimated productivity via measuring biomass at specific sites. Field and laboratory studies by Egan and Yarish (1990) show that kelp species are the most productive of the LIS macroalgae due to large size and rapid growth. At Flax Pond on Long Island, Brinkhuis (1983) found that salt marsh knotted wrack produced about 600 g (dry wt) m^{-2} $year^{-1}$. In comparison, estimates of above-ground production of cordgrass in Flax Pond ranged from approximately 600 to 1,300 g m^{-2} $year^{-1}$ (Houghton 1985). A Gas Research Institute and General Electric-sponsored biofuel study (Brinkhuis et al. 1983) growing five common LIS species in culture and comparing biomass and growth rates. This resulted in the recommendation of kelp as a biomass candidate, primarily due to its superior growth and productivity during the winter months. Graceful red weed, a warm water species, was also suggested as a candidate because of its high productivity during May to October. Green fleece, rockweed, and *Agardiella subulata* were additional possible candidates. Productivity by green fleece was similar to rockweed and kelp under these study conditions. Knotted wrack, dulse (*Palmaria palmata* and *P. palmaria*), and sea lettuce were ruled out. This study did not lead to commercial production because it was not economically advantageous at the time.

Some studies (Gerard 1999) showed an interesting relationship between cordgrass and knotted wrack. On southern New England and mid-Atlantic coasts, these species were mutually beneficial, and the combination of the vascular and non-vascular species resulted in high productivity for the marsh-estuarine system. It was thought that cordgrass ameliorated desiccation of knotted wrack, while a decaying layer beneath the mat of knotted wrack, provided nutrients for cordgrass (Brinkhuis 1976).

6.4.2.4 Trophic Interactions

As the base of the food chain, macroalgae directly and indirectly provide sustenance to a variety of animals, including humans (Lembi and Waaland 2007). For human use, kelp is used to make soup stock, or "dashi," and also eaten whole as a sea vegetable, or by pickling sections of the stipe as "sea pickles." It is also traditionally liquefied for use as a fertilizer in gardens (Chapman and Chapman 1980). Commercial cultivation or harvesting is not done in LIS as it is in many Asian and European nations, but individuals still gather and use the bounty from the shore. Irish moss is collected and used mostly to make traditional blancmange pudding, or to thicken stews. Dulse is dried and eaten as a snack or in various recipes; sea lettuce is used in "seaweed salad,"; and rockweed is an ingredient in the traditional New England clambake or brewed for tea. One additional human food that should be mentioned is *Porphyra* spp. Dried blades are flaked and used as a condiment. Certain species are gathered and pressed into sheets as nori to make sushi wrappers, but while this had been done commercially in Maine, it is presently not done in LIS or the Gulf of Maine (Chopin et al. 1999, 2001; He and Yarish 2006; Pereira and Yarish 2010; Yarish et al. 1999).

Aside from human consumption, there are herbivorous predators of these primary producers in various sizes and types. Finfishes, gastropod snails, and sea urchins are key consumers of macroalgae. The trophic relationship between kelp and urchins such as the purple sea urchin (*Strongylocentrotus purpuratus*) is well-known (e.g., Vadas 1977), and the chink shell is a voracious predator of kelp (Egan and Yarish 1990). Gastropods such as periwinkles feed on fleshy (e.g., sea lettuce) and even crustose algae such as *Ralfsia verrucosa*. Mesograzers such as copepods, amphipods, and polychaetes are less studied but important grazers of macroalgae (Lobban and Harrison 1994). However, in most situations, less than 20 % of the algal biomass consumed passes through herbivores; a greater amount becomes part of the detrital food web (Vadas 1985).

A study conducted at Avery Point examined mutually beneficial relationships between snails and their seaweed hosts (Stachowicz and Whitlatch 2005). Irish moss had greater abundance and far less fouling from ascidians when two common snails, *Anachis lafresnayi* and *Mitrella lunata*, were present than when they were absent. This indicates that the snails were feeding on the tunicates attached to Irish moss rather than the Irish moss itself, to the benefit of both.

Historical anecdotal reports praise graceful red weed and *A. subulata* as settlement surfaces for scallops in Niantic Bay and other locations (e.g., Goldberg et al. 2000). They are frequently found together, and observations that graceful red weed also tended to occur in hypoxic areas may be a reflection of its ability to take up N effectively.

Due to high balanced levels of protein, lipid, and carbohydrate, several of the common LIS algae have been used in aquaculture as nutrition for animals. A study at the National University of Galway (then University College Galway) showed that abalone preferred dulse in their diets, and that a mixed diet of dulse and kelp provided the best nutrition for these mollusks (Mai et al. 1992). Much remains to be investigated in terms of trophic transfer from algae.

6.4.2.5 Human Impacts

Kelp sampled near the Thames River contained high levels of metals including copper and cadmium (Shimshock et al. 1992). The ability of these algae to concentrate trace metals from the surrounding water suggests their use for biomonitoring.

In addition to indicating potential environmental problems, macroalgae can also assist in diminishing human impacts. *Porphyra* spp. has been investigated for potential as N scrubbers in nutrient bioremediation (e.g., Chopin et al. 1999, 2001; Carmona et al. 2001; McVey et al. 2002) and two native Sound species were deemed suitable for integrated multi-trophic aquaculture and bioremediation (Yarish et al. 1999; He and Yarish 2006).

Some threats are harder to document or observe. Van Patten et al. (1993) pointed out that LIS kelp at its southern limit could be eliminated by warming of 1–2 °C by reducing reproductive success. Reproductive effort in LIS was considerably less than that of kelp in New Hampshire or Nova Scotia. Davison et al. (2007) demonstrated in the laboratory that the interactions between N metabolism in Atlantic kelp and ultraviolet radiation are very complex, and need to be considered when evaluating the effects of anthropogenically increased UV on ocean productivity.

As climate change proceeds, increasing ocean acidification from the increasing levels of atmospheric carbon dioxide may particularly affect calcified species such as *Corallina officinalis,* due to the solubility of their high-magnesium calcite skeletons (Hall-Spencer et al. 2008). *C. officinalis* is found in rocky habitats and epiphytic on seagrass (Hemminga and Duarte 2000). Acidification may lead to profound changes in the food webs and diversity of seagrass meadows (Martin et al. 2008) because epiphytic coralline algae is primary colonizers of seagrass, and subsequently followed by many other species such as diatoms, sponges, foraminifera, worms, and other invertebrates (Corlett and Jones 2007). Bussell et al. (2007) found 125 species of invertebrates in communities associated with *C. officinalis* in a United Kingdom study, illustrating the richness of these associations. Martin et al. (2008) concluded that by the year 2100, calcium carbonate production by coralline algae could decrease by 50 %, with large consequences for local sediment budgets and biogeochemical cycles of carbon and carbonate in shallow coastal ecosystems.

Another human impact seen all too often is that of invasive species. Such species invasions are frequently brought about by human activities, e.g., the introduction of green fleece (*Codium fragile fragile*)to the east coast of North America in 1957 (Bouck and Morgan 1957) or earlier (Provan et al. 2008), probably via shipments of shellfish and ship hull transport (Carlton and Scanlon 1985). This highly invasive, buoyant species disrupts shellfish beds. Baitworms have been suggested as another, modern vector for distributing invasive species (Yarish et al. 2010a, b).

A globally spreading, large invasive red seaweed from Asia, *Grateloupia turuturu*, was first identified in LIS in 2004 (Millstone Environmental Laboratory 2009). An extensive, multi-state, multi-investigator Sea Grant study looked at the impacts and spread of this newcomer, which is well established at Millstone Point in Waterford, CT and as far east as Groton, CT. Although Irish moss, a competitor for habitat, is holding its own so far, future warming of LIS may facilitate the spread of *G. turuturu* and it may outcompete native species (Yarish et al. 2010a, b). The same study showed that *G. turuturu* was not a preferred food item for most herbivores, and that it grows quickly so could become dominant and change community structures. It reproduces well (Lin et al. in preparation) but decomposes quickly and thus is a poor source of food to surrounding systems (Janiak 2010). While there are good efforts at informing the public about the control of invasive species, more education, and more vigilance will be important for the future.

Several new invasive species in the genera *Porphyra* have been found in LIS since attention turned to it in the late 1990s as an economically important genus for the commercial sushi trade and other industries. Species were identified and described using DNA analyses (e.g., Broom et al. 2002; Klein et al. 2003 Neefus et al. 2000, 2008a, b). Very recently, Nettleton et al. (2012) found, by means of molecular screening, that the invasive alga *Gracilaria vermiculophylla* is in LIS. It has apparently gone undetected for years because it so closely resembles a related native species. This underscores the importance of DNA sequencing in future monitoring efforts.

In an EPA/Sea Grant-funded study of bait worm packaging as a potential vector of invasive species transport (Haska et al. 2012), investigators found that the seaweed packing used for baitworms yielded unintentional "stowaway" species. Microscopic examination and DNA analyses of purchased bait box contents discovered 13 species of macroalgae and 23 species of invertebrates. Two species of microalgae that are considered to be potentially toxic in bloom conditions, *Alexandrium fundyense* and *Pseudo-nitzschia multiseries*, were detected. This underscores the need for consideration of alternate packing materials and/or education on proper disposal methods of the seaweed packing used in bait boxes.

Finally, the careful records kept by the Millstone Environmental Laboratory (Keser et al. 2010) are insightful for suggesting what species of algae may become dominant in LIS with continued warming. We may expect a shift from kelp and other brown intertidal algae to more warm temperate red and brown alga such as the invasive species of *G. turuturu* and recently discovered *Gracilaria vermiculophylla*, as well as other warm temperature species of *Sargassum* spp.

6.4.2.6 Science Gaps and Management Implications

Brodie et al. (2009) pointed out that there are endangered algal species and even some that have become extinct. So little monitoring of the seaweeds is done at present that we might not even know whether important species are in serious decline. Brodie and co-workers summed it up by saying that "For the marine macroalgae, evidence of the impact of climate change, ocean acidification, and introduced species on native floras is often anecdotal and points to the need for long-term monitoring and scientific study to determine changes in abundance and distribution."

In the future, the use of macroalgae in integrated multitrophic aquaculture and bioremediation will need to grow. Further, given the global spread of invasive species and the uncertainties associated with large-scale ecosystem changes due to global warming, it will be essential to develop additional monitoring efforts and habitat surveys to have a basic understanding about what large- and small-scale ecological changes are happening in communities that depend on the macroalgae. If this does not happen, the old saying "You don't know what you've got till it's gone" could turn out to be true in the case of some valuable seaweeds and the fauna that depend on them.

6.5 Plankton

The earliest plankton studies in LIS date back 70 years with the first estimates of primary production (Riley 1941). During the 1950s, a series of studies on species composition and abundance of phytoplankton (Conover 1956), zooplankton (Deevey 1956), and primary production (Riley 1956) were carried out. During the 1980s, researchers measured micro- (Capriulo and Carpenter 1980) and mesozoo-plankton grazing (Dam Guerrero 1989; Dam and Peterson 1991, 1993), copepod phenology (Peterson 1986), and copepod egg production (Peterson and Bellantoni 1987). Starting in the 1990s, a more comprehensive spatial examination of plankton biomass and abundance took place (Capriulo et al. 2002), and has continued thanks to the Connecticut DEEP water quality monitoring program (Kaputa and Olson 2000; Dam et al. 2010). The program, with about 20 stations sampled at least at monthly intervals, covers the entire Sound. Monitoring programs for phytoplankton (Liu and Lin 2008) and zooplankton (Dam and McManus 2009), albeit at fewer stations, started during the first decade of the twenty-first century. Spatial studies of primary production were carried out in the first decade of the twenty-first century (Gobler et al. 2006; Goebel et al. 2006).

Here, we summarize information on both spatial and temporal patterns of phytoplankton and zooplankton in LIS. We also examine primary production and its fate, with emphasis on information available since the review by Capriulo et al. (2002). In the analysis, we take advantage of the extensive spatial coverage of the LIS Study (1988–1989) and the CTDEEP water quality monitoring program since 1991.

6.5.1 Phytoplankton

The seasonal cycle of phytoplankton in LIS was first investigated in the early 1950s (Conover 1956). In that study, and subsequent ones in the 1980s (Peterson 1986; Dam and Peterson 1991; Dam et al. 1994), sampling was done in CLIS and only for a few seasons. However, because there is a strong human population gradient from west to east along the shores of the Sound, there is a concomitant gradient in nutrient loading, plankton biomass and primary production (Bowman 1977; Wolfe et al. 1991; Lee and Lwiza 2008; Goebel et al. 2006). The first comprehensive spatial study in the Sound took place in the early 1990s: three nearshore stations that covered the eastern, central, and WLIS were sampled at monthly intervals for a period of 3 years (Capriulo et al. 2002).

6.5.1.1 Temporal and Spatial Patterns of Chlorophyll

In the CTDEEP spatial study of LIS, started in 1991, about 20 stations covering the entire Sound are sampled monthly (biweekly in the summer) for chlorophyll. The program divides the Sound into five regions, from west to east (Fig. 6.10): The West Narrows, the East Narrows, the West Basin, the Central Basin, and the East Basin (Kaputa and Olson 2000). Figures 6.11 and 6.12 show, respectively, time series of dissolved inorganic N (DIN) and chlorophyll at the extreme ends of the Sound—the West Narrows and the East Basin. During the 1988–2005 period, means of annual and DIN and chlorophyll were consistently higher, by a factor of three to five, in the West Narrows than in the East Basin (Dam et al. 2010). During that period, DIN concentrations were significantly greater in the West Narrows than in the Central and East Basins (Dam et al. 2010). Chlorophyll concentrations were significantly greater in the West Narrows than all other regions during some years, but in other years, chlorophyll concentrations in the West Narrows were only greater in the East Basin (Dam et al. 2010). In summary, there is indeed a dramatic decrease in nutrients and phytoplankton biomass from west to east in LIS, but this decrease is not monotonic. Most of the spatial regional differences occur among the extreme ends of the Sound, the Narrows, and the East Basin.

LIS also displays both seasonal and long-term patterns in phytoplankton biomass and nutrient concentrations (Figs. 6.11 and 6.12). The seasonal pattern is characterized by a chlorophyll maximum in late winter or early spring. A smaller chlorophyll peak occurs in early fall. Chlorophyll decreases during the summer, when nutrients are dramatically drawn down (Riley and Conover 1956; Peterson 1986; Capriulo et al. 2002). This pattern has not changed since the 1950s, and appears to be independent of location in the Sound (Figs. 6.11 and 6.12).

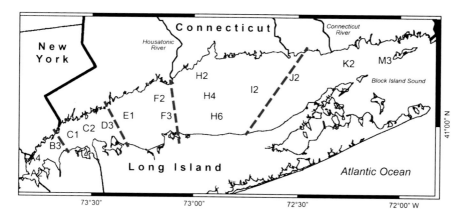

Fig. 6.10 Sampling stations for the CTDEEP Water Quality Monitoring Program. Stations are the same as those sampled in 1988–1989 by the Long Island Sound Study. *Blue dashed lines* indicate boundaries between regions (after Kaputa and Olson 2000)

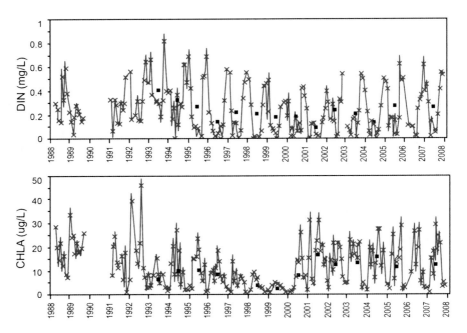

Fig. 6.11 Monthly means (*crosses*) and annual means (*squares*) in dissolved inorganic nitrogen, DIN (*top*) and chlorophyll (*bottom*) from 1988–2007 in the West Narrows surface waters. Data are from the Long Island Sound Study (1988–1989) and from the CTDEEP Water Quality Monitoring program (Dam et al. 2010). Annual means are shown only if 10 or more months were sampled in a particular year, and if the missing months were neither consecutive nor both January and December. Non-detectable data points were assumed to be 25 % of the detection limit. There is a hint of a linear decrease of DIN: slope $= -0.01$ mg L^{-1} year $^{-1}$, $r^2 = 0.24$, $p = 0.084$. The slope does not change even if the non-detectable points are assumed to be 0 % of the detection limit

From 1991 to 2005, significant decreases in annual mean and maximum DIN concentrations in surface waters were observed in the West Narrows. Mean DIN decreased during the 1990s to a minimum of less than 0.2 mg L^{-1} in 2000 (Fig. 6.11). Decreasing trends were observed for TDN and TN in that region for the same period (Dam et al. 2010). In contrast, there was no hint of a trend for DIN in the East Basin. The decreasing trend in DIN in the West Narrows is probably linked to the decrease in nutrient loading that has resulted from sewage-treatment improvements in that region.

The time-dependent pattern in chlorophyll concentration was different from the one for DIN. A dramatic decrease in chlorophyll was observed throughout LIS for the period 1988–1999, and no spring blooms were apparent during the years 1999 and 2000. Since 2000, chlorophyll has increased and remained fairly constant with time. While there is a spatial correlation between nutrients and chlorophyll biomass, the two variables appear less correlated in time, that is, from 1988 to 2007, there was a system-wide change in chlorophyll with time, but not in DIN. The

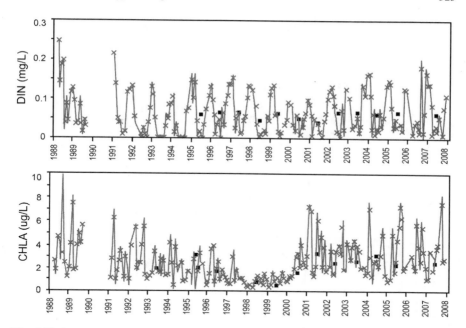

Fig. 6.12 Monthly means (*crosses*) and annual means (*squares*) in dissolved inorganic nitrogen, DIN (*top*) and Chlorophyll *a* (*bottom*) from 1988 to 2007 in the East Basin surface waters. Details as in caption of Fig. 6.11

chlorophyll pattern perhaps reflects some regional forcing, whereas the DIN patterns appear to reflect the differential nutrient loading from east to west.

Figures 6.11 and 6.12 show considerable interannual variability of phytoplankton biomass in LIS. However, there is no clear evidence of any trend in biomass from the early studies of the 1950s until the 2000s. The only fair comparison is the Central Basin since that is the region that was originally studied by G. Riley and co-workers. Figure 6.13 shows monthly chlorophyll concentrations in the Central Basin for the period 1988–2008. The range of observations for this period is similar to 1952–1954 (Riley and Conover 1956; Conover 1956) and 1992–1995 (Capriulo et al. 2002). Furthermore, a comparison of decadal means in the different seasons of the year throughout the 1950s, 1990s, and 2000s does not show any clear trends (Dam et al. 2010).

6.5.1.2 Phytoplankton Community Structure

Conover (1956) used settling chambers to show that diatoms dominate the phytoplankton in LIS, except perhaps in summer when dinoflagellates and other small flagellates prevail. Capriulo et al. (2002) made the same observation, but noticed that peak abundance was not in spring, but summer and fall (see Fig. 6.28 by Capriulo et al. 2002). Since the early 2000s, Lin and co-workers have analyzed data from the CTDEEP phytoplankton monitoring program. They did microscopic identifications for 10 representative stations in 2002, 2003, and 2007–2010, and molecular analysis for the <5 μm size-fraction at four of the stations in 2003. Although phytoplankton species number fluctuated somewhat interannually (Fig. 6.14), the community structure was relatively stable, both in time and space (Fig. 6.15). Diatoms contributed 61 % of the species richness, and dinoflagellates accounted for 26 %. Minor components included chrysophytes (2.6 %), raphidophytes (2.1 %), chlorophytes (3.1 %), cryptophytes (1.0 %), euglenophytes (0.5 %), and a number of unidentified species (3.7 %). Total species number exhibited two small peaks yearly, one in early spring and the other in late autumn, mainly attributable to diatom species (Fig. 6.15a). There was no discernible spatial trend in species richness along the Sound (Fig. 6.15b). These patterns resemble the observations from the early 1950s (Riley and Conover 1956; Harris and Riley 1956).

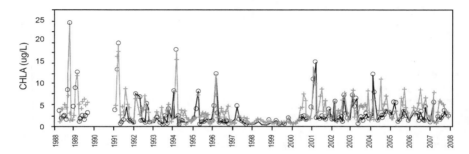

Fig. 6.13 Chlorophyll concentration (μg L^{-1}) in the Central Basin of Long Island Sound. *Circles* represent surface waters and *crosses* represent bottom waters. Details as in caption of Fig. 6.11

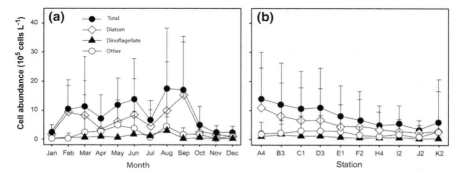

Fig. 6.14 Temporal and spatial variation of cell abundance of phytoplankton in LIS between 2002 and 2010. Shown are means and standard deviations (*error bars*). Data from the CTDEEP Water Quality Monitoring program (S. Lin, unpublished)

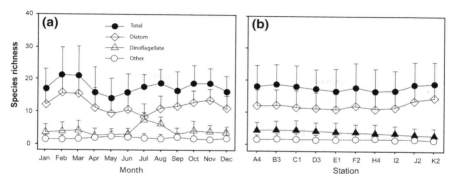

Fig. 6.15 Temporal (**a**) and spatial variation (**b**) of phytoplankton community structure in LIS between 2002 and 2010. Shown are means and standard deviations (*error bars*). Data from the CTDEEP Water Quality Monitoring program (S. Lin, unpublished)

A closer look at the 2002–2010 data showed that *Skeletonema costatum* was dominant in ELIS, *S. costatum* and *Thalassiosira* spp. in CLIS, and *Thalassiosira* spp., *Thalassiosira nordenskioldii*, and *Dactyliosolen fragilissimus* in WLIS. Such spatial species differentiation could be a result of responses to eutrophication (Cloern 2001). But as diatoms were the unassailably predominant group in the phytoplankton community (Conover 1956; Harris and Riley 1956; Capriulo et al. 2002; Liu and Lin 2008), apparently the effect of eutrophication in the main body of LIS is relatively low. However, it is well recognized that some inner bays have suffered severe eutrophication, where dinoflagellates such as *Prorocentrum minimum*, *Akashiwo sanguinea*, and *Alexandrium fundyense* can form blooms seasonally. For example, *A. fundyense* has formed blooms in Huntington Bay, NY, causing Paralytic Shellfish Poisoning outbreaks since 2006 (Hattenrath et al. 2010).

Although diatoms dominated the phytoplankton community throughout the year, seasonal variation in dominant species occurs. For 2002–2010, *Leptocylindrus minius*, *Scenedesmus* spp., *Thalassiosira* spp., and *S. costatum* were dominant in spring; *Thalassiosira* spp., *Thalassiosira nordenskioldii*, Thalassiosira *gravida*, *D. fragilissimus*, and *Prorocentrum triestinum*, in summer; *S. costatum* and *Thalassiosira* spp. in fall; and *Thalassiosira* spp. and *Asterionellopsis glacialis* in winter. Historically, *S. costatum* was the predominant species in winter, while in summer, water column stratification and low inorganic N favored other species, especially dinoflagellates, of which *Gonyaulax* spp. could make up ~50 % of the phytoplankton cell population (Harris and Riley 1956). In the last decade, a peak of dinoflagellate species number was noticeable in summer (Fig. 6.15a). *P. triestinum*, *P. minimum*, and *P. micans* appeared to replace *Gonyaulax* spp. as the dominant dinoflagellate species in WLIS in summer. Overall, *Thalassiosira* spp. and *S. costatum* were the predominant species in all seasons and almost the whole Sound. Recent molecular analyses have revealed several species out of the originally established *S. costatum* species; whether the observed *S. costatum* is composed of multiple species remains to be determined.

Very small phytoplankton, such as the ubiquitous cyanobacteria *Synechococcus* spp., has not been systematically studied in LIS, but they are known to be abundant, especially in summer. For example, Campbell (1985) measured abundances up to 1.8×10^8 cells L^{-1} in CLIS (Crane Neck shore station). At 1–2 fg chl cell^{-1} (Liu et al. 1999), this equates to 0.18–0.36 μg chl L^{-1}, most likely less than 10 % of the total phytoplankton biomass. Their role in primary production in LIS is not known.

6.5.1.3 Phytoplankton Abundance

Phytoplankton abundance in LIS varies seasonally, typically with a major peak in late winter or early spring, and a lower peak in the fall (Conover 1956; Harris and Riley 1956; Sun et al. 1994). During 2002–2010, the largest peak was apparent in the summer (Fig. 6.14a), in stark contrast to the work from the 1950s (Conover 1956), but similar to what was observed in the early 1990s in CLIS (Capriulo et al. 2002). The latter authors observed, however, the largest peak in phytoplankton abundance in the fall in WLIS. For 2002–2010, total abundance could reach about 4.4×10^5 cells L^{-1} (2009) to 1.5×10^6 cells L^{-1} (2007), of which diatoms contributed over half of the total biomass (~64 %), while dinoflagellate accounted for about 11 % (Fig. 6.14a). These figures are similar to those reported by Capriulo et al. (2002). Conover (1956) reported peaks in phytoplankton abundance an order of magnitude higher (up to ~30 million cells L^{-1}), but most of the observations fell within the range reported here.

As previously observed (Bowman 1977; Aller and Benninger 1981; Wolfe et al. 1991; Capriulo et al. 2002; Liu and Lin 2008), phytoplankton abundance was highest in the high nutrient and low S waters of the western Sound (Fig. 6.15b). As with the case for chlorophyll, most of the significant spatial differences in phytoplankton abundance arise from the extreme ends of LIS.

6.5.1.4 Primary Production

The few available measurements and estimates of primary production for LIS bracket the production between 400 and 854 g C m^{-2} year^{-1}. This two-fold range is not surprising given that different methods were employed in the studies, that sampling occurred in different regions, and that studies were done in three different decades. Riley (1956) provided a single estimate of annual gross primary production (470 g C m^{-2} year^{-1}) from the dark and light bottle technique. That study cites that Riley (1941) estimated production at 600–1,000 g C m^{-2} year^{-1}. Dam Guerrero (1989) used short ^{14}C incubations during mid-day, which yielded something between gross and net production. Measurements were integrated for the uppermost 10 m of the water column, and sampling frequency was weekly or biweekly from February to August. Sampling took place at a single station, near H6 (see Fig. 6.10), in CLIS. That study estimated primary production at 854 g C m^{-2} year^{-1}, with virtually identical estimates for the winter-spring and the summer periods. Goebel et al. (2006) derived gross primary production estimates from physiological parameters measured in oxygen-based photosynthesis-irradiance (P–I) incubations, integrated both through time of day and depth (uppermost 10 m). Sampling was biweekly during summer months and monthly during spring and fall. Eight stations from the central to WLIS were sampled; net primary production (NPP) averaged 400 ± 80 g C m^{-2} year^{-1}. Because algal respiration was assumed to be 50 % of gross production, the average gross production was 800 g C m^{-2} year^{-1}, similar to estimates from Dam Guerrero (1989).

The only study that has examined spatial variability in productivity in LIS is by Goebel et al. (2006). They observed a gradient of declining productivity from west to east, which correlates with a similar gradient in N loadings and chlorophyll biomass (see Fig. 6.4 in Goebel et al. 2006). However, productivity only varies by a factor of two, whereas N loadings vary roughly by an order of magnitude, while biomass varies by three- to five-fold. This implies that other factors must also constrain productivity. Indeed, nutrient addition assays indicate that nutrient limitation of primary production, either by N or Si, occurs in LIS (Gobler et al. 2006), but that this limitation varies with season and region. For example, WLIS appears N-limited during late spring and summer, whereas such limitation occurs in CLIS from spring until fall. Grazing, discussed later, is another potentially controlling factor. The magnitude of the gradient in mesozooplankton abundance (see next section) from east to west matches that of chlorophyll and primary production. Provided that individual grazing rates do not decrease from east to west, then zooplankton grazing should have an effect on the chlorophyll and primary production.

6.5.2 Zooplankton

The first survey of zooplankton abundance and seasonality in LIS was done in the early 1950s (Deevey 1956), and was entirely restricted to metazooplankton. (The term mesozooplankton applies to organisms >200 μm, and usually also refers to metazooplankton. The term microzooplankton applies to organisms <200 μm, and usually refers to protozooplankton. For example, most copepod nauplii and some copepodid stages are small enough to be microzooplankton, but are also metazooplankton. Here, we use the term meta- and mesozooplankton interchangeably, as well as micro- and protozooplankton.). The first protozooplankton (unicellular heterotrophs) survey was done in 1979–1980 (Capriulo and Carpenter 1983). Both of these surveys were mostly confined to CLIS. Work on copepod phenology, also in the central Sound, resumed in the 1980s (Peterson 1986). The first temporal-spatial survey for both metazooplankton and protozooplankton was carried in the 1990s (Capriulo et al. 2002), but was confined to three locations along the east–west axis of LIS (Capriulo et al. 2002). Since the early 2000s, a more comprehensive temporal-spatial survey of the zooplankton has been carried out by the CTDEEP monitoring program (Dam and McManus 2009). This program samples six stations along the east–west axis of the Sound. Here, we restrict comparisons among studies to

CLIS, where sampling locations are common to all studies. With respect to the metazooplankton, Deevey (1956) sampled monthly with a no. 10 silk net (~158 μm mesh) near the LISS/CTDEEP station H4 in the Central Basin from March 1952 to May 1953. Sampling was from bottom to surface waters, although it is unclear if the tows were oblique or vertical. Capriulo et al. (2002) sampled monthly from 1993 through 1995 (landward of station H2). Oblique tows from bottom to surface were done with a 202 μm mesh. Dam and McManus (2009) data are from the CTDEEP zooplankton monitoring program. Sampling was monthly or bimonthly during the summer months and consisted of vertical tows from bottom to surface with a 202 μm mesh net. Peterson (1986) sampled approximately weekly from February to November in the early to mid-1980s (near station H6). Tows were vertical from bottom to surface with a 202 μm mesh net. However, we do not include Peterson's data in all the comparisons because sampling did not take place throughout the entire year. Protozooplankton samples were collected from whole water samples, preserved and concentrated after settling, and then counted under inverted microscopes (Capriulo and Carpenter 1983; Capriulo et al. 2002; Dam and McManus 2009).

6.5.2.1 Metazooplankton

The seasonal cycle of metazooplankton abundance for the early 1950s (Deevey 1956) and the 2000s (Dam and McManus 2009) is shown in Table 6.2. The minimum abundance occurs in January and the maximum in June, regardless of the year, and abundance varies by one order of magnitude throughout the season. The seasonal pattern of total mesozooplankton biomass for 2002–2009 (Dam and McManus 2009) is similar, but with peaks between April and May (Fig. 6.16). The offset in peak abundance and biomass is likely due to the biomass dominance of

Table 6.2 Monthly mesozooplankton abundance (ind. m^{-3}) at station H4 (see Fig. 6.10 for station location)

Month	1952–1953 Abundance	2002–2004 Abundance	2008–2009 Abundance
January	10,530	11,459	17,010
February	14,665	10,332 ($n = 2$)	15,570
March	35,138 ($n = 2$)	15,802 ($n = 2$)	32,110
April	68,650 ($n = 2$)	93,556 ($n = 2$)	58,120
May	97,323 ($n = 2$)	66,728 ($n = 2$)	139,400
June	142,500	82,464 ($n = 2$)	204,420
July	83,840	32,332 ($n = 2$)	20,350
August	152,070	37,468 ($n = 3$)	23,720
September	80,575	21,759 ($n = 3$)	
October	19,290	8,123 ($n = 3$)	12,235 ($n = 2$)
November	32,735	3,363 ($n = 2$)	12,830
December	15,090	5,739 ($n = 2$)	11,420

Numbers in parentheses represent the number of months used to calculate a monthly mean. Abundance for 1952–1953 is from Deevey (1956). All other data are from Dam et al. (2010)

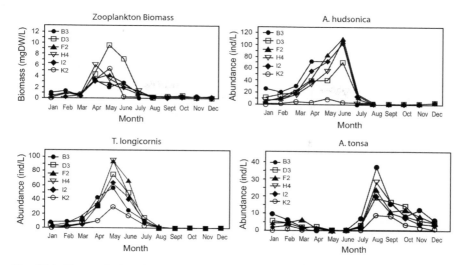

Fig. 6.16 Mesozooplankton in Long Island Sound. Shown are climatology plots for the period 2002–2009. The *upper left plot* is for total biomass (determined directly from aliquots of samples). The *other plots* are for abundance of the dominant zooplankton species. For station locations, see Fig. 6.10. Data from the CTDEEP zooplankton Monitoring program (Dam and McManus 2009)

the copepod *Temora longicornis* relative to *Acartia* spp. (Fig. 6.16). *T. longicornis*, a larger species than *Acartia* spp., peaked in May during this time period.

Copepods account for 80–90 % of the metazooplankton abundance in LIS (Dam and McManus 2009). The following taxa have been routinely reported for LIS: Arthropoda (copepods, mysids, crab larvae, amphipods, barnacle nauplii, and cladocerans); Annelida (polychaete larvae); Mollusca (gastropod and bivalve larvae); Echinodermata (sea star larvae); Chordata: *Oikopleura* sp.; Bryozoa; and Chaetognatha (e.g., the arrow worm, *Sagitta elegans*).

The seasonal cycle of zooplankton species in LIS is characterized by two distinct assemblages (Peterson 1986). The winter–spring assemblage is dominated by the copepods *Acartia hudsonica* (abundance) and *T. longicornis* (biomass). *Pseudocalanus* sp. is also present at this time, but in much lower abundance relative to the other two species. The summer-fall assemblage is dominated by the copepod *Acartia tonsa*. Two other species, *Paracalanus crassirostris* and *Oithona similis* are also abundant during this time period. Copepod species abundance data were examined for the 1950s (Deevey 1956), 1980s (Peterson 1986 and unpublished observations), 1990s (Capriulo et al. 2002) and 2000s (Dam and McManus 2009). *A. tonsa* peaked in late August 1952 at ~80,000 individuals m^{-3} (Deevey 1956). From 1982–1987, maximum *A. tonsa* abundance also occurred in August, but the peaks ranged from 2,000 to 25,000 individuals m^{-3} (Peterson 1986 and unpublished data). Similarly, from 1993 to 1995, *A. tonsa* peaks ranged from 3,000 to 25,000 individuals m^{-3} (Capriulo et al. 2002). The peak *A. tonsa* abundance from 2002 to 2009 ranged from <10,000 to 40,000 individuals m^{-3}, with peak abundance usually in August (Fig. 6.16; Dam and McManus 2009). In the previous century, the abundance of *A. hudsonica* peaked in March with 70,000 individuals m^{-3} in the 1950s (Deevey 1956), 15,000 individuals m^{-3} in the 1980s (Peterson 1986), and 30,000 individuals m^{-3} in the 1990s (Capriulo et al. 2002). During the 2000s, peak abundance ranged from 15,000 (March 2002–2003) to 300,000 m^{-3} (June 2003) (Dam and McManus 2009). The peak abundance for the first decade of this century, however, was in June (Fig. 6.16). This shift in peak abundance is intriguing, but remains unexplained. *T. longicornis* abundance peaked in June, with 60,000 individuals m^{-3} in the 1950s (Deevey 1956), a range from 10,000 to 45,000 individuals m^{-3} in the 1980s (Peterson 1986), and a mean of 40,000 individuals m^{-3} in the 1990s (Capriulo et al. 2002). During the 2002–2009 period, average peak abundance was in May and values ranged from 25,000 individuals m^{-3} to almost 100,000 m^{-3} (Fig. 6.16).

The seasonal patterns of abundance and the species composition of the zooplankton appear relatively unchanged since the 1950s, with the exception of a very recent change in the phenology of *A. tonsa*. Typically, this species is found in LIS from June to December. However, during 2008–2009 (Dam and McManus 2009) and in 2010 (unpublished observations), *A. tonsa* in WLIS was present from June to April. If this change persists, then we might be witnessing a broadening of the growth season of this species.

There are clear decadal differences in zooplankton abundance (Fig. 6.17). Mean annual zooplankton abundance during 2002–2004 was significantly lower than 1952–1953 (1-tailed t test, $p = 0.04$). Notice that most of the observations for 2002–2004 fall below the 1:1 line in Fig. 6.17. This is not the case for the comparison of 1952–1953 to 2008–2009. There was no difference in mean annual zooplankton abundance between these latter two periods (1-tailed t-test, $p = 0.29$).

Zooplankton abundance from the 2000s (Dam and McManus 2009) was generally higher than what was reported for the early 1990s by Capriulo et al. (2002). This could

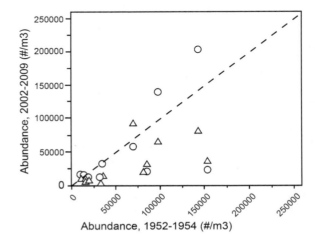

Fig. 6.17 Mean monthly zooplankton abundance (No. m^{-3}) at station H4 for 1952–1953 (Deevey 1956) versus 2002–2004 and 2008–2009 (Dam and McManus 2009). *Dashed line* indicates 1:1 relationship. *Triangles* years 2002–2004, *circles* years 2008–2009

be due to the locations of the stations in the latter study, which were in relatively shallow water. Also, the means reported in that study are geometric, whereas the ones in the former are arithmetic. Alternatively, this difference reflects the kind of decadal variability shown in Fig. 6.17. The lower zooplankton abundance during the 1990s could be explained by the dramatic decrease in phytoplankton biomass during that period.

Capriulo et al. (2002) examined zooplankton spatial distribution during the 1990s. They sampled three stations in relatively shallow water (about 10 m deep) between Stamford (western end) and Madison (eastern end), on the north side of LIS. Here, we highlight the data collected by the CTDEEP zooplankton monitoring program (Dam and McManus 2009), which has greater spatial (six stations from end to end in the Sound) and temporal (most of the last 10 years) coverage. The mean yearly total metazooplankton abundances show a monotonic pattern of decrease from west to east that is consistent from year to year, with differences of three- to five-fold between the extreme ends of the Sound (Table 6.3). However, the annual variability within stations is such that the only consistent statistically significant difference occurs between stations B3 and K2, the extreme ends of the Sound. This pattern, although not the details, is consistent with Capriulo et al. (2002) who also observed a decreasing gradient of zooplankton concentration from west to east. However, in that study the range of abundance varied by a factor of three, not five, and the stations were not as far apart. In any case, the decreasing trend of zooplankton abundance from west to east mirrors the gradient in phytoplankton biomass (Figs. 6.11 and 6.12) and abundance (Fig. 6.14), and may suggest that metazooplankton are generally food-limited in LIS. There is no indication that the high degree of eutrophication in WLIS is negatively affecting the zooplankton community, at least in terms of abundance.

6.5.2.2 Protozooplankton

The most common protozooplankton in the ocean are heterotrophic nanoflagellates, ciliates, and heterotrophic dinoflagellates. Currently, there is information on seasonal and spatial patterns of the first two groups in LIS. The first study

Table 6.3 Mean annual total mesozooplankton abundance for periods October 2008–October 2009, March 2007–February 2008, August 2003–August 2004, and August 2002–August 2003

Station	Abundance (ind.L^{-1})			
	2008–2009	2007–2008	2003–2004	2002–2003
B3	77.5	61.6	53.6	48.3
D3	41.9	48.6	38.4	34.5
F2	64.8	57.1	39.0	37.5
H4	46.9	55.2	33.2	33.9
I2	49.1	43.6	32.7	30.8
K2	19.4	16.4	11.4	11.1

Data from the CTDEEP zooplankton monitoring program (Dam and McManus 2009). See Fig. 6.10 for station locations of the seasonal cycle of protozooplankton in LIS was on ciliates, and reported abundances of ~270–2,000 L^{-1} for most of the year, with peak abundance of $1.3 \times 10^4 \, L^{-1}$ in summer in CLIS (Capriulo and Carpenter 1983). Tintinnids overwhelmingly dominated the ciliate community, with the exception of late spring (Capriulo and Carpenter 1983). At the time, 28 species of ciliates were reported. The list of ciliate species increased to 71 in the subsequent study of the early 1990s (Capriulo et al. 2002). Species richness decreased from west to east, with 94 % of the reported species found in WLIS and 62 % in ELIS. However, no information on abundance and seasonal cycles of ciliates was reported in that study.

In the CTDEEP monitoring program, ciliates are reported as three separate categories: tintinnids, other (naked) heterotrophic ciliates, and *Myrionecta rubra* (formerly known as *Mesodinium rubrum*). The latter is a mixotrophic ciliate (i.e., it eats phytoplankton but also photosynthesizes), but its main trophic mode is autotrophy, so it is effectively a phytoplankter. Examples of annual cycles of abundances of tintinnids, naked ciliates, and *M. rubra* are shown in Fig. 6.18. A summary of abundance for 2002–2009 is contained in Table 6.4. All three groups show broad peaks in abundance from spring through summer, but with great spatial and temporal variability (Fig. 6.18). Tintinnids in particular can vary by orders of magnitude from one station to the next and from one survey to the next. Naked or aloricate ciliates, comprised chiefly of oligotrichs (e.g., *Strombidium* spp.) and non-tintinnid choreotrichs (e.g., *Strobilidium* spp. and *Strombidinopsis* spp.), were usually more abundant than the lorica-bearing tintinnids. Although not shown here, small aloricate forms (10–15 μm), which are probably bacterivorous (Sherr et al. 1986), were sometimes abundant, especially in WLIS. These ciliates contribute relatively little biomass to the microzooplankton assemblage because of their small size (e.g., at their most abundant, ciliates less than 20 μm in diameter were about two-thirds of the total ciliate abundance, but only 2.3 % of the total biomass).

Both tintinnids and naked ciliates show a gradient from high abundance in WLIS to lower abundance in the east (Fig. 6.19). Most of this is driven by high abundance at station B3, the westernmost Sound, and low abundance at K2 (the easternmost Sound), which also revealed the highest contrast in the metazooplankton. As in the case of the metazooplankton, statistically significant differences among stations were only apparent between stations B3 and K2. *M. rubra* like autotrophic ciliates did not have a predictable pattern among stations.

Maximum ciliate abundance for 2002–2009 for LIS (Table 6.4) is similar to Capriulo and Carpenter (1983); however, mean abundance appears relatively low. This is probably because the former values include the easternmost stations, which typically have the lowest abundances.

McManus (1986) reported abundances of heterotrophic nanoflagellates that ranged from about 200 to 8,000 cells mL^{-1}, with a broad peak occurring during the summer in CLIS. Similar abundances were reported by Capriulo et al. (2002). Over their 3 year study, average abundance ranged between 10^2 and $6.4 \times 10^3 \, mL^{-1}$, two orders of magnitude greater than the ciliate abundance. The authors reported significantly higher concentrations in the westernmost waters

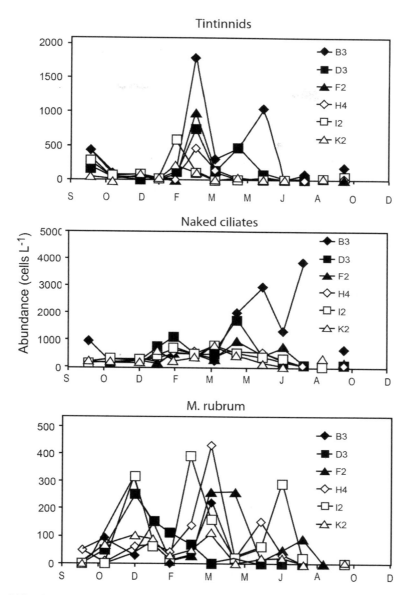

Fig. 6.18 Abundances of tintinnids (*top*), other heterotrophic ciliates (*middle*), and the autotrophic ciliate *Mesodinium rubrum* (=*Myrionecta rubra*) for 2008–2009. For station locations see Fig. 6.10. Data from the CTDEEP zooplankton Monitoring program

relative to central and ELIS. Peak abundance was during the summer, with considerable interannual variability in abundance. Beyond the two studies, there have been few measurements of these organisms in LIS. As bacterivores, they are a

Table 6.4 Descriptive statistics for abundance and biomass of ciliate microzooplankton from the CTDEEP monitoring program (August 2002–October 2009)

	Tintinnids No L^{-1}	Tintinnid biomass (μgC L^{-1})	Naked ciliates L^{-1}	Naked ciliates biomass (μgC L^{-1})	*Mesodinium- like* L^{-1}	*Mesodinium- like* biomass (μgC L^{-1})
Maximum	5,760	1,938	8,215	122	25,671	48
Minimum	0	0	0	0	0	0
Median	40	0	270	1	80	0
Mean	197.99	6.66	513.76	3.03	418.13	0.94
N	320	320	320	320	320	320
Standard deviation	572.14	108.31	794.43	9.17	1629.59	3.12

Data from Dam and McManus (2009)

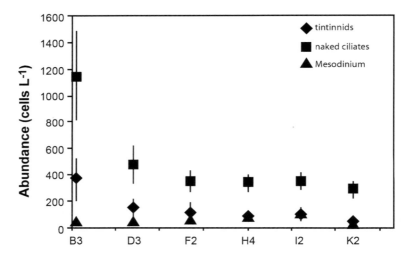

Fig. 6.19 Abundances of ciliates (individuals per L), plotted as annual averages by station for the period 2008–2009 (from Dam and McManus 2009). See Fig. 6.10 legend for station locations

component of the microbial food web, which cycles inorganic nutrients and mineralizes organic carbon (Taylor 1982).

6.5.3 Fate of Primary Production

The main loss terms of net productivity are grazing, sinking, advection, and degradation A portion of the sinking organic matter becomes buried in the sediment (see Cuomo et al., Chap. 4, and Varekamp et al., Chap. 5, in this volume), and some is used by the benthos. However, there have been no concerted studies that attempt a budget of the primary production and its loss terms at different times of the year or in different regions of LIS. Here, we highlight what is known about the main loss terms for the production. To our knowledge, the degradation term for phytoplankton biomass has not been studied.

6.5.3.1 Grazing

Microzooplankton grazing often accounts for a significant fraction of phytoplankton mortality, particularly in warmer months, in most aquatic environments (Calbet and Landry 2004, but see Caron et al. 2000). An early study conducted using seawater from the central basin of LIS and size-fractionated for grazing experiments (i.e., <35 μm for prey and >35 to <202 μm for predator additions) found that microzooplankton, which in those samples consisted primarily of tintinnids, grazed ~12 % day^{-1} of the standing stock of phytoplankton in November and between 14 and 41 % day^{-1} in June (Capriulo and Carpenter 1980). Although summer ambient seawater temperatures were warmer, in the laboratory T had no discernible influence on individual clearance rate (Capriulo 1982). Grazing percentage (P), however, is determined by both the abundance of grazers and individual feeding rates, and microzooplankton abundances were substantially higher in June than in November (6,700–10,500 organisms L^{-1} vs. 1,900–2,500 L^{-1}; Capriulo and Carpenter 1980).

York et al. (2011) measured microzooplankton grazing rates in WLIS during March and between June and July over 2 years using the dilution technique (Landry and Hassett 1982). Significant grazing (g day^{-1}), as measured by changes in total Chlorophyll a, was detected 50 % of the time in winter (mean $g \pm$ s.e. = 0.24 day^{-1} ± 0.12, n = 9) and 62.5 % in the summer (0.66 day^{-1} ± 0.19, n = 16). (The authors suggested that at those times when grazing was not detected, suspended, non-food particles may have interfered with the efficient removal of phytoplankton cells by the micrograzers). Grazing coefficients (i.e., g) can be used to estimate the grazing impact of microzooplankton on phytoplankton standing stock (% removed day^{-1}, as above), where $P = (1 - e^{-g}) \times 100$ (after Verity et al. 2002). Using this approach, the mean grazing impact (±s.e.) for WLIS based on the data of York et al. (2011) was 17.5 % day^{-1} ± 8.1 and 37.2 % day^{-1} ± 8.2 for March and June–July, respectively. Although the dilution technique is a different methodological approach, these estimates of P are remarkably similar to those of Capriulo and Carpenter (1980) for both the cold and warm-T periods. Furthermore, the seasonal shift in P in WLIS corresponded to change in microzooplankton biomass (mean ± s.e; March = 6.6 ± 1.1 μg C L^{-1}, and June–July = 30.4 ± 1.1 μg C L^{-1}; York et al. 2011), and represented a similar summer population increase (~4X) as previously noted in CLIS. As compared to metazooplankton (e.g., copepodites and adult copepods), micrograzers may be more efficient at consuming smaller phytoplankton cells and less able to consume larger items, such as chain-forming diatoms that often dominate during the spring bloom in LIS. For example, maximum prey sizes of 10 and 55 μm cell diameter were noted for two species of heterotrophic dinoflagellates (Hansen and Calado 1999 and references therein). Grazing coefficients trended higher on smaller phytoplankton cells as determined from changes in <10 μm Chlorophyll a (mean $g \pm$ s.e. $= 0.85$ day$^{-1} \pm 0.21$; York et al. 2011) compared to total Chlorophyll a in the dilution experiments, though the results were not statistically significant.

Less is known about the grazing impact of the total metazooplankton community on phytoplankton in LIS, and studies conducted to date have focused on a particular copepod species, albeit at times a dominant species such as *T. longicornis* during late winter and spring (Dam and Peterson 1991; Capriulo et al. 2002). A study of the *in situ* feeding of female *T. longicornis* suggested that this population alone could remove between <1 and 34 % day^{-1} of the phytoplankton standing stock in the central basin (Dam and Peterson 1993). The highest impacts tended to be found between mid- to late spring (~10–34 % day^{-1}) while lower values were in late winter and mid-summer (<1–2 % day^{-1}). Dam Guerrero (1989) estimated an average removal of 20 % of the primary production by *T. longicornis*. The grazing impacts were not significantly related to the population abundance of the copepods (Table 6.6 in Dam and Peterson 1993), but instead, likely due to confounding environmental influences on copepod feeding behavior. The grazing impact of a population is determined by both population abundance and individual ingestion rate. Individual ingestion rate of copepods (e.g., ng Chlorophyll a individual^{-1} h^{-1}) is often measured as a function of gut-pigment content and gut-clearance rate (Båmstedt et al. 1999). For *T. longicornis* females, several factors influenced individual gut-pigment content including phytoplankton size-structure and abundance (Dam and Peterson 1991, 1993). The gut-pigment content of *T. longicornis* females was highest during the spring bloom when the phytoplankton community was dominated by larger cells (>20 μm) and lowest in late spring and early summer when phytoplankton abundance was lower. The gut-clearance rate of these copepods was also positively related to seawater T (Dam and Peterson 1988).

The grazing impact of *T. longicornis* during spring in LIS may be on par or even higher than microzooplankton grazing. The situation in summer and fall is likely to be different. The calanoid copepod *A. tonsa* is a dominant member of the zooplankton community during summer and fall (Capriulo et al. 2002), and is typically an omnivore with a partial dietary dependence on heterotrophic prey, such as ciliates, to fuel reproduction (Gifford and Dagg 1991). Dam et al. (1994) found that phytoplankton (i.e., Chlorophyll a) ingestion and egg production rates (eggs female^{-1} day^{-1}) of *A. tonsa* were better correlated to the >10 μm size-fraction compared to total Chlorophyll a, and that maximum ingestion rates (Fig. 6.1 in Dam et al. 1994) occurred during the fall bloom in September. Using in situ abundance estimates and ingestion rates, Dam Guerrero (1989) estimated that *A. tonsa* removed on average 4 % of the primary production. In another Long Island estuary, Great South Bay, mesozooplankton dominated by *A. tonsa* in summer consumed 1–4 % day^{-1} of total primary production in summer (Lonsdale et al. 1996). The results indicate a substantial difference in the trophic role of these two copepod species in LIS. Whereas *T. longicornis* may play a substantial role in the control of phytoplankton in spring, the same is less probable for *A. tonsa* during summer and fall. Although carnivorous feeding behavior has not been investigated in LIS *per se*, it likely contributes to the reproductive success of both *T. longicornis* and *A. tonsa* (Gifford and Dagg 1991; Dam et al. 1994; Peterson and Dam 1996; Lonsdale et al. 1996).

Because the abundance and biomass of zooplankton are heavily dominated by *T. longicornis, A. tonsa,* and *A. hudsonica,* grazing estimates of these species are probably good proxies for total mesozooplankton grazing. To date, there is no single widely accepted technique to measure total mesozooplankton grazing akin to the dilution technique for microzooplankton grazing. Dam Guerrero (1989) used the downward flux of zooplankton fecal pellets captured by free-drifting sediment traps (Welschmeyer and Lorenzen 1985) to estimate mesozooplankton grazing impact during the spring season of 1987. He concluded that mesozooplankton removed ~25 % of the total primary production during that season. Total mesozooplankton grazing and *T. longicornis* grazing for the same period were significantly and positively correlated ($r = 0.58$), and grazing by *T. longicornis* accounted for 80 % of the total mesozooplankton grazing. No similar exercise has been carried out for the summer-fall period. Based on calculations of carbon required to satisfy metabolic demands of copepods, Riley (1956) estimated that zooplankton consumed 26 % of the annual primary production.

In summary, all indications are that during the winter and spring, micro- and mesozooplankton each remove about 25 % of the primary production, for a total of 50 %. During the summer and fall, microzooplankton grazing is more than 50 % of the primary production and mesozooplankton grazing is much less. During the LISICOS (Long Island Sound Integrated Coastal Observing System) study in the Narrows regions of the Sound (westernmost end of LIS) in July 1995, microzooplankton grazing was estimated to remove up to 75 % of the primary production (York et al. 2011), whereas mesozooplankton removed <20 % of the production (H. Dam, unpublished).

6.5.3.2 Sinking and Horizontal Export

Phytoplankton sinking from the water column to the benthos and horizontal export via advection can be significant loss terms for primary production. However, both of these terms are poorly constrained for LIS. For example, during the spring 1987, phytoplankton sinking flux, as measured by free-drifting sediment traps, was estimated to be 2.6-fold greater than mesozooplankton grazing (Dam Guerrero 1989). Since mesozooplankton grazing was estimated at the time to be 25 % of the primary production, by this accounting, loss due to sinking was the equivalent of 65 % of the primary production. Because all other loss terms (microzooplankton grazing, horizontal export) would only account for the remaining 10 %, the sinking loss term was probably overestimated. During the 1995 LISICOS study of the Narrows region of the Sound, phytoplankton sinking flux was measured with both moored and free-drifting sediment traps, and estimated to account for 3–30 % of the primary production. By difference from grazing (see above) and sinking, losses due to horizontal export had to be <20 %. Strikingly, community respiration exceeded primary production by about an order of magnitude (J. Kremer, personal communication). Therefore, a mass balance could not be achieved. A similar exercise has not yet been attempted for the spring season, where a significant fraction of the annual production takes place. In any case, the unbalanced carbon budget from the LISICOS studies suggests that portions of the Sound might be highly heterotrophic. This implies that a large source of terrestrial or oceanic, not estuarine, carbon must exist to sustain the respiratory demands in the water column. Sediment studies indicate a varying and occasionally large influx of terrestrial organic carbon (Chap. 5, Fig. 5.41; Varekamp et al. 2004, 2010). In summary, there is a great deal of uncertainty as to the role of sinking and horizontal export of the primary production. Integrated studies in which primary production and its loss terms which are simultaneously measured are required.

6.5.4 Harmful Algal Blooms

Harmful algal blooms (HABs) occur when a single species of phytoplankton grows to a density that has a direct negative impact upon other organisms or an ecosystem through production of toxins, mechanical damage, or by other means. HABs often are associated with shellfish poisoning syndromes in human consumers of contaminated molluscs, economic losses to coastal communities, and commercial fisheries, and HAB-associated invertebrate, fish, bird, and mammal mortalities. These events are sometimes referred to as "red tides" because the microalgae that most frequently cause HABs, dinoflagellates, often contains a reddish pigment (peridinin), and thus can discolor waters red when blooms occur. Some dinoflagellate blooms that color waters red are harmless. Because blooms caused by other groups of algae with different coloration (green and brown) can also be harmful, the term "harmful algal blooms," or HABs, have been adopted by scientists to describe these diverse events. All algal blooms are not harmful or toxic, and many HABs do not impart any discoloration to the water.

Globally, the phytoplankton communities within many coastal ecosystems have undergone phase shifts in recent decades, with species associated with HABs becoming more prevalent (Hallegraeff 1993; Glibert et al. 2005; Heisler et al. 2008). The number of species, duration, intensity, and distribution of HABs has all increased, accelerating negative impacts on human health, fisheries, and economies. Some parts of LIS are prime examples of regions undergoing such changes with regard to HABs. Prior to 2006, algal blooms in LIS were viewed as nuisances and contributors to hypoxia. Since that time, toxic, and harmful algal blooms have occurred annually within some parts of the estuary. For example, since 2006, blooms of the saxitoxin-producing dinoflagellate *Alexandrium fundyense* have emerged in Northport and Huntington Bays along the north shore of Long Island

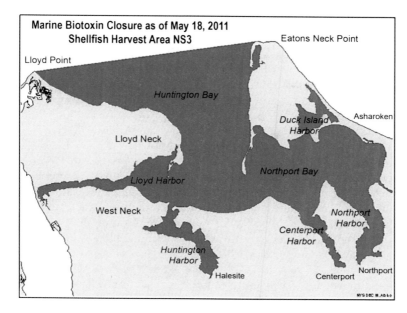

Fig. 6.20 Shellfish beds (~10,000 acres) in Huntington and Northport Bays closed by the NYSDEC due to PSP production by *A. fundyense* in 2011. This region was never closed due to PSP prior to 2006, but has been closed in 2006, 2008, 2009, 2010, and 2011 due to saxitoxin contamination of shellfish during *A. fundyense* blooms

every spring. These blooms persist for ~2 months (late April through June), and achieve cell densities $>10^6$ cells L^{-1} and water column saxitoxin concentrations $>2.4 \times 10^4$ pmol STX eq. L^{-1} (Hattenrath et al. 2010). During the blooms, shellfish have become highly toxic to humans (>1.4 mg STX eq. $100 \ g^{-1}$ shellfish tissue), resulting in the closure of nearly 10,000 acres of shellfish beds in Northport and Huntington Bays during five of the past 6 years since 2006 (Fig. 6.20; Hattenrath et al. 2010).

Hattenrath et al. (2010) investigated factors promoting *A. fundyense* blooms in Northport Bay. Densities of benthic *A. fundyense* cysts at the onset of blooms generally have been orders of magnitude lower than levels needed to account for observed cell densities, indicating that in situ growth of vegetative cells is responsible for peak bloom densities. Experimental enrichment of bloom water with biologically available N, particularly ammonium, significantly increased densities of *A. fundyense* and saxitoxin concentrations relative to unamended control treatments (Hattenrath et al. 2010). The $\delta^{15}N$ signatures (12–23 ‰) of particulate organic matter (POM) during blooms were similar to those of sewage (10–30 ‰), and both toxin and densities of *A. fundyense* were significantly correlated with POM $\delta^{15}N$ ($p < 0.001$; Hattenrath et al. 2010). These findings suggest that *A. fundyense* growth was supported by a source of wastewater, such as the sewage treatment plant that discharges into Northport Harbor or on-site septic systems.

Warmer than average atmospheric temperatures in the late winter and spring of 2008 and a cooler May contributed to an extended period of water column temperatures optimal for growth of *A. fundyense* (12–20 °C), and thus may have contributed to the larger and longer bloom in 2008 (Hattenrath et al. 2010). Altogether, this evidence suggests that sewage-derived N loading, coupled with above-average spring temperatures, can promote intense and toxic *A. fundyense* blooms in embayments within LIS. The persistence of a dense cyst bed (>500 cysts cm^{-3}) in Northport Harbor sediment since the onset of these blooms is likely responsible for the lasting nature of these events, but perhaps not their interannual intensity.

Blooms of *A. fundyense* recently have been most intense in Northport and Huntington Bays, but surveys of LIS from 2008 to 2011 have demonstrated that undetected blooms of *A. fundyense* had occurred in multiple nearshore regions of New York and Connecticut, with low abundances of these species present even within open water regions of LIS. For example, in 2009, a bloom of *A. fundyense* in LIS near the Mattituck Inlet was denser than the blooms in Northport and Huntington Bays. Moreover, elevated levels of *A. fundyense* have now been confirmed at more than 25 sites across New York and Connecticut, demonstrating that these HABs are not an isolated phenomenon in a single harbor. The widespread distribution of *A. fundyense* highlights the potential for this toxic dinoflagellate species to expand across LIS if the suitable conditions such as those currently present in Northport (cyst beds, high N loads, and poor flushing) are replicated elsewhere.

The dinoflagellate *A. fundyense* is not the only species of toxic phytoplankton that blooms in LIS. In 2008, another dinoflagellate, *Dinophysis acuminata*, began forming large and persistent annual blooms (>10^6 cells L^{-1}) in Northport Bay. These blooms have recurred annually since then, and the species forming them has been confirmed as *Dinophysis acuminata* by scanning electron microscopy. This species has been responsible for diarrhetic shellfish poisoning (DSP) events around the world (Yasumoto et al. 1980; Hallegraeff and Lucas 1988; Campbell et al. 2010), and the blooms in 2008 and 2010 generated the toxins okadaic acid and DTX-1 (Steve Morton, NOAA, personal communication), both of which are the causative agents of DSP-syndrome and are federally regulated by the US Food and Drug Administration. Blooms of *Dinophysis* have been reported on the east coast of the US (Maranda and Shimizu 1987; Tango et al. 2002, although DSP-causing toxins in shellfish had not been reported above the action level in the US until recently. Campbell et al. (2010) reported that in 2008, a bloom of *D. acuminata* exceeding 10^5 cells L^{-1} occurred in the Gulf of Mexico causing DSP-toxicity in oysters and the closure of shellfish beds. Although the co-occurrence of *D. acuminata* blooms and other dinoflagellates in LIS suggests that they are promoted by similar factors, there have been few field studies investigating factors that promote *Dinophysis* blooms. Currently, the factors promoting blooms of *D. acuminata* in LIS are unknown. Moreover, the occurrence of DSP in LIS represents a serious development, as the NYSDEC at this time is monitoring for *D. acuminata* or DSP toxins in the water column or shellfish.

Another dinoflagellate species with harmful and ecosystem-disruption effects that blooms episodically in LIS is *Prorocentrum minimum*. This species has been reported in LIS as a spring-summer bloom-forming phytoplankter for as long as records have been published, although it may be listed under species pseudonyms *Exuviella minimum*, *E. marie-lebouriae*, or *Prorocentrum marie-lebouriae* (Riley 1956b; Riley and Conover 1967; Capriulo et al. 2002). This species was recognized only recently as a "harmful" species (Heil et al. 2005). Possible associations with shellfish poisoning in human consumers have not been verified, and a toxin responsible for harmful effects upon shellfish and other invertebrates has not been characterized, leading some to speculate that the bloom effects are attributable to excessively high biomass loading leading to hypoxia (Landsberg 2002).

Association of *P. minimum* blooms with poor performance of larval and post-set shellfish in the Milford Aquaculture Laboratory's shellfish hatchery in the 1950s and 1960s was documented (Davis and Chanley 1956), leading to studies exploring the possibility that trichocysts released by the dinoflagellate cells were responsible for molluscan responses (Ukeles and Sweeney 1969). A particularly widespread (Throgs Neck to New Haven) and long-lived (May–August) bloom of *P. minimum* in 1987 were associated with massive fish kills and caused arrested growth and mortality in hard clams planted near Stamford and Milford as part of a growth study conducted by the Milford Lab (Wikfors and Smolowitz 1993). Subsequent experimental studies revealed that *P. minimum* varies greatly in "toxicity" to shellfish, with declining populations more bioactive (Wikfors 2005). Most bivalve species are able to protect themselves against *P. minimum* exposure for at least a few days by rejecting captured cells in pseudofeces or through post-ingestive rejection within hemocyte-coated fecal strands (Hégaret et al. 2008). This dinoflagellate likely has episodic, negative effects upon farmed and wild populations of bivalves annually within LIS embayments, but there is no obvious trend in occurrence since the 1950s; therefore, this organism could be considered a stable component of the LIS plankton ecology.

The ichthyotoxic dinoflagellate *Cochlodinium polykrikoides* is a catenated, bloom-forming species that has become increasingly common in coastal zones around the world (Kudela and Gobler 2012). Historically, blooms of *C. polykrikoides* have been reported most frequently in Asia, with South Korea alone reporting more than $100 million in annual fisheries losses attributable to this species during the 1990s. Since 2004, this species has been forming dense blooms in the waters of eastern Long Island, including the Peconic estuary that exchanges with ELIS (Gobler et al. 2008; Kudela and Gobler 2012). To the north of this region, blooms have been observed recently in Point Judith Pond, RI (Hargraves and Maranda 2002) as well as in Narragansett Bay (S.E. Shumway, personal observation), Martha's Vineyard and Nantucket, MA (G. Wikfors, personal observation) although scientific reports of the latter sites have yet to be published. This species has been known to be transported hundreds of km across the waters of Asia (e.g., Korean to Japan; Malaysia to the Philippines) (Kudela and Gobler 2012). It seems likely that such transport has already occurred across ELIS. Furthermore, it seems plausible that such transport could introduce blooms into new, coastal regions of LIS in the future.

Similar to *P. minimum*, the toxic diatom *Pseudo-nitzschia multiseries* appears to have been a component of the LIS phytoplankton for as long as records have been kept, but under the pseudonym *Nitzschia seriata* (Riley 1956b; Riley and Conover 1967). Several species in this genus produce domoic acid, a glutamic-acid analog that can be transferred through filter-feeding shellfish to human seafood consumers in whom it can cause Amnesic Shellfish Poisoning (ASP). *Pseudo-nitzschia* spp. generally are components of the mid-to-late spring bloom, but seldom dominate the assemblage. Currently, there are no reports of domoic-acid poisoning in humans or wildlife in LIS and no clear trends in extent or intensity of occurrence.

A final phytoplankton species that forms HABs near LIS is *Aureococcus anophagefferens*; a pelagophyte causing destructive brown tides in northeastern and mid-Atlantic US estuaries for more than 25 years, with blooms occurring most frequently in Long Island waters. Large ($>10^6$ cells mL^{-1}) blooms effectively shade the benthos, causing reduction in eelgrass beds (Cosper et al. 1987). Brown tide caused recruitment failure and starvation in bay scallop (*Argopecten irradians*) populations (Bricelj and Kuenstner 1989) and high rates of mortality in bivalve molluscs during blooms (Bricelj et al. 2001; Greenfield and Lonsdale 2002) generally have been attributed to cessation of feeding and starvation (Gainey and Shumway 1991). Blooms have occurred consistently within Long Island south-shore estuaries since 1985, but occurred in the Peconic estuary on eastern Long Island from 1985 to 1995 only (Gobler et al. 2005). Quantification of *A. anophagefferens* by the Suffolk County Department of Health Services in harbors along the New York coastline of LIS revealed the widespread presence of *A. anophagefferens* cells, with maximal densities of 5×10^3 cells mL^{-1}—an order of magnitude lower than the levels known to inhibit shellfish feeding (Bricelj et al. 2001). Given the departure of blooms from the neighboring Peconic estuary and the low levels of *A. anophagefferens* typically present in LIS, brown tides are one of the HABs less likely to occur in LIS in the near future.

At greatest risk of being impacted by a possible change in the HAB ecology of LIS is its $80 million molluscan shellfish fishery, chiefly consisting of wild caught hard clams and bottom-cultivated eastern oysters (*Crassostrea virginica*). Shellfish-harvest closures because of fecal-indicator bacteria are common in LIS, but until the recent *A. fundyense* blooms on the north shore of Long Island, the shellfish industry in LIS has been largely immune to the PSP closures that plague northern New England shellfisheries. The state resource agencies in New York and Connecticut responsible for assuring the safety of shellfish landings, NYSDEC and the Department of Agriculture in Connecticut, have monitoring and control programs approved by the US Food and Drug Administration to detect the presence of toxigenic algae in the water and/or toxins in shellfish tissues. These surveillance programs will need to keep pace with changing conditions in the phytoplankton ecology of LIS to maintain the economic benefits of a thriving shellfishery in LIS.

6.5.5 Science Gaps and Management Implications

There is remarkable agreement in the spatial patterns of abundance and biomass of phytoplankton, protozooplankton, and metazooplankton in LIS; there is a decreasing gradient from west to east, but it is mostly underlain by differences at the extreme ends of the Sound. This gradient has some implications for the two major problems in LIS: eutrophication and hypoxia. Clearly, there is some degree of correlation between these two variables in the Sound. Western LIS, with its high nutrient concentration and high phytoplankton biomass, experiences summer-time hypoxia, whereas the eastern Sound does not (Lee and Lwiza 2008; O'Donnell et al. 2008). However, the linkage is complex. Nutrient concentrations, specifically DIN, declined during the 1990s in the western Sound (Fig. 6.11), with concomitant decreases in chlorophyll (Fig. 6.11), yet summer-time hypoxia pervasively occurred in WLIS during this period (Lee and Lwiza 2008). Thus, further work is necessary to elucidate the mechanistic linkages among nutrients, phytoplankton biomass, and hypoxia (see Sect. 6.7.1).

There is considerable interannual (e.g., chlorophyll, Figs. 6.11 and 6.12) and decadal (e.g., metazooplankton, Fig. 6.17) variabilities in the plankton in LIS. Such variabilities have implications for resource limitation of consumers; for example, the low zooplankton abundance during the early 2000s might have reflected the dramatic decrease of chlorophyll that occurred during 1998–2000, which may have lowered zooplankton abundance (and hence future recruitment) at that time. Unfortunately, we do not have zooplankton abundance data for this period. Changes in chlorophyll and zooplankton took place throughout the Sound, suggesting regional rather than local forcing. Finally, there is no clear evidence from water column studies that LIS is now more eutrophic than 60 years ago, as both nutrient and chlorophyll levels do not seem to have changed much since then. Nevertheless, benthic foraminiferal assemblages have changed dramatically since the 1960s (Sect. 6.5.2).

Capriulo et al. (2002) hypothesized that the high abundance of copepods in WLIS could be a factor in the development of hypoxia due to the production of fecal pellets that would quickly sink to the bottom and decompose. This hypothesis was tested recently. Fecal pellet production rates, zooplankton standing stocks, the downward flux of fecal pellets and the standing stock of fecal pellets in the water column were measured in LIS in July 2005 and March 2006 (Loglisci 2007). She estimated that pellet carbon flux was the equivalent of 2.3–45.2 % of the total particulate organic carbon flux. Further, she estimated that if the entire sinking pellet pool degraded on the seafloor, it would account for up to 88 % of the total sediment oxygen demand in March, but only 18 % in July. Thus, while fecal pellets can represent a significant drawdown of DO during the winter, this is not a time when hypoxia is a problem. On the other hand, when hypoxia is prevalent, zooplankton pellets account for at best ~20 % of the drawdown of DO.

Finally, in terms of developing management strategies for hypoxia in LIS, a significant lack of knowledge and understanding of the fate of the primary production, including the relative amount of organic matter buried and preserved in the sediment. There are very large imbalances between sources and sinks of carbon in WLIS, where hypoxia is common. A concerted effort is required to address this gap in knowledge.

6.6 Deep Water Benthos

6.6.1 Deep Water Benthic Habitats and Macrofaunal Communities

The sea floor of LIS is primarily comprised of sedimentary environments that form physical habitat mosaics of varying spatial extent, scale, and complexity (Fig. 6.21). These soft-sediment environments are dotted with patches of hard substrates such as boulder fields and rocky outcrops, adding to the overall diversity of seafloor habitats in LIS. Superimposed on the physical habitat structure are biogenic structures such as tubes, pits, burrows, shell hash, and structures by habitat-forming species such as mussels and hydrozoans (Fig. 6.22). The relative mix of overall habitat structure and complexity varies across spatial scales and sediment patch type. Communities in these habitats are typically dominated by polychaete annelids, various crustacean taxa such as amphipods, and bivalves and gastropods. This section focuses on the soft-sediment habitats and communities of the deeper reaches (>3–4 m depth) of LIS, reviews our current state of knowledge of these communities, provides an assessment of what aspects of LIS benthos need to be better understood, and suggests the types of studies that may provide the necessary information.

6.6.1.1 Studies Prior to 1985

Prior to 1985, our knowledge of the benthic ecology of the deep-water portions of LIS was based on a handful of Sound-wide surveys and several datasets in specific portions of LIS (Fig. 6.23). Sanders (1956) and McCall (1975) investigated the benthos of the central basin of LIS, revealing the general nature and composition of these communities, primarily those inhabiting mud and sandy-mud habitats. McCall (1977), Rhoads et al. (1978), and others (see Cuomo et al., Chap. 4, in this volume) conducted more detailed studies of the mud communities in the central

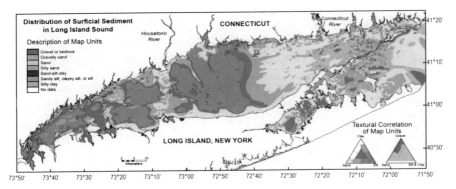

Fig. 6.21 Distribution of seafloor habitat/patch types in LIS based on sediment texture. Map is from Poppe et al. (2001)

Fig. 6.22 Image of small-scale biogenic and physical features that increase habitat complexity on portions of LIS. Note mussels forming clumps on sediment and a portion of a boulder to the right with various fouling organisms. At the base of the boulder is an accumulation of shell hash. Largest mussels are ~2–3 cm in length for scale

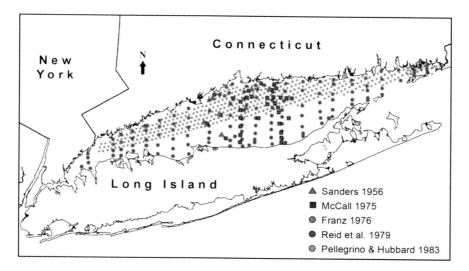

Fig. 6.23 Benthic community surveys in the deeper water portions of LIS conducted prior to 1985. See text for details

basin particularly with respect to animal sediment relations, how they affect sediment biogeochemistry, and their responses to disturbances and subsequent recovery/successional dynamics. Portions of this work were tied to the study of dredged material disposal sites by the Army Corps of Engineers as part of the Disposal

Area Monitoring System, DAMOS (see Sect. 6.6.1.4 later in this chapter). This research led to a widely applied model of soft-sediment community recolonization and succession (Fig. 6.24), and also technological developments to study soft-sediment communities, specifically the development of the REMOTS camera for sediment profile imaging (see Rhoads and Germano 1982), and the potential to survey benthic conditions and communities more rapidly than conventional sampling. This model is used almost exclusively to assess and predict benthic community responses to human disturbances. There were also some surveys in Fishers Island Sound (Zajac 1998a).

The characteristics of soft-sediment infaunal communities in LIS have been reviewed previously by Zajac (1998b) and Zajac et al. (2000a, b) based on broad-scale surveys conducted from the mid-1950s through the 1980s, primarily studied by Sanders (1956), McCall (1975), Reid et al. (1979), and Pellegrino and Hubbard (1983). These analyses showed that benthic communities in LIS exhibit significant variation at several spatial scales. There is a gradient in species richness

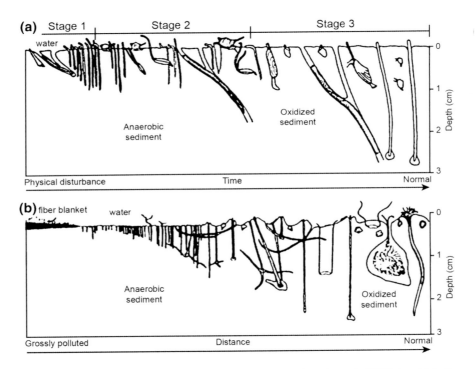

Fig. 6.24 a Model of soft-sediment succession proposed by Rhoads et al. (1978) in which disturbance is followed by recolonization of stage I species comprised of opportunistic species which live in the upper few cm of the sediment. Eventually a "climax" Stage 3 community is reestablished comprised of deeper dwelling more K-selected type species. The successional model stages have similarities to responses of infauna along an organic pollution gradient as proposed by Pearson and Rosenberg (1978) shown in (**b**)

and community structure along the west to east axis of the Sound. Species richness is relatively low in the western and central basins of LIS, although there are
areas that appear to have higher richness; these coincide with coarser sediments
(Fig. 6.25). Species richness starts to increase more regularly across habitats on
the eastern side of the central basin and in the area of the Mattituck Sill, is somewhat lower beyond this transition zone among sedimentary environments (see
Lewis, Chap. 2, in this volume), and then increases sharply east of the Connecticut
River. The large-scale, east to west gradient likely reflects to some degree a larger
potential species pool at the eastern end of LIS and the connection to the open
coastal waters of Block Island Sound and the Atlantic Ocean. In central and WLIS,
lower species richness may reflect a smaller pool of potential species that have
entered the Sound proper, but also a smaller set that can successfully maintain
populations. Patch and smaller-scale spatial differences in species richness may be
related to the sediment characteristics of specific patches with lower richness in
muddy sediments and higher richness in sandy and coarser sediments (Fig. 6.26).
Furthermore, species richness may be affected by small scale, physical, and biogenic habitat characteristics (e.g., Hewitt et al. 2005) but interactions between
small-scale habitat structure and species richness are not well-known for LIS.

Analysis of data collected by Pellegrino and Hubbard (1983), Zajac (1998b),
and Zajac et al. (2000a, b) showed that benthic community structure in LIS varies
from west to east, exhibiting several spatial transitions that are related to the large-
scale sedimentary and benthic landscape features of the LIS sea floor (Fig. 6.27).
Based on classification (clustering) analysis, 11 community types were recognized
and several of these exhibited sub-community groups. In the depositional, muddy
habitats of the western and central basins of LIS, three main types of benthic

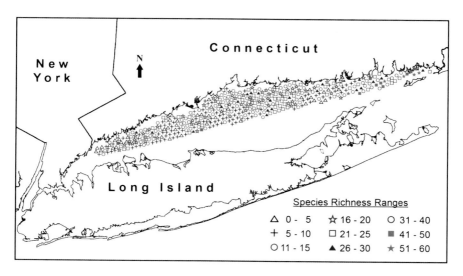

Fig. 6.25 Spatial patterns of species richness in LIS based on data presented in Pellegrino and
Hubbard (1983). Values represent the total number of species per 0.04 m^2 grab sample

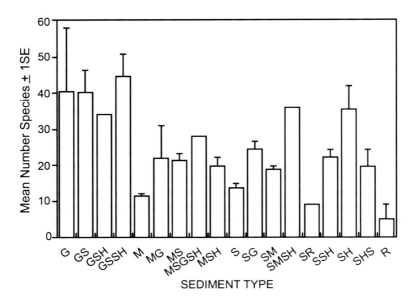

Fig. 6.26 Relationship between species richness and sediment type in LIS. The values represent the mean number of species per sample at stations with the same bulk sediment characteristics identified by Pellegrino and Hubbard (1983) Sediment type designations are as follows: *G* Gravel; *GS* Gravel, Sand; *GSH* Gravel, Shell; *GSSH* Gravel, Sand, Shell; *M* Mud; *MG* Mud, Gravel; *MS* Muddy Sand; *MSGSH* Muddy sand, Gravel, Shell; *MSH* Mud, Shell; *S* Sand; *SG* Sand, Gravel; *SM* Sandy mud; *SMSH* Sandy mud, Shell; *SR* Sand, Rocks; *SSH* Sand, Shell; *SH* Shell; *SHS* Shell, Sand; *R* Rocky

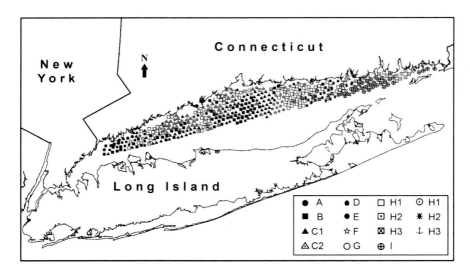

Fig. 6.27 Spatial distribution of benthic communities identified via clustering analysis by Zajac et al. (2000a, b) of data provided in Pellegrino and Hubbard (1983)

communities were recognized, dominated by the polychaetes *Nephtys incisa* and *Cistenoides gouldii*, and the bivalves *Mulinia lateralis*, *Nucula annulata* and *Pitar morrhuana*, differing mostly in terms of relative number of these dominant species. The communities correspond to a mud assemblage identified by Reid et al. (1979) that spanned the central and western basins of LIS and the Narrows in the far WLIS, and are similar to the community described by Sanders (1956) in the central basin. Another community type dominated by the polychaetes *Clymenella zonalis* and *Mediomastus ambiseta* and the bivalve *M. lateralis* was also found in some muddy areas, especially along transitional areas among different sedimentary environments in east-CLIS. Several types of benthic communities were found in more hydrologically dynamic areas with coarser sediments. A community type dominated by several polychaetes including *Asabellides oculata* and *Spiophanes bombyx* and the bivalve *Tellina agilis* was found in the transition between the eastern and central basins and along the bathymetric highs separating the central and western basins and Narrows (Fig. 6.27). A community dominated by the polychaetes *Cirratulis grandis*, *C. cirratus*, *Prionospio heterobranchia*, *P. tenuis*, and the amphipod *Aeginnia longicornis* was found in sandy sedimentary environments in the eastern basin. Areas characterized by primarily coarse-grained sediments contained communities dominated by several species of tubiculous amphipods and polychaetes. These analyses were based on data collected only in the Connecticut waters of LIS and as such may not provide an accurate depiction of deeper water benthic assemblages along the southern axis of LIS. We can extrapolate that benthic communities are similar in the portions of specific patches that were not sampled by Pellegrino and Hubbard (1983) to the extent that sedimentary habitats are not excessively different. However, Reid et al. (1979) identified a sand assemblage in shallower water along the north shore of Long Island, suggesting more variation in community structure in the transitional areas among deeper water muddy basins and sandy areas.

In summary, studies conducted prior to about 1985 provided a general understanding of composition and structure of the seafloor communities of LIS, as well as a disturbance/response model that could be used for environmental assessment, and a better conception, in part, of the temporal dynamics of these communities.

6.6.1.2 Studies from 1985 to 2010

Between 1985 and 2010, there were several important developments in the study of seafloor habitats of LIS and their ecology. Central among these was the integration of geologic surveys of the LIS sea floor and studies of the Sound's benthic ecology, which provided the basis for understanding the structure and dynamics of the LIS benthos relative to characteristics of the benthic landscape in LIS (see Lewis, Chap. 2, in this volume). Between the late 1980s and mid-1990s, collaboration among researchers at the USGS, CTDEEP, and the University of New Haven began to integrate seafloor mapping and benthic community studies in several deep-water sections of LIS (Figs. 6.28 and 6.29). This work revealed that

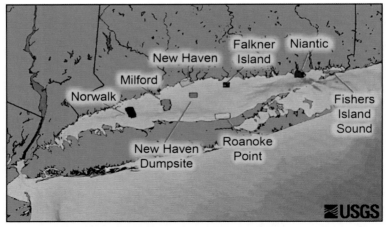

Map showing location of Long Island Sound sidescan sonar mosaics. Select the
mosaic area name from the table below to view the associated metadata file.

Fig. 6.28 Locations of USGS side scan mosaic study areas in LIS. See also USGS website for
up to date overview of USGS and NOAA seafloor mapping in LIS

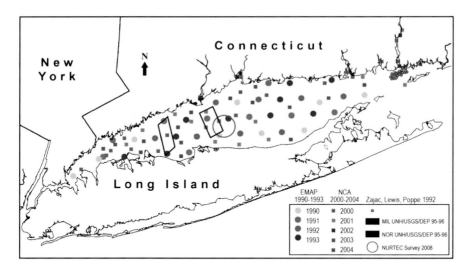

Fig. 6.29 Location of benthic surveys and studies in LIS conducted between 1985 and 2010
(some studies may not be shown); does not include inshore/intertidal studies

seafloor habitat/patch structure can be an important determinant of benthic community structure at small ($\leq m^2$), meso (10–100 s of m^2), and large scales (>100 s
of m^2), and seafloor mapping has provided the ability to analyze benthic community structure in habitats that could not be easily located. This includes the

transition zones between different types of sedimentary and hard substrate habitats, and sedimentary habitats that have varying geomorphological characteristics (Zajac 1996, 1998a; Zajac et al. 2000a, 2003). These studies also suggested that benthic community responses to disturbances such as hypoxia may be spatially complex (Fig. 6.30) and that successional stages in the Rhoads et al. (1978) model (Fig. 6.24) may not hold in all areas of LIS (Zajac 1998a, 1999, 2001). Maung (2010) recently showed that benthic responses to hypoxia are complex relative to geochemical dynamics at the sediment–water interface.

Concurrent advancements in mapping benthic habitats and integrating them with ecological studies were made in the nearshore habitats of LIS over this period (see Sect. 6.2, Littoral Zone). Building on the increase in seafloor and habitat mapping in the Sound and the growing appreciation of the value of integrated benthic mapping and ecological studies, Auster et al. (2009; see EPA LISS website) developed a Habitat Classification Model for LIS, which can be applied as a framework for continued integration of these types of studies.

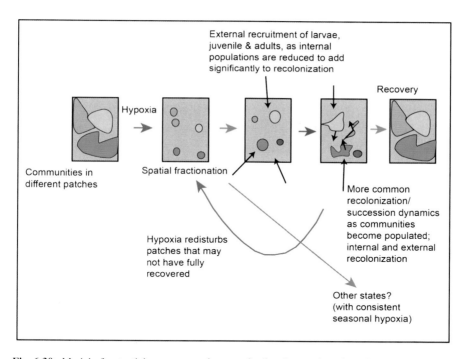

Fig. 6.30 Model of potential responses to large-scale disturbances in soft-sediment benthic landscapes. Because of increased spatial variation in remnant community structure, initial recolonization patterns may also be variable and not follow the patterns predicted by models developed from small-scale experimental work. There may be some temporal lag as to when more typical successional patterns are evident as biotic patches reach critical abundance levels so as to make significant contributions to recolonization of the overall disturbed portion of the benthic landscape. If hypoxia is recurrent, there may be an overall, long-term shift in community structure. Modified from Zajac (1998a, b)

The seafloor mapping data in conjunction with several of the benthic study datasets have provided the framework for assessing the application of marine and coastal spatial and conservation planning in LIS. Neely and Zajac (2008) showed that potential conservation planning scenarios in LIS would be constrained and shaped by the strong physical, chemical, and biological gradients in LIS, and conservation scenarios would need to recognize and incorporate such gradients. They showed that separate candidate conservation areas would be needed in the western and eastern portions of LIS due to the spatial heterogeneity of richness and community type distributions.

There have also been several Sound-wide benthic survey/monitoring studies, specifically the EPA Environmental Monitoring and Assessment Program (EMAP) and the National Coastal Assessment (NCA) program (Fig. 6.29). These studies sought to assess the environmental quality of estuarine and coastal areas of the US and as such their focus was regional. The data collected included benthic community structure, sediment chemical composition, and other environmental characteristics as well as sediment toxicity. Using the EMAP data from LIS, Schimmel et al. (1999) assessed the relative degree of impacted versus non-impacted sites based on a benthic index comprised of a species diversity measure and the abundances of tubificid and spionid annelids. Their analyses suggested that the western deep-water portions of LIS were impacted, as were some deep water sites in the eastern portion of LIS, and many harbors and large embayments.

6.6.1.3 Science Gaps and Management Implications

Although there has been a relatively substantial effort to study the benthos of LIS, there remain significant gaps in our knowledge. In particular, the temporal dynamics of these communities are not well understood, both with respect to seasonal fluctuations and interannual dynamics. Many of the benthic surveys noted above sampled at specific stations just once and as such it is difficult to interpret changes in data obtained decades apart without a good understanding of shorter-term fluctuations. Zajac and Whitlatch (in preparation) analyzed the benthic data collected as part of the EPA EMAP and the NCA programs which span 3-year periods about a decade apart (early 1990s and early 2000s), and found that there were several significant changes in benthic abundances and species richness of the central and western basins of LIS. However, Zajac (1998a) found that in these areas, there is significant seasonal and interannual variation in benthic community structure over a 2-year period that was approximately of the same magnitude of the decadal changes found by Zajac and Whitlatch (in preparation). Although the EMAP and the NCA programs sampled over 3–4 years, samples were only taken once a year and not in the same locations because the focus of these programs was to assess regional status of estuarine conditions, not within-estuary dynamics. As such, we are missing data critical not only to understanding the basic dynamics of benthic systems in LIS, but also to assessing temporal changes that may result from of human and natural disturbances and longer term climate change. There have been no studies of the deep water (nor shallow water) benthic communities that have assessed seasonal and year to year changes for periods greater than 2 years, nor studies that have assessed recovery following disturbance for more than that time span. Benthic disturbance succession models (Fig. 6.24) predict that recovery periods may be on the order of the life span of the species that characterize endpoint, "climax" communities. At least in the deep-water mud environments of LIS, these potentially include species such as *N. incisa*, maldanid polychaetes, and bivalves such as false quahogs (*Pitar morrhuanus*) and hard clams, and are among those expected in climax communities. All have potential life spans >3 years. Studies of these communities and populations of key species are necessary to parse the various spatial and temporal patterns of variability so that long-term trends can be identified. We effectively have no information on successional dynamics and endpoint communities in the coarser-grained habitats of LIS, and indeed these may be quite different from that characterized in Fig. 6.23 (see Zajac 2001 for an extended discussion of this point). There have been few follow-up studies after certain types of impacts such as pipeline construction that have assessed long-term impacts of such disturbances. Finally, and quite surprisingly, there is no long-term, consistent monitoring of benthos inside and out of hypoxic zones, which along with concomitant physical and chemical monitoring, are critical to understanding the long-term impact of seasonal hypoxia and how this may interact with changing climate, such as increasing water temperatures that may exacerbate seasonal hypoxia.

Food-web dynamics are a key for understanding the overall ecology and ecosystem dynamics of various environments in LIS. To that end, data on biomass and productivity of all food-web components are necessary. Unfortunately, very few data of this type (e.g., Carey 1962; Kroeger 1997) are available for benthic species in LIS, and as such comprise a significant unknown in terms of having LIS specific data that can be incorporated into food-web models (Zajac et al. 2008). Future benthic studies in LIS should routinely quantify the biomass and, to the extent possible, productivity of dominant taxa.

Currently, there is significant effort to continue seafloor habitat mapping studies in LIS. The production of seafloor maps showing detailed bathymetry, sediment type, and other features needs to be paired with benthic ecological studies of areas being mapped, as seafloor maps without associated ecological information will potentially provide little value for understanding and managing these environments. A greater focus on systematic mapping and conducting ecological surveys of inshore benthic environments along both the Connecticut and New York coasts of LIS is also strongly warranted.

Many of these informational needs could be addressed by establishing sites that would be consistently monitored within the context of a "natural experiment" design (Fig. 6.31). The locations of the benthic monitoring areas would be based on the premise of a "natural experiment," testing spatial difference in community structure at multiple scales and allowing temporal changes (as might occur with climate change) to be assessed in a variety of benthic habitat types across the estuarine gradient in LIS. Within these locations, coordinated physical and chemical monitoring and habitat mapping would allow use of an ecosystem-based approach.

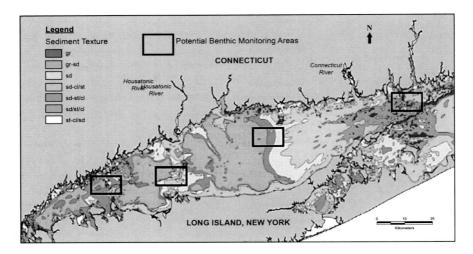

Fig. 6.31 Potential, long-term benthic monitoring areas in Long Island Sound. Each area is comprised of several types of sedimentary habitats and together employ a natural gradient "experimental" design. Their location at the transitions among different regions of the Sound would integrate with the known distributions of benthic community types

A similar shallow water set of monitoring areas could also be established. The areas shown in Fig. 6.31 are suggested sites including a variety of sedimentary habitats repeated along the west-east gradient of LIS and would allow for understanding benthic ecology relative to sediment type and to the gradients of physical and chemical conditions in the Sound.

Effective stewardship of the LIS benthos will depend on a much better understanding of all its ecological components. The deep water benthic habitats of LIS are one of these critical components, yet our understanding of the resident organisms, their population and community dynamics, and their contributions to ecosystem function lags far behind the gains we have made over the past 20–25 years with respect to our knowledge of the physical and chemical nature of the estuary, and certain pelagic components. As such it is critical that a more concerted and systematic effort is placed in research so as to close this gap in our overall understanding of LIS.

6.6.1.4 Benthic Disturbance

Overview and Natural Disturbance

Nearshore and subtidal benthic habitats in LIS are frequently affected by anthropogenic and natural disturbances: physical, chemical (anoxia, contamination, and acidification), thermal (exposure to unusually low or high air or water temperatures), or biological (settlement or growth of larvae or adults in sufficient density to alter surface or subsurface conditions). Disturbance of relatively stable sediments or hard-bottom substrates is thought to be sufficiently important to ben-thic habitat structure and function that it provides the driver for several models of benthic succession and community structure (Fig. 6.32) (Sanders 1969; Pearson and Rosenberg 1978 Rhoads et al. 1978; Rhoads and Germano 1986; Zajac 2001). The frequency and degree of disturbance vary widely but generally are expected to decrease with increasing water depth. Shallow water and intertidal habitats experi-ence high frequency disturbance from waves, tides, and storms. Subtidal habitats

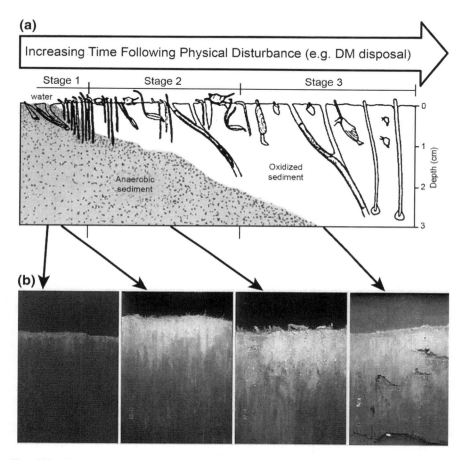

Fig. 6.32 The drawing at the *top* (from Fig. 6.24) illustrates the development of infaunal suc-cessional stages over time following a physical disturbance. The SPI images *below* the drawing provide examples of the different successional stages. a Highly-reduced sediment with a very shallow RPD layer (contrast between *light colored* surface sediments and *dark underlying* sedi-ments) and little evidence of infauna. **b** Numerous small polychaete tubes are visible at the sedi-ment surface (Stage 1) with a slightly deeper aRPD compared to the previous image. **c** A mixture of polychaete and amphipod tubes occurs at the sediment surface (Stage 2). **d** Numerous burrow openings and feeding pockets (voids) at depth within the sediment are evidence of deposit-feed-ing, Stage 3 infauna (from Germano et al. 2011)

may experience tidal current exchange, i.e., seasonal disturbance from storms or T, but most disturbances occur with periodicities measured in months or years. Most subtidal habitats in LIS are well protected from frequent natural disturbance with the exception of areas scoured by tidal currents near The Race or on shallow ridges or shoals (e.g., Stratford Shoal, Long Sand Shoal). However, the tidal exchange of sediment across the sediment–water interface can be quite substantial in many of these subtidal habitats.

The seasonal influx of suspended sediment into the Sound has been estimated to be 9.3×10^8 kg year^{-1} (Farrow et al. 1986; Rhoads 1994), equivalent to an annual sediment mass accumulation rate of 0.05 g cm^{-2} year^{-1} over 1792 km^2 of muddy sediment (Bokuniewicz and Gordon 1980). This estimate compares with long-term sedimentation rates determined by ^{210}Pb profiles from the center of LIS of 0.05 g cm^{-2} year^{-1} and radiocarbon dating results of 0.077 g cm^{-2} year^{-1} (Benoit et al. 1979). Varekamp et al. (2010) measured mass accumulation rates in cores from LIS (using ^{210}Pb, ^{137}Cs, and ^{14}C) of 0.01–0.05 g cm^{-2} year^{-1} before the 1800s; rates increased to approximately 0.25 g cm^{-2} year^{-1} in the mid-1800s. The deposition rate may be higher (0.1 g cm^{-2} year^{-1}) in the western basin (Bokuniewicz 1988), but Varekamp et al. (2010) did not confirm this observation. In addition to the influx of suspended sediment, the central and western basins are subject to tidal resuspension and deposition of large volumes of sediment (Rhoads et al. 1984; Knebel and Poppe 2000). Most of this resuspended sediment is trapped below the pycnocline in summer and creates a transitory near-bottom turbidity zone (Rhoads et al. 1984). Average near-bottom turbidity values of 5 mg L^{-1} result in an estimate of 2.5×10^8 kg or 27 % of the annual supply in suspension with higher values in spring and early summer (Bokuniewicz 1988; Rhoads 1994). The flux of sediment due to suspension and redeposition appears to be much higher than the net long-term sedimentation (Rhoads 1994). Using McCall's sediment trap data, Rhoads calculated that 1×10^{12} kg of fine sediment was resuspended annually or the equivalent of 1,000 times the long-term sedimentation rate (McCall 1977; Rhoads 1994). These measured and estimated rates suggest that the benthic environments in the central and western basins experience very high fluxes of fine-grained sediments and that very little of the net influx of sediment is removed from the resuspension cycle (Rhoads 1994). The benthic community in these environments is therefore exposed and presumably adapted to relatively high exchange of sediment across the sediment–water interface, despite the apparently protected conditions (McCall 1977). During certain storm events, the resuspension levels may be much higher, but there is little evidence that storm events are a major source of disturbance except in shallow nearshore habitats.

Dredging and Dredged Material Disposal

Dredging affects channel floor habitats, with minimal loss of suspended sediments to surrounding habitats (Bohlen et al. 1979; Wilber et al. 2007). Material removed during dredging is frequently placed on the sea floor in open water habitats with immediate, short-term effects on the benthic community (Germano et al. 1994; Bolam and Rees 2003). The annual average dry weight of dredged material placed in LIS has been estimated at 4.1×10^8 kg year^{-1} or less than half of the annual sedimentation rate (Rhoads 1994). Of this amount, dispersal losses from passage through the water column and resuspension after placement are estimated at 6 %, or about 3 % of the annual non-disposal sediment input (Rhoads 1994).

Historically, dredged material disposal in the Sound occurred at many sites located just outside harbors adjacent to the recognized channel (Fredette et al. 1993). As a result, some benthic habitats still retain traces of placement of isolated piles of harbor sediment (Poppe et al. 2001, 2010; ENSR 2007).

Disposal is now confined to designated areas between 3.43 and 6.86 km^2 (1–2 nautical miles^2 in deeper areas of LIS (Fredette et al. 1993). Disposal of dredged sediments is permitted for materials that are determined to be suitable for unconfined open water disposal based on biological based testing protocols (Fredette and French 2004; EPA/USACE 1991). Disposal material typically consists of seasonal placement of 10–500,000 cubic yards (7.65–382,300 m^3) of harbor sediments at buoys by releasing the material from a split-hull barge at the surface. Each barge contains between 382 and 2,294 m^3 (500–3,000 yd^3 of water-laden sediment. Recent sea floor imaging studies, experimental placement, and laboratory experiments have clarified the physical processes involved in placement of dredged material in open water (Fig. 6.33; Valente et al. 2012; ENSR 2007). The sediment released from the bottom of the barge falls rapidly to the sea floor, entraining some water but retaining a coherent mass until it contacts the sea floor. Upon contact, the vertical force of the bolus of water-entrained sediment is transferred to lateral forces and rapidly spreads in a circular pattern to form a low mound or crater shape on the sea floor (Fig. 6.34). Depending on the water depth, barge volume, water content of the dredged material, and the seafloor surface, the mound or crater is between 50 and 300 m in diameter. This process of sediment placement creates a disturbed sediment surface that consists of a coherent layer of dredged material 10–200 cm thick in the center of the mound, thinning to mixed layers of ambient sediment and fresh dredged material. At the outer margin of the placement feature, layers 1–2 mm thick of fresh dredged material can be detected with sediment profile imaging techniques (Germano et al. 2011). This process of placement is usually repeated 10–250 times at a single disposal buoy during a disposal season (October–May), resulting in a shallow mound 1–5 m thick and 200–1,500 m in diameter (Figs. 6.33 and 6.34; Valente et al. 2012; ENSR 2007).

After placement of sediments on the sea floor, the resulting mound remolds over a period of months due to consolidation (Silva et al. 1994; Poindexter-Rollings 1990), bioturbation, erosion, and deposition from near-bottom flow processes (Rhoads 1994). As a result, the volume of the mound will decrease and the surface will smooth and begin to converge with conditions of adjacent ambient sediments. The degree of convergence will depend on water depth, physical conditions of the placement site, and the differential between placed and ambient sediments (Rhoads 1994). Management of placement sites for containment of dredged material dictates creation of stable mounds in depositional environments (Fredette and French 2004).

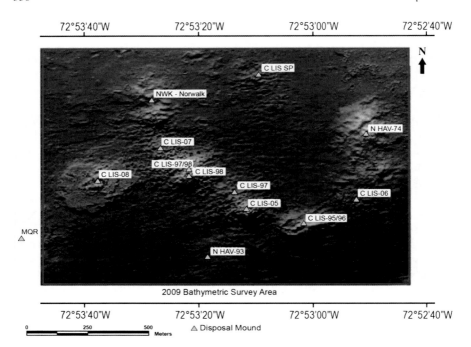

Fig. 6.33 Hillshaded multibeam bathymetry of a portion of the Central Long Island Sound Disposal Site surveyed in 2009. Individual mounds (e.g., CLIS-05) represent accumulated disposal of dredged material for one or more disposal seasons (October–May). Relief is exaggerated to highlight low relief features that result from impact of dredged material on the seafloor (from Valente et al. 2012)

Mounds containing dredged material unsuitable for unconfined placement due to elevated levels of contaminants are engineered to meet regulatory guidelines under which the contaminants are made inaccessible (Fredette and French 2004; EPA/USACE 1991). Studies of the stability of these engineered or "capped" mounds (mounds constructed to isolate the unsuitable material beneath a sediment "cap"; Palermo 1991) have demonstrated that pore water exchange of contaminants with the overlying ambient water is well below background levels (Bokuniewicz 1989; Murray et al. 1994). Longer term studies of mounds have demonstrated that once the rapid consolidation phase has been completed (ca. 1 year), the surface layers of sediment (10–30 cm) are the only horizons available to interact with biological resources (Fredette et al. 1992; Murray et al. 1994).

The consequence to benthic habitats of the placement of dredged material has been studied in LIS for over 40 years (Fredette and French 2004; Valente 2004). The nature of the impacts can vary depending on the composition of the dredged material and the habitat at the disposal site (Bolam et al. 2006). A structured monitoring approach of disposal impacts has been utilized in LIS since 1977 as part of the Disposal Area Monitoring System, or DAMOS (Germano et al. 1994; Fredette and French 2004). Based on the results of DAMOS studies and the understanding

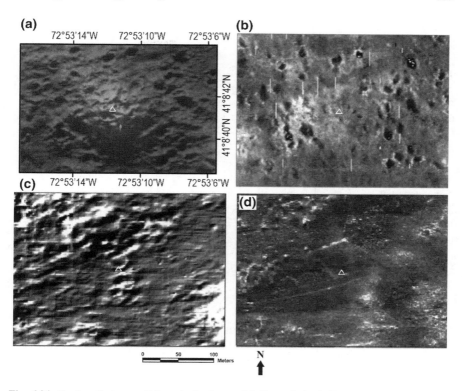

Fig. 6.34 Surface features within a dredged material disposal site in Long Island Sound before and after several disposal seasons. Images are centered over the location of a mound (CLIS-05) formed from October 2005 to May 2006. **a** Hillshaded multi-beam bathymetry from 2009, 3.5 years after mound formation. **b** Backscatter mosaic (sidescan imagery) from multibeam survey from 2009. **c** Hillshaded multi-beam bathymetry from 2005 prior to mound formation. **d** Sidescan sonar mosaic from 1997 (from Poppe et al. 2001). Individual ring features and impact craters are from single disposal events with split-hull disposal barges (from Valente et al. 2012; ENSR 2007)

of the physical processes described above, the benthic disturbance that results from placement of dredged material in open water habitats in LIS is a remobilization of surface sediments, burial of surface sediments and benthic infauna, and introduction of disturbed sediments with high organic loads into discrete areas (Germano et al. 1994). Virtually all benthic infauna are smothered in layers that exceed 15 cm. The ability to escape a given depth of burial is related to the life habits of the fauna (Kranz 1974; Maurer et al. 1986; Kjeilen-Eilertsen et al. 2004); strong burrowing deposit feeders can escape from 10 cm or more of burial (Jackson and James 1979; Bellchambers and Richardson 1995), but attached epifaunal suspension feeders cannot survive more than 1 cm (Kranz 1974). This means that some organisms can burrow up through thin layers of fresh sediment, but many will not. The center region of the disposal mounds formed in the Sound is usually sufficiently thick to catastrophically bury all organisms that cannot move quickly (Germano et al. 1994).

An "apron" of thinner material can extend 100–500 m beyond the bathymetrically detectable margin of the mound [accumulations greater than 10 cm can be reliably detected with multibeam surveys, greater than 20 cm with single beam fathometers (Fredette and French 2004; Carey et al. 2012)]. In the apron of thinner deposition, the introduction of high organic loads provides a surge in potential food supply for deposit feeders and rapid bioturbation usually obliterates the distinct layer within months (Germano et al. 2011).

The surface of the mound, including the apron, attracts high settlement densities of small surface deposit feeders (polychaetes, amphipods, bivalves, and meiofauna). This response has been documented by numerous monitoring studies in LIS and is consistent with the successional model of Rhoads et al. (1978) and Pearson and Rosenberg (1978). The nature and rate of recolonization can be strongly influenced by the timing of disturbance relative to seasonal pulses of settlement and growth of larvae (Zajac and Whitlatch 1982; Wilber et al. 2007). The successional model of response to physical disturbance from placement of dredged materials (Rhoads et al. 1978) has been tested with observation of disposal mounds in LIS since 1982 with the use of sediment profile imagery (Germano et al. 2011). Sediment profile imaging (SPI) utilizes a cross-sectional image of the upper 20 cm of the sediments to observe visual evidence of organism-sediment interactions. A phenomenological model (Rhoads and Germano 1982, 1986) has been used to interpret the ecological effects of dredged material in LIS (Germano et al. 1994) and minimize the impacts of disturbance (Fredette 1998; Fredette and French 2004).

The infaunal successional model (Rhoads and Germano 1986) posits that stage 1 organisms (small, tube-dwelling surface deposit feeders) appear within days or weeks of physical disturbance or deposition of a fresh layer of dredged material. If no further disturbance occurs, these stage 1 organisms are replaced by infaunal deposit feeders (stage 2) and eventually by larger infaunal deposit feeders (stage 3), many that feed in a head-down orientation that creates distinctive feeding voids (Germano et al. 2011). The establishment of this mature community may take months to years to complete and results in a deepening of the bioturbated mixed sediment layer and convergence with the surrounding benthic habitat conditions, depending on factors such as the spatio-temporal structure of the species pool (Zajac 2001). Potential variation in the rate of succession is illustrated by recent results collected from a disposal mound 5 months after cessation of disposal in 2009 (Fig. 6.35).

Benthic disturbance from dredged material disposal in LIS has immediate effects on sessile epifauna and infauna (Germano et al. 1994, 2011). The management approach to dredged material disposal in LIS includes biological testing of sediments and active management of disposal to segregate materials determined to be unsuitable for unconfined open water placement (Fredette and French 2004; Carey et al. 2006). During the development of the management approach, dredged material known to contain elevated levels of metals and PAHs was placed at the Central Long Island Sound Disposal Site in 1983 at several locations (capped and uncapped) as an experiment and monitored extensively

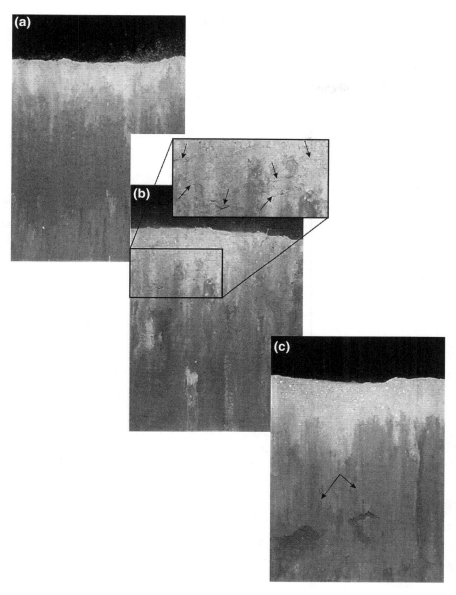

Fig. 6.35 Sediment profile images collected in October 2009 from a disposal mound (CLIS-08) formed in from October 2008 to May 2009 at the Central Long Island Sound Disposal Site. **a** Image illustrating dredged material colonized by tube-building worms (Stage 1). **b** Image illustrating a transitional successional status from Stage 1 to 2. Small Stage 1 worm tubes are visible at the sediment surface, and numerous small tunnels produced by burrowing Stage 2 meiofauna (e.g., crustaceans and bivalves) occur just below the surface (*arrows inset*). **c** Image showing a large vertical burrow and feeding voids (*arrows*) resulting in a Stage 3 successional designation (from Valente et al. 2012)

for more than 20 years (Germano and Rhoads 1984; Myre and Germano 2007). Short-term biological effects were observed after placement of unconfined dredged material (Myre and Germano 2007). This joint USEPA/USACE Field Verification Program was designed to field-verify existing test methods for predicting the environmental consequences of dredged material. The biological testing and resulting management approaches (sequestering dredged material with evidence of biological effects beneath a "cap" layer of material without significant biological effects) have contributed to the lack of observable long-term ecological effects from disposal activities in LIS. Apart from alterations of habitat due to introduction of different grain-size composition or changes in sediment transport conditions due to elevation of the sea floor, there is no evidence of long-term effects on benthic processes or habitat conditions (Germano et al. 2011).

6.6.2 Benthic Foraminifera

Foraminifera are unicellular, heterotrophic eukaryotes in the super-group Rhizaria, characterized by a branching, anastomizing network of granular reticulopodia (Adl et al. 2005). Many species have a proteinaceous theca, but many others make a shell (test) by agglutinating mineral grains in an organic or mineral matrix or by secreting $CaCO_3$. Tests may consist of one or many chambers. Foraminifera are marine, living from brackish coastal regions to the deepest ocean trenches (Pawlowski and Holzmann 2008).

There are about 50 species of living planktonic foraminifera and several thousand benthic ones (Murray 1991, 2006, 2007). Foraminifera are part of the meiobenthos, i.e., mostly between 63 and 1,000 μm in diameter. Benthic foraminifera are most diverse (hundreds of morphological species) along the lower continental shelf (Culver and Buzas 1982; Gooday 1993). Estuaries usually contain a few tens of species at most, coastal salt marshes and mangrove forests about 15 species, with 5–10 dominant species (Murray 1991, 2006; Scott et al. 2001; Javaux and Scott 2003).

Foraminifera are ubiquitous in the marine realm, their tests are easily fossilizable, and their small size makes it possible to obtain statistically valid data using relatively small samples (see e.g., Jorissen et al. 2007). Planktonic and benthic foraminiferal fossil assemblages thus have been used widely to reconstruct environmental changes on timescales from millions of years (see e.g., Thomas 2007) through historical times. In coastal regions, profound changes in foraminiferal fauna occurred partly in response to anthropogenic changes (see e.g., Alve 1995, 1996; Alve and Murray 1995; Culver and Buzas 1995; Karlsen et al. 2000; Scott et al. 2001; Platon et al. 2005, Murray 2006; Sen Gupta and Platon 2006; Nikulina et al. 2008; Gooday et al. 2009). In LIS, foraminifera have been used to study salt marsh ecology and reconstruct relative sea level rise, and eutrophication.

6.6.2.1 Foraminifera in Coastal Salt Marshes

The distribution and relative abundance of foraminifera in intertidal marsh settings are primarily controlled by tidal emergence of the marsh surface (Scott and Medioli 1978, 1980; Scott et al. 2001). Exposure time is controlled by the elevation of the salt marsh surface above mean sea level and the tidal range, so that marsh foraminifera are distributed in vertical zones, similar to marsh vegetation (Scott and Medioli 1978, 1980; Scott et al. 2001; Horton and Edwards 2006). The zones can be expressed in vertical distance from mean high water (MHW). In high and middle marsh, benthic foraminifera live on the marsh surface or in the topmost 2.5 cm of the sediment, whereas in the low marsh they are dominantly found in the upper 5 cm but with a significant number of living specimens down to 15 cm (Saffert and Thomas 1997; Edwards et al. 2004a).

Dominant foraminifera in coastal salt marshes are agglutinants, constructing tests from mineral and/or biogenic grains (e.g., quartz, mica, diatom frustules) (Bradshaw 1968; Scott et al. 2001; de Rijk 1995). The most common species along the coasts of LIS are *Trochammina macrescens*, *T. inflata*, *Tiphotrocha comprimata*, *Haplophragmoides manilaensis*, *Miliammina fusca*, and *Arenoparrella manilaensis*. Less common are *Siphotrochammina lobata*, *Balticammina pesudomacresens*, *Polysaccamina ipohalina*, *Pseudothurammina limnetis*, *Ammotium salsum*, *Ammoastuta inepta*, and *Textularia earlandi*. *T. macrescens*, *T. inflata*, and *T. comprimata* are the most abundant in middle to high marsh regions, with *H. manilaensis* increasing in relative abundance at lower S, and *Arenoparrella mexicana* close to creeks (Scott and Medioli 1978; Thomas and Varekamp 1991; Nydick et al. 1995; de Rijk 1995; Saffert and Thomas 1997; de Rijk and Troelstra 1999; Scott et al. 2001; Edwards et al. 2004a). Lower in the intertidal zone, *M. fusca* is more common, with *A. salsum* or *A. inepta* at lower salinities. On lower intertidal mudflats, the calcareous species *Ammonia parkinsoniana*, *Elphidium* spp., and *Haynesina germanica* may be found occur (Fig. 6.36).

Coastal marshes are sensitive to changes in sea level (e.g., Donnelly and Bertness 2001). Foraminiferal assemblages are indicative of the elevation of the marsh surface above mean sea level, so that fossil assemblages in core samples can be used to indicate the position of each sample with regard to mean sea level when the foraminifera were living. By dating the samples, we can reconstruct the rate of sea level rise over the time of deposition of the samples (Scott and Medioli 1980; Thomas and Varekamp 1991; Varekamp et al. 1992; Nydick et al. 1995; Varekamp and Thomas 1998; Scott et al. 2001; Edwards et al. 2004b; Horton and Edwards 2006). Marsh foraminifera have become the dominant method for evaluating sea level variation (Armstrong and Brasier 2005). The accuracy of these estimates depends on the width of the foraminiferal zones; thus, on the tidal range, and on the exact location of the living foraminifera on or in the sediments (Saffert and Thomas 1997).

With recent rapid sea level rise, New England salt marshes underwent dramatic shifts, with the low marsh cordgrass replacing the high marsh salt hay (Thomas and Varekamp 1991; Varekamp et al. 1992; Warren and Niering 1993; Nydick et al. 1995; Varekamp and Thomas 1998; Thompson et al. 2000; Donnelly and Bertness 2001; Edwards et al. 2004b; Fitzgerald et al. 2008). Ultimately, loss of salt marsh is expected, and has been observed (Donnelly and Bertness 2001). Benthic foraminifera thus can be used to monitor the ability of salt marshes to adapt to increasing rates of sea level rise.

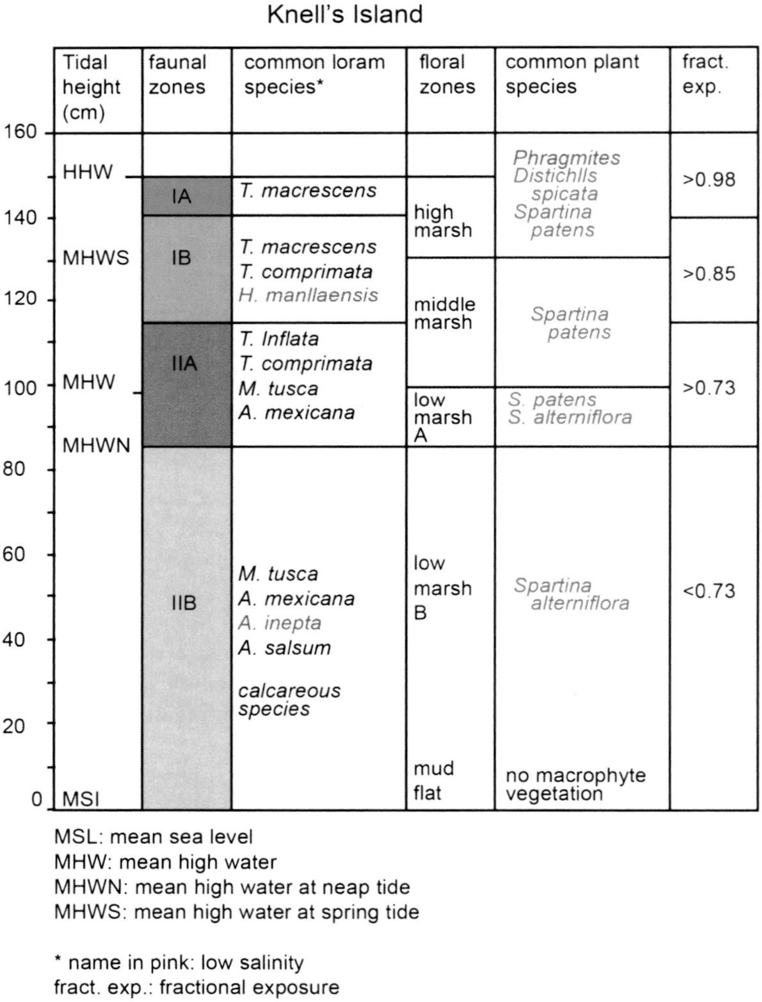

Knell's Island

Tidal height (cm)	faunal zones	common foram species*	floral zones	common plant species	fract. exp.
160					
HHW — 140	IA	*T. macrescens*	high marsh	*Phragmites Distichlis spicata Spartina patens*	>0.98
MHWS — 120	IB	*T. macrescens T. comprimata H. manllaensis*	middle marsh	*Spartina patens*	>0.85
MHW — 100 MHWN — 80	IIA	*T. Inflata T. comprimata M. tusca A. mexicana*	low marsh A	*S. patens S. alterniflora*	>0.73
60 — 40 — 20	IIB	*M. tusca A. mexicana A. inepta A. salsum calcareous species*	low marsh B	*Spartina alterniflora*	<0.73
0 MSI			mud flat	no macrophyte vegetation	

MSL: mean sea level
MHW: mean high water
MHWN: mean high water at neap tide
MHWS: mean high water at spring tide

* name in pink: low salinity
fract. exp.: fractional exposure

Fig. 6.36 Foraminiferal zonation and plant zonation within the intertidal zone of the salt marshes at Knell's Island, Housatonic River

6.6.2.2 Subtidal Foraminifera

Foraminifera from WLIS and New York Harbor were first described by Shupack (1934), followed by more detailed studies in the late 1940s (Parker 1952), and an exhaustive study in the 1960s (Buzas 1965), with additional data analysis by Murray (1976). The morphology of the dominant *Elphidium* spp. was studied by Buzas (1966), Miller et al. (1982), and Buzas et al. (1985). Thomas et al. (2000) described assemblages collected in the 1990s. Thomas et al. (2004a, b, 2009, 2010; Thomas and Varekamp 2011) compared data on assemblages from grab samples (i.e., recently living) with assemblage data on core samples representing the last millennia (Thomas et al. 2004a, b, 2009, 2010; Varekamp et al. 2010).

The LIS foraminiferal assemblages are "marginally marine," with a low species richness and diversity (e.g., Murray 1991; Scott et al. 2001). Abundant species have high tolerances for fluctuations in T, oxygen, S, and environmental pollution (e.g., Moodley and Hess 1992; Alve 1995; Culver and Buzas 1995; Gooday et al. 2009). Foraminifera are common to abundant in most of LIS, with the exception of non-depositional areas, i.e., east of 72.6°W (Knebel and Poppe 2000).

The most frequent species in LIS (Fig. 6.37) (Parker 1948; Buzas 1965; Thomas et al. 2000) are the calcareous hyaline *Elphidium excavatum, E. incertum,*

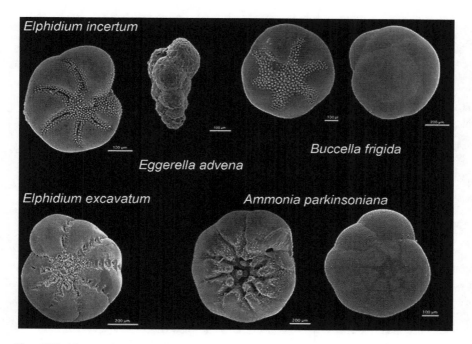

Fig. 6.37 Most common benthic foraminiferal species in Long Island Sound. Both *ventral* and *dorsal views* are given for the two trochospiral species, *Bucella frigida* and *Ammonia parkinsoniana*

Buccella frigida, and *A. parkinsoniana* (formerly described as *A. beccarii,* or *A. beccarii* s.), and the agglutinated species *Eggerella advena* and *Trochammina squamata.* Rare species (<5 %) include the agglutinated *Rheophax scorpiurus, R. nana,* and *Textularia earlandi,* the calcareous hyaline *Lagena* spp., *Dentalina* spp., *Nonion commune, Buliminella elegantissima, Fursenkoina fusiformis, Bolivina* spp., *Hopkinsina* spp., and *Polymorphina novangliae,* the epiphytic species *Cibicides lobatulus* and *Cornuspira planorbis,* and the calcareous porcellaneous *Quinqueloculina* spp. The rare species are more common in ELIS than toward the west, and they were more common in the 1960s (Buzas 1965) and earlier (Parker 1948) than they are today (Thomas et al. 2000). Species characteristic for coastal salt marsh environments are found in LIS close to the shore, but not as living specimens.

In LIS, Buzas (1965) recognized an assemblage with >60 % *E. excavatum* at depths from 3 to 19 m (average 11 m), an assemblage with common *E. excavatum,* >9 % *B. frigida* and <19 % *E. advena* at 15–33 m (average 26 m), and a third assemblage with >19 % *E. advena* at 19–39 m depth (average 30 m). *A. parkinsoniana* was rare (<5 %). The most abundant species remained the same in the 1990s (Thomas et al. 2000), but assemblages differed profoundly from these in the 1960s by the establishment of an east–west rather than a depth zonation, with *A. parkinsoniana* becoming common and often dominant (up to 85 %) in westernmost LIS (Thomas et al. 2004a, b, 2009, 2010; Thomas and Varekamp 2011). This is a highly significant change. Core studies show that *E. excavatum* was the dominant foraminiferal species in shallow water settings since the establishment of LIS (Thomas et al. 2004a, b, 2011; Varekamp et al. 2010), and that *A. parkinsoniana* began to increase in abundance in the late 1960s to early 1970s in westernmost LIS, spreading eastward (Thomas et al. 2004a, b). In addition, in the 1990s, the relative abundance of *E. advena* in the deeper part of LIS declined from >19 % (Buzas 1965) to a few percent at most (Thomas et al. 2000, 2004a, b, 2010). Core studies show that it was more abundant before the 1800s, declined in abundance in the early twentieth century, increased shortly in the early 1960s, followed by a continued decline (Thomas et al. 2004a, b; Thomas and Varekamp 2011) (Fig. 6.38).

Buzas (1965) suggested that the assemblages might be controlled by food supply, with the *E. excavatum*-dominated faunas in the photic zone feeding on benthic algae and diatoms. Murray (1991) described *E. excavatum* as an herbivore, and *E. advena* as a detritivore (Alve 1995). *E. excavatum* consume diatoms and can sequester their chloroplasts, thus becoming functional photosynthesizers, giving them competitive advantage over heterotrophs in sunlit waters (Lopez 1979; Lee et al. 1988; Lee and Lee 1989; Lee and Anderson 1991; Bernhard and Bowser 1999; Pillet et al. 2011). *E. excavatum* also sequesters chloroplasts in aphotic waters, where they might help in survival during hypoxia (Cedhagen 1991), and increasing feeding efficiency (e.g., Bernhard and Bowser 1999).

The ecology of *E. incertum* is not well-known; it flourishes in quiet waters without strong current activity. Little is known about the ecology of *B. frigida,* which is common in fine-grained, organic-rich sediments (Murray 1991, 2006).

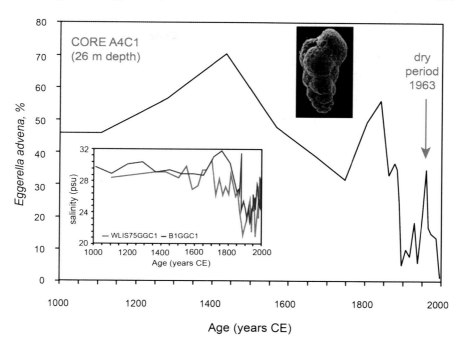

Fig. 6.38 Relative abundance of *Eggerella advena* (shown in Scanning Electron Micrograph) for the last 1,000 years, as observed in Core A4C1 (western LIS). *Inset* shows reconsgtructed salinity for the last 1,000 years for cores WLIS 75GGC1 (The Narrows) and B1GGC2 (Outside the mouth of the Housatonic River)

The agglutinated *Eggerella advena* survives under low oxygen and severely polluted conditions, but is relatively sensitive to low S (Murray 1991; Alve 1995). *A. parkinsoniana* is cosmopolitan, omnivorous (Bradshaw 1957), survives wide swings in T and S, and is abundant in highly polluted regions (Murray 1991, 2006; Alve 1995).

6.6.2.3 Causes of Foraminiferal Faunal Change

The foraminiferal faunal changes in core records in the early to middle 1800s consist of increased relative and absolute abundance of *E. excavatum*, possibly caused by eutrophication which led to increased diatom productivity from the earliest nineteenth century on (Varekamp et al. 2010). The decline in relative abundance of the agglutinated species *E. advena* at the same time may have been caused by decreasing S (Varekamp et al. 2004, 2009, 2010), as documented by stable isotope studies of benthic foraminiferal tests. A short-term increase in S (and the relative abundance of *E. advena*) occurred during the drought years in the 1960s (Buzas 1965).

Abundance of *Ammonia* spp. related to *Elphidium* spp. is commonly expressed as the *Ammonia-Elphidium* index (A–E index, Fig. 6.39), which has high values in eutrophied, hypoxic waters (e.g., Chesapeake Bay and the Gulf of Mexico; Sen Gupta et al. 1996; Karlsen et al. 2000; Platon et al. 2005; Sen Gupta and Platon 2006; Gooday et al. 2009). Increased abundance of *Ammonia* spp. thus has been tentatively explained as caused by hypoxia. We see correlation between hypoxia and high abundance of *Ammonia* spp., but do not interpret this as causation: both *E. excavatum* and *A. parkinsoniana* are extremely resilient under hypoxia, and survive full anoxia for several days (Moodley and Hess 1992). Regional rise in water temperatures could be a contributing factor, because *Ammonia* spp. is more common in warm waters than *Elphidium* spp. (Murray 1991, 2006), and need several weeks of temperatures above 20 °C for efficient reproduction (Bradshaw 1957). It is, however, improbable that warming is the main factor causing the rise in abundance of *A. parkinsoniana*, because temperatures are not very different in western and ELIS.

A more probable cause for the decrease in abundance of *E. excavatum* may be a decrease in overall bulk diatom productivity, as seen in biogenic silica (S) records (Varekamp et al. 2010), possibly resulting from low N/Si values during severe eutrophication. This may have led to declining relative abundances of the specialized diatom consumer *E. excavatum* (Varekamp et al. 2005). Limited genetic

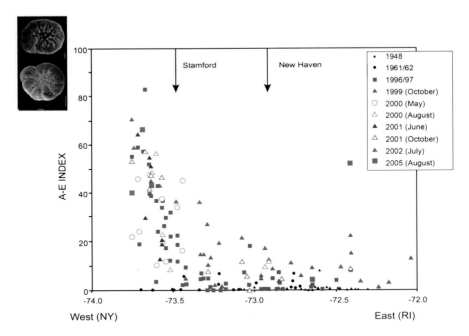

Fig. 6.39 A–E index (percentage *Ammonia parkinsoniana* normalized to the sum of the percentages of *Ammonia* and *Elphidium* species) for foraminiferal assemblages in grab samples collected in various years; all samples plotted versus longitude (E–W). Upper Scanning Electron Micrograph: *Ammonia parkinsoniana, ventral side*; Lower Scanning Electron Micrograph: *Elphidium excavatum*

evidence on the genus *Ammonia* in LIS, however, allows the possibility that the recent increase in relative abundance was caused by an invasive cryptospecies (Holzmann and Pawlowski 1997; Hayward et al. 2004).

In conclusion, benthic foraminifera have been used to document ecosystem changes over time, and may be used in the future to monitor ecosystem response due to climate change and N input (Gooday et al. 2009). In Europe, a standardized protocol has been developed that uses benthic foraminifera to monitor status of coastal marine ecosystems (Schoenfeld et al. 2012).

6.7 Fish, Shellfish, and Wildlife

6.7.1 Finfish Communities and Fisheries

The estuarine waters of LIS are nursery and feeding grounds for over 100 species of resident and migratory finfishes (Gottschall and Pacileo 2010). The variety of bottom types, water depths, currents, and tide ranges found in different parts of LIS creates a myriad of habitats for a large array of species. This diversity of species has supported ecosystem functions and commercial harvest for centuries.

Occupation of the region around LIS by native people has been verified for the past 12,000 years, although 2,000–3,000 years ago permanent settlements along the coast in bays and estuaries were maintained by hunter-gatherer-fisher cultures (Bernstein 2006). First nation's people were primarily dependent on fish and shellfish close to shore. Remains of shell middens suggest heavy use of hard clam and oyster, and lesser consumption of soft clam, bay scallop, and some knobbed whelk (*Busycon carica*) (Witek 1990), that appears to have been close to a year-round harvest (Lightfoot and Cerrato 1988). Early accounts describe fishing through use of torchlight with spears and clubs, nets, hooks, bow and arrow, traps, and weirs in the LIS region (Banks 1990). Weirs were constructed using wooden stakes driven into the river bottom, occasionally supported by rock walls in areas of swift current, where water level receded in low tides (Banks 1990). Stakes were interwoven with branches to create a barrier that confined fish such that they could be captured by dipnet or spear (Banks 1990). Relicts of these structures remain today in many of the river mouths that drain to the Sound.

Anadromous fishes provided the greatest density of available food for societies in the Hudson Valley (Fisher 1983), and capture of anadromous species in Connecticut is supported by a number of stake weirs in the Housatonic River and their common use throughout the eastern US (Lutins 1992). Thus, the primary fish species captured by native cultures around LIS would have included, in order of likely importance: American shad, American eel (*Anguilla rostrata*) and sea lamprey (*Petromyzon marinus*), and also, blueback herring (*Alosa aestivalis*) and Atlantic salmon (*Salmo salar*). Long Island Native Americans were also known as skilled whalers, although it is not clear if pre-contact populations simply scavenged whales when they became stranded (Merwin 1993).

In the early 1600s, Dutch colonists settled around New York (New Amsterdam at the time) and along the Sound's shoreline, discovering abundant fish and mammals in the nursery areas of the numerous marshes and small tributaries (see Weingold and Pillsbury, Chap. 1, in this volume). Lucrative fur and fish trading enterprises were initiated, with the beaver (*Castor canadensis*) a prime target. Interactions between the first nations and colonists initially were peaceful, but war over these valuable resources was inevitable. The Dutch West Indies Company maintained control over much of the fish trade around the region, of which details are scant, until the English took over in 1664. Trade documents indicate that the Dutch imitated the native cultures' consumption of herring, possibly fueled by religious beliefs that limited meat in their diet, and an export monopoly of pickled herring that allowed the Dutch to enjoy substantial wealth (Matthews 1927). However, even though the primary fish species exported by the Dutch was herring, local consumption by the colonists was probably the same fish and shellfish species as native cultures enjoyed. Initially, dead or stranded whales were a coveted resource and provided revenue and materials for entire towns if found (Ross 1902).

Initially, much of the region was farmland and in order to boost production, fish were added to the fields in the fall. Atlantic menhaden (*Brevoortia tyrannus*) was of critical importance in this regard. Starting in the 1600s, menhaden were taken in small quantities from small bays, inlets, and river mouths in shallow water using simple gear (Hathaway 1910). While menhaden were primarily used as fertilizer here, they were consumed by poorer inhabitants in the southern US (Hathaway 1910). Menhaden were an important fertilizer on Long Island by the 1700s, with the claim that it doubled land value (Ross 1902). Macroalgae, seagrass, and marsh grasses were also used (Ross 1902), and mussels (blue and ribbed) were recorded as used for fertilizer in the early twentieth century (Matthews 1927). Menhaden form vast schools during migrations between northern feeding and summer spawning habitats, which could easily be landed by haul seines drawn ashore by horses. Millions of pounds were harvested every year.

During the 1650s, Long Island whalers became engaged in the expanding commercial rendering of oil such that between 1670 and 1850 whaling was considered "the great trade on the Long Island coast" (Ross 1902). Cold Spring Harbor had 9–10 ships engaged in whaling in 1843 although the number of whales harvested from LIS is unknown. By the late 1800s, whaling trips had become a multiyear effort, with ships sailing as far as the Pacific Ocean (Ross 1902). In 1850, the extraction of oil from menhaden was developed on an industrial scale. Until 1870, the menhaden harvest was relatively small, but the need for oil combined with the decline of whales created a new fishery that persists until the present time. The ease with which the menhaden could be caught with purse seines and quickly brought to nearby factories fueled a very lucrative fish oil industry that replaced whale oil as a critical Long Island commodity. In 1881, over 200 million fish were harvested and rendered into over a million gallons of fish oil in Southport alone (Ross 1902). From 1885 to 1910, annual US menhaden catches ranged from about a quarter million in 1892 to over one million pounds in 1903 (Hathaway 1910). In 1910, menhaden were still considered a coast-wide resource but had disappeared from the Gulf of Maine (Hathaway 1910). Modeling of the adjacent Great South Bay system demonstrates that they played a critical ecosystem role at the turn of the century in the Long Island region, both as a filter feeder and forage fish for top predators (Nuttall et al. 2011). Thus, over the past 100 years, the roles of both menhaden and top predators such as the sand tiger shark were dramatically reduced within inshore bays, and loss of migratory species significantly changed energy flows within and among food webs (Nuttall et al. 2011).

The past century saw the advent of industrialization and with it unprecedented human impacts on the shores and waters of the Sound. To date, 25–35 % of the Sound's tidal wetlands have been lost due to dredging, filling, and development (LISS 1994, 2003). The loss of wetlands increases from east to west, directly related to pressures of urbanization (LISS 2003). Increasing environmental perturbations related to industrial pollution began to have an effect on the Sound's animal populations. Poor water quality accelerated the mortality of the Sound's oysters such that the annual harvest declined from 10 million pounds in 1957 to less than 1 million pounds by 1960 (Bureau of Commercial Fisheries 1963).

Several diadromous species also declined, primarily due to industrial dams but also declining water quality (Limburg and Waldman 2009; Hall et al. 2011). There are over 1,000 dams in the Connecticut River watershed, although only six mainstem dams play a role in anadromous fish declines and restoration. Mainstem dams at Enfield in Connecticut, and Holyoke and Turner's Falls in Massachusetts embody historic fish passage issues. The first mill dams were erected throughout Atlantic coastal watersheds during the early 1600s; however, the technology required to span larger rivers was not available until the late 1700s. Dams provided power to industry but more importantly produced navigable waterways by bypassing flow through lock systems and increasing water depth above the dam sufficiently to allow transport of goods (Cumbler 1991).

The success and failure of fish passage projects since 1800 have ebbed and flowed in response to local litigation due to the effects of floods and spring freshets (Ducheney et al. 2006). In 1866, the Commonwealth of Massachusetts enacted legislation requiring free passage of shad and salmon on the Connecticut River. In 1873, a wooden fish ladder was constructed on the Holyoke Dam, but no fish used the ladder and in 1895 the Massachusetts legislature passed an act that exempted Holyoke Water Power from the need of maintaining a fishway. The current dam at Holyoke was built in 1900 and in 1938 the act exempting fishway maintenance was repealed. In 1940, a second ineffective fish ladder was constructed. In 1955, a fish elevator at Holyoke Dam provided some passage, which was further improved in the 1970s when a fish ladder was also constructed at Turners Falls. The Enfield Dam fell into disrepair in the late 1970s and is presently breached, thus allowing full passage, although many people are petitioning for its reconstruction.

Fish passage targets identified by the Connecticut River Atlantic Salmon Commission and required in the 1999 Holyoke Dam FERC (Federal Energy Regulatory Commission) license include one million American shad, one million blueback herring, 6000 Atlantic salmon, 500 shortnose sturgeon (*Acipenser brevirostrum*), and an unspecified number of American eel (Ducheney et al. 2006).

Salmon were considered plentiful in 1797, but few remained in 1808 and none by 1820 (Cumbler 1991). In 1849, seines could still catch over two thousand shad in a single sweep, but shad declined after 1820 and disappeared completely from the upper river in the nineteenth century. During the late 1800s, artificial propagation of shad became widespread. Between 1896 and 1910, rearing ponds provided 57 million shad fry stocked into the Connecticut River (Moss 1965). Seth Green, a pioneer in fish farming, was commissioned to hatch shad and in 1868 released an estimated 40 million shad fry into the Connecticut River (Moss 1965).

Restoring salmon runs have been far more difficult because Atlantic salmon were extirpated from state waters in 1820. A native stock had to be reconstructed using wild and farmed salmon from Maine and Canada that were released into the Connecticut River. Returning fish have contributed to the distinct Connecticut River strain since 1979, and hundreds of millions of young parr and smolt have been released into the river since 1970 (USASAC 2011). However, to date, the number of returning salmon captured at various river monitoring locations has never exceeded 530 (USFWS 2011).

6.7.1.1 Shellfish Aquaculture

Historically, natural oyster beds were a prominent feature along the coast of Connecticut, both shores of Long Island, all of Manhattan Island, and up the Hudson River as far as the saltwater extended (Matthews 1927). Long Island Sound was among the first areas to be heavily exploited for its wild shellfish resources; by the late 1700s, oyster catches were already declining (EOBRT 2007). Oyster Bay enacted fishing seasons for oysters in 1801 suggesting that supplies were being threatened, and in 1807 allowed ownership of bottom ground for the production of oysters (Ross 1902). Dredges were introduced in some areas in the early 1800s, and catches declined even more rapidly despite strong demand (Kurlansky 2006; EOBRT 2007). By 1855, natural oyster reefs had been reduced to such little value that they were used only for the seed they produced (Matthews 1927) and East River oystermen began to use clean shells to catch the seed of oysters, so that they could be transplanted for growth to market (Matthews 1927). During the same period, a fishery for bay scallops developed in the Sound, with a large fleet dredging the bottom during fall and winter months. Oyster production was limited due to low demand after peak landings in 1910 until 1938 when production was impacted by a major hurricane that covered many beds with sediment and destroyed harvest vessels (MacKenzie 1997). Beds were again destroyed by a hurricane in 1950 and inadequate supplies limited oyster production until the mid-1980s (MacKenzie 1997).

Investment in pollution abatement projects since the 1970s has aided a rebound of aquaculture in the Sound (LISS 2003). Production was also greatly increased by massive private and government cultch plantings during 1988–1991 (MacKenzie 1997). The majority of the oysters in the region are currently produced in Connecticut waters. From 1990 to 1996, an average of 6,43,000 bushels of eastern oysters and 1,29,000 bushels of hard clams were harvested annually from cultured beds in Connecticut waters alone, worth $26–49 million in market value annually (CT Department of Agriculture, Division of Aquaculture landings data). In 1997–1998, a disease die-off in the Sound's oyster beds caused the annual oyster harvest in Connecticut to drop from 5,26,000 bushels in 1996 to 56,000 in 2001 (CT Department of Agriculture, Division of Aquaculture landings data).

Eastern oysters display a wide range of survival strategies because they inhabit a naturally variable environment; they are both colonizers and ecosystem engineers and have a high reproductive potential (EOBRT 2007). In the years following the die-off, CT DOA Aquaculture Division staff and federal aquaculturists at the NOAA laboratory in Milford, CT worked with industry to introduce disease resistant strains of seed oysters (Rawson et al. 2010). Oyster harvest rates have slowly increased, but remain well below the peak harvests. In recent years, the industry has turned to hard clams to fill the gap left by the loss of the oyster harvest, as well as the loss of the American lobster (*Homarus americanus*) harvest, in WLIS. By 2007, nearly 50 million hard clams and over 13 million oysters were harvested in Connecticut. Shellfish aquaculture in LIS also concentrates on hard clams and oysters. Oyster restoration efforts are increasingly using cultch planting and reserves (EOBRT 2007). The problems of single-species aquaculture are now well recognized, given their susceptibility to catastrophic episodes of disease and predation. Multi-species operations have been recently tested, combining cultures of blue mussels, razor clams, macroalgae (e.g., *Porphyra* spp.), and bay scallops (Getchis 2005).

6.7.1.2 Lobster Trap Fishery

Until 1999, the wild harvest of American lobster in LIS was second only to shellfish aquaculture in total revenue. From 1880 to 1892, 0.6–0.7 million kg of lobster were landed annually and represented the highest state landings from Connecticut until 1983 (Blake and Smith 1984). Annual landings declined by 50–90 % through the 1950s (Blake and Smith 1984), and began to slowly increase in the late 1960s as a trawl fishery and subsequent trap fishery moved offshore and encountered abundant lobster in offshore grounds (General Dynamics 1968). By the late 1980s, trap landings were further bolstered with the replacement of the traditional wooden lathe trap with the vinyl clad wire trap which proved much more durable and efficient. The advent of this gear upgrade, and regulations prohibiting the targeted harvest of lobster by trawlers in the Sound due to the damage this gear causes to lobsters (Smith and Howell 1987), changed the fishery from principally a part-time or mixed-species endeavor to a full time single-species industry. By the late 1990s, landings from LIS peaked at 5.3 million kg, supporting over 1,000 license holders and generating $36 million in ex-vessel value (CTDEP 2000; NYS DEC unpublished data). Historically, 50–60 % of this harvest occurred in western Sound waters (CTDEP 2000). A massive die-off that occurred in the fall of 1998 and intensified in 1999 impacted the productive western Sound waters most severely with landings declining by more than a third in 1 year (CTDEP 2000). Individual ports suffered even greater losses, particularly in The Narrows, where landings in the port of Greenwich fell 70 %, while Norwalk declined 48 %, and Stamford dropped 43 % in 1999. A decade later, much of the lobster harvest activity was gone in these ports and the remnant lobster population was concentrated in the deeper waters of the central basin and The Race (see Sect. 6.8.4, Lobster Mortality Events). The rank of NY's lobster harvest in pounds and dollars has continued to decline since the die-off. Currently, lobster ranks ninth in dollars and sixteenth in pounds in seafood harvest in NY.

6.7.1.3 Finfish Trawl and Gill-net Fishery

Springtime signals American shad migration from the Sound up Connecticut's major rivers to spawn. The commercial shad fishery takes advantage of the migration "run" by tending drift gill-nets in the rivers' current to entrap the fish. This fishery peaked in the mid-1800s but continues today with a handful of drift gill-net fishermen. The Connecticut River shad population, the largest in the state, fell to only 16,000 spawning adults in 1967 when a federally coordinated restoration program was started by the US Fish and Wildlife Service. About 30 years later, after rigorous water quality programs were put in place and fishways allowed the fish to migrate past major dams to preferred spawning grounds, the population reached 300,000 spawning adults (USFWS 2011). Automated counters installed in the Holyoke Dam fishway now provide a quick means of monitoring the spawning population which has been stable at 156,000–168,000 fish from 2006 to 2010 (USFWS 2011) after the fishery harvested 30,000–50,000 (CTDEP MFIS 2010) fish for traditional springtime shad bakes.

Although landings from the trawl fishery earn less than a quarter of the annual revenues that come from aquaculture and the lobster fishery, trawl-caught finfish have been the main source of locally harvested seafood since the invention of coal-fired steam engines in the early twentieth century (CTDEP MFIS 2010). By the early 1900s, diverse net and long-line fisheries exploited the swift currents in The Race targeting bluefish, mackerel (*Scomber scombrus*), and menhaden (Ross 1902). It is difficult to precisely track harvest of individual finfish species solely from LIS because catches landed in LIS ports do not necessarily mean the fish were caught within its waters. Much of the Sound's harvest has been landed at out-of-state ports since the 1960s (Blake and Smith 1984), and harvest from elsewhere is routinely landed at ports around the Sound. However, landings data are an essential tool used by fisheries scientists and managers to track trends in harvest and provide evidence of ecosystem dynamics.

Connecticut DEP landing reports were highest in the mid-1990s and have declined over the past decade (Fig. 6.40; CTDEP MFIS 2010). Since the 1960s, commercial finfish landings have been dominated by summer and winter flounder, and these two species are also the most valuable (CTDEP MFIS 2010). Historically, winter flounder was also among the most abundant species in LIS,

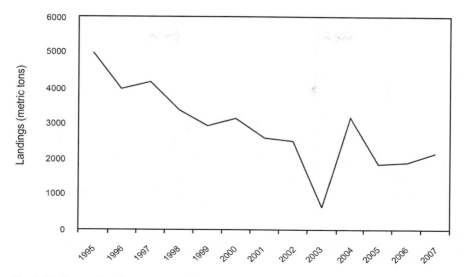

Fig. 6.40 Connecticut Department of Energy and Environmental Protection commercial fishery landings data, 1995–2007

but this species has experienced a drastic decline in both landings and value since the late 1990s (CTDEP MFIS 2010). Conversely, scup was traditionally considered of low value, but since 2003 demand and value of this species have increased (Fig. 6.41; CTDEP MFIS 2010).

The economic value of Connecticut commercial finfish and shellfish landings has been monitored by CTDEP since 1978. Data for Connecticut show that the ex-vessel value of total landings steadily increased from a few million dollars before 1980 to a record high value of over $61 million in 1992, boosted by increasingly large shellfish landings (CTDEP MFIS 2010). Despite declines in total weight harvested, the overall value of total landings has generally increased, providing over $30 million to harvesters annually since 2005 (CTDEP MFIS 2010). These figures represent only the value of the landings at the dock. The total economic impact of commercial fishing on the Sound's coastal communities is far greater.

6.7.1.4 Recreational Fisheries

Recreational fishing became an important pastime beginning in the late 1800s and into the early 1900s, as fishing and hunting clubs cropped up, targeting some 200 species of salt- and freshwater fishes (Ross 1902). In more recent years, recreational fishing activity in LIS has remained steady at around 1.5 million trips each year since 1981 (Fig. 6.42). Recreational landings peaked in the early 1980s and have fallen by nearly 50 %, likely due to a combination of regulatory restrictions and declines in some important resident species. The most popular species

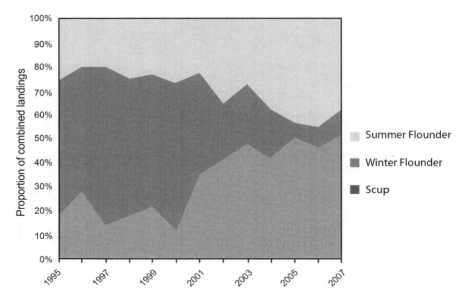

Fig. 6.41 Proportion of commercial landings for summer flounder, winter flounder and scup, 1995–2007

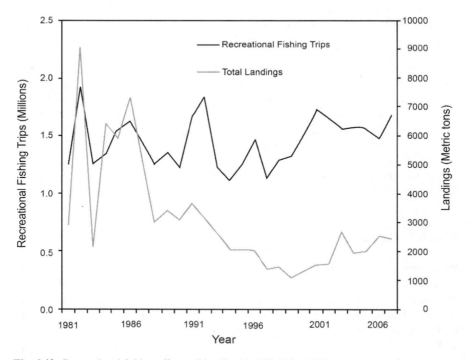

Fig. 6.42 Recreational fishing effort and landings in LIS 1981–2007

targeted in the Sound are bluefish, striped bass, scup, summer flounder, and tautog (*Tautoga onitis*, also commonly called blackfish). For several species, the reported catches from recreational fisheries are considerably greater than commercial landings. For striped bass and bluefish, in particular, the recreational fishing is almost an order of magnitude greater than the commercial landings, showing that these species primarily experience recreational harvest (MacLeod 2009). In recent years, anglers caught over 10 million fish of all species each year, but kept less than a third of their catch (MacLeod 2009; NYS DEC unpublished data). Altobello (1992) estimated that the economic value generated by recreational and commercial fishing activities in LIS contributed $1.2 billion per year to the regional economy.

6.7.1.5 Trends in Abundance, Diversity, and Community Structure

Components of the Sound's finfish community have been documented for centuries in commercial and recreational catch records as well as research studies, but the most comprehensive data set, in spatial coverage and duration, is the CTDEEP LIS Trawl Survey (LISTS). The survey has been conducted each spring and fall since 1984 and utilizes a stratified-random sampling design so all depths and bottom types are sampled proportionately to their area (Gottschall and Pacileo 2010). Forty 1 × 2 nautical mile sampling sites were chosen at random each month for a total of 200 sites per year. Sites are sampled using a 14 m otter trawl with a 51 mm cod end towed for 30 min. The survey is used to identify species distribution and abundance on spatial and temporal scales and provides the primary data for management and scientific investigations related to the Sound's finfishes and shellfish.

Seasonal Trends

The overall abundance of finfishes captured by the survey has remained relatively stable since 1984 with an increase observed in the fall and a decrease in the spring (Fig. 6.43). In the first 5 years (1984–1989) of the survey, catch rates in spring and fall were nearly identical; however, in the last 5 years (2003–2008), they were almost an order of magnitude different. Over the time series (1984–2008), the most commonly captured species in the fall were butterfish (*Peprilus triacanthus*), scup, long-finned squid (*Loligo pealei*), weakfish (*Cynoscion regalis*), bluefish, bay anchovy (*Anchoa mitchilli*), windowpane flounder (*Scophthalmus aquosus*), winter flounder, and little skate (*Leucoraja erinacea*) (Fig. 6.44a). Of the top ten finfish species, significant increases in abundance were observed for butterfish ($r = 0.4$, $p < = 0.042$), scup ($r = 0.6$, $p < = 0.001$), weakfish ($r = 0.6$, $p < = 0.001$), and anchovy ($r = 0.5$, $p < = 0.013$), while significant decreases were observed in abundance for windowpane flounder ($r = -0.5$, $p < = 0.005$) and winter flounder ($r = -0.5$, $p < = 0.022$). The top ten species captured in the spring included winter flounder, scup, windowpane flounder, American lobster, butterfish, little

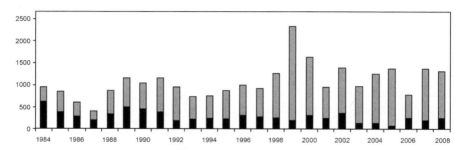

Fig. 6.43 LIS Trawl Survey mean abundance of finfish in spring (*gray*) and fall (*black*) cruises, 1984–2008

skate, long-finned squid, Atlantic herring (*Clupea harengus*), fourspot flounder, and red hake (*Urophycis chuss*) (Fig. 6.44b). Of the ten most abundant finfish species increasing trends were observed in scup ($r = 0.5$, $p < = 0.009$) and butterfish ($r = 0.54, p < = 0.005$), while decreasing trends were observed for winter flounder ($r = -0.7, p < = 0.001$), windowpane flounder ($r = -0.8, p <= 0.001$), little skate ($r = -0.4, p <= 0.033$), fourspot flounder ($r = -0.6, p < = 0.001$), and red hake ($r = -0.4, p < = 0.037$).

In addition to changes in individual species, large-scale changes have been observed in multi-species groups representing different habitat use and ecological guilds. All of the finfish species captured by the survey (Tables 6.5 and 6.6) were grouped into species that are bottom tending (epibenthic, $n = 36$), species that prefer the upper water column (pelagic, $n = 35$), and species that utilize both benthic and pelagic habitats (demersal, $n = 24$). Spring catches have been dominated by epibenthic species, although their abundance has declined significantly since 1984 ($r^2 = 0.67$, $p < 0.001$; Fig. 6.45). Over the same period, the abundance of demersal species significantly increased ($r^2 = 0.24$, $p = 0.007$) though with large variance. The abundance of pelagic species showed no trend. However, the bottom tending gear used in the survey may have under-sampled these species and thus reduced the statistical power to detect trends. Fall abundance data showed a greater increase in demersal species ($r^2 = 0.44$, $p < 0.001$) than is seen in spring catches, while fall abundance of the other two groups showed no trend. Epibenthic species continue to be about twice as abundant in the survey as demersal and pelagic species.

In order to identify important sub-groups in the trends observed for demersal, epibenthic, and pelagic species, the abundance trends of flatfish, skates, and shellfish were estimated. The aggregate abundance trends for the 30 most common species were analyzed for the following groups: demersal (scup, silver hake (*Merluccius bilinearis*)), tautog, spot (*Leiostomus xanthurus*), black sea bass ((*Centropristes striata*), weakfish), skate (little skate and winter skate (*Leucoraja ocellata*)), shellfish (American lobster), epibenthic (striped sea robin (*Prionotus evolans*)), red hake, northern sea robin (*Prionotus carolinus*, also commonly called gurnard), and flatfish (windowpane flounder, winter flounder, fourspot

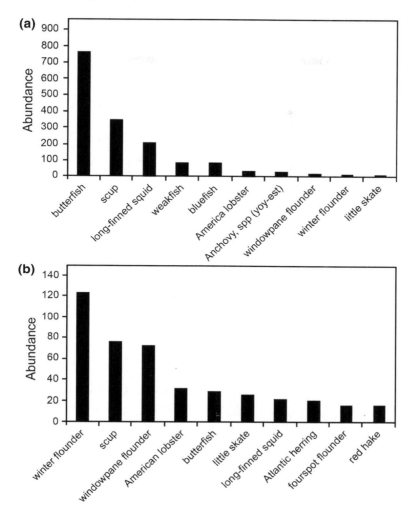

Fig. 6.44 Top ten most abundant species captured in the fall (**a**) and spring (**b**) LIS trawl survey. *Note* Of shellfish only American lobster has been reported for the duration of the survey so the relative order of the top ten species is biased towards finfish

flounder, hogchoker (*Trinectes maculatus*)), summer flounder, smallmouth flounder (*Etropus microstomus*). In both seasons flatfishes and skate have declined, while demersal species have increased, especially in the fall (Fig. 6.46). The increase in demersal species is driven largely by scup, while decreases in flatfishes are driven by winter flounder, windowpane founder, and fourspot flounder. Epibenthic abundance decreased in the fall and increased in the spring, while American lobster increased in the 1980s until a decline in the 1990s.

Table 6.5 Definitions of the T adaptation and depth groups used to classify finfish species captured in Long Island Sound Trawl Survey

Adaptation groups

Adapted to cold temperate waters

More abundant north of Cape Cod, MA than south of New York

Behaviorally adapted to cold temperatures, including subfreezing

Preferred T range approximately 3–15 °C

Spawns at lower end of T tolerance

Adapted to warm temperate waters

More abundant south of New York than north of Cape Cod, MA

Behaviorally avoids temperatures <10 °C

Preferred T range approximately 11–22 °C

Spawns at higher end of T tolerance

Adapted to subtropical/tropical waters

Rare north of Chesapeake and occasional or rare in the mid-Atlantic

Strays captured north of mid-Atlantic are usually juveniles

Not tolerant of temperatures <10 °C

Spawns only south of New York Bight

Depth groups

Epibenthic (E)

Found exclusively or almost exclusively on the bottom

Feeds almost entirely on benthic prey

If fished, taken only by bottom gear such as otter trawl nets

Demersal (D)

Associated with the bottom or bottom structure but may use water column occasionally

Feeds primarily on bottom organisms; uses bottom sediments in reproduction

If fished, taken primarily by bottom tending gear or mid-water gear such as pound nets

Pelagic (P)

Uses the entire water column or primarily surface waters; eggs and larvae develop entirely in surface waters

Feeds primarily on surface prey or a mix of benthic/surface prey

If fished, taken primarily by off-bottom or surface gear such as drift gill-nets or long lines

Classifications are based on information taken from Collette and Klein-MacPhee (2002) and Murdy et al. (1997)

The decline in spring abundance was not accompanied by a decline in overall finfish species diversity (Fig. 6.47). Diversity in spring catches has remained fairly steady over the 25-year time series, averaging about 11 species per sample, while diversity in fall catches has increased, rising from about 11 species per sample to 13.

Spatial Trends

The LISTS survey is designed to randomly sample 12 spatial habitats: three bottom types (sand, mud, and transitional between the two) at four 9.1 m water depths. Average catches over the time series among these 12 habitat types ranged from 27 fish per sample at deep sand sites, primarily found mid-Sound west of the

Table 6.6 Classification of 95 species of finfish captured in the LIS Trawl Survey by adaptation and depth group

Common name	Scientific name	Adaptation group	Depth group
alewife	*Alosa pseudoharengus*	Cold	P
American plaice	*Hippoglossoides platessoides*	Cold	E
Atlantic Herring	*Clupea harengus*	Cold	P
Atlantic Sturgeon	*Acipenser oxyrinchus*	Cold	D
barndoor skate	*Dipturus laevis*	Cold	E
Atlantic cod	*Gadus morhua*	Cold	D
cunner	*Tautogolabrus adspersus*	Cold	D
fawn cusk-eel	*Lepophidium profundorum*	Cold	E
fourspot flounder	*Hippoglossina oblonga*	Cold	E
goosefish (monkfish)	*Lophius americanus*	Cold	E
grubby	*Myoxocephalus aeneus*	Cold	E
haddock	*Melanogrammus aeglefinus*	Cold	D
little skate	*Leucoraj erinacea*	Cold	E
lumpfish	*Cyclopterus lumpus*	Cold	E
Atlantic mackerel	*Scomber scombrus*	Cold	P
ocean pout	*Zoarces americanus*	Cold	E
northern pipefish	*Syngnathus fuscus*	Cold	E
pollock	*Pollachius virens*	Cold	P
rockling	*Enchelyopus cimbrius*	Cold	E
red hake	*Urophycis chuss*	Cold	E
rock gunnel	*Pholis gunnellus*	Cold	E
rainbow smelt	*Osmerus mordax*	Cold	P
Atlantic salmon	*Salmo salar*	Cold	P
longhorn sculpin	*Myoxocephalus octodecemspinosus*	Cold	E
spiney dogfish	*Squalus acanthias*	Cold	P
searaven	*Hemitripterus americanus*	Cold	E
Atlantic seasnail	*Liparis atlanticus*	Cold	E
Atlantic tomcod	*Microgadus tomcod*	Cold	D
winter flounder	*Pseudopleuronectes americanus*	Cold	P
whiting (silver hake)	*Merluccius bilinearis*	Cold	D
windowpane	*Scophthalmus aquosus*	Cold	E
winter skate	*Leucoraja ocellata*	Cold	E
yellowtail flounder	*Limanda ferruginea*	Cold	E
American shad	*Alosa sapidissima*	Warm	P
Atlantic silverssides	*Menidia menidia*	Warm	P
blueback herring	*Alosa aestivalis*	Warm	P
tautog (blackfish)	*Tautoga onitis*	Warm	D
bluefish	*Pomatomus saltatrix*	Warm	P
Atlantic bontio	*Sarda sarda*	Warm	P
black seabass	*Centropristis striata*	Warm	D
butterfish	*Peprilus triacanthus*	Warm	P
clearnose skate	*Raja eglanteria*	Warm	E
conger eel	*Conger oceanicus*	Warm	E
Atlantic croaker	*Micropogonias undulatus*	Warm	D
American eel	*Anguilla rostrata*	Warm	E
gizzard shad	*Dorosoma cepedianum*	Warm	P
hogchoker	*Trinectes maculatus*	Warm	E

(continued)

Table 6.6 (continued)

Common name	Scientific name	Adaptation group	Depth group
hickory shad	*Alosa mediocris*	Warm	P
menhaden	*Brevoortia tyrannus*	Warm	P
northern kingfish	*Menticirrhus saxatilis*	Warm	D
naked goby	*Gobiosoma bosci*	Warm	E
northern searobin	*Prionotus carolinus*	Warm	E
scup (porgy)	*Stenotomus chrysops*	Warm	D
northern puffer	*Sphoeroides maculatus*	Warm	E
striped cusk-eel	*Ophidion marginatum*	Warm	E
lined sea horse	*Hippocampus erectus*	Warm	E
summer flounder	*Paralichthys dentatus*	Warm	E
sea lamprey	*Petromyzon marinus*	Warm	D
smooth dogfish	*Mustelus canis*	Warm	D
smallmouth flounder	*Etropus microstomus*	Warm	E
spotted hake	*Urophycis regia*	Warm	E
spot	*Leiostomus xanthurus*	Warm	D
stripped searobin	*Prionotus evolans*	Warm	E
striped bass	*Morone saxatilis*	Warm	P
oyster toadfish	*Opsanus tau*	Warm	E
white pearch	*Morone americana*	Warm	D
weakfish	*Cynoscion regalis*	Warm	D
bigeye scad	*Selar crumenophthalmus*	Sub-tropical	P
bigeye	*Priacanthus arenatus*	Sub-tropical	P
crevalle jack	*Caranx hippos*	Sub-tropical	P
planehead filefish	*Monacanthus hispidus*	Sub-tropical	P
Lizardfish	*Synodus foetens*	Sub-tropical	D
lookdown	*Selene vomer*	Sub-tropical	P
Atlantic moonfish	*Selene setapinnis*	Sub-tropical	P
northern sennet	*Sphyraena borealis*	Sub-tropical	P
orange filefish	*Aluterus schoepfi*	Sub-tropical	P
Atlantic round herring	*Etrumeus teres*	Sub-tropical	P
roughtail stringray	*Dasyatis centroura*	Sub-tropical	E
banded rudderfish	*Seriola zonata*	Sub-tropical	D
rough scad	*Trachurus lathami*	Sub-tropical	P
sanbar shark	*Carcharhinus plumbeus*	Sub-tropical	D
sharksucker	*Echeneis naucrates*	Sub-tropical	P
spanish mackerel	*Scomberomorus maculatus*	Sub-tropical	P
African pompano	*Alectis ciliaris*	Tropical	P
blue runner	*Caranx crysos*	Tropical	P
dwarf goatfish	*Upeneus parvus*	Tropical	D
glasseye snapper	*Priacanthus cruentatus*	Tropical	D
gray triggerfish	*Balistes capriscus*	Tropical	D
mackerel scad	*Decapterus macarellus*	Tropical	P
red connetfish	*Fistularia petimba*	Tropical	P
round scad	*Decapterus punctatus*	Tropical	P
red goatfish	*Mullus auratus*	Tropical	D
short bigeye	*Pristigenys alta*	Tropical	P
striped burrfish	*Chilomycterus schoephi*	Tropical	D
yellow jack	*Caranx bartholomaei*	Tropical	P

See Table 6.5 for complete definitions

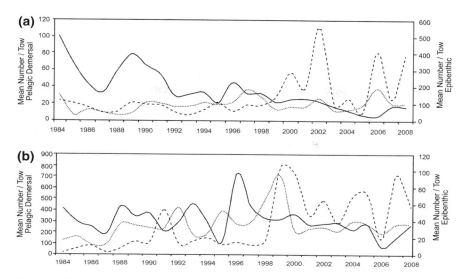

Fig. 6.45 LIS Trawl Survey mean finfish abundance by habitat type in spring (**a**) and fall (**b**) cruises, 1984–2008. Epibenthic (*solid line*), demersal (*dashed line*), and pelagic (*dotted line*) species trends are shown. The negative trend in epibenthic species abundance is significant in spring (r-square = 0.67, $p < 0.001$). The positive trend in demersal species abundance is significant in spring (r-square = 0.24, $p = 0.007$) and fall (r-square 0.44, $p < 0.001$)

Connecticut River mouth, upward to an average of 99 fish per sample at mud sites in the western basin at mid-depths of 20–30 m. This spatial pattern has remained fairly constant over the 25-year time series.

Although the spatial distribution of the finfish community has not changed, research trawl catch data indicate individual species shifts and a general trend toward decreased flatfish abundance that appears to have stabilized after 1999. A multivariate analysis was completed that only considers species that were caught in over 40 % of tows in order to remove the influence of highly migratory species whose affinity to a specific habitat or species group cannot be well supported (see Jordaan et al. 2010). Bootstrapping analysis and principal component analysis (PCA) were applied using a correlation matrix, with 16 and 17 species in the analysis of spring and fall data, respectively. The three principal components (PCs) account for 48 and 45 %, and the top five PCs explain 63 and 58 % of the total variance for spring and fall survey data, respectively. PCA scores for individual sites were then interpolated using inverse-distance weighting and contrasting species groups shaded in light (negative scores outside one standard deviation) and dark (positive scores outside one standard deviation). Thus, species groups can be visualized as three sets (PC1, PC2, and PC3) of contrasting species groups (Fig. 6.50). A further challenge is that many species are at historic low abundances and have undergone further rapid population declines during the time period of the LISTS survey (Limburg and Waldman 2009). The results indicate increasing habitat differentiation and organization from spring to fall (Fig. 6.50), which is shared

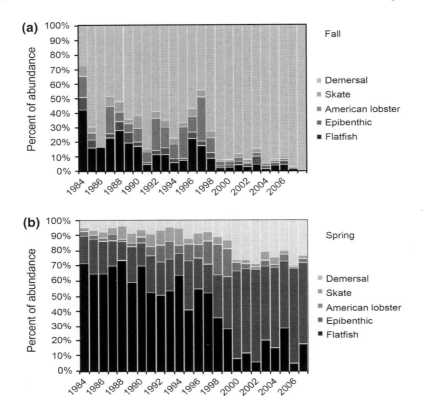

Fig. 6.46 Trends in the percent abundance of the top 30 most abundant species captured in the LIS trawl survey grouped by demersal, skate, American lobster, epibenthic and flatfish. *Note* American lobster was the only shellfish reported because it has been consistently recorded for the entire duration of the survey

by other temperate systems (Jordaan et al. 2011). Temperate regions face a fall-winter overturn, forced by cooling surface waters and storms that deconstruct ecological structure followed by an annual re-organization.

These research trawl catch data can also be used to track individual species habitat shifts using multivariate (principal component) analysis (see Jordaan et al. 2010). Finfish species caught in at least 40 % of survey tows were analyzed using this approach and results indicate increasing habitat differentiation and organization from spring to fall (Fig. 6.48), a result shared by other temperate systems (Jordaan et al. 2011). Temperate regions face a fall-winter overturn, forced by cooling surface waters relative bottom T, and storms/ice conditions that deconstruct ecological structure followed by an annual re-organization. Biological activity often peaks with reproduction and population expansion during spring blooms, and this is followed by a summer growth season. In the fall, species spatially segregate into western, west-central, east-central, and eastern groups, and also along

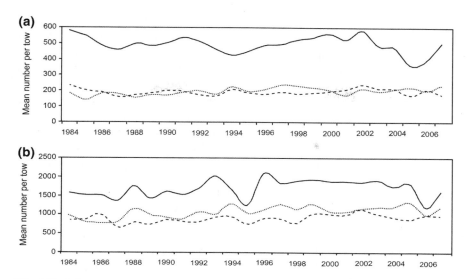

Fig. 6.47 LIS Trawl Survey finfish diversity by habitat type in spring (**a**) and fall (**b**) cruises, 1984–2008. Epibenthic (*solid line*), demersal (*dashed line*), and pelagic (*dotted line*) species trends are shown

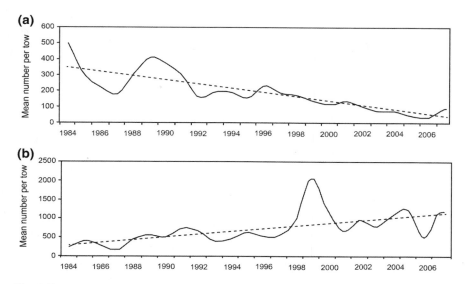

Fig. 6.48 LIS Trawl Survey abundance of cold-adapted finfish in spring (**a**) and warm-adapted finfish in fall (**b**) cruises, 1984–2008. Significant regression slopes (*dashed lines*) are shown; cold-adapted r-square = 0.49 and warm-adapted r-square = 0.60; $p < 0.001$

depth contours, to create a mosaic of assemblage groups (Fig. 6.48). The ecological structure identified is consistent with expectations that the westernmost Sound experiences lower oxygen levels and other environmental characteristics that differ from the central basin, a remnant of the post-glaciation lake. The Race region in the easternmost Sound, in contrast, experiences higher water velocity and less anthropogenic impacts other than fishing pressure. The ecological structures of both the eastern and western regions do not appear to relate to depth, suggesting more homogeneous habitat characteristics than the central regions. These results do not include analyses of species dynamics such as predator–prey relationships and their interaction with climate and fishing. A full ecosystem modeling exercise would better elucidate key ecological processes.

Community assemblage structure and species distributional changes have been linked to fishing pressure (Fogarty and Murawski 1998), density-dependent waves (Fauchald et al. 2006), food availability (Olsson et al. 2006), and environmental drivers that impact the condition of individuals (Lucey and Nye 2010; Nye et al. 2009; Frisk et al. 2008, 2010). Hypoxia has been linked to decreased species abundance in the Sound, with longfin squid and bluefish showing the most sensitivity (Howell and Simpson 1994). Variation of T can cause changes in stratification, hypoxia duration and severity, disease prevalence, and other indirect effects. Trends in the LISTS show that cold-adapted species have declined in abundance, while warm-adapted species have increased, suggesting that warming temperatures may be altering the LIS finfish community. The decrease of cold-adapted species has been especially dramatic during the spring (Figs. 6.48 and 6.49). In contrast, warm-adapted and southern migrant species adapted to sub-tropical and tropical conditions have increased.

6.7.1.6 Science Gaps and Management Implications

Long Island Sound is a productive marine ecosystem that has provided an abundance of resources to the suburban and urban populations of Connecticut and New York for over two centuries and to first nation people for 12,000 years. Available data suggest that the system is characterized by species switches important to local fisheries

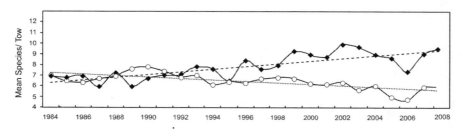

Fig. 6.49 LIS Trawl Survey mean diversity of cold-adapted finfish in spring (*open circles*) and warm-adapted finfish in fall (*solid diamonds*), 1984–2008. Significant regression slopes (*dashed lines*) are shown; cold-adapted r-square = 0.49 and warm-adapted r-square = 0.60; $p < 0.001$

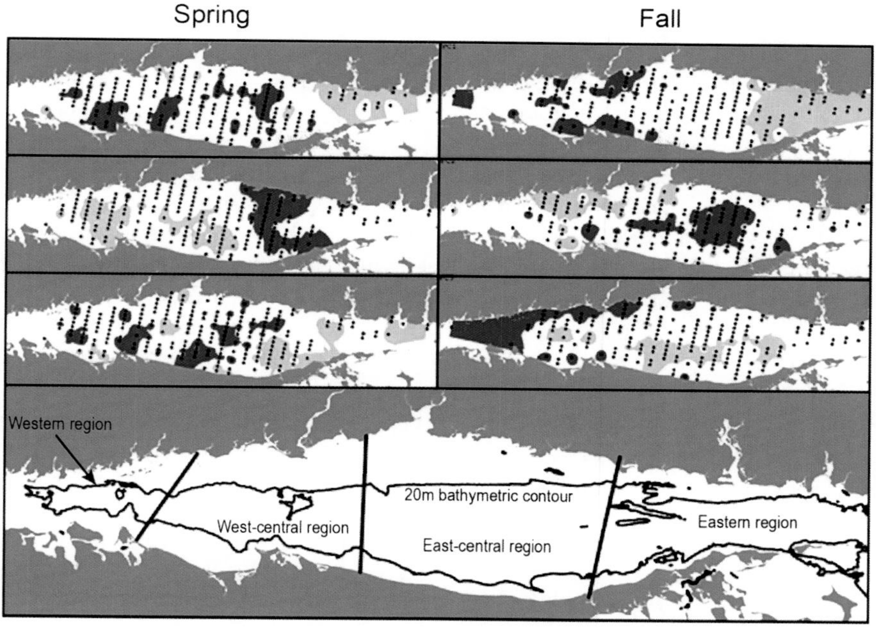

Fig. 6.50 Results of a multivariate analysis of the Long Island Sound Trawl Survey (LISTS) data. Divisions in species groups fall along bathymetric and inner and outer Sound gradients. PCA scores for individual sites were interpolated using inverse-distance weighting; sites *shaded* in *light* had negative scores outside one standard deviation, while sites *shaded* in *dark* had positive scores outside one standard deviation. Sites with scores within one standard deviation are *not shaded*. Thus, species groups can be visualized as three sets (PC1, PC2 and PC3) of contrasting species groups for the spring and fall data sets. *Dots* represent LISTS station locations. An overall structure map is shown with bathymetry marked at the 20 m contour

and shifts in finfish community composition. Perhaps the first major shift occurred with the over-harvest of whales and other marine mammals in the eighteenth century, followed by the dramatic decline in anadromous species in the nineteenth century, and the early twentieth century coast wide over-harvest of Atlantic menhaden. More recently, LIS has seen dramatic declines in winter flounder and American lobster and increases in warm-adapted species. Climate change and hypoxia have been proposed as modern drivers of change, suggesting that the twenty-first century will also be characterized by large-scale biological changes in LIS.

The LIS finfish community has responded to a warming trend by shifting from one in which cold-adapted species are numerically dominant, particularly in spring, to one in which warm-adapted species are increasingly abundant, especially in fall. Most of these warm-adapted species have historic ranges centered off the mid-Atlantic and a northward shift of these populations to LIS is consistent with trends and forecasts at larger biogeographic scales (Murawski 1993; Nye et al. 2009; Hare et al. 2010). Temperature has been identified as an important factor in reducing the contribution of cold-adapted species and increasing the role of warm-adapted species in LIS. However, causes of distributional trends in finfish populations can come from many sources and be complex combinations of variables. Mechanisms of community change in LIS remain elusive as a result of a paucity of finfish studies.

Additional mechanisms need to be explored in the Sound to gain insight into the shifts in abundance observed in the survey and in fisheries data. Long Island Sound has a long history of seafood harvest that has been well documented as a driver of habitat and community change in the western Atlantic (Fogarty and Murawski 1998; Frisk et al. 2011a). The relationship between fishery harvest and system change can be direct, such as the removal of traditional groundfishes in the western Atlantic, but can also be indirect and difficult to detect. For example, Atlantic menhaden has recently been identified as a keystone species in the nineteenth century Great South Bay, New York ecosystem (Nuttall et al. 2011). The loss of this keystone species in addition to several top predators marked the beginning of a century-long decline in system maturity of Great South Bay (Nuttall et al. 2011).

Deciphering community and species shifts in temporal and spatial distribution through the use of seasonal surveys, conducted annually during relatively fixed periods, is a complex and difficult challenge (Dunton et al. 2010; Frisk et al. 2008). The Sound has diverse finfish fauna (Briggs and Waldman 2002) that exhibit a wide range of migration behaviors (Latham 1917). Seasonal variation in the abundance of species collected in surveys can be a function of changes in timing of migration due to environmental factors and decadal patterns in species distribution (Dunton et al. 2010; Frisk et al. 2008). Ecosystem modeling exercises using time-varying parameters such as have been applied to Chesapeake Bay and Delaware Bay (Christensen et al. 2009; Frisk et al. 2011b) would be instructive in relating trends observed in surveys and harvest to the biology of the Sound's natural resources.

6.7.2 Wildlife: Marine Reptiles, Birds, and Mammals

6.7.2.1 Marine Reptiles

Sea turtles have been reported in the northeast states including LIS since the beginning of the last century (Murphy 1916; Babcock 1919; DeSola 1931). Surveys for abundance and distribution of sea turtles in LIS have not been conducted with any consistency. The lack of survey data has fostered inaccurate assumptions related to the presence or absence of sea turtles in LIS. Information on animals inhabiting LIS has come from stranding data, opportunistic sightings reported by mariners, and the general public (citizen scientists) (DiGiovanni et al. 2000). These sightings have been supplemented by the occurrence of stranded animals alive and dead over the last 3 decades. These data were used to assess the presence of animals in the LIS. Few *in situ* tracking and diet studies were conducted in the last 2 decades, limiting our ability to determine habitat usage in these areas.

The only tracking data for sea turtles are from studies conducted in the Peconic Bay Estuary during the late 1980s and from animals rehabilitated and satellite tagged by the Riverhead Foundation for Marine Research and Preservation (RFMRP) starting in 2002. Satellite tracking of rehabilitated animals has revealed that sea turtles are using the waters around Long Island on a regular basis (DiGiovanni et al. 2009).

Long Island Sound is a seasonal foraging habitat to three species of sea turtles: Loggerhead (*Caretta caretta*), Kemp's ridley (*Lepidochelys kempii*), and Atlantic green (*Chelonia mydas*). These animals enter the coastal waters of the Northwest Atlantic after spending their early years surviving in near surface waters of the North Atlantic Gyre and then move into the coastal environment of the northwest Atlantic. It is in these bays and estuaries that they begin to develop their benthic lifestyle (Morreale and Standora 2005).

Few studies have addressed the extent to which sea turtles use LIS as a habitat. Diet studies on sea turtles were conducted during the late 1980s and early 1990s by Burke and co-workers on sea turtles captured in the Peconic Estuary and life history and habitat data were collected (Morreale and Standora 1992). These studies revealed a diet of spider crabs (*Libinia emarginata*) and green crabs for Kemp's ridley sea turtles (Burke et al. 1994) and for loggerhead sea turtles (Burke et al. 1993, 1994). The Atlantic green sea turtle is an omnivore during its early developmental years, but as it enters the coastal habitat of the western North Atlantic it becomes more of an herbivore, feeding on green fleece and brown algae. Nesting of sea turtles has not been reported in LIS for any of the species encountered.

Sea turtles begin to show up in the eastern bays and estuaries during late spring or summer. They remain in the area throughout the summer beginning their southward or offshore migration from New York waters in mid-October. If these animals do not leave early enough in the fall, they can become cold stunned (hypothermic) and wash up on the beaches around New York and New England.

Cold stunning of sea turtles is a major problem for the Kemp's ridley, loggerhead, and green sea turtles. These animals wash up on north-facing beaches of LIS (Morreale et al. 1992; Burke et al. 1991). Historically, the sea turtle most affected by cold stunning is the Kemp's ridley. Reports from the early years of the last century described cold-stunned animals washing up in LIS (Latham 1969). They are thought to have been Kemp's ridley sea turtles. In recent years, cold-stunned Atlantic green sea turtles are becoming more prevalent (DiGiovanni unpublished data). This shift could be attributed to more animals visiting the waters of eastern Long Island.

6.7.2.2 Birds

Long Island Sound is important for three main groups of birds: those found at sea, on coastal beaches and mudflats, and in tidal marshes. Systematic surveys of offshore birds are largely lacking, and with just a few exceptions, comprehensive information on abundance and distribution is limited. More is known about coastal species and detailed studies have been conducted on several beach-nesting and tidal marsh birds.

Most open water species, such as waterfowl, loons, and grebes, are common during non-breeding periods, with few lingering into summer. Congregations of several thousand greater scaup (*Aythya marila*, also commonly known as blue-bill) are common during winter, especially toward the western end of the Sound. Detailed studies of their diets (Cronan 1957) and contaminant levels (Cohen et al. 2000) have been conducted. White-winged scoter (*Melanitta fusca*) *and* brant (*Branta bernicla*) also occur in large flocks, especially during migration, but most other sea ducks are less numerous. Common and red-throated loons (*Gavia immer* and *G. stellata* respectively), and horned grebes (*Podiceps auritus*) are moderately prevalent along the Sound's coasts from fall until spring.

Great cormorants (*Phalacrocorax carbo*) *also* occur primarily during winter, but the closely related double-crested cormorant (*Phalacrocorax auritus*) occurs year-round, with abundance peaks during spring and fall migration and a growing breeding population. Concerns have been raised over the adverse effects of cormorant increases on their prey, but a study of alewives suggests that cormorant-caused mortality is largely compensatory and is unlikely to impact populations at a regional scale (Dalton et al. 2009).

Although gulls occur year-round, both abundance and variety are greatest from fall until spring. Large congregations occur from late March-early April, when thousands of migrant gulls feed on offshore plankton. These flocks are dominated by ring-billed, herring, and Bonaparte's gulls (*Chroicocephalus philadelphia*), and several other species as well. Herring and greater black-backed gulls (*L. marinus*) both nest in small numbers along the coast. Ring-billed gulls over-summer, and there is an influx of laughing gulls (*Leucophaeus atricilla*) from farther south in late summer. Several species of terns summer in LIS. The federally endangered roseate tern (*Sterna dougallii*) nests in small numbers on Faulkner's Island and in much larger numbers on Great Gull Island (Spendelow et al. 1995). At both sites, the colonies are dominated by greater numbers of common terns (*S. hirundo*), which also nest in small numbers elsewhere along the coast.

Pelagic species are much less common, although some—e.g., northern gannets (*Morus bassanus*) and razorbills (*Alca torda*)—seem to be using the Sound more often than previously and have become regular non-breeding visitors, especially in ELIS. During summer, small numbers of Wilson's storm-petrels (*Oceanites oceanicus*), which breed in the southern hemisphere during the austral summer, can also be found.

The Sound's shoreline is important for a suite of beach-nesting birds, and for non-breeding shorebirds and gulls. Among the nesting species, piping plovers are perhaps the best studied, and the species of greatest conservation concern. Federally listed under the US Endangered Species Act, this species and least terns are the focus of management programs to fence nesting areas and discourage human disturbance. Other beach-nesting species include American oystercatcher

(*Haematopus palliatus*) and black skimmers (*Rynchops niger*), both of which appear to be increasing. The primary threats to all of these species are trampling of the eggs by people and elevated predation levels due to human activities (Warnock et al. 2001). Nest flooding on high spring tides adds a more natural cause of breeding failure, albeit one that is likely exacerbated by the reduced area of suitable nesting habitat due to coastal development, and that is likely to become a growing problem with rising sea levels.

Locally non-breeding shorebirds, especially sanderling (*Calidris alba*), forage on beaches seeking prey in wet sand as each wave retreats. Other species, including purple sandpipers (*C.maritima*) and ruddy turnstones (*Arenaria interpres*), use rocky shores and jetties where they eat mollusks, crustaceans, and the like. At high tide, these species are joined by a variety of other shorebirds displaced from the mudflats where they forage by probing wet mud for invertebrates. High tide roosts—such as those that form on the sand bars at Milford Point, CT—can contain thousands of semipalmated sandpipers (*Calidris pusilla*) and numerous other species. Although these roost sites are used only for short periods of the day, the scarcity of disturbance-free sites can make them especially important. The shorebird concentrations at such sites also make them a magnet for predators such as peregrine falcons (*Falco peregrinus*) and merlins (*F. columbarius*).

The last group of birds that depend on LIS's habitats are found in tidal marshes. The primary breeding species include American black ducks (*Anas rubripes*), clapper rails (*Rallus longirostris*), willets (*Tringa semipalmata*), and both seaside and saltmarsh sparrows. Several species of herons, egrets, as well as glossy ibis (*Plegadis falcinellus*), which mostly nest in colonies on nearshore islands, fly to marshes to forage on small fishes and invertebrates. Breeding ospreys (*Pandion haliaetus*) reverse this commute, nesting primarily on human-built platforms placed in coastal marshes, but flying offshore to hunt for fish. During migration, various other species of shorebirds and ducks are also found.

Historically, habitat for all these birds has declined as tidal marshes was developed and modified through mosquito ditching, tidal restriction, upland development, and invasion by the introduced common reed (Bertness et al. 2009; Crain et al. 2009). Tidal marsh habitats today are better protected and considerable restoration work has occurred (Warren et al. 2002). Nonetheless, the rising sea levels associated with a warming climate poses a new threat with potentially dire consequences. The saltmarsh sparrow, for example, is already prone to frequent nest flooding and even small increases in relative sea level could seriously impact reproduction and increase extinction risk (Bayard and Elphick 2011). Over time, all species that nest in the high marsh are likely to experience increased nest flooding and ultimately most tidal marsh habitat is projected to convert to either low marsh or mudflat (Hoover 2009).

6.7.2.3 Mammals

Historically, numerous marine mammals have used LIS throughout the seasons. These include two groups: cetaceans and pinnipeds. Systematic surveys throughout LIS for marine mammals have not been conducted annually. The only continuous monitoring of marine mammals is done during the winter and spring, and focuses on pinnipeds. These surveys, conducted by the Riverhead Foundation for Marine Research and Preservation since 1997, have documented an increase in harbor seals (*Phoca vitulina*) and gray seals (*Halichoerus grypus*) in the waters around Long Island (DiGiovanni et al. 2009). The peak season for seals in New York waters is winter, with animals arriving from northern waters of New England and Canada in the fall and leaving in late spring. Recent data suggest that this season is expanding and more animals are staying in the area year-round. Pinniped surveys conducted over the last 2 decades have revealed an increase in seal population from hundreds of animals in the early 1990s to thousands during the last count during the winter of 2011.

Due to the lack of survey data for cetaceans, data collected via strandings, opportunistic sightings, and the NOAA Sighting Advisory System for North Atlantic right whales (*Eubalaena glacialis*) have been used as baseline. Only if animals were encountered on a regular basis were they thought to frequent LIS. Cetaceans are encountered throughout the year and include bottlenose dolphins (*Tursiops truncatus*), common dolphins (*Delphinus delphis*), harbor porpoise (*Phocoena phocoena*), and North Atlantic right whales. The North Atlantic right whale has been encountered in the eastern portion of LIS. Harbor porpoise have been recovered throughout LIS by the rescue program and sightings have occurred as far west as the East River and Flushing Bay. These reports were collected by the Riverhead Foundation's Sighting Program and verified by photos or by biologists from the Foundation. Bottlenose dolphins were reported in WLIS in Huntington Harbor in 2009. This sighting event identified over 100 animals and lasted more than a week.

As these populations were reduced through natural and anthropogenic causes, their encounter rates have decreased. Since the introduction of the Marine Mammal Protection Act in 1972 we have started to see an increase in occurrences throughout LIS, from Sheffield Island, Pelham Bay Park, and the north shore of Long Island in the west to Little Gull Island to the east. Over the last 2 decades, harbor seal abundance has increased in these areas. Since 2001, gray seals have been seen on western Long Island haul-out sites and have increased over the last decade (DiGiovanni et al. 2009).

6.8 Cross-Cutting Themes and Summary

In this section we examine some of the environmental problems facing the biota of LIS. These issues, including the development of seasonal hypoxia, invasions of non-native species, biological changes due to regional patterns of warming and increased CO_2, and a catastrophic die-off of an ecologically and economically important species, are complex and multi-faceted. These are clearly not the only challenges facing the biota of LIS, but the nexus of their scientific complexity and their economic and management significance highlights their importance.

6.8.1 Hypoxia: Biological Contributions and Ecological Consequences

6.8.1.1 Biological Contributions to Hypoxia: Processes

Significant portions of nearshore marine regions, including LIS, experience persistent seasonal hypoxia, changing the structure and function of marine communities and subsequently altering nutrient cycles, reducing energy flow to higher trophic levels, and compressing habitats (Rabalais and Turner 2001; Diaz and Rosenberg 2008; Stramma et al. 2008). The expanse and dynamics of persistent oxygen minimum zones (OMZs) are closely related to human activity and climate change (Hendy and Pedersen 2006). Increasing stratification, reduced ventilation, and declining gas solubilities due to warming are expected to shift the global balance between C and O_2, thereby expanding O_2-depleted regions (Keeling and Garcia 2002; Rabalais et al. 2010; Keeling et al. 2010; Falkowski et al. 2011).

Hypoxia is defined as maintenance of DO concentrations below either 2 or 3 mg O_2 L^{-1} (62 or 94 μM O_2) in bottom waters (Pavela et al. 1983; Parker and O'Reilly 1991; Welsh and Eller 1991). Many regulatory agencies define hypoxia as <3 mg O_2 L^{-1} because juveniles and adults of many benthic invertebrate and fish species experience physiological stress below this concentration (Ritter and Montagna 1999). Seasonal hypoxia is common to numerous estuaries, such as LIS, Chesapeake Bay, and the Neuse River (Gooday et al. 2009), and develops when consumption rates of DO exceed its resupply rates to bottom waters through physical mixing and transport processes (ventilation). Seasonal hypoxia in bottom waters has been observed in the western narrows of LIS (WLIS) since the early 1970s and some suggest that mild hypoxia dates back to the pre-colonial era (Parker and O'Reilly 1991). Geochemical and micropaleontological core observations show that more severe hypoxia did not pre-date the middle nineteenth century (Thomas et al. 2004a, b, 2009, 2010; Varekamp et al. 2009, 2010). Ever since systematic monitoring commenced in 1987, LIS summer hypoxia has varied interannually between 78 and 1,000 km^2 (30–390 mi^2) in areal extent and 33–82 days in duration (http://longislandsoundstudy.net). Over the last 3 decades, hypoxia has occurred annually and usually initiates near Throgs Neck Bridge and propagates eastward to at least the New York-Connecticut border through late summer (LISS 1994). Although monitoring by state and local government agencies dates as far back as 1914, historical trends are unclear because of relatively low spatial and temporal sampling resolution and analytical limitations of early monitoring programs.

Processes controlling hypoxia in the LIS ecosystem have received comparatively little direct study and have been assumed to be much like those of Chesapeake Bay and other major estuaries by modelers and resource managers. Most estuaries originate from a major river and minor tributaries discharging directly to the sea through a drowned river valley, a glacial cut or through a bar-built lagoon. Their circulation is controlled by freshwater discharge rates, geomorphology, and ocean tides in a relatively straightforward manner. However,

LIS is distinct from most other estuaries. Today, most freshwater inputs to LIS are orthogonal to the estuary's primary axis, mainly from the Connecticut, Housatonic, and Thames Rivers and wastewater treatment facilities (WWTFs). The unusual hydrography and geomorphology of LIS, including three sub-basins formed by two north–south sills, impose limited tidal exchange with the ocean and long hydraulic residence times (O'Donnell et al., Chap. 3, in this volume). These attributes make LIS susceptible to strong seasonal physical stratification, especially when vertical mixing by wind stress and thermohaline circulation is minimal (Welsh and Eller 1991; Vigil 1991; Anderson and Taylor 2001).

In the isolated bottom waters of LIS during late summer and early fall, DO concentrations are drawn down by biological and chemical oxygen utilization. Dissolved oxygen drawdown in bottom waters is fueled by labile organic matter and reduced inorganic materials that are presumably controlled by riverine and nonpoint sources, and WWTF discharges. Monitoring programs reveal that discharges from WWTFs, which include combined sewer overflows (CSOs), supply to LIS an average of 61 % of the system's total organic carbon (TOC) loading, totaling ~137,000 tons year^{-1}, and 73 % of N nutrient loading, amounting to 53,000 tons year^{-1}; the remainder derives from nonpoint sources (CTDEP and NYSDEC 2000). WWTF discharges to LIS are subjected to secondary treatment, a process which reduces biochemical oxygen demand (BOD) by 92 % relative to the untreated influent, while only removing ~37 % of total N (NRC 1993). Consequently, BOD in effluents from New York City wastewater treatment facilities (WWTFs) declined by about 60 % during the latter half of the twentieth century (O'Shea and Brosnan 2000). Therefore, labile organic matter directly fueling BOD in the receiving waters today is probably only a small fraction of the reported TOC discharge, most of which may be relatively refractory and slowly remineralized. High precipitation events may promote episodic spikes in BOD, because influent from CSOs can temporarily overwhelm WWTFs, reducing their hydraulic retention times, and cause incompletely treated high-BOD effluent to enter LIS (Loucks and Johnson 1991; St. John et al. 1991). However, over most of hypoxia onset and maturation, marine productivity due to inputs of inorganic nutrients, not TOC loadings, are believed to be the primary driver of oxygen consumption in LIS bottom waters as evidenced in management and modeling efforts (Welsh and Eller 1991; Welsh 1995). Over time scales of centuries, the proportions of autochthonous and allochthonous organic matter (OM) have varied considerably (Varekamp et al. 2009, 2010).

Excessive macronutrient (N, P, Si) loadings generally stimulate phytoplankton communities to rapidly proliferate, barring other limitations, such as light exposure, micronutrients, or inhibitory chemicals. Several fates are possible for flourishing phytoplankton cells: they can divide, be eaten, lyse after succumbing to viral infection, or sink from surface waters. Above the compensation depth during the day, phytoplankton are net producers of DO, sometimes driving it to supersaturation in surface waters. However, like all aerobes, photosynthetic organisms must also respire continuously. In fact, algal respiration, R_a, in LIS's euphotic zone, is estimated to consume 5–52 % of gross primary production (GPP) during the day (Goebel et al. 2006). In LIS surface waters, R_a has been estimated to contribute on average 40–50 % to total daytime plankton community respiration (R_c) (Anderson and Taylor 2001; Goebel and Kremer 2007). During the night, euphotic zone residents become net sinks for DO, respiring away stored energy reserves. Furthermore, viable phytoplankton residing below the compensation depth for prolonged periods, due to vertical mixing or sinking, continue to respire until energy reserves are totally dissipated. Viable phytoplankton can sink from surface waters as individual cells or amorphous aggregates, depending upon species and physiological status (Kiørboe et al. 1996; Lomas and Moran 2011). Some can also be packaged in metazoan fecal pellets in an undigested state and sink rapidly to depth while presumably still respiring (Dam et al. 1995).

In addition to their own endogenous respiration, phytoplankton fuel respiration of other trophic levels when they: fall prey to protistan and metazoan herbivores; release extracellular organic carbon (EOC); decompose; are defecated by herbivores; or are lysed by viruses. Heterotrophic prokaryotes (bacteria and archaea) and fungi are the beneficiaries of the organic substrates produced by these processes. Vertical export of phytoplankton-derived organic matter to bottom waters through sinking of cells, aggregates, and fecal pellets, through mesozooplankton migrations, and mixing provides a major mechanistic link between inorganic nutrient loadings to surface waters and oxygen consumption in bottom waters (Welsh 1995; Anderson and Taylor 2001). The efficiency with which inorganic nutrient loadings are translated into bottom water oxygen consumption is dependent upon interactions of several physical, chemical, and biological variables including T, wind stress, organic matter quality, and plankton community structure (Welsh and Eller 1991; Vigil 1991; Michaels and Silver 1988; Wilson et al. 2008). Unicellular organisms dominate community respiration (R_c) because of their overwhelming abundances and high intrinsic metabolic rates. Over 98 % of planktonic R_c could be attributed to organisms with diameters <30 μm in coastal mesocosms (Williams 1981). Thus, the focus of studies on the genesis of hypoxia has historically been on microorganisms (e.g., Thomas et al. 2000; Gooday et al. 2009).

6.8.1.2 Dissolved Oxygen Dynamics

A previously unpublished study from G. Taylor's group (Stony Brook University) examined DO dynamics with respect to nutrient inventories, primary production, and bacterial activities at a single site in WLIS (station A4 near Execution Rocks: 40°52.3′N 73°44.1′W) through the summers of 1997 and 1998. This study was conducted concurrently with a parallel lower Hudson River estuary study; station and methodological details for both studies can be found in Taylor et al. (2003a, b). During this study hypoxic bottom waters (<3.0 mg O_2 L^{-1} or <25 % sat) were observed only in August of both years with depressed DO concentrations (<5.5 mg O_2 L^{-1} or <67 % sat) in the samples immediately preceding (Fig. 6.51a). Bottom DO rebounded almost completely by mid-October in both years. DO in the mixed layer was near saturation throughout these observations (78–117 % sat) and

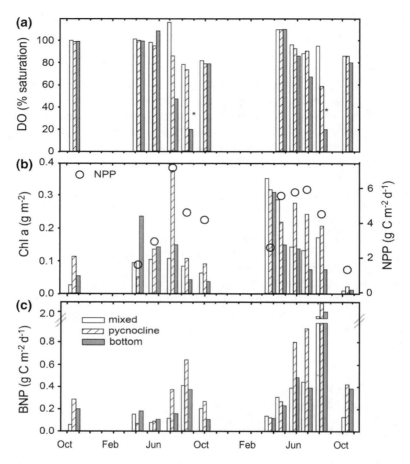

Fig. 6.51 Temporal variations in dissolved oxygen concentrations, DO (**a**), total Chlorophyll *a* inventories, Chl*a*, and net primary production, NPP determined by ^{14}C-bicarbonate assimilation (**b**), and bacterial net production, BNP, determined by ^{3}H-thymidine incorporation into DNA and microscopic cell sizing (**c**) observed at station A4 in western LIS between 21 Oct 1996 and 14 Oct 1998. Mean DO concentrations for surface, pycnocline and bottom layers were determined from continuous profiles provided by a YSI polarographic electrode on the sampling rosette. Biomass and production measurements (**b**, **c**) were determined from discrete Niskin bottle samples collected from the middle of mixed layer, pycnocline and bottom layer and integrated over entire layer by simple trapezoidal interpolation. * = hypoxia (DO < 3.0 mg O_2 L^{-1})

was usually intermediate (59–109 % sat) within the pycnocline. In both years, highest rates of primary production occurred in the weeks prior to and during detection of hypoxia (Fig. 6.51b). Chlorophyll *a* (Chl*a*) inventories within the pycnocline and bottom waters tended to be highest in the spring and decline through summer and fall, paralleling the decline of DO below the mixed layer (Fig. 6.51a, b). Within the three layers, heterotrophic bacterial production (BNP), estimated from ^{3}H-thymidine incorporation into DNA and cell sizing measurements, tended to increase through the summer and reached annual maxima when hypoxia was established, then declined into the fall (Fig. 6.51c). In 10 of the 12 observations, the majority of heterotrophic production resided in waters below the mixed layer, similar to Chl$a$ distributions. Patterns evident in Fig. 6.51 are consistent with a scenario in which spring and early summer phytoplankton blooms in the mixed layer are exported to depth passively by sinking or actively via herbivory, where they decompose to fuel microheterotrophic metabolism, drawing down bottom water DO in mid-to-late summer. These analyses also suggest that the euphotic zone (1 % of surface irradiance) extended into the pycnocline when waters were relatively clear which appeared to be the case in five of our 12 observations. Thus, some undefined fraction of primary production and DO production occurred in deeper, stratified waters on these occasions.

Temperature affects seasonal DO dynamics by contributing to stratification, oxygen solubility, and rates of microbial metabolism. Solar insolation warms surface waters from spring through summer to establish physical stratification. Previous studies in WLIS assert that declines in bottom DO may be more dependent upon changes in thermal stratification than changes in point and nonpoint loadings (Torgersen et al. 1997; O'Shea and Brosnan 2000; Anderson and Taylor 2001; Wilson et al. 2008; O'Donnell et al., Chap. 3, in this volume). Oxygen is less soluble in warmer water. DO concentrations observed in pycnocline and bottom water samples of the study reported here were more strongly correlated with T ($r = -0.82$, $p \ll 0.0001$, $n = 22$) than any other measured variable, explaining 67 % of the variance in DO. Rates of chemical reactions and metabolism of poikilotherms vary in direct proportion to water T. Bacterioplankton metabolism appears to possess Q_{10} responses lying between two and three (Shiah et al. 2000). Thus, a portion of the seasonal BNP increases observed at station A4 may have been driven by warming waters (Fig. 6.51c). About 40 % of the variance ($r = 0.64$; $p < 0.001$, $n = 24$) in bacterial specific growth rates (BNP/bacterial biomass $= \mu = $ div day^{-1}) was explained simply by T, while the remaining variance is likely explained by labile substrate availability. Because measuring labile substrate availability is impracticable, measurements of ^{14}C-glucose, ^{14}C-glycolate, and ^{14}C-acetate turnover (k, % day^{-1}) were used as proxies for bacterial substrate assimilation activity. Below the mixed layer, turnover of all three substrates was highly correlated with BNP; k_{glu} vs. BNP, $r = 0.73$, $p \ll 0.0001$; k_{gly} vs. BNP, $r = 0.61$, $p < 0.05$; k_{ace} versus BNP, $r = 0.86$, $p \ll 0.0001$. Thus, between 37 and 74 % of variance in BNP can be explained by labile substrate turnover. Labile substrate turnover probably does not vary totally independently from T because processes liberating substrates from biogenic polymers, i.e., enzymatic hydrolysis, are T-sensitive (Taylor et al. 2009). Thus, covariance among BNP, k_{glu}, k_{gly}, k_{ace}, and T is expected.

Water column oxygen consumption was also measured in an automated respirometer within the mixed layer, pycnocline, and bottom waters (Taylor et al. 2003a). Community respiration below the mixed layer varied between 0.16 and 4.0 μM O$_2$ h^{-1} ($=1.8$ μM O$_2$ h^{-1}) and on average was 1.4 times slower than in the mixed layer. Rates overlap with those reported by Goebel and Kremer (2007) for direct measurements at eight LIS stations and depth-dependent trends were similar.

However, R_c values reported in both these studies are significantly faster than those (0.12–0.36 μM O_2 h^{-1}) derived from a model developed for a nearby station (E10) (Hydroqual 1995). Using R_c measured in pycnocline and bottom water samples and DO saturation concentrations calculated from T and S, potential DO turnover rates (k_{Rc}) were estimated (Fig. 6.52). Below the mixed layer at station A4, k_{Rc} varied between 2 and 40 % day^{-1}, meaning that if these waters were initially DO-saturated and not ventilated, then planktonic R_c alone would totally deplete DO in 2.5–64 days (=13 days). Clearly, turnover of actual DO inventories in bottom waters was faster at the time of sampling because DO was undersaturated in most instances (Fig. 6.52). Also apparent from this analysis is a requirement for significant physical ventilation from vertical and lateral transport to prevent onset of anoxia (Torgersen et al. 1997).

Adding estimates of benthic oxygen demand do not significantly alter rates of subsurface DO turnover. To illustrate, the highest published LIS benthic oxygen demand (~47 mmol m^{-2} day^{-1}; Mackin et al. 1991) only contributes 1.5–16.6 % (=6 %) to observed total k_{Rc} in bottom waters. Chemical oxygen demand (COD) was previously evaluated in formalin-killed controls and found to be consistently <10 % of total R_c (Taylor et al. 2003a). Thus, results from both study years indicate that metabolism of suspended microbial communities usually accounts for >85 % of the total oxygen demand below the mixed layer, with benthic communities and COD accounting for the remainder, similar to findings published

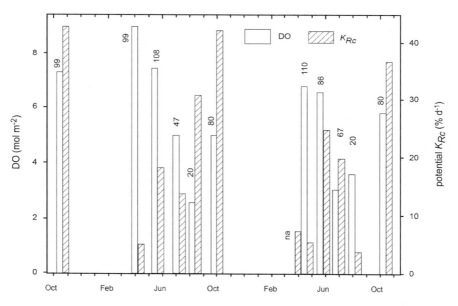

Fig. 6.52 Dissolved oxygen inventories observed in waters underlying the mixed layer and potential respiratory turnover rates (k_{Rc}) of saturating DO levels in these waters. k_{Rc} based on community respiration rates (R_c) measured in pycnocline and bottom water samples divided by molar DO saturation calculated from temperature and S. Numbers above bars present observed % saturation at time of sampling

elsewhere (Welsh and Eller 1991; Anderson and Taylor 2001; Goebel and Kremer 2007). However, sediment communities may periodically account for higher proportions of R_c in the aftermath of seasonal blooms (Aller 1994). The principal sources of nutrients and labile OM that support R_c can be episodic and may be associated with rainfall patterns (Anderson and Taylor 2001).

Nitrification is another DO-consuming biological process that is seldom considered in coastal eutrophication and hypoxia studies. It is a chemoautotrophic process transforming ammonium to nitrite then to nitrate, consuming O_2 in both steps and producing biomass from dissolved inorganic carbon (DIC). This process is performed by β- and γ-proteobacteria and ammonia-oxidizing archaea (Francis et al. 2005). Ammonium, the energy donor in nitrification, can be a significant fraction of the total N in WWTF effluents. In the lower Hudson River estuary, for example, elevated levels of dark $DI^{14}C$ assimilation (DDA) coincided with high WWTF-derived ammonium loadings, suggesting that nitrification was locally important in carbon, N, and oxygen dynamics (Taylor et al. 2003a). The DDA measured in surface water samples at station A4 strongly co-varied with R_c, PO_4^{3-}, total inorganic N ($TIN = NO_2^- + NO_3^- + NH_4^+$), and extracellular organic carbon (EOC) produced by phytoplankton (Table 6.7). A portion of DDA may be attributable to anaplerotic reactions of all metabolizing cells, but observed DDA exceeds maximum predicted anaplerotic CO_2 fixation for phyto- and bacterioplankton combined by as much as 17-fold and averages threefold (Taylor et al. 2001). Thus, excess DDA is plausibly driven by nitrification. All inorganic N species are summed as TIN here because their individual distributions with depth and time were extremely coherent. The potential contribution of nitrification to hypoxia, to microbial production entering food webs, and to N transformations can be cautiously inferred from these

Table 6.7 Bivariate Pearson correlation matrix for community respiration (R_c), Chlorophyll a concentrations (Chl$a$), phaeopigments (Pheo), net primary production (NPP), extracellular organic carbon production (EOC), dark DIC assimilation (DDA), bacterial biomass (BBiom), bacterial net production (BNP), ortho-phosphate (PO_4^{3-}), nitrite + nitrate + ammonium (TIN), and particulate and dissolved organic matter (POC, DOC) observed within the mixed layer of station A4

	Chl$a$	Pheo	NPP	EOC	DDA	BBiom	BNP	PO_4^{3-}	TIN	POC	DOC
R_c	ns	ns	ns	**0.61**	<u>**0.73**</u>	ns	0.56	<u>**0.76**</u>	<u>**0.96**</u>	ns	ns
Chl$a$		ns	ns	ns	ns	**0.78**	ns	ns	ns	<u>**0.87**</u>	ns
Pheo			ns	ns	ns	ns	ns	ns	ns	**0.67**	ns
NPP				ns	ns	ns	ns	ns	ns	ns	ns
EOC					<u>**0.96**</u>	ns	ns	**0.61**	**0.66**	ns	0.54
DDA						ns	ns	<u>**0.74**</u>	<u>**0.78**</u>	ns	0.57
BBiom							0.60	ns	ns	<u>**0.75**</u>	**0.58**
BNP								ns	ns	ns	ns
PO_4^{3-}									<u>**0.84**</u>	ns	ns
TIN										ns	ns
POC											ns

Statistical significance levels are specified as follows: >90 %, **>95 %** and <u>**>99 %**</u>. "ns" = <90 % ($n = 12$ samples)

Table 6.8 Bivariate Pearson correlation matrix for community respiration (R_c), Chlorophyll *a* concentrations (Chl*a*), phaeopigments (Pheo), surface net primary production (NPP), bacterial biomass (BBiom), bacterial net production (BNP), ortho-phosphate (PO_4^{3-}), nitrite + nitrate + ammonium (TIN), particulate and dissolved organic matter (POC, DOC) and S observed below the mixed layer of station A4 unless otherwise noted

	Chl*a*	Pheo	NPP	BBiom	BNP	PO_4^{3-}	TIN	N:P	POC	DOC	Salinity
R_c	ns	**0.57**	−0.38	ns	0.44	**0.62**	**0.84**	**0.58**	ns	ns	0.46
Chl*a*		ns	ns	0.50	ns	−0.46	−0.47	ns	**−0.74**	−0.35	−0.45
Pheo			ns	ns	ns	ns	ns	ns	**0.56**	ns	ns
NPP				0.49	ns	ns	−0.42	**−0.65**	ns	−0.36	ns
BBiom					0.75	ns	ns	−0.49	0.41	ns	**−0.55**
BNP						**0.60**	0.46	−0.52	ns	ns	−0.42
PO_4^{3-}							**0.79**	ns	ns	0.41	0.47
TIN								**0.68**	ns	0.48	**0.55**
N:P									ns	ns	ns
POC										ns	ns
DOC											0.39

Statistical significance levels are specified as follows: >90 %, **>95 %** and **>99 %**. "ns" = <90 % (*n* = 12 pycnocline + 12 bottom water samples)

results. However, direct measurements using ^{15}N tracers and ancillary molecular approaches (e.g., functional gene/transcript quantification) will be necessary to accurately assess the importance of nitrification in this system. Studies of δ^{15}N in organic matter and DIN may help elucidate the importance of nitrification in LIS as well (Varekamp et al., Chap. 5, Sect. 5.7.3, in this volume).

Temporal variations in inorganic nutrients (PO_4^{3-}, TIN) explained more variance in R_c at station A4 than other metrics considered (Tables 6.7 and 6.8). Variations in inventories of particulate and dissolved organic carbon (POC, DOC), phytoplankton, and bacterioplankton and NPP did not explain observed variance in R_c in either the upper or lower water column. However, bacterial production measurements did co-vary with R_c in both layers, but only explained up to ~30 % of the variance (Tables 6.7 and 6.8). Correlations between R_c and pheopigments (+) and NPP (−) in pycnocline and bottom water samples were consistent with R_c lagging behind surface productivity and responding to herbivorous activity and delivery of phytodetritus. However, sampling resolution was far too coarse to resolve potentially lagged responses. About 12 sampling campaigns at a single station were clearly inadequate to fully elucidate the relationships between R_c and a host of controlling variables.

6.8.1.3 Biological Consequences of Hypoxia

Effects on Invertebrates

Long Island Sound has been subject to the effects of hypoxic events and other disturbances on benthic community composition (Rhoads et al. 1978). The role of oxygen was an important component of research into a catastrophic die-off of LIS lobsters in 1999 (see Sect. 6.8.4). Otherwise, the effects of hypoxia on invertebrates have been surprisingly little studied in LIS. Therefore, much of this section relies on information from other bodies of water, though in many cases the species examined occur in LIS.

All free-living animals require oxygen, but species vary widely in their tolerance to hypoxia. Many coastal benthic species are adapted to survive hypoxic conditions as long as it is not severe hypoxia or anoxia lasting too long. Many species can survive mild hypoxia for 1–2 weeks (Diaz and Rosenberg 1995). Differences in tolerance to hypoxia, combined with behavioral and physiological responses to hypoxia and associated stresses, characterize interspecific effects and ecological consequences.

Mobile animals such as fishes generally move away from hypoxic water. Benthic invertebrates are generally much less mobile than nekton, with the exception of some epifaunal predators (shrimp, mysids, blue crabs). Infauna respond to hypoxic exposure by moving closer to the sediment surface, and under severe hypoxia, crawl out of the sediment (Diaz and Rosenberg 2008). Sensitivity to hypoxia is typically highest in fishes, followed by crustaceans, polychaetes, with bivalves commonly being the most tolerant (Gray et al. 2002). At 2 mg O_2 L^{-1}, benthic fauna in the Adriatic Sea exhibited escape responses (with epifauna moving horizontally and infauna moving upward in the sediment); at 0.5 mg O_2 L^{-1}, infauna started to emerge and began to die (Riedel et al. 2008). This behavior is consistent with other studies, and underscores the importance of oxygen concentration on behavior and survival (Diaz and Rosenberg 1995, 2008).

Exposure to periodic hypoxic conditions is common and natural for many coastal benthic animals, including those living in intertidal zones and in organic rich muds. Inhabitants of these environments exhibit a wide range of physiological adaptations to tolerate hypoxic periods. Animals that cannot escape must compensate for reduced oxygen to maintain metabolic rate, reduce their metabolic rate, or use anaerobic pathways to process energy (Burnett 1997). Many species respond to moderate hypoxia by increasing ventilation or heart rate to acquire sufficient O_2, but at lower O_2 concentrations, net metabolic rates decline. Some species induce production of respiratory pigments during exposure to hypoxia (Mangum 1997). Many benthic invertebrates that tolerate hypoxia use a variety of anaerobic pathways, with metabolic rates typically reduced under these conditions (Grieshaber et al. 1994). One exception is the transient opportunistic bivalve *Mulinia lateralis*; animals collected from Port Jefferson Harbor maintained high metabolic rate even under anoxic conditions (Shumway et al. 1993). Anaerobic metabolic pathways are crucial mechanisms for tolerating hypoxic episodes, but because animals reduce metabolic rates and do not feed during these periods, growth and secondary production are reduced during severe hypoxia.

Exposure to low oxygen levels in marine environments is typically accompanied by other potential stressors such as sulfide, ammonium, carbondioxide, and elevated T. Marine benthic fauna are invariably exposed to elevated levels of sulfide and ammonium, both of which are toxic to aerobes. The 1999 die-off of lobsters in LIS may have been due to such interacting stressors. Sulfide is the product of anaerobic decomposition of organic matter by bacteria that use sulfate as the electron acceptor. Because of the high concentration of sulfate in seawater, bacterial sulfate reduction accounts for a significant fraction of organic matter decomposition in coastal marine muds (Jørgensen 1977). Under hypoxic conditions, sulfide and ammonium build up in sediment porewater and in bottom water because their oxidation is reduced. Many studies that have investigated tolerance to hypoxia have not measured the additive or interactive effects of exposure to elevated concentrations of these toxic compounds (Vaquer-Sunyer and Duarte 2010). Benthic infauna in muds are chronically exposed to micromolar concentrations of porewater sulfide and most of them exhibit considerable tolerance to it. Benthic invertebrates are able to live in sulfide-rich sediment in part by minimizing exposure through ventilation of overlying water into their tubes and burrows (Aller 1984). They have efficient detoxification mechanisms to oxidize sulfide that is taken up into their tissues, but in most cases these mechanisms require oxygen (Völkel and Grieshaber 1994). Thus, under hypoxic conditions, they are doubly exposed to the direct stresses of low oxygen level and elevated sulfide exposure. If not all internal sulfide is detoxified, cytochrome c oxidase is inhibited, resulting in sulfide-dependent anaerobic metabolism (Grieshaber and Völkel 1998).

Some species are quite tolerant to sulfide exposure. The enrichment opportunistic polychaete *Capitella teleta* (previously described as *Capitella* sp. I) (Blake et al. 2009) survives and burrows actively at 7 mM sulfide (Wada et al. 2006). Their larvae may use sulfide as a settlement cue (Cuomo 1985, but see Dubilier 1988) and can survive up to 2 mM sulfide (Dubilier 1988). As described later, *C. teleta* rapidly colonizes organically enriched sediments having high concentrations of porewater sulfide.

Sulfide is toxic to aerobic organisms because it binds to hemoglobin and cytochrome c oxidase (Somero et al. 1989). Some benthic animals may also use sulfide, either through sulfide-oxidizing bacterial symbionts (Fisher 1990; Cavanaugh et al. 2006), feeding on free-living sulfide-oxidizing bacteria (Tsutsumi 1990; Tsutsumi et al. 2001), or even directly as an energy source (Doeller et al. 1999; Parrino et al. 2000; Tielens et al. 2002). But at the high sulfide concentrations found in organically enriched muds of LIS, sulfide is directly toxic, causing animals to shift to less efficient anaerobic pathways at higher oxygen levels than they would otherwise. Hypoxic conditions as typically defined (3 mg O_2 L^{-1}) are likely to underestimate the combined impact of low oxygen and sulfide exposure (Vaquer-Sunyer and Duarte 2008). Survival during hypoxia is reduced significantly (average 30 % higher mortality than hypoxia alone) during sulfide exposure, suggesting that hypoxia effects in natural environments are not due to low oxygen alone (Vaquer-Sunyer and Duarte 2010).

Elevated temperatures reduce survival time and increase the lethal oxygen level (Vaquer-Sunyer and Duarte 2011). Given that the most severe hypoxia occurs in the late summer when water temperatures peak, it is likely that climate change may exacerbate hypoxia-T interactions. But in WLIS, warming trend is primarily exhibited in surface waters above the pycnocline, which are not susceptible to hypoxia (Wilson et al. 2008). The bottom waters of WLIS have trended toward slightly lower summer temperatures because the remnant of the cold pool is isolated by earlier stratification. Thus, during most summers, bottom water temperatures may not be exacerbating the effects of exposure to sulfide and ammonium in LIS.

Hypoxic bottom waters are also elevated in CO_2 resulting from the mineralization of organic carbon. Though CO_2 is not toxic in the sense that sulfide and ammonium are, elevated levels can affect organisms in many ways. Elevated concentrations of CO_2 (hypercapnia) produce acidosis in organisms (Burnett 1997). Temperature-induced acidosis has been shown to be deleterious to lobsters in LIS (Dove et al. 2005).

Effects on Food Webs

Predator–prey interactions are affected in complex ways by hypoxia, depending on its severity and spatial and temporal patterns. During chronic hypoxia, hypoxic water may be a refuge from predators that are less tolerant to low oxygen. For example, seasonally hypoxic Narragansett Bay supports a large population of hard clams because its main predators—sea stars, fish, and several crab species—avoid hypoxic conditions that the hard clam can tolerate (Altieri 2008).

When bottom waters become episodically hypoxic, prey may be more susceptible to predators. In Chesapeake Bay, periods of hypoxia result in higher predation on the bivalve *Macoma balthica* because it moves toward the sediment surface, making it more vulnerable to predation. Predators move from shallower, more oxic water into recovering hypoxic waters to prey on benthic animals faster than the prey can respond to the recovered conditions by burrowing deeper into the sediment (Long and Seitz 2008). This appears to occur in many cases during periodic hypoxia and recovery (Diaz and Rosenberg 2008).

Although predators are typically more sensitive to hypoxia, there is considerable variation among species. Lobsters (in LIS) (Howell and Simpson 1994) and blue crabs are relatively tolerant to hypoxia compared to other mobile epibenthic predators, although blue crabs exhibit high individual variability due to differences in hemocyanin structure, possibly due to acclimation history (Bell et al. 2010). Among water column predators, gelatinous animals appear to be more tolerant to hypoxia than fish (Grove and Breitburg 2005). The ctenophore *Mnemiopsis leidyi* uses hypoxic bottom water but its growth and reproduction are reduced at 2.5 mg O_2 L^{-1}. Growth of the scyphomedusan *Chrysaora quinquecirrha* is not affected by hypoxia but it avoids hypoxic water. In general, hypoxic conditions result in more energy flow to bacteria and to gelatinous zooplankton, reducing energy going to higher trophic levels such as finfishes.

Early life stages are often more sensitive to hypoxia than adults. Hypoxia can increase mortality of fish embryos. Bay anchovy eggs that sink into hypoxic waters experience higher mortality than those in normoxic conditions (Breitburg et al. 2009a, b; see Keister et al. 2000).

Disturbance, Defaunalization, and Ecological Succession

Severe hypoxia or anoxia results in mass mortality of all exposed animals in the seabed and bottom water. Upon return to more oxic conditions, the sediment is sequentially colonized by a relatively predictable succession of animals. The model for the ecological succession of the benthos was developed by seminal work in LIS (McCall 1977; Rhoads et al. 1978; Rhoads and Germano 1982, 1986) and in Swedish waters (Pearson and Rosenberg 1978) following both spatial and temporal responses to disturbances including hypoxia, pollution, and dredged sediment disposal. The first organisms to colonize sediments defaunated by severe hypoxia are small, tube-dwelling polychaetes, typically spionids and capitellids. These species, characteristically *C. teleta* and *Streblospio benedicti* in LIS, have opportunistic life histories characterized by small size, short life span, rapid growth, and high reproductive output. The *C. teleta* epitomizes the enrichment opportunist that rapidly colonizes organically enriched muds once oxic conditions in bottom water return. Sediment porewater has high sulfide concentrations at this stage, making it a geochemically harsh habitat to colonize. But the sediment does not harbor competitors and it is food-rich. *C. teleta* and *S. benedicti* have high tolerances for sulfide (Llanso 1991). Under these conditions, food is abundant but oxygen may be limiting, and growth of these colonizers is sensitive to oxygen concentration (Forbes and Lopez 1990). High metabolic potential of these opportunistic species makes them sensitive to oxygen level. The *C. teleta* requires high concentrations of labile organic matter to grow. At higher oxygen levels, it has high metabolic rates, independent of the amount of food available, indicating a metabolic system "poised" for aerobic respiration. High population growth rate occurs under conditions of high food and low but increasing oxygen concentrations (Forbes et al. 1994). The *C. teleta* can persist in very low population densities (<100 individuals m^{-2}) during unfavorable periods (Tsutsumi 1990). They may survive these periods by shrinking or degrowing (Forbes and Lopez 1990). When conditions become favorable, surviving worms grow rapidly.

M. lateralis also rapidly colonizes disturbed sediment, and has been known to be a transient opportunist in LIS for over 5 decades (Sanders 1956; Levinton 1970; Valente et al. 1992). This suspension feeder exhibits a physiological strategy similar to *C. teleta*, poised for rapid growth under favorable environmental conditions but exhibiting reduced survival under environmental stress (Shumway 1983; Shumway et al. 1993). It survives more poorly than many other infaunal bivalves under anoxic and sulfidic conditions. Surprisingly, its metabolic heat dissipation remains high even under anoxic conditions. Its deficient ability to regulate metabolism results in poor survival when environmental conditions worsen, such as periods of low food or low oxygen. *M. lateralis* is well represented in shell death assemblages, but only sporadically in living benthic communities (Levinton 1970).

Physiological and Behavioral Responses of Fishes

Long Island Sound provides productive habitats for young fish and crustaceans. The bays, harbors, banks, and wetlands provide a variety of protected nursery areas for early growth and rapid development of important species (Weinstein 1979; McEnroe et al. 1995; Able and Fahey 1998; Smith and Able 2003). Development of hypoxia, through its sublethal and lethal effects on vulnerable early life stages, degrades effective nursery habitats, diminishing their size and number. Reduction of essential habitats may decrease access to rich-food resources critical to the survival and success of early juvenile populations.

Hypoxia can affect fishes directly by altering metabolic rates and their physiological capabilities for tolerance and acclimation, and indirectly through altered behavior that affects abundance and distribution; or through reduced feeding and reduced prey availability (Eby et al. 2005; Pihl et al. 1992; McEnroe 1991; Neuenfeldt 2002). Migrations of important anadromous species, including salmon, shad, and Atlantic sturgeon (*Acipenser oxyrinchus*) can be blocked by hypoxic waters (Maes et al. 2007; Weisberg et al. 1996; Albert 1988). Severe hypoxia periodically produces fish kills in LIS. Hypoxia effects on fishes have been the subject of recent reviews (USEPA 2000; Richards et al. 2009).

Hypoxia may reduce growth (Bejda et al. 1992; Cech et al. 1984; Stierhoff et al. 2003, 2009b), impair endocrine system function, and disrupt reproduction (Wu et al. 2003; Wu 2002; Landry et al. 2007). Hypoxia is also associated with decreased physical fitness and consequent vulnerability to predation (Roussel 2007), lowered immunity to disease and increased susceptibility to pollutants (Mellergaard and Nielsen 1990; Sniesko 1973; Plumb et al. 1976; Walters and Plumb 1980; Lloyd 1961). As a consequence, hypoxia acts to decrease fitness, leading to mortalities through decreased abilities to avoid, or tolerate, environmental stressors and to loss of health with vulnerability to predation, disease, and toxicant effects (Heath 1987; Pihl et al. 1991; Eby and Crowder 2002; Eby et al. 2005; Tyler and Targett 2007; McEnroe 1991).

A summary of effects of hypoxia on physiology and behavior for LIS fishes is listed in Table 6.9.

Tolerance

Fishes were the most sensitive group of marine animals in an analysis of sublethal and lethal indicators of hypoxia (Vaquer-Sunyer and Duarte 2008). Tolerance to hypoxia for species that occur in LIS has been described by Poucher and Coiro (1997) and Miller et al. (2002) and was reviewed by USEPA in 2000 and 2003 by Breitburg et al. (2001, 2003). Early life stages are especially sensitive to low

Table 6.9 Effects of low dissolved oxygen on fish physiology and behavior (compiled by M. McEnroe)

Fish species	Lab or field	Effects of low dissolved oxygen (DO)	Reference
Teleosts			
Atlantic croaker (*Micropogonias undulates*)	Field	Juvenile croaker present at DO of 1–2 mg O_2 L^{-1}, rarely in trawls at DO <1 mg O_2 L^{-1}	Bell and Eggleston (2005)
	Field	Chesapeake Bay absent catches <1.4 mg O_2 L^{-1}	Pihl et al. (1991), Eby et al. (2005)
	Lab	Avoid 2.0 mg O_2 L^{-1}	Wannamaker and Rice (2000)
	Field	Growth decreased >50 % in habitats with intermittent hypoxia	Eby et al. (2005)
Atlantic menhaden (*Brevoortia tyrannus*)	Lab	96 h LC 50 of 1.04 mg O_2 L^{-1}	USEPA (2000)
	Lab	1.2 mg O_2 L^{-1} for 24 h, no mortality, 25 °C	Shimps et al. (2005)
	Lab	30–40 % mortality at 1.2 mg O_2 L^{-1} for 24 h, 30 °C	
	Lab	100 % mortality in <6 h at 0.6 mg O_2 L^{-1}	
	Lab	100 % survival at 1.5 mg O_2 L^{-1} for 2 weeks	
	Lab	More tolerant than spot	McNatt and Rice (2004)
	Field	Most will avoid 1 and 2 mg O_2 L^{-1}	Shimps et al. (2005)
	Field	Avoidance threshold 2.6 mg O_2 L^{-1} threshold Neuse R	Wannamaker and Rice (2000)
	Field	Absent Cape fear collections <1.4 mg O_2 L^{-1}	Eby and Crowder (2002)
	Lab	Threshold of avoidance was 2–4 mg O_2 L^{-1}	Schwartz et al. (1981)
	Lab	Threshold for growth reduction of 60 % DO < 1.5 mg O_2 L^{-1}	Wannamaker and Rice (2000)
		At 30 °C; higher growth than at 25 °C, reduced 63 % at 1.5 mg O_2 L^{-1} and 25 °C	Wannamaker and Rice (2000)
Atlantic silversides (*Menidia menidia*)	Lab	Growth decreased 47 % after 28 days at 3.9 mg O_2 L^{-1}	Poucher and Coirio (1997)

(continued)

Table 6.9 (continued)

Fish species	Lab or field	Effects of low dissolved oxygen (DO)	Reference
Bay anchovy (*Anchoa mitchilli*)	Field	Present at 1.5 mg O_2 L^{-1}	Robinette (1983)
	Lab	96 h LC_{50} 1.6 mg O_2 L^{-1} for larvae	Chesney and Houde (1989)
	Field	Bay anchovy stay in well-oxygenated surface waters, make brief forays to hypoxic bottom water to feed	Ludsin et al. (2009)
Bluefish (*Pomatomus saltatrix*)	Field	Avoided < 4 mg O_2 L^{-1}	Middaugh et al. (1981)
	Field	Absent NY Bight when low DO	Oliver et al. (1989)
Hogchocker (*Trinectes maculatus*)	Lab	24 h LC_{50} = 0.5 mg O_2 L^{-1}	Pihl et al. (1991)
	Lab	Survive 1.0 mg O_2 L^{-1} for 10 days	
	Lab	Die within 1 day at 0.4 mg O_2 L^{-1}	
	Field	Chesapeake present in trawls at low DO < 1.4 mg O_2 L^{-1}	
Inland Silversides (*Menidia beryllina*)	Lab	96 h LC_{50} = 1.0 mg O_2 L^{-1}	USEPA (2000)
	Lab	Larval 96 h LC_{50} = 1.4 mg O_2 L^{-1}	Miller et al. (2002)
Mummichog (*Fundulus heteroclitus*)	Lab	Juveniles no mortality 9 days at 1.0 mg O_2 L^{-1} at 25 °C	Stierhoff et al. (2003)
	Lab	Maintained growth to 3.0 mg O_2 L^{-1}	Stierhoff et al. (2003)
	Lab	Growth decreased 60 % at 1.0 mg O_2 L^{-1} with surface access; reduced 90 % w/o surface access	Stierhoff et al. (2003)
	Lab	No effect of diel hypoxia (1–11 mg O_2 L^{-1}) on growth	Stierhoff et al. (2003)
	Lab	No avoidance of 1 versus 4 mg O_2 L^{-1}	Wannamaker and Rice (2000)
Red drum (*Sciaenops ocellatus*)	Lab	Larval 96 h LC_{50} = 1.8 mg O_2 L^{-1}	Miller et al. (2002)
Spot (*Leiostomus xanthurus*)	Lab	24 h LC_{50} values of 0.7 mg O_2 L^{-1}	USEPA (2000)
	Lab	1.2 mg O_2 L^{-1} for 24 h, no mortality, 25 °C	Shimps et al. (2005)

(continued)

Table 6.9 (continued)

Fish species	Lab or field	Effects of low dissolved oxygen (DO)	Reference
	Lab	30–40 % mortality at 1.2 mg O_2 L^{-1} and 30 °C for 24 h	
	Lab	100 % mortality in <6 h at 0.6 mg O_2 L^{-1}	
	Lab	Juveniles low avoidance threshold of $\leq$1.0 mg O_2 L^{-1}	Wannamaker and Rice (2000)
	Field	No negative trend in abundance with DO	Bell and Eggleston (2005)
	Field	Chesapeake present in trawls at low DO < 1.4 mg O_2 L^{-1}	Pihl et al. (1991)
	Field	Move into hypoxic waters	Pihl et al. (1992)
	Lab	Growth reduced 31 % at 25C and 89 % at 30 °C after 14 days at DO $\leq$ 1.5 mg O_2 L^{-1}	McNatt and Rice (2004)
Striped bass (*Morone saxatilis*)	Lab	Reduced growth of juveniles in hypoxia (~ 4.7–5.3 mg O_2 L^{-1})	Cech et al. (1984)
	Lab	DO $\leq$ 4.0 mg O_2 L^{-1} reduced feeding and growth	Brandt et al. (1998) (reviewed in USEPA 2003; Brandt et al 2009)
	Field	DO $\leq$3.0–4.0 mg O_2 L^{-1} reduced feeding in fish 2 and 4 years old	Chittenden (1971) Breitburg et al. (1994)
		DO $\leq$3.0 mg O_2 L^{-1} reduced swimming, agitated	
Summer flounder (*Paralichthyes dentatus*)	Lab	24 h LC_{50} = 1.59 mg O_2 L^{-1} for juveniles at 24–25 °C	USEPA (2000)
	Field	Absent in trawls DO < 2, decline in abundance between DO 2–4 mg O_2 L^{-1}	Bell and Eggleston (2005)
	Lab	Growth of juveniles reduced 25 % at $\leq$3.5 mg O_2 L^{-1} and 50–60 % at 2.0 mg O_2 L^{-1} at 20 and 25 °C for 14 days	Stierhoff et al. (2006)

(continued)

Table 6.9   (continued)

Fish species	Lab or field	Effects of low dissolved oxygen (DO)	Reference
	Lab	Reduced growth in diel cycling (2–11 mg O_2 L^{-1}) regime At $\geq$25 °C growth was reduced at 5.0 mg O_2 L^{-1}, At 30 °C, 90 % reduction in growth at 2.0 mg O_2 L^{-1}	Stierhoff et al. (2006) Stierhoff et al. (2006)
	Field	Reduced growth rates in fish in creek with DO cycle of 0.1–17 mg O_2 L^{-1}; significantly reduced growth between 3.5 and 5.0 mg O_2 L^{-1}	Stierhoff et al. (2006)
	Field	Growth rate decrease in wild-caught fish after $\geq$120 days with mean DO rarely dropping below growth protective criteria of 4.80 mg O_2 L^{-1}, temperatures 16–32 °C	Stierhoff et al. (2009a)
Weakfish (*Cynoscion regalis*)	Lab	Avoidance threshold of 1 mg O_2 L^{-1}	Steirhoff et al. (2009c)
	Lab	Growth unaffected by constant DO of 2.0 mg O_2 L^{-1} for 7 days, nor by diel cycling hypoxia of 2–11 mg O_2 L^{-1} at 20, 25 and 30 °C	Steirhoff et al. (2009c)
	Field	Growth rate decrease in wild-caught juvenile fish mean DO rarely dropping below growth protective criteria of 4.80 mg O_2 L^{-1}, temperatures 16–32 °C for >120 days	Steirhoff et al. (2009a)
	Field	Juvenile weakfish leave when DO < 2 mg O_2 L^{-1} and quickly return when DO rises >2 mg O_2 L^{-1}	Tyler and Targett (2007)
	Lab	Reduced activity at DO < 2.8 mg O_2 L^{-1}; prior exposure to hypoxia altered swimming behavior	Brodie et al. (2009)

(continued)

Table 6.9 (continued)

Fish species	Lab or field	Effects of low dissolved oxygen (DO)	Reference
Winter flounder (*Pleuronectes americanus*)	Lab	24 h LC_{50} value of 1.4 mg O_2 L^{-1}	USEPA (2000)
	Lab	At 20C juvenile growth reduced 50 and 60 % at 5.0 and 3.5 mg O_2 L^{-1}; growth zero at 2.0 mg O_2 L^{-1}. At 25 °C poor growth at all DO levels, lost weight at 2.0 mg O_2 L^{-1}	Stierhoff et al. (2006)
	Lab	Significant growth reduction in juveniles at 2.2 mg O_2 L^{-1}	Bejda et al. (1992)
	Field	Growth depression of juveniles occurred at a mean DO of 5.0 mg O_2 L^{-1} after 49 days growth in field	Stierhoff et al. (2009a)
	Field	Juvenile growth declined in relation to time spent at DO $\leq$ 2.3 mg O_2 L^{-1} in Narragansett Bay	Meng et al. (2008)
Chondrosteans			
Atlantic sturgeon (*Acipenser oxyrinchus*)	Lab	High mortality of YOY at 3 mg O_2 L^{-1} for 10 days at 26 °C	Secor and Gunderson (1998)
	Lab	Juveniles Avoid DO < 3 mg O_2 L^{-1}	Niklitschek (2001), Niklitschek and Secor (2005)
	Lab	Juveniles exhibited threefold decrease in growth in hypoxia (3 mg O_2 L^{-1}) versus normoxia (7 mg O_2 L^{-1}) at 26 °C	
	Lab	Juveniles decreased growth at 40 % oxygen saturation at 20 and 27 °C, approximately 3.3 and 2.9 mg O_2 L^{-1}, respectively	Secor and Gunderson (1998)
	Lab		Niklitschek (2001), Secor and Niklitschek (2001)
Shortnose sturgeon (*A. brevirostrum*)	Lab	24 h LC_{50} values of 2.7 and 2.2 mg O_2 L^{-1} at 77 and 104 days post-hatch (dph) at 20 °C, which rose to 3.1 mg O_2 L^{-1} at 27 °C	Campbell and Goodman (2004)

(continued)

Table 6.9 (continued)

Fish species	Lab or field	Effects of low dissolved oxygen (DO)	Reference
	Lab	Juveniles avoid DO < 3 mg O_2 L^{-1}	Niklitschek (2001), Niklitschek and Secor (2005)
	Lab	Decreased growth of juveniles at 40 % oxygen saturation at 20 and 27 °C, approximately 3.3 and 2.9 mg O_2 L^{-1}, respectively	Niklitschek (2001) Secor and Niklitschek (2001)

DO (Douderoff and Shumway 1970; Davis 1975; Chapman 1986; Rombough 1988; Breitburg 2002; Miller et al. 2002). In general, lethal oxygen concentrations declined with development from larvae to post-larvae and juveniles. Species mean LC_{50} (median lethal concentration) were greatest for larvae (1.4–3.3 mg O_2 L^{-1}), intermediate (1.0–2.2 mg O_2 L^{-1}) for post-larvae, and least (0.5–1.6 mg O_2 L^{-1}) for juveniles (Miller et al. 2002) and adults (Breitburg et al. 2001; Pihl et al. 1991; Poucher and Coiro 1997; USEPA 2000, 2003). Among sensitive early life stages were bay anchovy eggs 12 h LC_{50} of 2.7 mg O_2 L^{-1} (Chesney and Houde 1989) and striped bass larvae with a 12 h LC_{50} of 2.4 mg O_2 L^{-1} (Miller et al. 2002).

Many marine fishes use nearshore waters and wetlands of LIS as nursery habitats for feeding and rapid growth critical to the success of juveniles (Weinstein 1979; Able and Fahey 1998; McEnroe et al. 1995). Early life stages of mummichog, silverside (*Menidia menidia*), bay anchovy, American eel, tautog, winter flounder, weakfish, bluefish,, and striped bass all inhabit marshes and creeks that become warm and hypoxic in summer (Talbot and Able 1984; Kimble and Able 2007; McEnroe et al. 1995).

Hypoxia produces a variety of reactions in fish eggs and larvae. It may delay hatching (Voyer and Hennecky 1972), stimulate hatching (DiMichele and Powers 1984), reduce hatching success or induce deformities (Shumway et al. 1964). Hypoxia can reduce larval growth rate (Rombough 1988), leading to protracted exposure to predators (Giorgi 1981). For species that spawn in deeper water, effects on eggs and larvae will depend on their location in the water column. Striped bass eggs sink below the pycnocline where they can be exposed to hypoxia (Keister et al. 2000). Fish larvae have very high specific metabolic rates and high oxygen requirements. Most swimming activity of larvae is supported by aerobic metabolism, as is the capture and digestion of prey. Gills may be absent or poorly developed in early larval stages, so larvae are especially sensitive to periodic hypoxia as may occur during night-time minima in shallow waters. In many fishes, gill filaments become fully functional at metamorphosis, and this is accompanied by production of blood hemoglobin (De Silva and Tytler 1973). As gills develop, oxygen uptake efficiency and hypoxia tolerance increase. For example, in Atlantic herring larvae, incipient lethal DO values declined from 3.1 to 2.9 mg O_2 L^{-1} as gills developed, and to 2.2 mg O_2 L^{-1} at metamorphosis; equivalent decline in lethal DO was found for a larval flatfish: 2.7, 2.5, 1.7 mg O_2 L^{-1}, respectively (De Silva and Tytler 1973, Niklitschek and Secor 2009).

In general, juvenile and adult fishes have similar ranges of tolerance. For juvenile fishes, 96 h LC_{50} ranged from 0.6 to 1.6 mg O_2 L^{-1}; for adult fishes, the LC_{50} range was 0.5–1.6 mg O_2 L^{-1}. Sensitive species with $LC_{50's}$ 1.1–1.6 mg O_2 L^{-1} include striped bass, northern pipefish (*Syngnathus fuscus*), winter flounder, scup, Atlantic menhaden, and summer flounder. More tolerant were fourspine stickleback (*Apeltes quadracus*), windowpane flounder, Northern sea robin, hogchoker, and spot with LC_{50} 0.5 to >1.0 mg O_2 L^{-1} (Breitburg et al. 2001; Pihl et al. 1991; Poucher and Coiro 1997; USEPA 2000).

Physiological Consequences

Under normoxic conditions, the maximal rate a fish can use oxygen is limited by capacities of the gills and cardiovascular system to take up oxygen and distribute it to the tissues. However, the potential demands for oxygen by all the organs and tissues in aggregate exceed the maximal rate of oxygen uptake (Guderly and Pörtner 2010). The limitation of maximum aerobic capacity of fishes causes conflicts among demands of life activities that require aerobic energy and require partitioning of energy between those demands.

Farrell and Richards (2009) stated that for "all fishes, the zone of environmental hypoxia is characterized by a progressive loss of physiological functions as the hypoxic state deepens and aerobic scope declines." They suggested that with deepening hypoxia that loss of principal aerobic functions might start with reduction of locomotion, followed by impairment of reproduction, and later the slowing or cessation of growth. Fish species vary widely in the onset of such sublethal physiological effects; thresholds ranged from 10 to 2 mg O_2 L^{-1} (Vaquer-Sunyer and Duarte 2008).

Hypoxia occurs in the summer when water temperatures are high, so for ectotherms, metabolic rates, and oxygen requirements are highest when DO levels are lowest. Elevated T tends to raise the DO threshold for onset of sublethal effects (Cech et al. 1984, 1990; Schurmann and Steffensen 1992; Schurmann and Steffensen 1997; Secor and Gunderson 1998; Claireaux et al. 2000). As a result, hypoxia tolerance may be severely limited at elevated T, and survival may decline rapidly (Schurmann and Steffensen 1997; Shimps et al. 2005; Secor and Gunderson 1998; Campbell and Goodman 2004).

Avoidance

Fishing surveys have reported fish avoidance of hypoxic areas. Avoidance was usually observed as a decline in abundance and horizontal distribution in the catches of several species, accompanied by a decrease in species diversity. Thresholds of DO for avoidance have commonly been reported in the range of 3–1.5 mg O_2 L^{-1} (Howell and Simpson 1994; Baden et al. 1990; Pihl et al. 1991; Rabalais and Turner 2001; Eby and Crowder 2002). LIS trawl surveys found catches-per-unit-effort were lower in trawl samples taken at <2–3 mg O_2 L^{-1}, and at lowest DO concentrations sampled, many species were absent (Howell and Simpson 1994). In other estuaries, similar avoidance thresholds of 2–3 mg O_2 L^{-1} have been reported for juvenile menhaden, spot, summer flounder, and weakfish (Eby and Crowder 2002; Bell and Eggleston 2005; Tyler and Targett 2007), while Atlantic croaker (*Micropogonias undulatus*), hogchoker, and spot appeared to have lower avoidance thresholds, between 1 and 2 mg O_2 L^{-1} (Bell and Eggleston 2005; Pihl et al. 1991).

In deep waters with chronic hypoxia, fishes moved nearshore to shallower, more oxic waters (Pihl et al. 1991; Breitburg et al. 1992; Eby and Crowder 2002, 2004; Bell and Eggleston 2005). Fishes may show a stronger avoidance response to chronic hypoxia than to episodic hypoxia (Bell and Eggleston 2005).

Laboratory experiments have confirmed that estuarine larval and juvenile fishes also move away from hypoxic water. Avoidance thresholds vary greatly among species, from 4 to <1 mg O_2 L^{-1} (Deubler and Posner 1963; Breitburg 1994; Wannamaker and Rice 2000; Weltzien et al. 1999). Avoidance thresholds are given in Table 6.9. For some species, apparent avoidance thresholds reported for the field seem to be higher than those determined in the laboratory (Tyler and Targett 2007; Stierhoff et al. 2009a; Brady and Targett 2010), although menhaden had lower avoidance thresholds in the field (Wannamaker and Rice 2000; Eby and Crowder 2002; Schwartz et al. 1981).

Somatic Growth

Hypoxia has been directly associated with reduced growth in fishes, including a number of species commonly inhabiting LIS, and particularly for juvenile stages that inhabit nursery areas of the estuary (Stierhoff et al. 2003; Poucher and Coiro 1997; Cech et al. 1984; Hales and Able 1995; Bejda et al. 1992; Miller et al. 1995; Stierhoff et al. 2009b, c; McNatt and Rice 2004; Secor and Gunderson 1998). The USEPA (2000) made an extensive review of the minimal oxygen levels requirements for growth and survival of estuarine and marine fishes. Among LIS species, larval stages of crustaceans and fishes had higher thresholds for impairment of growth (4.7 mg O_2 L^{-1}) than did older juveniles, 3.3 mg O_2 L^{-1} (Miller et al. 1995; Poucher and Coiro 1997). The USEPA advisory criterion for protection of growth in marine species is 4.8 mg O_2 L^{-1} (USEPA 2000). Here, we focus on reports that were made subsequent to the USEPA (2000) review. Thresholds for growth reduction are summarized in Table 6.9.

Poor feeding under hypoxic conditions has been associated with much of the reduction of growth in many species at low DO (Ripley and Foran 2007; Chabot and Dutil 1999; Stierhoff et al. 2009b, c; Niklitschek 2001). Digestion requires increased aerobic metabolism (Brett and Groves 1979), so hypoxia tends to restrict feeding and slows digestion, thus providing nutrition inadequate for normal growth.

Some fishes avoid waters in the DO range associated with inhibition of growth (Breitburg 2002; USEPA 2000). The USEPA (2000) review also reported DO thresholds for fish avoidance behavior and concluded that for most species, the DO critical for growth inhibition occurred at approximately the same DO level at which avoidance behavior was reported. That critical DO concentration for growth inhibition was two to three times greater than the DO for the lethal LC_{50} criterion. More recent studies on juvenile spot, winter flounder, summer flounder, weakfish, and mummichog have found that growth may be impaired at DO concentrations higher than levels inducing avoidance responses (McNatt and Rice 2004; Stierhoff et al. 2009b, c; Wannamaker and Rice 2000; Stierhoff et al. 2003). Thus, active avoidance might not be sufficient to prevent hypoxia-related growth impairment.

Growth of striped bass and winter flounder were impacted at relatively high oxygen concentrations close to 5.0 mg O_2 L^{-1} (Cech et al. 1984; Breitburg 1994). Growth of summer flounder and Atlantic silversides were impacted at 3–4 mg O_2 L^{-1} (Stierhoff et al. 2006; Poucher and Coiro 1997). Juvenile Atlantic sturgeon growth was depressed three-fold at 3 mg O_2 L^{-1} (Secor and Gunderson 1998) and growth rates of Atlantic and shortnosed sturgeons were reduced at 40 % saturation (Niklitschek 2001; Secor and Niklitschek 2001; reviewed in USEPA 2003). Juvenile weakfish, Atlantic menhaden, and spot are relatively more tolerant with lower thresholds for growth reduction (DO < 2.0 mg O_2 L^{-1}).

At elevated temperatures, growth limitation has been commonly reported to occur at higher DO levels in winter flounder and summer flounder (Stierhoff et al. 2009a; Stierhoff et al. 2006), spot (McNatt and Rice 2004), and Atlantic sturgeon (Secor and Gunderson 1998), but not for juvenile weakfish (Stierhoff et al. 2009c).

Many larval and juveniles fishes that inhabit nearshore marshes experience fluctuating diel regimes of oxygen concentration with typical variation from hypoxia during night and in early morning, to hyperoxia during afternoon (Sanger et al. 2002). In laboratory studies, exposure of juvenile winter flounder, summer flounder, and Atlantic menhaden to cycles of oxygen concentration typical of such diel variation reduced growth when the oxygen minima achieved were similar to those that lead to decreased growth at constant low DO (Bejda et al. 1992; Steirhoff et al. 2006; McNatt and Rice 2004), and Table 6.9. In episodic hypoxia, growth of juvenile Atlantic croaker was decreased by half (Eby et al. 2005). In contrast, mummichog growth appeared not affected by cycling widely from 1 to 11 mg O_2 L^{-1}, though growth decreased in constant deep hypoxia at 1.0 mg O_2 L^{-1} (Steirhoff et al. 2003).

Summer flounder experiencing conditions of protracted diel hypoxia in the wild had impaired growth at higher oxygen concentrations than those reported for fish exposed to hypoxia in the laboratory (Stierhoff et al. 2009b). Similar results were found for weakfish; in laboratory studies, no growth reduction was found, even at low DO of 2.0 mg O_2 L^{-1} (Stierhoff et al. 2009c). The growth rate of wild-caught fish, however, was apparently impacted by diel fluctuation in oxygen, despite the mean

DO rarely dropping below growth protective criteria of 4.8 mg O_2 L^{-1} (Stierhoff et al. 2009b). Reduced growth of fishes in habitats with diel hypoxia has often been attributed to reduction in habitat area and abundance of prey, leading to increased competition (Eby et al. 2005) or to increased energetic costs of migration (Perez-Dominguez et al. 2006). Juvenile fishes may utilize shallow hypoxic waters because such areas may have fewer piscine predators than adjacent deeper habitats, and may provide shallow refugia for young fish (Paterson and Whitfield 2000; Sogard 1994).

Reproduction

Hypoxia effects on reproduction on LIS fishes have not been investigated, but hypoxia is known to affect reproductive and endocrine systems profoundly. In adult fishes, gonadal maturation is a complex aerobic growth process and the maturing gonad is typically one of the largest tissues requiring oxygen. Hypoxia disrupts endocrines associated with reproduction (Wu 2009; Wu et al. 2003; Thomas et al. 2007, 2011; Cheek et al. 2009). Declines occurred in trihydroxy-progesterone, FSH, LH, and GnRH in Atlantic croaker under hypoxic conditions (Thomas et al. 2006, 2007). Similarly, there was a significant fall in concentrations of plasma sex steroids in hypoxia-exposed mummichogs and gulf killifish (*Fundulus grandis*) (Landry et al. 2007).

Initiation of adult spawning may be delayed or inhibited by hypoxic exposure (Landry et al. 2007). One-month exposure to hypoxia (1.7–2.7 mg O_2 L^{-1}) reduced fertility and hatching success in eggs of Atlantic croaker (Thomas et al. 2006, 2007). Similar findings were reported for Gulf killifish, with females producing fewer eggs (Landry et al. 2007). Recent field studies of gulf killifish and Atlantic croaker inhabiting marsh creeks with diel hypoxia show characteristics of endocrine disruption and reproductive impairment (decline in gonad size and lower plasma concentration of reproductive hormones) similar to findings in laboratory studies (Cheek et al. 2009; Landry et al. 2007). Thomas and Rahman (2012) reported that in Atlantic croaker collected from chronically hypoxic waters in the Gulf of Mexico, 19 % of females had ovarian masculinization and aromatase suppression; these aberrations of normal reproduction, similar to those induced in the laboratory, suggest that chronic hypoxia causes the disturbances producing masculinization. Further suggestion of the role of hypoxia in masculinization is the greater numbers of males collected from hypoxic waters of the Gulf of Mexico (Shang et al. 2006; Thomas and Rahman 2012). Hypoxia also affects sex differentiation and can even lead to development of a single sex in zebrafish (*Danio rerio*) (Shang et al. 2006).

Compensations for Environmental Hypoxia

Fishes may be unable to avoid hypoxia but they may enhance oxygen uptake or decrease oxygen consumption; they often do both. Fishes have developed a variety of morphological, physiological, and behavioral adjustments to improve their abilities to extract oxygen in such impacted environments. However, few such adaptations have been studied for fishes of LIS.

A specialized behavior of some species is aquatic surface respiration (ASR) in which fish use the surface water, which usually has highest concentrations of oxygen in the water, for gill respiration. Mummichogs adopt ASR behaviors with increasing frequency in progressive hypoxia (McEnroe and Allen 1995; Stierhoff et al. 2003). Similar ASR behavior is adopted by other fishes in LIS (M. McEnroe, personal observation). Aquatic surface respiration allows fishes to survive and grow in habitats that are otherwise limiting (Weber and Kramer 1983; Stierhoff et al. 2003) but may also expose them to increased avian predation (Kramer et al. 1983; Kramer 1987; Kersten et al. 1991).

Many fishes increase ventilation rate in low DO (Pihl et al. 1991; Wannamaker and Rice 2000; Taylor and Miller 2001; Perry et al. 2009; Crocker and Cech 1997). As DO goes down further, ventilation rate then declines (Larsson et al. 1976; Perry et al. 2009; McEnroe and Kroslowitz 1997). Oxygen uptake during ventilation is directly related to the surface area of the gill (Hughes 1984). Some fishes living in hypoxic water have been reported to possess a greater gill surface area (GSA) than conspecifics inhabiting normoxic water (Chapman et al. 2000; Chapman and Hulen 2001; Timmerman and Chapman 2004) and in some species GSA can be rapidly altered in response to environmental oxygen (Sollid et al. 2003).

Oxygen uptake is controlled by blood hemoglobin (Hb) affinity for oxygen, and the amount of oxygen that can be carried in the blood depends directly on its hemoglobin (Hb) content in the red blood cells (rbc). Under hypoxia, many teleosts elevate blood Hb concentration by increasing the rbc number and thus hematocrit (Hct) (Hall et al. 1926; Grant and Root 1952; Petersen and Petersen 1990; Stierhoff et al. 2003). Menhaden caught in low DO waters had greater Hb content of blood than did fishes caught in normoxic waters (Hall et al. 1926). Windowpane flounder from the hypoxic WLIS had significantly greater volume of red cells than flounder from the eastern Sound, possibly in response to exposure to low DO (Dawson 1990). Mummichogs significantly increased Hct in acute hypoxia (Stierhoff et al. 2003), but Hct declined in chronic hypoxia (Greaney et al. 1980). Hemoglobin in blood of fishes living in hypoxic water generally has greater affinity for oxygen, and low P_{50} (O_2 concentration resulting in 50 % saturation of hemoglobin) compared with species from normoxic habitats (Campagna and Cech 1981; Weber 1988; Weber and Jensen 1988; Jensen et al. 1993). Hypoxia-tolerant mummichogs have Hb with very high affinity, followed by monkfish (*Lophius americanus*, also commonly called goosefish), oyster toadfish (*Opsanus tau*), and sea robins (DiMichele and Powers 1984; Green and Root 1933).

Hypoxia tolerance correlates with both gill surface area and high blood Hb affinity for oxygen (Richards 2011) although both these factors are also related to lifestyle activity (Grey 1954; Hall et al. 1926) and phylogeny (Richards 2011).

6.8.1.4 Science Gaps and Management Implications

Hypoxia in LIS results from complex interactions among geomorphologic, hydrographic, meteorological, allochthonous chemical input, and biological properties of the system. Prolonged stratification and eutrophication appear to be essential ingredients for development of hypoxic bottom waters. Results presented here and in recent publications support Welsh's (1995) hypothesis that inorganic nutrient loadings to LIS, primarily from municipal wastewater treatment facilities (WWTFs), and photosynthetic production fuel the biological respiration that drives the system to hypoxia. Approximately, 85 % of the oxygen consumption in the bottom waters of WLIS is attributable to planktonic microorganisms with the remainder due to larger planktonic organisms, benthic respiration, and chemical oxygen demand. The relative contributions to planktonic respiration due to direct remineralization of WWTF-derived TOC or oxidation of WWTF-derived ammonium through nitrification deserve investigation.

As a result of treatment facility upgrades, total inorganic N loadings from New York City WWTFs in the twenty-first century are considerably lower than they were in the 1980s and 1990s (Wilson et al. 2008 and references therein), but hypoxia has shown no clear indication of abating in duration, intensity, or areal extent in recent years (http://longislandsoundstudy.net/category/status-and-trends/water-quality). Although nutrient input has been reduced, current input is still sufficient to drive bottom waters in WLIS to seasonal hypoxia. The physical controls of stratification and ventilation have responded to climate change, and may counteract the response to reduced N loadings. An improved model of oxygen dynamics is a first-order management issue for LIS; it will require more thorough monitoring of physical, chemical, and biological processes, along with research to elucidate the roles of relatively understudied processes such as nitrification, allochthonous organic matter, and organic matter stored in the sediments as contributors to hypoxia.

The ecological effects of severe seasonal hypoxia on benthic communities are relatively well-known in LIS, at least over short ecological time scales of months to approximately 1 year. Considerably less is known about the combined effects of hypoxia and other stressors, including exposure to elevated levels of sulfide and ammonium, elevated, warmer water, and high CO_2 concentrations. There is considerable need for a long-term program to monitor the benthos in seasonally hypoxic areas and normoxic areas for comparison.

Low DO affects the physiology of fishes by limiting the aerobic scope available for swimming, feeding, growth, and reproduction. Early life stages are especially sensitive, and hypoxic conditions reduce inhabitability of essential nursery habitats of LIS. USEPA (2000) water quality criteria to protect marine species (4.8 mg O_2 L^{-1}), if achieved throughout LIS, would provide adequate protection for most species inhabiting the Sound. Notwithstanding improvements made, seasonal hypoxia in LIS does not meet those criteria. The combined effects of hypoxia and other factors such as T and pH on fishes in LIS deserve further study. These factors may act synergistically to exacerbate effects of hypoxia on the physiology of fishes; they critically reduce capabilities for uptake and transport of oxygen, fitness, and survival. During recent decades, T increases and pH decreases have occurred in regional waters. Such changing factors could require future reconsideration of the water quality criteria for DO.

6.8.2 Biological Invasions

Introduced species are often prominent members of LIS communities (Table 6.10): at summer's end, a buoy pulled up along the shore may be covered with the Asian green fleece (C. fragile fragile) and the Asian tunicate Styela clava, a habitat likely once dominated by the native blue mussel. Marina float edges support the Japanese sea anemone Diadumene lineata by the millions, a few centimeters above profuse populations of additional invasive sea squirt species Botrylloides violaceus and Didemnum vexillum from Asia and Botryllus schlosseri from Europe. Along rocky shores, the most common snail is the European common periwinkle, and the most common crab is the Asian shore crab. Between 1995 and 2010, new non-native species were detected at the rate of about one new species per year. Nevertheless, the scale and extent—the depth and breadth—of invasions in the Sound have often been lower on the environmental radar.

A question typically posed relative to recently glaciated environments such as LIS is whether "everything is really introduced anyway," implying that adding additional species by human-mediated transport simply extends a process already in play. The young age (less than 20,000 years) of, for example, newly deglaciated rocky intertidal shores and their natural recolonization by characteristic organisms such as barnacles, mussels, crabs, and seaweeds, appear to underscore an "it's invaded anyway" scenario, and that distinguishing human-mediated activities from natural recolonization is simply a matter of terminology, not process.

Several critical ecological and evolutionary aspects pertain here. Biological invasions consist of two separate and distinct phenomena: Range expansions and introductions. Range expansions are the movement of species over time by "natural" means, such as current-mediated dispersal of larvae or rafted adults. These movements occur largely along corridors, such as continental margins and island chains, often in response to long-term changes in ocean and shoreline conditions. Introductions are the transport of species by human-mediated vectors (such as shipping) and occur largely across barriers such as ocean basins and continents. These latter events are sui generis, having no precedent in Earth history at the time scales involved. We can now move virtually any species anywhere in the world in 24 h.

Thus, introductions are not simply accelerating natural processes. The vast majority of species involved would not have arrived eventually, simply given enough time (beyond invoking time scales beyond human concern in environmental mediation and management, such as hundreds of millions of years and the movement of continents). Natural recolonization of

Table 6.10 Examples of marine and estuarine invasions from Long Island Sound (compiled by J.T. Carlton)

Taxon/species	Origin	First occurrence in Southern New England/ New York	Vector (to Atlantic North America)	Remarks	Reference(s) documenting or reviewing first discovery or other occurrences in southern New England
CNIDARIA					
Anthozoa					
Diadumene lineata (= *Haliplanella luciae*) (Orange-striped sea anemone)	NWP	1892	SF	A clonal species that can achieve extraordinary abundance in fouling on floats	Verrill (1898)
MOLLUSCA					
Gastropoda					
Littorina littorea (European common periwinkle)	Europe	1875	BR/IR	Abundant on rocky intertidal shores, mud-sand flats, pilings, and salt marsh edges	Carlton (1982), Blakeslee et al. (2008)
Bivalvia					
Teredo navalis (shipworm)	SP?	<1860s	SB		Carlton (1992)
CRUSTACEA					
Isopoda					
Ianiropsis sp.	NWP?	1999	SF/BW	A tiny, abundant isopod in fouling communities	Pederson et al. (2005)
Amphipoda					
Caprella mutica	NWP	1998	SF/BW	Now most abundant caprellid in Long Island Sound fouling communities	Boos et al. (2011)
Microdeutopus gryllotalpa	Europe	1871	SF		Myers (1969), Bousfield (1973)
Decapoda					

(continued)

Table 6.10 (continued)

Taxon/species	Origin	First occurrence in Southern New England/ New York	Vector (to Atlantic North America)	Remarks	Reference(s) documenting or reviewing first discovery or other occurrences in southern New England
Brachyura					
Carcinus maenas (European shore crab)	Europe	<1817	SF/BR	Say (1817) reported this crab based upon a specimen collected by then 17-year old Titian Peale, later a well-known American naturalist and artist. Less common intertidally since the arrival of *Hemigrapsussanguineus* in the 1990s	Carlton and Cohen (2003)
Hemigrapsus sanguineus (Asian shore crab)	NWP	1992	BW	Abundant in rocky intertidal; also in salt marshes and subtidal fouling communities	McDermott (1998)
Caridea					
Palaemon macrodactylus (Oriental shrimp)	NWP	2001	BW	Common in brackish water fouling communities	Warkentine and Rachlin (2010)
BRYOZOA					
Cheilostomata					
Membranipora membranacea	Europe	1990	BW	Common on kelp blades	Berman et al. (1992)
Bugula neritina	SP?	1985	SF	Appearing in New England as permanently established populations in the 1980s. Very abundant in summer fouling communities	McGovern and Hellberg (2003)

(continued)

Table 6.10 (continued)

Taxon/species	Origin	First occurrence in Southern New England/ New York	Vector (to Atlantic North America)	Remarks	Reference(s) documenting or reviewing first discovery or other occurrences in southern New England
Bugula simplex	Mediterranean	1871	SF	In pre-1960s New England literature as *Bugula flabellata*, a distinct European species not in North America	Ryland and Hayward (1991), Ryland et al. (2011)
KAMPTOZOA					
Barentsia benedeni	Europe	1977	SF	No doubt present much earlier than the first records reflect, and much more widely distributed. Forms dense fouling mats in brackish water	Canning and Carlton (2000)
CHORDATA **Ascidiacea**					
Botryllus schlosseri	Europe (WP?)	1838	SF	"This is the most handsome and conspicuous compound ascidian found on the eastern coast of the United States" (Van Name 1945). Abundant in fouling communities	Verrill et al. (1873)
Botrylloides violaceus	NWP	1974	SF		Whitlatch and Osman (2000), Pederson et al. (2005)
Styela clava	NWP	1973	SF	Often very abundant in fouling communities; reaching 15 cm and more in height	Whitlatch and Osman (2000); Pederson et al. (2005)

(continued)

Table 6.10 (continued)

Taxon/species	Origin	First occurrence in Southern New England/ New York	Vector (to Atlantic North America)	Remarks	Reference(s) documenting or reviewing first discovery or other occurrences in southern New England
Styela canopus (=*S. partita*)	NWP	*Circa* 1870	SF	In the 1800s and early 1900s *S. canopus* was described as occurring "in summer in large masses on the piles of wharves" (Van Name 1912) in southern New England, along with the native ascidians *Aplidium constellatum* and *Perophora viridis* and the cryptogenic ascidian *Didemnum candidum* [as *D. lutarium*]; both *S. canopus* and *D. candidum* have become rare since the arrival during and since the 1970s of *Styela clava*, *Ascidiella aspersa*, *Botrylloides violaceus*, and other new fouling organisms	Pederson et al. (2005)
Ascidiella aspersa	Europe	*Circa* 1983	SF	Common in fouling communities	Whitlatch and Osman (2000), Pederson et al. (2005)
Diplosoma listerianum	NWP? (widely stated as Europe)	1970s	SF	Common in fouling communities	Whitlatch and Osman (2000), Pederson et al. (2005)
Didemnum vexillum	NWP	2000	SF	Often abundant in fouling assemblages, and one of the few non-native species found in the open ocean, now occurring in abundance on Georges Bank	Kott (2004), Pederson et al. (2005), Lambert (2009)

(continued)

Table 6.10 (continued)

Taxon/species	Origin	First occurrence in Southern New England/ New York	Vector (to Atlantic North America)	Remarks	Reference(s) documenting or reviewing first discovery or other occurrences in southern New England
ALGAE					
Chlorophyta (green algae)					
Codium fragile fragile (= Codium fragile tomentosoides)	NWP	1957	SF (not commercial oysters)	Abundant in sublittoral and common in low intertidal	Carlton and Scanlon (1985), Mathieson et al. (2008a, b)
Rhodophyta (red algae)					
Grateloupia turuturu (= G. doryphora)	NWP	1996	SF	First detected in Narragansett Bay, and now becoming more common in Long Island Sound	Marston and Villalard-Bohnsack (2002), Mathieson et al. (2008a, b, c)
Antithamnion hubbsii (= A. pectinatum auctt.; = A. nipponicum)	NWP	1986	SF/BW	Often abundant intertidally	Marston and Villalard-Bohnsack (2002)
Dumontia contorta (= D. incrassata)	Europe	1928	SF		Mathieson et al. (2008b)
"Heterosiphonia" japonica	NWP	2009	SF/BW	Thousands of specimens washed ashore in Rhode Island after Hurricane Bill, signaling the presence of this species on the Atlantic coast	Schneider (2010)
Bonnemaisonia hamifera (= Asparagopsis hamifera; = Trailliella intricata)	NWP	1927	SF		Mathieson et al. (2008a, b)
Neosiphonia harveyi (= Polysiphonia harveyi)	NWP	1847	SF	Common intertidally	McIvor et al. (2001), Mathieson et al. (2008a, b)
Lomentaria clavellosa	Europe	1971	SF		Mathieson et al. (2008a, b)

First occurrence is a date of first report or the date of first collection but not necessarily the date of introduction. A given species may have been collected earlier elsewhere on the Atlantic coast

Origins: NWP Northwest Pacific Ocean, SP South Pacific Ocean, WP Western Pacific Ocean

Vectors: SB Ship boring, SF Ship fouling, IR Intentional release, BR Ballast Rocks, BW Ballast water habitats rendered historically abiotic (by whatever process) characteristically derives from adjacent donor biotas and the recolonizing species concerned usually have, in turn, a long evolutionary history in the region (Sorte et al. 2010). In contrast, human-mediated invasions typically bring species from distant regions that lack the longer term (millennia to millions of years) evolutionary integration and roots of indigenous or endemic taxa. Species brought by ships or other means from European or Pacific theaters are not part of a "natural" continuing post-glacial colonization process, and can (and do) fundamentally change our understanding of the biological, biogeographic, and ecological histories of our shores.

Human-mediated invasions of marine and estuarine organisms into LIS have likely been occurring for nearly 500 years. Giovanni da Verrazano, sailing by LIS in 1524 aboard *La Dauphine* (De Vorsey 2007), may have been the first European since the Vikings to bring non-native species across the Atlantic to American shores. However, aside from the archeological record (which generally provides a resource for only hard-bodied or hard-shelled organisms), scientific records of the modern day marine life of LIS date only from the late 1700s and early 1800s, and these early works are fragmentary at best (for example, Herbst 1782–1804; Schoepf 1788; Rafinesque 1817, 1819). A more comprehensive picture begins to emerge by the late 1800s, but for many groups, there are few records until well into the mid-twentieth century.

A result of these historical lacunae is that many species that may have been introduced between the 1500s and 1800s, a nearly 400 year period, that are now common to both Europe and North America are assumed, without evidence, to be naturally occurring amphiatlantic taxa. These often include well-known ship-fouling species, an attribute that by itself does not mean that they were introduced. Instead, pending a detailed rendering of their historical biogeography, genetics, biology, and ecology, many taxa now have to be removed from "native" lists and treated as cryptogenic (Carlton 1996), as further discussed below.

As a result of the historiography of LIS biodiversity, our modern understanding of the scale of introductions (and thus their importance in the ecological history) of LIS is highly constrained by our limited knowledge of the early history of invasions. The recent resolution that the common native red alga *Polysiphonia harveyi*, found in 1847 in Stonington, CT is in fact an Asian species (now known as *Neosiphonia harveyi*) introduced on ship hulls (Table 6.10), speaks to our often translucent windows into the past. Our lack of knowledge is not restricted to long-ago eras: our sense of the pulse of changes in biodiversity in the Sound is also limited by the absence of any modern extensive, Sound-wide standard monitoring or observation program. Sometime between the late 1990s and 2010, a large (up to 6 cm in length) shrimp from Asia, *Palaemon macrodactylus*, spread throughout LIS and into Narragansett Bay without notice. Carlton (2009) has reviewed the many additional reasons for the potential underestimation of the number of non-native species. Despite the long-term density of marine biologists from Woods Hole to New York, and despite the recent completion of a decade's global effort focused on censusing marine life, at the beginning of the twenty-first century, we have no thorough, scholarly atlas of either the historical or modern biota of LIS.

We thus do not know the number of non-native species in LIS. About 70 largely macroscopic non-indigenous species are recognized in the 275 km inland sea corridor from the Hudson River estuary to Buzzards Bay (J.T. Carlton, unpublished data). The actual number of non-native species in southern New England may be two or three times of that. Examples are presented here (Table 6.10) where some of the more conspicuous and prominent invertebrate invaders include: Sea anemones, mollusks, crustaceans, bryozoans, and ascidians. Non-native but largely freshwater fishes may enter the brackish or marine waters of the Sound, such as the mosquito fish (*Gambusia affinis*) and the brown trout (*Salmo trutta*)(Gordon 1974; Mills et al. 1997; Briggs and Waldman 2002). Certain non-native salt marsh plants are common, such as the introduced genotype of the common reed (Saltonstall 2002). The non-native mute swan (*Cygnus olor*) is a common omnivore in salt water throughout the Sound (Allin and Husband 2003).

Poorly known are many common and often predominant taxa such as protists, diatoms, foraminiferans, sponges, numerous species of hydroids, nematodes, flatworms, oligochaetes, polychaetes, cheilostome and ctenostome bryozoans, nudibranchs, copepods, amphipods, isopods, and tanaids. Examples of cryptogenic species of some of these groups are shown in Table 6.11. Many of these species are not rare: the hydroid *Obelia dichotoma* (in earlier literature as *O. commissuralis*) has long been a prominent member of regional fouling communities, Nutting (1901) noting it as "abundant in Woods Hole [Cape Cod, Massachusetts].... growing on piling of wharves and on submerged timbers." The bryozoan *Amathia vidovici* forms prominent fouling masses in LIS in the summer, and has been present in New England since at least the 1850s (Verrill et al. 1873, as a "new species," *Vesicularia dichotoma*). Its biogeographic history remains unknown, as does the history of many other bryozoans.

Most of the recognized invasions of LIS are mollusks, crustaceans, ascidians, and macroalgae, groups that are either larger in size, more tractable taxonomically, or have attracted interest by specialists (such as ascidians and certain common filamentous algae). Approximately, half of the non-native species in LIS are European and approximately half have their origin or probable origin in the Pacific Ocean. Of interest is that while nearly three-quarters of the 14 species of non-native algae recognized from southern New England come from the Northwest Pacific Ocean, all of the introduced mollusks (nine species) of clear provenance originate from the Atlantic Ocean. No introduced mollusks from Japan or the Northwest Pacific in general are yet established in LIS and adjacent waters (J.T. Carlton, unpublished data). The reasons for this contrast remain unknown.

Absent major historical importations of foreign oysters (such as occurred on the Pacific coast of North America and in France), shipping has been the primary vector for non-native species invasions in LIS. Ship-mediated vectors include hull fouling, internally fouled seawater systems (piping and sea chests), ballast water, solid ballast (such as rocks), and the shipworm- and gribble-bored hulls of wooden vessels. Early invasions included a guild of classic European harbor-associated fouling organisms, such as the green crab, and, not shown in Table 6.10, the salt marsh snail *Myosotella myosotis* (present by the 1830s) and the hydroid

Ectopleura dumortieri (present since about the 1850s) and likely many of the other hydroids in a holding pattern in the cryptogenic bin in Table 6.11. Probably arriving from Europe, but possibly not native there, were the sea squirt *Botryllus schlosseri* and the shipworm *Teredo navalis*, both now suspected of originating in the Pacific Ocean. Some of these may be pre-colonial or colonial-era invasions commencing as early as the 1500s and 1600s. Between the 1870s and 1890s,

Table 6.11 Cryptogenics: Examples of species in Long Island Sound that may be native or introduced (compiled by J.T. Carlton)

Porifera	*Protohydra leuckartii*
	Sarsia tubulosa
Halichondria bowerbanki	Polychaeta
	Arabella iriçolor
Haliclona canaliculata	*Autolytus prolifer*
	Brania clavata
Leucosolenia sp.	*Eumida sanguinea*
	Fabricia sabella
Lissodendoryx isodictyalis	*Harmothoe imbricata*
Scypha sp.	*Lepidonotus squamatus*
	Loimia medusa
	Myxicola infundibulum
	Nicolea zostericola
Hydrozoa (hydroids)	*Phyllodoce maculata*
Amphinema dinema	Bryozoa
	Electra spp.
Amphinema rugosum	*Conopeum* spp.
Campanularia hincksii	
	Cryptosula pallasiana
Clytia hemisphaerica	*Bowerbankia* spp.
	Amathia vidovici
Dynamena pumila	*Anguinella palmata*
	Amphipoda
Ectopleura larynx	*Caprella penantis*
	Caprella equilibra
Ectopleura crocea	Tanaidacea
Gonothyraea loveni	*Leptochelia savignyi*
	Tanais dulongii
Halecium halecinum	Ascidiacea
	Ciona intestinalis
Laomedea calceolifera	*Didemnum candidum*
	Perophora viridis
Obelia dichotoma	
Obelia bidentata	
Opercularella lacerata	
Pennaria disticha	
Plumularia setacea	
Protohydra leuckartii	

coincident with a global surge in shipping (Carlton and Cohen 2003; Carlton et al. 2011), the bryozoan *Bugula simplex*, the sea anemone *Diadumene lineata*, and the hydroid *Gonionemus vertens* all first appeared in southern New England.

What followed appears to be a fairly quiescent period of invasions for the next 50 years, marked by apparently few invasions (or by fewer studies). However, starting in the 1950s, invasions recommenced and in fairly rapid-fire succession prominent fouling organisms began to arrive green fleece in 1957; the sea squirts *S. clava* (1973), *B. violaceus* (1974), *D. listerianum* (1970s), and *A. aspersa* (1983), and the bryozoan *Bugula neritina* (1985). The 1990s–2000s continued to mark an era of numerous invasions in southern New England, including the bryozoan *Membranipora membranacea* (from earlier inoculations in the Gulf of Maine), the Asian shore crab (from earlier inoculations to the south), the alga *Grateloupia turuturu*, the amphipod *Caprella mutica,* and the isopods *Ianiropsis* spp. and *Synidotea laevidorsalis*, the sea squirt *D. vexillum*, the shrimp *P. macrodactylus*, the alga *Gracilaria vermiculophylla* and the alga *Heterosiphonia japonica,* and others (Table 6.10; Pederson et al. 2005).

Finally, as Stachowicz et al. (2002) have noted, new invasions of LIS may be mediated by warming waters. These invasions consist of two guilds: Species from overseas that were historically impeded from colonizing New England by previously too-cold temperatures and species from southern United States waters that are steadily making their way northward (Carlton 2010; Carlton et al. 2011). Harbingers of a future Sound include southern elements that now make only a transient appearance, but may 1 day overwinter and become part of the established biota. Among these are the bryozoan *Zoobotryon verticillatum*, which can form startlingly large masses (two meters long by one meter wide) fouling marina floats and docks in southern California, and the golf-ball size sea squirt *Styela plicata*, which can become equally prominent. Both have been found sparingly in LIS in the 2000s, as waif populations (*Z. verticillatum*) or as individuals (*S. plicata*). While *Z. verticillatum*is unlikely to be native to southern US waters, and *S. plicata* is native to Asia, future invasions will include native species as well, already reflected in the 1980s establishment of the well-known southern clam *Rangia cuneata* in oligohaline waters of the Hudson River (Carlton 1992).

6.8.3 Effects of Climate Change

6.8.3.1 Coastal Acidification

The process of fossil fuel combustion that promotes increasing concentrations of anthropogenic CO_2 leading to declining levels of pH and CO_3^{2-} in the ocean has generally been referred to as ocean acidification (Doney et al. 2009). Estuaries experience lower pH nearshore-specific processes that can further enhance dissolved CO_2 concentrations. Many estuaries are "net heterotrophic" due to terrestrial, riverine, and wetland supplements of allochthonous carbon (Gattuso et al. 1998;

Ram et al. 2003; Taylor et al. 2003a; Koch and Gobler 2009; Gattuso and Hansson 2011), which can lead to waters that are supersaturated with CO_2. Episodic discharge of acidic river waters into the Gulf of Maine can depress aragonite Ω (saturation) values with potentially negative consequences for calcifying organisms such as soft-shelled clams (Salisbury et al. 2008). Additionally, upwelling can result in coastal waters with CO_2 levels exceeding 1,000 ppm and decreased levels of CO_3^{2-} (Feely et al. 2008; Salisbury et al. 2008; Doney et al. 2009). Anthropogenic nutrient loading in coastal zones (Howarth 2008) has increased algal blooms (Beman et al. 2005), and subsequent heterotrophic degradation of bloom-derived organic matter may result in substantial declines in pH and increases in CO_2. This may especially be the case in temperate coastal zones, including LIS, during summer when the net heterotrophic nature of these systems can be maximal (Blight et al. 1995; Ram et al. 2003; Thomas et al. 2004a, b). Sediments in estuaries are typically the most heterotrophic part of the system due to the accumulation of sinking organic matter and generally have the highest CO_2 concentrations. The microbial degradation of this organic matter results in the flux of CO_2 at the sediment–water interface (Rasmussen and Jørgensen 1992) and has the same chemical effects as atmospheric fluxes of CO_2, reducing the seawater pH and carbonate ion availability. Estuaries on Long Island and Chesapeake Bay have experienced pCO_2 concentrations between 500 and 1,500 ppm and pH levels as low as 7.6 (Talmage and Gobler 2009; Waldbusser et al. 2011). Coastal zones are regions that may regularly experience high levels of CO_2 and may already periodically experience decreased CO_3^{2-} availability.

The decrease in CO_3^{2-} availability may threaten marine organisms with calcifying parts. One group of marine organisms affected by increasing CO_2 levels are bivalve molluscs (e.g., Miller et al. 2007). Filter-feeding bivalves are considered ecosystem engineers in coastal waters due to the filtration services they provide (Colson and Sturmer 2000; Gutierrez et al. 2003). Their filtration has the potential to control eutrophication and harmful algal blooms (Officer et al. 1982; Cerrato et al. 2004), thus increasing light penetration (Newell and Koch 2004), which has been shown to benefit submerged aquatic vegetation in a LIS estuary (Carroll et al. 2008; Wall et al. 2008). As such, these organisms can have major effects on ecosystem structure and function (Raillard and Menesguen 1994; Grant 1996; Arnold et al. 2002).

In recent decades, wild populations of shellfish native to LIS, such as the hard clam, Eastern oyster, and bay scallop, have been under increasing pressure from overfishing, loss of habitat, hypoxia, and harmful algal blooms, and their populations have experienced precipitous declines (Jackson et al. 2001; Myers et al. 2007). New York estuaries offer prime examples of such declines, as NYSDEC reports on shellfish landings for the hard clam fishery in Great South Bay and the bay scallop fishery in the Peconic Estuary have shown declines by more than 99 % since the early 1980s. Factors cited as contributing to these precipitous declines include overharvesting (Kraeuter et al. 2008), reduced reproductive success, predation (Kraeuter 2001), harmful algal blooms (Greenfield et al. 2004; Bricelj and MacQuarrie 2007), and a changing food supply (Greenfield et al. 2005; Lonsdale et al. 2007).

Shellfish produce calcareous shells and the production of these shells depends on the same calcification processes and availability of CO_3^{2-}. Studies of $CaCO_3$ secreting bivalves have found that sediments undersaturated with respect to aragonite ($\Omega = \sim0.3$) can cause enhanced mortality of juvenile hard clams (at 0.2, 0.3, 1, and 2 mm size classes) (Green et al. 2004). Elevated (~740 ppm) CO_2 concentrations decreased calcification rates in blue mussels and oysters (a decrease of 25 and 10 %, respectively, over present levels (Gazeau et al. 2007). Berge et al. (2006) described decreased growth and metabolic rates in the blue mussel at pH levels of 7.4. Similarly, in the Mediterranean mussel (*Mytilus galloprovincialis*), a reduction in seawater pH to 7.3 decreased rates of oxygen consumption, increased N excretion indicating the net degradation of proteins, and reduced growth (Michaelidis et al. 2005).

Bivalve larvae are critical to the population dynamics of the adult populations, as any decline in larval populations can have profound implications for future shellfisheries (Caley et al. 1996; Gosselin and Qian 1997; Carriker 2001; Cragg 2006; Miller et al. 2007). They may also be highly sensitive to increased CO_2 concentrations. Green et al. (2009) demonstrated that increasing surface sediment aragonite Ω from 0.25 to 0.53 can increase the settlement of soft shell clams by three-fold, suggesting that recently settled larvae may be sensitive to CO_3^{2-} availability (Green et al. 2009). Experimentally enhanced CO_2 has been shown to decrease the development rate of Pacific oyster (*Crassostrea gigas*) larvae (Kurihara et al. 2007) and the Mediterranean mussel (Kurihara et al. 2008). Carbon dioxide can play a central role in influencing the survival and physiology of ecologically valuable bivalve larvae; experiments on Long Island showed that hard clams, eastern oysters, and bay scallops displayed dramatic declines in survivorship and delayed metamorphosis under elevated CO_2 concentrations projected for the twenty-first century and beyond (Talmage and Gobler 2009). The eastern oyster was the most resilient of the three with lowered growth and survival under elevated CO_2, but significant decreases in survival only under the highest CO_2 treatments ($\sim1,500$ ppm). The hard clam and bay scallop larvae both displayed increased growth, survival, and more robust and rapid development when grown under pre-industrial (~250 ppm) CO_2 concentrations compared to current levels (~390 ppm) (Talmage and Gobler 2010). Coastal ecosystems already experience elevated levels of CO_2 (Salisbury et al. 2008; Talmage and Gobler 2009; Waldbusser et al. 2011), in part due to decomposition of the organic matter present in estuaries (Gattuso et al. 1998; Paerl et al. 1998; Thomas et al. 2004a, b). Bivalve larvae spawned into such environments will experience significant reductions in their survival rates. As coastal oceans acidify over the next two centuries, there may be selection pressure for bivalves to become more resistant to high CO_2. Acidification of LIS is nearly inevitable, and may already be occurring in areas of increased eutrophication and freshwater inputs.

6.8.3.2   Temperature Effects During Winter and Spring

The winter-spring phytoplankton bloom in LIS has historically dominated annual primary production (Riley and Conover 1967). It provides a significant source of organic matter for both pelagic and benthic components of the food web. Timing and magnitude of spring blooms are controlled by several factors including incident radiation, wind and tidal stress, and water T (Iriarte and Purdie 2004).

Temperate coastal systems like LIS typically exhibit a mismatch between the spring bloom and zooplankton grazing because low temperatures restrict zooplankton activity at the time of the bloom (Smetacek 1984; Keller et al. 2001), so that much of the bloom production is deposited ungrazed on the seabed where this fresh phytodetritus fuels a detritus-based benthic and demersal food web (Graf et al. 1982; Rudnick and Oviatt 1986). Benthic faunal utilization of this material is typically also delayed in temperate systems because of low T, and as bottom water warms in spring, microbial and faunal activities have been shown to rapidly increase in LIS (Cheng and Lopez 1991; Gerino et al. 1998).

There is a long-term trend of winter warming in LIS and surrounding region (Lwiza 2008). Slightly elevated temperatures during the winter-spring dramatically affect the magnitude and fate of the spring bloom (Keller et al. 1999, 2001; Oviatt 2004; Oviatt et al. 2002). In Narragansett Bay, winter conditions ~2 °C warmer than average (3–4 °C instead of ~1–2 °C) resulted in a weakened phytoplankton bloom caused by higher abundance of grazing planktonic copepods. This slightly warmer water (4 °C) allowed copepods to grow fast enough to graze phytoplankton production. Thus, slightly warmer temperatures have a major impact on phytoplankton production that is mediated by the response of zooplankton to T. Under warmer conditions, more production goes to pelagic components of the food web, and less to benthic and demersal components. It is not known whether benthic fauna and sediment microbial communities respond similarly to warming of the water column, but benthic communities definitely respond to decreased input of phytodetritus (Nixon et al. 2009). The long-term reduction in the Narragansett Bay winter-spring bloom over the past half century has been caused by warmer winters along with changes in wind mixing and cloud cover, resulting in significantly lower input of organic matter to the seabed and reduced benthic metabolism (Nixon et al. 2009; Smith et al. 2010). The region has also experienced reduction in cold-water demersal species and increase in warm water pelagic fishes. Warmer winter temperature increases predation by the sand shrimp (*Crangon septemspinosa*) on winter flounder larvae, demonstrating the effect of climate change on predator–prey interactions. Increases in crabs (especially *Cancer* spp.) have also occurred (Collie et al. 2008). Long Island Sound probably responds similarly to Narragansett Bay with respect to winter T shifts, but it has been relatively little studied.

Sensitivity of phytoplankton and zooplankton to meteorological and oceanographic conditions such as slight changes in winter T may have ramifications for higher trophic levels and may determine relative success, composition, and diet of apex species. The effects of bottom-up control of food webs are complicated by several factors, including species replacement and top-down processes, but a shift in carbon and energy flow through different components of the food web is likely to have great impact on higher trophic level predators such as striped bass, weakfish, and bluefish (Oviatt 2004; Frederiksen et al. 2006, 2007).

The benthos of broad, shallow continental shelves in temperate regions experience a wide annual T range; benthos in LIS, for example, are subjected to an annual T range from ~0 to 22 °C. Temperature is likely to control benthic nutrition, especially for detritivores, because it affects both microbial degradation rates and macrofaunal responses. On temperate continental shelves, phytodetritus deposition resulting from the spring bloom often occurs when bottom water temperatures are low, limiting benthic response. Both benthic microorganisms and macrofauna appear to be limited by low T during this period (when T typically ranges from 0 to 4 °C), but they do not respond equally (Hines et al. 1982; Kristensen et al. 1992; Gerino et al. 1998). Graf et al. (1982) showed that microorganisms respond immediately to phytodetritus deposition but most mesothermic benthic macrofauna (an exception is *M. balthica*) do not respond until bottom water T increases a few degrees. This time lag in macrofaunal response may occur weeks after deposition of the spring bloom. Thus, freshly deposited phytodetritus may be partially mineralized before T increases to allow for macrofaunal feeding.

Although colder water species such as blue mussels are able to feed at temperatures as low as 1 °C (Thompson 1984), many mesothermic macrobenthos typical of LIS do not appear to feed at all at or below ~6 °C. The deposit-feeding bivalves *Nucula proxima* and *Yoldia limatula*, two dominants in CLIS (Gerino et al. 1998), cease feeding at 6 °C (Bender and Davis 1984; Cheng and Lopez 1991) and eastern oysters stop filtering at ~5 °C (Loosanoff 1958; Pomeroy et al. 2006). Bioturbation in temperate regions is controlled by T (Gerino et al. 1998; Goedkoop et al. 1997; Kristensen et al. 1992); bioturbation by *Neanthes virens* is inconsequential below 6 °C. Benthic foraminifera do not calcify (and thus grow) during the coldest ~5 months of the year in LIS (Varekamp et al. 2010).

Heterotrophic microbial activity is also affected by T. Temperature changes too often in temperate environments to select for psychrophilic bacteria (Isaksen and Jørgensen 1996; Jørgensen, personal communication). Nevertheless, mineralization of phytodetritus begins immediately upon deposition, though the rate is sensitive to T (e.g., Shiah and Ducklow 1994; Gerino et al. 1998).

The benthos in environments that experience low seasonal variation in T (deep sea, polar, and tropical regions) exhibit no such time lag between phytodetritus deposition and macrofaunal response; animals respond immediately to fresh food (Stead and Thompson 2003, 2006; Stead et al. 2003; Witte et al. 2003; Hudson et al. 2004; Moodley et al. 2005; McMahon et al. 2006). It is likely that the effect of T-controlled time lag in macrofaunal response in temperate regions ramifies through the food web.

6.8.4 Lobster Mortality Events

The American lobster fishery has been an important component of New England's economy and culture for hundreds of years. Abundance has waxed and waned over this time; LIS supported a large increase in the lobster population in the 1990s.

However, in the fall of 1999, the lobster population in LIS experienced an unprecedented mortality event from which it has not recovered. State and federal landings data show that prior to the die-off, Connecticut and New York commercial landings ranged from 7 to 11.7 million pounds annually, with an ex-vessel value of $18–40 million. Participation in the fishery peaked in 1998 when over 1200 residents of the two states bought licenses to fish commercially. By 2002, fewer than 900 residents purchased licenses and by 2009 that number declined to 592. Commercial landings plummeted to about 1 million pounds by 2009, with an ex-vessel value under $3.5 million.

6.8.4.1 Historic Mortality Events

The American lobster ranges from maritime Canada to Virginia; south of Long Island the species is found only offshore in deep cool water. Long Island Sound marks the southern extent of their distribution in warm nearshore waters. As the animals approach the limit of their T tolerance, the likelihood of being exposed to stressful conditions increases. Reports of localized small-scale mortality events in the Sound are common, especially during the fall when water temperatures are highest. Periodic larger mortality events have also occurred. For example, outbreaks of gaffkemia (caused by the bacteria *Aerococcus viridans homari*), also known as wasting or red tail disease, occurred in the fall of 1990, 1991, and 1993.

6.8.4.2 The "Die-off" in 1999

Beginning in late summer and early fall of 1999, reports of large numbers of dead and dying lobsters of all sizes came in from lobster fishers in the western Sound. Reports of dead and lethargic lobsters in the central and eastern basins of LIS increased soon after. In western ports, commercial landings declined by over 90 % in fall 1999 compared to their 1995–1998 average. Reductions in fall landings for ports east of Norwalk, CT ranged from 64 to 91 %. The declaration by the US Department of Commerce of a fisheries disaster in early 2000 fueled extensive research and data collection by many academic, federal, and state agencies from 2001 to 2005. These studies provided substantial evidence that a combination of physical factors pushed the Sound's lobster population far out of equilibrium with its environment in the fall of 1999. Lobsters were subjected to increasingly hostile conditions that overwhelmed their immune systems. Those unable to either cope physically or move into a better environment ultimately died. Although many of these adverse conditions were not new in the Sound, collectively they had never been as severe. The analogy of the "perfect storm" seems the best way to capture the sequence of events. The next sections list a chronology of the most important events, published in greater detail by Connecticut Sea Grant (Pearce and Balcom 2005; Balcom and Howell 2006).

Warm Waters and Hypoxic Conditions

Water temperature affects all of life history processes of American lobster including growth, maturity, spawning, egg, and larval development. American lobsters
are capable of detecting T changes of 1 °C (Jury and Watson 2000), demonstrate
a thermal preference of 12–18 °C, and will avoid temperatures >19 °C (Crossin
et al. 1998). Water temperatures >28 °C cause mortality to adult lobsters within
48 h and more quickly when dissolved oxygen is reduced below 6.4 mg L^{-1}
(McLeese 1956). Prolonged exposure to water T above 20 °C causes physiological
stress as indicated by hemolymph acidosis (Dove et al. 2005), increased respiration rate (Powers et al. 2004), and depressed immune response (Dove et al. 2005;
Steenbergen et al. 1978). It has also been linked to increased incidence of disease
including epizootic shell disease (Glenn and Pugh 2006), and excretory calcinosis
(Dove et al. 2004).

There has been a widespread increase in the spatial range and duration of water
temperatures above 20 °C in the coastal waters of southern New England. The
best illustration of this phenomenon is the marked increase in the number of days
each year when the mean bottom water T remains above 20 °C, as measured at the
intakes of the Millstone Power Station in ELIS (Fig. 6.53).

Bottom water temperatures were 1–2 °C warmer than average for a number of months during 1998–1999 under drought conditions. This T increase
exceeded the upper tolerance threshold for lobsters (20 °C) for more than 83
"degree days." Temperatures recorded in deep-water areas of the Sound were
higher than 21 °C by late summer 1999; shallow areas recorded temperatures
higher than 23 °C. As the water warmed up, oxygen solubility decreased.
Hypoxic conditions formed in the WLIS for about 50 days, which was not out
of the ordinary. The worst period occurred in the first week of August and most
of the hypoxia had dissipated by August 21. Lobsters are known to "herd" or
crowd in high numbers near margins of hypoxic zones where oxygen concentrations are slightly less than 2 mg O_2 L^{-1}, their normal respiratory threshold.
Because of these two factors—water T and hypoxia—an already dense population of lobsters was crowded into the few remaining areas of the Sound cool
enough and oxygenated enough to sustain them. Those lobsters were then
caught in shrinking areas of tolerable conditions and had no escape when their
"islands" disappeared.

In August 1999, the water column was stratified by both T and S. On August
28, the region felt the effects of Hurricane Dennis to the south, coupled with
strong winds from a northern cold front. These clashing weather conditions caused
rapid and complete vertical mixing of the water column, bringing warm surface
water to the bottom. As a result, deeper bottom water temperatures increased
by 1 °C in 6 h (from 21 to 22 °C) in some locations. When tropical storm Floyd
passed through on September 16, 1999, dropping more than three inches of rain,
the gush of freshwater runoff that followed the storm resulted in haline stratification, trapping the warm water on the bottom.

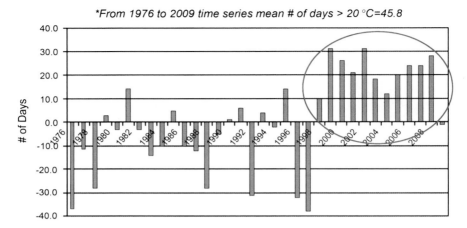

Fig. 6.53 Anomalies from the mean number of days >20 °C of the Millstone Power Station bottom temperature, 1979–2009. Anomalous high values in the year since the die-off (1999–2009) are *circled*. Daily mean temperature is computed from continuous recordings at the Power Station intakes. Data provided courtesy of Dominion Nuclear Connecticut (DNC 2010)

Role of Pesticides

The unusually large freshet from storm Floyd also may have flushed pesticides into the Sound from coastal areas of New York and Connecticut where they had been sprayed to combat West Nile virus, a new threat in the area. It is clear that lobsters in WLIS would have suffered extensive mortality in 1999 in the absence of any pesticide effects even though subsequent laboratory studies showed that these pesticides (methoprene and resmethrin) are lethal at low concentrations to larval, juvenile, and adult lobsters (De Guise et al. 2005; Walker et al. 2005). These authors demonstrated that sublethal effects, such as elevated stress hormone level and depressed immune responses, were noted after chronic exposure to as little as one-hundredth of the lethal dose. However, field sampling throughout the LIS watershed showed that concentrations of these pesticides measured in surface waters at 27 sites (39 total samples) from June to August 2001 and 2002 were lower than the levels found to be acutely toxic to larval and juvenile lobsters (the most sensitive stages) except in two cases in wetlands on Long Island immediately after spraying (Zulkosky et al. 2005). It cannot be ruled out that in limited areas of the far western Sound exposure to pyrethroid pesticides may have further weakened some lobsters, making them even more susceptible to disease or the inhospitable environmental conditions (De Guise et al. 2005). Lobster larvae, if they were present in these same areas at the time of exposure, also could have suffered. However, larval abundance in the western Sound peaked as usual in early July, according to CTDEP sampling (Giannini and Howell 2010), while most pesticide applications were made later in the summer and fall.

Several models were developed to describe the effects of transport, dilution, and degradation of pesticides washing into the perimeter of the Sound from August to October 1999 on the concentration distributions (Landeck Miller et al. 2005, Wilson et al. 2005). Model loadings were based on CT DEP and NYS DEC records of pesticide applications, including land-locked areas, and so the total mass of pesticides introduced to the models were upper bounds. Differences in the spatial resolution of the models, and the locations and delivery rates to coastal cells, led to substantial differences in the maximum concentrations and distributions that the models predicted. The models agreed that concentrations attenuated very rapidly with distance from the coast and that highest values were confined to areas of the western Narrows (New York waters off the Bronx, and Nassau and Suffolk Counties). They also agreed that predicted maxima (~1 PPB) were an order of magnitude below the estimated threshold for lethal impact on lobsters. However, the uncertainties in source locations and the degradation rates used in the models and the neglect of interactions with particles leaves open the possibility that pesticide may have been an added stressor to lobsters in those areas.

The Final Effects

The cumulative effects of meteorological and oceanographic factors stressed the lobsters to a point where their immune systems could not cope with rapidly changing and increasingly lethal conditions. During the last months of 1999, thousands of lobsters, crabs, and sea stars were collected dead, or in their weakened state they became infected with parasitic amoebae and then died (Mullen et al. 2005). Sustained above-average bottom water T was the driving force. Other factors including exposure to elevated levels of bottom water sulfide and ammonium collectively compounded the physiological stress of high T and resulted in the rapid onset of lobster mortality (Draxler et al. 2005; Robohm et al. 2005; Cuomo et al. 2005).

Mortality events following the 1999 "die-off" have been less widespread although better documented. Sampling of the commercial catch by CTDEEP and NYS DEC staff has recorded mortalities in all seasons but primarily in the fall. Observations of the Connecticut commercial catch since 1983 show an increase in the incidence of dead lobsters in the 5 years following the die-off; incidence increased from a very rare event (<1 % of the monthly observed catch) to a common one (up to 7 % of the monthly observed catch), with individual trips recording mortalities as high as 46 %.

6.8.4.3 Abundance Trends and Changes in Distribution

The CTDEEP trawl survey data indicate that conditions throughout LIS were favorable for recruitment of small young lobsters in the early 1990s, allowing the lobster stock to increase to historic high abundance (Gottschall and Pacileo 2010). This enhanced survival of young ended abruptly with the die-off. Similar trends were seen in Rhode Island and southern Massachusetts waters, as well as in offshore canyons south of Long Island (ASMFC 2009). Research survey catches in these areas show that abundance of small lobsters increased through the 1990s followed by a decline to record lows. These trends suggest that regional environmental factors were at first favorable and then very unfavorable for lobster production through the 1990s and continuing into the following decade.

Catch data from a lobster trap survey conducted by NYS DEC in WLIS during 2003 showed that lobster catches dropped sharply when bottom dissolved oxygen fell below 4 ppm and water temperatures rose above 18 °C. This drop was particularly noted for egg-bearing females. In 2003, bottom water temperatures recorded in this survey were above 19 °C from mid-August through September. By 2007, bottom water temperatures were above 19 °C from mid-July through September. The highest catch rate in all 4 years (2003–2007) of the survey was in the deepest water depths sampled (31–35 m).

Catch data from the trawl survey conducted by CTDEEP since 1984 (Gottschall and Pacileo 2010) also indicate that there has been a shift in lobster distribution in the central and western basins of LIS. At sites with muddy bottom sediment, preferred habitat for lobster in this area, catches have shifted from shallow inshore waters to deeper mid-Sound waters. From 1984 to 1991, the mean catch at sites less than 9 m deep was comparable to the mean catch at sites deeper than 27 m. However, the mean catch in 2000–2008 at shallow sites was less than half the mean for deep sites (Fig. 6.54). It appears that loss of optimal nearshore habitat may be causing the dwindling stock to contract spatially into deeper water. In WLIS, the potential expansion of chronic hypoxia under conditions of higher T compounds the physical effects of both factors.

6.8.4.4 Fishing and Natural Mortality

Assessment of the effect of harvest removals on the Sound's lobster population is complicated by the increase in natural losses made obvious during the die-off in 1999 and following years. Catch patterns in all southern New England trawl surveys carried out in state waters by Connecticut, Rhode Island, and Massachusetts and by the National Marine Fisheries Service in offshore federal waters indicates that non-fishing (natural) mortality increased approximately two-fold beginning in 1997 (ASMFC 2010). Since 1999, harvest removals (pounds landed) and harvest effort (traps fished and licenses sold) in this entire area have decreased substantially; however, lobster abundance has not rebounded to even near-average levels. Although lobster stocks north of Cape Cod are thriving, the reproductive potential, and abundance of the southern New England stock remains low.

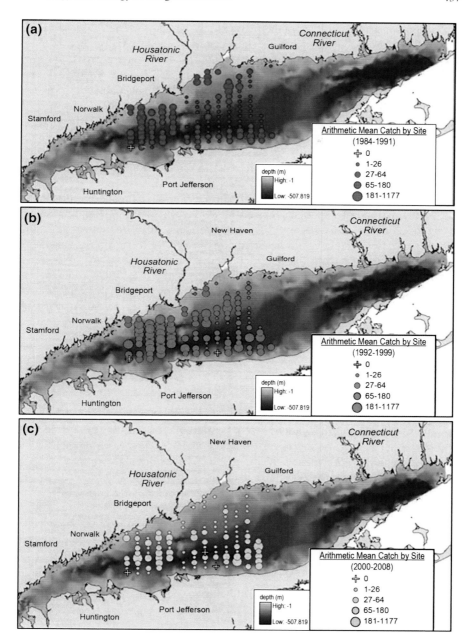

Fig. 6.54 Special distribution was American lobster catches in the western and central basins from the Long Island Sound Trawl Survey. Arithmetic mean catches are shown for mud bottom sites only over three time periods: **a** 1984–1991, **b** 1992–1999, and **c** 2000–2008

Some of the losses in the southern stock may be due to "shell disease" or the bacterial breakdown of the lobster's shell forming pits or lesions. First noted in 1997, shell disease has been observed throughout ELIS, Narragansett Bay and Rhode Island Sound, and Buzzards Bay and Vineyard Sound. Each year, it is observed in about 15–30 % of the population, with the highest percentage in egg-bearing females. An index of abundance of newly settled lobsters, measured in Narragansett Bay beginning in 1992, matched subsequent recruitment to the fishery for several years but broke down after 1997 when shell disease first became prevalent in Rhode Island waters (Gibson and Wahle 2005). Researchers have hypothesized that the supply of new recruits is greatly reduced by the increased mortality due to shell disease.

The history of disease outbreaks and mortality events paints a dire picture for the southern stock of American lobster. However, apparently devastating natural events often have positive as well as negative effects. All of the lethal and sublethal factors described above have exerted strong selective forces on the genetics of the southern New England lobster stock. Those animals that survive to maturity and successfully reproduce may be better suited to their changed environment than those that did not. And if conditions do not change too rapidly, this lobster stock may adapt so that they will be poised once again to take advantage of favorable conditions whenever they may occur.

6.9 Concluding Remarks

The biological communities of LIS have experienced a wide variety of impacts, including cultural eutrophication, hypoxia, introduction of non-native species, and climate change. In some regards, the consequences have been dramatic, with strong longitudinal gradients in plankton productivity, changes in critical habitats such as tidal marshes, dramatic declines in important species such as eelgrass, lobster and winter flounder, restructuring of benthos by periodic hypoxia, and the increasing importance of southern nekton species over northern species. In other ways, some aspects have not shown significant temporal patterns. There is no evidence that LIS is more eutrophic now than it was 60 years ago, although benthic foraminiferal assemblages changed fundamentally from those during the last 10,000 years, starting in the late 1960s (Thomas et al. 2001; Varekamp et al. 2010).

Our understanding of the biology of LIS is based on decades of research combined with sustained monitoring efforts. Some Long-term trends have been observed, but others may obscured by interannual variations. Observing systems measuring physical, chemical (e.g., T, S, currents, oxygen), and biological parameters (e.g., chlorophyll) have played a crucial role in understanding the processes affecting the biological community of LIS. Better spatial coverage of LIS with observing systems is strongly warranted. Monitoring of plankton, seagrass, and nekton is sufficiently robust (with some caveats regarding nekton monitoring discussed below) to characterize temporal changes in biomass and community structure at temporal scales ranging from seasonal to multidecadal. Other components of biological community of LIS have been monitored much more sporadically. So little monitoring of the seaweeds is done at present that we might not even know whether important species are in serious decline. Monitoring of benthic animal communities has been infrequent enough to conflate seasonal, interannual, and longer term trends. Regular monitoring that compares regions experiencing seasonal hypoxia to normoxic regions is also lacking, and shallow water regions are seriously neglected.

We strongly recommend continued and expanded monitoring efforts that cover critical habitats and biological communities in LIS. Plankton (including HAB-forming species) and nekton monitoring have been very informative, but without monitoring of the other organisms, it is very difficult to understand or predict how the food web responds to changes. Food-web dynamics (biomass, productivity, trophic linkages) are relatively poorly known, with significant gaps in understanding trophic linkages between primary producers and apex predators. It should be possible to take advantage of interannual variations in production and trophic linkages as natural experiments to better understand the workings of the LIS food web and how it will respond to longer term changes. For example, we know that much of the long-term trend in warming in LIS is due to warmer winters, and interannual variations in winter minimum water temperatures span the range of warming over the past several decades. Slightly warmer winters have been implicated in dramatic changes in phytoplankton production and its fate, yet little is known about how these changes ramify through the food web. These natural experiments need to be done at multiple spatial and temporal scales, and include multiple habitat types.

The value of monitoring is difficult to overstate but we must be cognizant of its limitations. One of the most important examples of this limitation is the connection between nutrient concentrations and the development of seasonal hypoxia in WLIS. Monitoring has demonstrated the congruence of spatial patterns of nutrients and hypoxia development, but even though N input in this region has been reduced, seasonal hypoxia has not abated. An improved model of oxygen dynamics is a first-order management issue for LIS; it will require more thorough monitoring of physical, chemical, and biological processes along with research to elucidate the role of relatively understudied processes such as nitrification, the roles of allochthonous organic matter, and organic matter stored in the sediments as contributors to hypoxia. We have a much better understanding of the factors controlling phytoplankton production than we do of its fate; rates of organic matter mineralization, especially in WLIS, are inadequately understood.

Another example of the limits of interpreting monitoring data comes from the fish surveys. These valuable seasonal surveys have shown long-term changes in relative and absolute abundance of different fish species, with cold-water species (dominant in spring) declining and warm water species (dominant in summer and fall) increasing. Nonetheless, the fixed seasonal schedule of these surveys has made interpretation of underlying causes difficult because many of these fish species are highly (and differentially) migratory, so their abundance in LIS is highly seasonal.

In summary, the biology of LIS is quite well characterized and understood in some regards and much less so in others. Expanded monitoring, research, and modeling are necessary to address current issues and to anticipate and address for future challenges.

Acknowledgments   Glenn Lopez would like to thank Hans Dam, Bob Whitlatch, and Roman Zajac for efforts in planning and organizing this chapter, all the coauthors for their contributions, the editors for their comments and corrections, and Robert Christian for his valuable review.

J. Vaudrey would like to recognize the contribution and support provided by the Department of Marine Sciences at the University of Connecticut, Connecticut Sea Grant (Kremer: R/ER-23), and the Connecticut Department of Environmental Protection (Kremer: FY2003 EPA 319 NPS #03-33 and FY2004 LISS Enhancement Project #AG0606257). Thanks to Dr. A. Branco and 15 Coastal Studies majors (University of Connecticut) for field assistance, 1999–2004.

C. Pickerell would like to recognize the contribution and support of the Sound Futures Fund of the National Fish and Wildlife Foundation for funding all of CCE's eelgrass restoration work in LIS. Without this support, none of this work would have been possible. Additional partners within the Sound Futures Fund have included the LIS Study, the National Oceanic and Atmospheric Administration, and Shell Oil Company. Save the Sound, CT also funded some of the early eelgrass work in LIS. Members of the CCE team instrumental in developing and refining site selection and planting methods for LIS include S. Schott and K. Petersen Manzo. This work would not have been possible without their hard work and dedication.

C. Yarish wishes to thank The Connecticut Department of Environmental Protection's Long Island Sound Research Fund (CWF-314-R), M. DiGiacomo-Cohen and R. Lewis of the Long Island Sound Resource Center and R. Rozsa for assistance with historical data for Long Island Sound seagrasses; to many undergraduate and graduate students from the UConn Department of Marine Sciences and the late R.A. Cooper, Director Emeritus of the Marine Science Technology Center; to A. Calabrese, Director Emeritus, Milford Laboratory, (NMFS, NOAA, DOC), Milford, Connecticut; and to M. Keser, J. Foertch and J. Swenarton, Millstone Environmental Laboratory, Millstone Power Station, Dominion Resources Services, Inc., Waterford, Connecticut.

H. Dam would like to thank the CTDEEP and the Office of LISS for funding that provided much of the data and analysis for the section on plankton.

Ellen Thomas thanks CT Sea Grant and the LISS EPA office for funding, and many Wesleyan undergraduate students for their hard work.

Chris Elphick thanks Patrick Comins, Frank Gallo, and Greg Hanisek for a careful reading of the text and the Center for Environmental Science and Engineering at the University of Connecticut for providing CSE with office space during the preparation of this manuscript.

Rob DiGiovanni extends his appreciation to Kimberly Durham and Julika Wocial for their help in preparing this document and providing historical background on the marine mammal and sea turtle rescue program.

Gordon Taylor is indebted to the many students and colleagues who have participated in the fieldwork and generously shared their insights into hypoxia. This contribution was supported in part by Hudson River Foundation Grant 012/96A.

Appendix 6.1

Definitions of the temperature adaptation and depth groups used to classify finfish species captured in Long Island Sound Trawl Survey. Classifications are based on information taken from Collette and Klein-MacPhee (2002) and Murdy et al. (1997).

ADAPTION GROUPS
Adapted to Cold Temperate Waters:

- more abundant north of Cape Cod, MA than south of New York
- behaviorally adapted to cold temperatures, including subfreezing
- preferred T range approximately 3–15 °C
- spawns at lower end of T tolerance

Adapted to Warm Temperate Waters:

- more abundant south of New York than north of Cape Cod, MA
- behaviorally avoids temperatures <10 °C
- preferred T range approximately 11–22 °C
- spawns at higher end of T tolerance

Adapted to Sub-tropical/Tropical Waters:

- rare north of Chesapeake and occasional or rare in the mid-Atlantic
- strays captured north of mid-Atlantic are usually juveniles
- not tolerant of temperatures <10 °C
- spawns only south of New York Bight

DEPTH GROUPS
Epibenthic (E)

- found exclusively or almost exclusively on the bottom
- feeds almost entirely on benthic prey
- if fished, taken only by bottom gear such as otter trawl nets

Demersal (D)

- associated with the bottom or bottom structure but may use water column occasionally
- feeds primarily on bottom organisms; uses bottom sediments in reproduction
- if fished, taken primarily by bottom tending gear or mid-water gear such as pound nets

Pelagic (P)

- uses the entire water column or primarily surface waters; eggs and larvae develop entirely in surface waters
- feeds primarily on surface prey or a mix of benthic/surface prey
- if fished, taken primarily by off-bottom or surface gear such as drift gill-nets or long lines

References

Able KW, Fahey MP (1998) The first year in the life of estuarine fishes in the Middle Atlantic Bight. Rutgers University Press, New Brunswick

Addy CE, Johnson R (1947) Status of eelgrass along the Atlantic coast during 1947. In: Proceedings, Northeastern Game Conference, pp 73–78

Adl SM, Simpson ABG, Farmer MA, Andersen RA, Anderson OR, Barta JR, Bowser SA, Brugerolle G, Fensome RA, Fredericq S, James TY, Karpov S, Kugrens P, Krug J, Lane CF, Lewis LA, Lodge J, Lynn DH, Mann DG, McCourt RM, Mendoza L, Moestrup Ø, Mozley-Standridge SE, Nerad T, Shearer CA, Smirnov AV, Spiegel FW, Taylor MFJR (2005) The new higher level classification of Eukaryotes with emphasis on the taxonomy of Protists. J Eukaryot Microbiol 52:39–451

Alber M, Swenson EM, Adamowicz SC, Mendelssohn IA (2008) Sudden marsh dieback: an overview of recent events in the US. Est Coast Shelf Sci 80:1–11

Albert R (1988) The historical context of water quality management for the Delaware estuary. Estuaries, Coasts 11:99–107

Aller RC (1984) The importance of relict burrow structures and burrow irrigation in controlling sedimentary solute distributions. Geochim Cosmochim Acta 48(10):1929–1934

Aller RC (1994) The sedimentary Mn cycle in Long Island Sound: its role as intermediate oxidant and the influence of bioturbation, O_2, and C_{org} flux on diagenetic reaction balances. J Mar Res 52:259–295

Aller RC, Benninger LK (1981) Spatial and temporal patterns of dissolved ammonium, manganese, and silica fluxes from bottom sediments of Long Island Sound, USA. J Mar Res 39:295–314

Allin CC, Husband TP (2003) Mute swan (Cygnus olor) impact on submerged aquatic vegetation and macroinvertebrates in a Rhode Island coastal pond. Northeast Nat 10:305–318

Altieri AH (2008) Dead zones enhance key fisheries species by providing predation refuge. Ecol 89(10):2808–2818

Altobello MA (1992) The economic importance of Long Island Sound's water quality dependent activities. University of Connecticut. Report to the US Environmental Protection Agency for the Long Island Sound Study, 41 pp

Alve E (1995) Benthic foraminiferal responses to estuarine pollution: an overview. J Foram Res 25:190–203

Alve E (1996) Benthic foraminiferal evidence of environmental change in the Skagerrak over the past six decades. Norske Geologiske Undersoknung, Bull 430:85–93

Alve E, Murray JW (1995) Benthic foraminiferal distribution and abundance changes in Skagerrak surface sediments: 1937 (Höglund) and 1992/1993 data compared. Mar Micropaleontol 25:269–288

Anderson SS (1972) The ecology of Morecambe Bay: II. Intertidal invertebrates and factors affecting their distribution. J Appl Ecol 9:161–178

Anderson TH, Taylor GT (2001) Nutrient pulses, plankton blooms and hypoxia in western Long Island Sound. Estuaries 24:228–243

Anisfeld SC, Hill TD (2011) Fertilization effects on elevation change and belowground carbon balance in a Long Island Sound tidal marsh. Estuaries Coasts 35:201–211

Armstrong H, Brasier MD (2005) Microfossils. Blackwell Publishing Ltd, Malden, pp 143–187

Arnold WS, Marelli DC, Parker M, Hoffman P, Frischer M, Scarpa J (2002) Enhancing hard clam (*Mercenaria* spp.) population density in the Indian River Lagoon, Florida: a comparison of strategies to maintain the commercial fishery. J Shellfish Res 21:659–672

ASMFC (2010) Recruitment failure in the Southern New England lobster stock. Technical Report to the Lobster Management Board, April 2010. http://www.asmfc.org/speciesdocum ents/lobster

Atlantic States Marine Fisheries Commission (ASMFC) (2009) American lobster stock assessment report for peer review. Stock Assessment Report 09-01. http://www.asmfc.org/species documents/lobster

Auster PJ, Heinonen KB, Witharana C, McKee M (2009) A habitat classification scheme for The Long Island Sound Region. Final Report, EPA Long Island Sound Study

Babcock HL (1919) Turtles of New England. Mem Boston Soc, Nat Hist 8:325–431

Baden SP, Loo LO, Pihl L, Rosenberg R (1990) Effects of eutrophication on benthic communities including fish: Swedish west coast. Ambio 19(3):113–22

Balcom N, Howell P (2006) Responding to a resource disaster: American lobsters in Long Island Sound, 1999–2004. Connecticut Sea Grant Report CTSG-06-02

Båmstedt U, Nejstgaard JC, Solberg PT, Høisaeter T (1999) Utilisation of small-sized food algae by *Calanus finmarchicus* (Copepod, Calanoida) and the significance of feeding history. Sarsia 84(1):19–28

Banks M (1990) Aboriginal weirs in southern New England. Bull Archaeol Soc Connect 53:73–84

Batiuk RA, Orth RJ, Moore KA, Dennison WC, Stevenson JC (1992) Chesapeake Bay submerged aquatic vegetation habitat requirements and restoration targets: a technical synthesis. Virginia Inst of Marine Science, Gloucester Point

Batiuk RA, Bergstrom P, Kemp M, Koch EW, Murray L, Stevenson JC, Bartleson R, Carter V, Rybicki NB, Landwehr JM, Gallegos CL, Karrh L, Naylor M, Wilcox D, Moore KA, Ailstock S, Teichberg M (2000) Chesapeake Bay submerged aquatic vegetation water quality and habitat-based requirements and restoration targets: a second technical synthesis. US Environmental Protection Agency, Annapolis

Bayard TS, Elphick CS (2011) Planning for sea-level rise: quantifying patterns of Saltmarsh Sparrow (*Ammodramus caudacutus*) nest flooding under current sea-level conditions. Auk 128:393–403

Beck MW, Heck KL Jr, Able KW, Childers DL, Eggleston DB, Gillanders BM, Halpern B, Hays CG, Hoshino K, Minello TJ, Orth RJ, Sheridan PF, Weinstein MP (2001) The identification, conservation, and management of estuarine and marine nurseries for fish and invertebrates. Bioscience 51:633–641

Bejda AJ, Phelan BA, Studholme AL (1992) The effect of dissolved oxygen on the growth of young-of-the-year winter flounder, *Pseudopleuronectes americanus*. Env Biol Fishes 34:321–327

Bell GW, Eggleston DB (2005) Species-specific avoidance responses by blue crabs and fish to chronic and episodic hypoxia. Mar Biol 146:761–770

Bell GW, Eggleston DB, Noga EJ (2010) Molecular keys unlock the mysteries of variable survival responses of blue crabs to hypoxia. Oecologia 163:57–68

Bellchambers LM, Richardson AMM (1995) The effect of substrate disturbance and burial depth on the venerid clam, *Katelysis scalarina* (Lamark, 1818). J Shellfish Res 14:41–44

Beman JM, Arrigo KR, Matson PA (2005) Agricultural runoff fuels large phytoplankton blooms in vulnerable areas of the ocean. Nature 434:211–214

Bender K, Davis WR (1984) The effect of feeding by *Yoldia limatula* on bioturbation. Ophelia 23:91–100

Benoit GJ, Turekian KK, Benninger LK (1979) Radiocarbon dating of a core from Long Island Sound. Est Coast Mar Sci 9:171–180

Berge JA, Bjerkeng B, Pettersen O, Schaanning MT, Oxnevad S (2006) Effects of increased sea water concentrations of CO_2 on growth of the bivalve *Mytilus edulis* L. Chemosphere 62:681–687

Berman J, Harris L, Lambert W, Buttrick M, Dufresne M (1992) Recent invasions of the Gulf of Maine: three contrasting ecological histories. Conservation Biol 6:435–441

Bernhard JM, Bowser SS (1999) Benthic foraminifera of dysoxic sediments: chloroplast sequestration and functional morphology. Earth Sci Rev 46:149–165

Bernstein DJ (2006) Long-term continuity in the archaeological record from the coast of New York and Southern New England, USA. J Island Coastal Archaeol 1(2):27–284

Berrill M, Berrill D (1981) A Sierra Club naturalist's guide to the North Atlantic Coast. Sierra Club Books, San Francisco

Bertness MD (1999) The ecology of Atlantic Shorelines. Sinauer Associates, Sunderland 417 pp

Bertness MD (2007) Atlantic shorelines: natural history and ecology. Princeton University Press, Princeton 431 pp

Bertness MD, Gaines SD, Hay ME (eds) (2001) Marine community ecology. Sinauer Associates, Sunderland, 550 pp

Bertness MD, Silliman BR, Holdredge C (2009) Shoreline development and the future of New England salt marsh landscapes. In: Silliman BR, Grosholz ED, Bertness MD (eds) Human impacts on salt marshes. University of California Press, Los Angeles, pp 137–148

Bintz JC, Nixon SW, Buckley BA, Granger SL (2003) Impacts of temperature and nutrients on coastal lagoon plant communities. Estuaries 26:765–776

Blake M, Smith E (1984) A marine resources management plan for the State of Connecticut. Department of Environmental Protection, Bureau of Fisheries, 244 pp

Blake JA, Grassle JP, Eckelbarger EJ (2009) *Capitella teleta*, a new species designation for the opportunistic and experimental *Capitella* sp. I, with a review of the literature for confirmed records. Zoosymposia 2:25–53

Blakeslee AMH, Byers JE, Leser MP (2008) Solving cryptogenic histories using host and parasite molecular genetics: the resolution of *Littorina littorea*'s North American origin. Mol Ecol 17:3684–3696

Blight SP, Bentley TL, Lefevre D, Robinson C, Rodrigues R, Rowlands J, Williams PJL (1995) Phasing of autotrophic and heterotrophic plankton metabolism in a temperate coastal ecosystem. Mar Ecol Prog Ser 128:61–75

Boer WF (2007) Seagrass-sediment interactions, positive feedbacks and critical thresholds for occurrence: a review. Hydrobiologia 591:5–24

Bohlen WF, Cundy DF, Tramontano JM (1979) Suspended material distribution in the wake of estuarine channel dredging operations. Est Coast Mar Sci 9:699–711

Bokuniewicz HJ (1988) Sedimentation of fine-grained particles in Long Island Sound: a review of evidence prior to 1987. Special report 83, Reference 88-2, EPA National Estuaries Program, 36 pp

Bokuniewicz HJ (1989) Behavior of sand caps on subaqueous dredged-sediment disposal sites. In: Hood DW, Schoener A, Park PK (eds) Oceanic processes in marine pollution, vol 4., Scientific monitoring strategies for ocean waste disposalKreiger Publishing Co, Malabar, pp 221–229

Bokuniewicz HJ, Gordon RB (1980) Sediment transport and deposition in Long Island Sound. Adv Geophys 22:69–106

Bokuniewicz H, Tanski JJ (1983) Sediment partitioning at an eroding coastal bluff. Northeastern Geol 5:73–81

Bolam SG, Rees H (2003) Minimising the impacts of maintenance dredged material disposal in the coastal environment: a habitat approach. Environ Manag 32(2):171–188

Bolam SG, Rees HL, Somerfield P, Smith R, Clarke KR, Warwick RM, Atkins M, Garnacho E (2006) Ecological consequences of dredged material disposal in the marine environment: a holistic assessment of activities around the England and Wales coastline. Mar Poll Bull 52(6):415–426

Boos K, Ashton GV, Cook EJ (2011) The Japanese skeleton shrimp *Caprella mutica* (Crustacea: Amphipoda): a global invader of coastal waters. In: Galil BS, Clark PF, Carlton JT (eds)

In the wrong place-Alien marine crustaceans: distribution, biology, and impacts. Springer, Dordrecht, pp 129–156

Bos AR, Bouma TJ, de Kort GLJ, van Katwijk MM (2007) Ecosystem engineering by annual intertidal seagrass beds: sediment accretion and modification. Estuar Coast Shelf Sci 74:344–348

Bouck GE, Morgan E (1957) The occurrence of *Codium* in Long Island Waters. Bull Torrey Bot Club 84:384–387

Bouma TJ, De Vries MB, Low E, Peralta G, Tánczos IC, Van De Koppel J, Herman PMJ (2005) Trade-offs related to ecosystem engineering: a case study on stiffness of emerging macrophytes. Ecology 86:2187–2199

Bousfield EL (1973) Shallow-water Gammaridean Amphipoda of New England. Cornell University Press, New York, pp 312

Bowman MJ (1977) Nutrient distributions and transport in Long Island Sound. Est Coast Mar Sci 5:531–548

Bradshaw JS (1957) Laboratory studies on the rate of growth of the foraminifer '*Streblus*' *beccarii* (Linne) var. *tepida* (Cushman). J Paleontol 31:1138–1147

Bradshaw JS (1968) Environmental parameters and marsh foraminifera. Limnol Oceanogr 13:26–28

Brady DC, Targett TE (2010) Characterizing the escape response of juvenile summer flounder *Paralichthys dentatus* to diel-cycling hypoxia. J Fish Biol 77:137–152

Brady DC, Tuzzolino DM, Targett TE (2009) Behavioral responses of juvenile weakfish, Cynoscion regalis, to diel-cycling hypoxia: swimming speed, angular correlation, expected displacement and effects of hypoxia acclimation. Can J Fish Aquat Sci 66(3):415–424

Brandt SB, Demers E, Tyler JA, Gerken MA (1998) Fish bioenergetic modelling: Chesapeake Bay ecosystem modelling program (1993–1998). Report to the Chesapeake Bay Program, US Environmental Protection Agency, Chesapeake Bay Program Office, Annapolis, MD

Brandt SB, Gerken M, Hartman KJ, Demers E (2009) Effects of hypoxia on food consumption and growth of juvenile striped bass (Morone saxatilis). J Exp Mar Biol Ecol 38:(1) pp S143–S149

Breitburg DL (1994) Behavioral responses of fish larvae to low dissolved oxygen concentrations in a stratified water column. Mar Biol 120:615–625

Breitburg DL (2002) Effects of hypoxia, and the balance between hypoxia and enrichment, on coastal fishes and fisheries. Estuaries 25:886–900

Breitburg DL, Adamack A, Rose KA, Kolesar SE, Decker B, Purcell JE, Keister JE, Cowan HH (2003) The pattern and influence of low dissolved oxygen in the Patuxent River, a seasonally hypoxic estuary. Estuaries 26(2A):280–297

Breitburg DL, Craig JK, Fulford RS, Rose KA, Boyton WR, Brady D, Ciotti BJ, Diaz RJ, Freidland KD, Hagy JD III, Hart DR, Hines AH, Houde ED, Kolesar SE, Nixon SW, Rice JA, Secor DH, Targett TE (2009a) Nutrient enrichment and fisheries exploitation: interactive effects on estuarine living resources and their management. Hydrobiologia 629:31–47

Breitburg DL, Hondorp DW, Davias LW, Diaz RJ (2009b) Hypoxia, nitrogen and fisheries: integrating effects across local and global landscapes. Ann Rev Mar Sci 1:329–350

Breitburg DL, Pihl, L, Kolesar SE (2001) Effects of low dissolved oxygen on the behavior, ecology, and harvest of fishes: A comparison of the Chesapeake Bay and the Baltic-Kattegat Systems. In : Coastal Hypoxia: Consequences for the living resources and ecosystems. Coastal and Estuarine Studies 58, Rabelais, NN and RE Turner, eds. American Geophysical Union, Washington, DC

Breitburg DL, Steinberg N, DuBeau S, Cooksey C, Houde ED (1994) Effects of low dissolved oxygen on predation on estuarine fish larvae. Mar Ecol Prog Ser 104(3):235–246

Brett JR, Groves TDD (1979) Physiological energetics. In: Hoar WS, Randall DJ, Brett JR (eds) Fish physiology, Vol VIII: bioenergetics and growth. Academic Press, New York, p 786

Bricelj VM, Kuenstner SH (1989) Effects of the "brown tide" on the feeding physiology and growth of bay scallops and mussels. In: Cosper EM, Bricelj VM, Carpenter EJ (eds) Novel phytoplankton blooms: causes and impacts of recurrent brown tides and other unusual blooms. Springer, Berlin, pp 491–509

Bricelj VM, MacQuarrie SP (2007) Effects of brown tide (*Aureococcus anophageﬀerens*) on hard clam *Mercenaria mercenaria* larvae and implications for benthic recruitment. Mar Ecol Prog Ser 331:147–159

Bricelj VM, MacQuarrie SP, Schaffner RA (2001) Differential effects of *Aureococcus anophageﬀerens* isolates ("brown tide") in unialgal and mixed suspensions on bivalve feeding. Mar Biol 139:605–615

Briggs PT, Waldman JR (2002) Annotated list of fishes reported from the marine waters of New York. Northeast Nat 9:47–80

Brinkhuis BH (1976) The ecology of salt marsh fucoids. Mar Biol 34:325–333

Brinkhuis BH (1983) Seaweeds in New York waters: a primer. New York Sea Grant Institute, Stony Brook , 21 pp

Brinkhuis BH, Davies D, Hanisak D, Herman H, Macler B, Peterson W (1982) In: Squires D, McKay L (eds) Marine biomass: New York State species and site studies. Report to New York State Energy Research and Development Authority Gas Research Institute. ERDA Report 82–87, 139 pp

Brinkhuis BH, Breda VA, Mariani ED, Tobin S, Macler BA (1983) New York biomass program-culture of Laminaria saccharina. J Mariculture Soc 14:360–379

Brodie J, Andersen RA, Kawachi M, Millar AJK (2009) Endangered algal species and how to protect them. Phycologia 48(3):423–438

Broom JE, Nelson WA, Yarish C, Jones WA, Aguilar Rosas R, Aguilar Rosas LE (2002) A reassessment of the taxonomic status of *Porphyra suborbiculata, Porphyra carolinensis* and *Porphyra lilliputiana* (Bangiales, Rhodophyta) based on molecular and morphological data. Eur J Phycol 37:227–235

Bureau of Commercial Fisheries (1963) A program of fishery research and services, North Atlantic Region. US Department of the Interior, Fish and Wildlife Service, Bureau of Commercial Fisheries, Washington DC, 20 pp

Burke VJ, Standora EA, Morreale SJ (1991) Factors affecting strandings of cold stunned juvenile Kemp's ridley and loggerhead sea turtles in Long Island, New York. Copeia 1991:1136–1138

Burke VJ, Morreale SJ, Standora EA (1994) Dietary composition of Kemp's ridley sea turtles in the waters of New York. Fish Bull 92:26–32

Burnett LE (1997) The changes of living in hypoxia and hypercapnia aquatic environments. Amer Zool 37:633–640

Bussell JA, Lucas I, Seed R (2007) Patterns in the invertebrate assemblage associated with *Corallina officinalis* in tide pools. J Mar Biol Assn U K 87:383–388

Buzas MA (1965) The distribution and abundance of Foraminifera in Long Island Sound. Smithsonian Inst Miscellaneous Coll 149(1):1–88

Buzas MA (1966) The discrimination of morphological groups of *Elphidium* (foraminifera) in Long Island Sound through canonical analysis and invariant characters. J Paleontol 40:585–594

Buzas MA, Culver SJ, Isham LB (1985) A comparison of fourteen Elphidiidid (Foraminiferida) taxa. J Paleontol 59:1075–1090

Calbet A, Landry MR (2004) Phytoplankton growth, microzooplankton grazing, and carbon cycling in marine systems. Limnol Oceanogr 49:51–57

Caley MJ, Carr MH, Hixon MA, Hughes TP, Jones GP, Menge BA (1996) Recruitment and the local dynamics of open marine populations. Ann Rev Ecol Syst 27:477–500

Campagna VA, Cech JJ Jr (1981) Gill ventilation and respiratory efficiency of Sacramento Blackfish. *Orthodon microlepidotus* in hypoxic environments. J Fish Biol 19:581–591

Campbell L (1985) Investigations of marine, phycoerythrin-containing *Synechococcus* spp. (Cyanobacteria): distribution of serogroups and growth rate measurements. Ph.D. dissertation, Stony Brook, State University of New York

Campbell JG, Goodman LR (2004) Acute sensitivity of juvenile shortnose sturgeon to low dissolved oxygen concentrations. Trans Am Fish Soc 133(3):772–776

Campbell L, Olson RJ, Sosik HM et al (2010) First harmful *Dinophysis* bloom in the US is revealed by automated imaging flow cytometry. J Phycol 46:66–75

Canning MH, Carlton JT (2000) Predation on kamptozoans (Entoprocta). Invert Biol 119:386–387

Capriulo GM (1982) Feeding of field collected tintinnid micro-zooplankton on natural food. Mar Bio 71:73–86

Capriulo GM, Carpenter EJ (1980) Grazing by 35 to 202 μm micro-zooplankton in Long Island Sound. Mar Biol 56:319–326

Capriulo GM, Carpenter EJ (1983) Abundance, species composition and feeding impact of tintinnid microzooplankton in central Long Island Sound. Mar Ecol Prog Ser 10:277–288

Capriulo GM, Smith G, Troy R, Wikfors G, Pellet J, Yarish C (2002) The planktonic food web structure of a temperate zone estuary, and its alternation due to due to eutrophication. Hydrobiologia 475(476):263–333

Carey AG (1962) An ecological study of two benthic animal populations in Long Island Sound. PhD dissertation, Yale University, 61 pp

Carey DA, Lewis D, Wolf S, Greenblatt M, Fredette TJ (2006) Long term stability of capped dredged material disposal mounds: Stamford, New Haven North and Cap Site 2 in Long Island Sound. Proceedings of the 26th annual Western Dredging Association Technical Conference, San Diego, CA. Https://westerndredging.org/index.php/information/proceedings-presentations/category/43-26th-annual-weda-conference?download=395:13-carey-long-term-stability-of-capped-dredged-material-disposal-mounds

Carey DA, Hickey K, Myre PL, Read LB, Esten ME (2012) Monitoring surveys at the historical Brenton Reef disposal site 2007 and 2009. DAMOS Contribution 187, US Army Corps of Engineers, New England District, Concord, MA, 134 pp

Carlton JT (1982) The historical biogeography of *Littorina littorea* on the Atlantic coast of NorthAmerica, and implications for the interpretation of the structure of New England intertidal communities.Malacological Rev 15:146

Carlton JT (1992) Introduced marine and estuarine mollusks of North America: an end-of-the-20th-century perspective. J Shellfish Res 11:489–505

Carlton JT (1996) Biological invasions and cryptogenic species. Ecology 77:1653–1655

Carlton JT (2009) Deep invasion ecology and the assembly of communities in historical time. In: Rilov G, Crooks JA (eds) Biological invasions in marine ecosystems. Springer, Berlin, pp 13–56

Carlton JT (2010) Invertebrates, marine. In: Simberloff D, Rejmanek M (eds) Encyclopedia of biological invasions. University of California Press, Berkeley

Carlton JT, Cohen AN (2003) Episodic global dispersal in shallow water marine organisms: the case history of the European shore crabs *Carcinus maenas* and *Carcinus aestuarii*. J Biogeography 30:1809–1820

Carlton JT, Scanlon JA (1985) Progression and dispersal of an introduced alga: *Codium fragile* ssp. *tomentosoides* (Chlorophyta) on the Atlantic coast of North America. Bot Mar 28:155–165

Carlton JT, Newman WA, Pitombo FB (2011) Barnacle invasions: introduced, cryptogenic, and range expanding Cirripedia of North and South America. In: Galil BS, Clark PF, Carlton JT (eds) In the wrong place–Alien marine crustaceans: distribution, biology, and impacts. Springer, Dordrecht, pp 159–213

Carmona R, Kraemer GP, Zertuche JA, Chanes L, Chopin T, Neefus C, Yarish C (2001) Exploring *Porphyra* species for use as nitrogen scrubbers in integrated aquaculture. J Phycol 37(3):10. http://dx.doi.org/10.1111/j.1529-8817.2001.jpy37303-18.x

Caron DA, Dennett MR, Lonsdale DL, Moran DM, Shalapyonok L (2000) Microzooplankton herbivory in the Ross Sea, Antarctica. Deep-Sea Res 11(47):3249–3272

Carriker MR (2001) Embryogenesis and organogenesis of Veligers and early juveniles. In: Kraeuter JN, Castagna M (eds) Biology of the hard clam. Elsevier Science, New York, pp 77–115

Carroll J, Gobler CJ, Peterson BJ (2008) Resource-restricted growth of eelgrass in New York estuaries: light limitation, and alleviation of nutrient stress by hard clams. Mar Ecol Prog Ser 369:51–62

Cavanaugh CM, McKiness ZP, Newton ILG, Stewart FJ (2006) Marine chemosynthetic symbioses. Prokaryotes 1:475–507

Cech JJ Jr, Mitchel SJ, Wragg TE (1984) Comparative growth of juvenile white sturgeon and striped bass–effects of temperature and hypoxia. Estuaries 7(1):12–18

Cech JJ, Mitchell SJ, Castleberry DT, McEnroe M (1990) Distribution of California stream fishes: influences of environmental temperature and hypoxia. Environ Biol Fishes 29:95–105

Cedhagen T (1991) Retention of chloroplasts and bathymetric distribution in the sublittoral foraminiferan *Nonionellina labradorica*. Ophelia 33:17–30

Cerrato RM, Bokuniewicz HJ (1986) The benthic fauna at four potential containment/wetlands stabilization areas. Marine Sciences Research Center Special Report No 73. State University of New York, Stony Brook, p 117

Cerrato RM, Holt LC (2008) North Shore bays benthic mapping: ground truth studies. Marine Sciences Research Center Special Report No 135. Marine Sciences Research Center, School of Marine and Atmospheric Sciences, Stony Brook University, Stony Brook, NY, 184 pp

Cerrato RM, Caron DA, Lonsdale DJ, Rose JM, Schaffner RA (2004) Effect of the northern quahog *Mercenaria mercenaria* on the development of blooms of the brown tide alga *Aureococcus anophagefferens*. Mar Ecol Prog Ser 281:93–108

Chabot D, Dutil J-D (1999) Reduced growth of Atlantic cod in non-lethal hypoxic conditions. J Fish Biol 55:472–491

Chambers RM, Meyerson LA, Saltonstall K (1999) Expansion of *Phragmites australis* into tidal wetlands of North America. Aquat Botany 64:261–273

Chapman VJ, Chapman DJ (1980) Seaweeds and their uses. Chapman and Hall, New York

Cheek AO, Landry CA, Steele SL, Manning S (2009) Diel hypoxia in marsh creeks impairs the reproductive capacity of estuarine fish populations. Mar Ecol Prog Ser 392:211–221

Cheng IJ, Lopez GR (1991) Contributions of bacteria and sedimentary organic matter to the diet of *Nucula proxima*, a deposit-feeding protobranchiate bivalve. Ophelia 34:157–170

Chesney EJ, Houde ED (1989) Laboratory studies on the effect of hypoxic waters on the survival of eggs and yolk-sac larvae of the bay anchovy, *Anchoa mitchilli*, p.184–191. In ED Houde, EJ Chesney,TA Newberger, AV Vazquez, CE Zastrow, LG Morin. HR Harvey, and JW Gooch, (eds.). Population Biology of the Bay Anchovy in Mid-Chesapeake Bay. Final Report to Maryland Sea Grant Ref. No. (UM-CEES) CBI, 89–141, Solomons, Maryland

Chopin T, Yarish C, Wilkes R, Belyea E, Lu S, Mathieson A (1999) Developing *Porphyra*/salmon integrated aquaculture for bioremediation and diversification of the aquaculture industry. J Appl Phycol 11:463–472

Chopin T, Buschmann AJ, Halling C, Troell M, Kautsky N, Neori A, Kraemer GP, Zertuche-Gonzalez JA, Yarish C, Neefus C (2001) Integrating seaweeds into mariculture systems: a key towards sustainability. J Phycol 37(6):975–986

Christensen V, Walters CJ, Ahrens R, Alder J, Buszowski J, Christensen LB, Cheung WWL, Dunne J, Froese R, Karpouzi V, Kaschner K, Kearney K, Lai S, Lam V, Palomares MLD, Peters-Mason A, Piroddi C, Sarmiento JL, Steenbeek J, Sumaila R, Watson R, Zeller D, Pauly D (2009) Database-driven models of the world's largest marine ecosystems. Ecol Model 220:1984–1996

Churchill AC, Cok AE, Riner MI (1978) Stabilization of subtidal sediments by the transplantation of the seagrass Zostera marina L. New York Sea Grant Report, NYSSGR-RS-78-15, Albany, NY

Claireaux G, Webber DM, Lagardère J-P, Kerr SR (2000) Influence of water temperature and oxygenation on the aerobic metabolic scope of Atlantic cod (*Gadus morhua*). J Sea Res 44:257–265

Cloern JE (2001) Our evolving conceptual model of the coastal eutrophication problem. Mar Ecol Prog Ser 210:223–253

Cohen JB, Barclay JS, Major AR, Fisher JP (2000) Wintering scaup as biomonitors of metal contamination in federal wildlife refuges in the Long Island region. Arch Environ Contam Toxicol 38:83–92

Collette B, Klein-MacPhee G (eds) (2002) Bigelow and Schroeder's fishes of the Gulf of Maine, 3rd edn. Smithsonian Inst Press, Washington, DC

Collie JS, Wood AD, Jeffries HP (2008) Long-term shifts in the species composition of a coastal fish community. Can J Fish Aquat Sci 65:1352–1365

Colson S, Sturmer LN (2000) One shining moment known as Clamelot: the Cedar Key story. J Shellfish Res 19:477–480

Connecticut Department of Environmental Protection (CT DEP) (2000) Impact of 1999 lobster mortality in Long Island Sound. Report to the Natl Marine Fisheries Service (US Commerce Dept) in support of disaster assistance under the Sustainable Fisheries Act, 55 pp

Connecticut Department of Environmental Protection and New York State Department of Environmental Conservation (2000) A total maximum daily load analysis to achieve water quality standards for dissolved oxygen in Long Island Sound

Conover SAM (1956) Oceanography of Long Island Sound, 1952–1954. IV. Phytoplankton. Bull Bingham Oceanogr Collect 15:62–112

Corlett H, Jones B (2007) Epiphyte communities on *Thalassia testudinum* from Grand Cayman, British West Indies: their composition, structure, and contribution to lagoonal sediments. Sed Geol 194:245–262. doi:10.1016/j.sedgeo.2006.06.010

Cosper EM, Dennison WC, Carpenter EJ, Bricelj VM, Mitchell JG, Kuenstner SH, Colflesh DC, Dewey M (1987) Recurrent and persistent "brown tide" blooms perturb coastal marine ecosystem. Estuaries 10:284–290

Cottam C (1933) Disappearance of eelgrass along the Atlantic Coast. Plant Dis Rep 17:46–53

Cragg SM (2006) Development, physiology, behaviour, and ecology of scallop larvae. In: Shumway SE, Parsons GJ (eds) Scallops: biology, ecology, and aquaculture. Elsevier, Amsterdam, pp 45–122

Crain CM, Gedan KM, Dionne M (2009) Tidal restrictions and mosquito ditching in New England marshes. In: Silliman BR, Grosholz ED, Bertness MD (eds) Human impacts on salt marshes. University of California Press, Berkeley, pp 149–169

Crocker CE, Cech JJ Jr (1997) Effects of environmental hypoxia on oxygen consumption rate and swimming activity in juvenile white sturgeon, *Acipenser transmontanus* in relation to temperature and life intervals. EnvIron Biol Fish 50:383–388

Croker RA (1970) Intertidal sand macrofauna from Long Island, New York. Chesapeake Sci 11:134–137

Cronan JM Jr (1957) Food and feeding habits of the scaups in Connecticut waters. Auk 74:459–468

Crossin G, Al-Ayoub S, Jury S, Howell W, Watson W (1998) Behavioral thermoregulation in the American lobster *Homarus americanus*. J Exp Biol 201:365–374

CT DEP (Connecticut Department of Environmental Protection) (2000) Impact of 1999 lobster mortalities in Long Island Sound. CT DEP Marine Fisheries Office, Old Lyme, CT, 47 pp

CT DEP MFIS (Connecticut Department of Environmental Protection Marine Fisheries Information System) (2010) Marine Fisheries Office, Old Lyme, CT

Culver SJ, Buzas M (1982) Recent benthic foraminiferal provinces between Newfoundland and Yucatan. Geol Soc Am Bull 93:269–277

Culver SJ, Buzas M (1995) The effects of anthropogenic habitat disturbance, habitat destruction, and global warming on shallow marine benthic foraminifera. J Foram Res 25:204–211

Cumbler JT (1991) The early making of an environmental consciousness: fish, fisheries commissions and the Connecticut River. Env History Rev 15(4):73–91

Cuomo MC (1985) Sulphide as a larval settlement cue for *Capitella* sp. I. Biogeochem 1:169–181

Cuomo C, Zinn GA (1997) Benthic invertebrates of the Lower West River. In: Casagrande DG (ed) Restoration of an urban salt marsh: an interdisciplinary approach, pp 152-161, Center for Coastal and Watershed Systems, Bull No 100. Yale School of Forestry and Environmental Studies, New Haven, p 270

Cuomo C, Valente R, Dogru D (2005) Seasonal variations in sediment and bottom water chemistry of western Long Island Sound: implications for lobster mortality. J Shellfish Res 24(3):805–814

Dalton CM, Ellis D, Post DM (2009) The impact of double-crested cormorant (*Phalacrocorax auritus*) predation on anadromous alewife (*Alosa pseudoharengus*) in south-central Connecticut, USA. Can J Fish Aquat Sci 66:177–186

Dam Guerrero HG (1989) The dynamics of copepod grazing in Long Island Sound. PhD Dissertation, State University of New York, Stony Brook, NY

Dam HG, McManus GB (2009) Final report: monitoring mesozooplankton and microzooplankton in Long Island Sound, National Coastal Assessment, reporting period: March 2007–April 2008

Dam HG, Peterson WT (1988) The effect of temperature on the gut clearance rate constant of planktonic copepods. J Exp Mar Bio Ecol 123:1–14

Dam HG, Peterson WT (1991) In situ feeding behavior of the copepod *Temora longicornis* effects of seasonal changes in chlorophyll size fractions and female size. Mar Ecol Prog Ser 71:113–123

Dam HG, Peterson WT (1993) Seasonal contrasts in the diel vertical distribution, feeding behavior, and grazing impact of the copepod *Temora longicornis* in Long Island Sound. J Mar Res 51:561–594

Dam HG, Peterson WT, Bellantoni DC (1994) Seasonal feeding and fecundity of the calanoid copepod *Acartia tonsa* in Long Island Sound: is omnivory important to egg production? Hydrobiologia 292(293):191–199

Dam HG, Roman MR, Youngbluth MJ (1995) Downward export of respiratory carbon and dissolved inorganic nitrogen by diel-migrant mesozooplankton at the JGOFS Bermuda time-series station. Deep-Sea Res Pt I 42:1187–1197

Dam HG, O'Donnell J, Siuda AN (2010) A synthesis of water quality and planktonic resource monitoring data for Long Island Sound. Final report EPA grant number LI-97127501

Davis JC (1975) Minimal dissolved oxygen requirements of aquatic life with emphasis of Canadian species: a review. J Fish Res Board Canada 32:2295–2332

Davis HC, Chanley PE (1956) Effects of some dissolved substances on bivalve larvae. Proc Natl Shellfisheries Assoc 46:59–74

Davison IR, Jordan TL, Fegley JC, Grobe C (2007) Response of *Laminaria saccharina* (Phaeophyta) growth and photosynthesis to simultaneous ultraviolet radiation and nitrogen limitation. J Phycol 43:636–646

Dawson MA (1990) Blood chemistry of the Windowpane flounder *Scophthalamus aquosus* in Long Island Sound: geographical, season, and experimental variations. Fish Bull, US 88:429–437

De Guise S, Maratea J, Chang E, Perkins C (2005) Resmethrin immunotoxicity and endocrine disrupting effects in the American lobster (*Homarus americanus*) upon experimental exposure. J Shellfish Res 24(3):781–786

De Rijk S (1995) Agglutinated foraminifera as indicators of salt marsh development in relation to Late Holocene sea level rise. PhD dissertation, Vrije University, Amsterdam

De Rijk S, Troelstra S (1999) The application of a foraminiferal actuo-facies model to salt-marsh cores. Palaeogeogr Palaeoclim Palaeoecol 149:59–66

De Silva, CD and P Tytler. 1973. The influence of reduced environmental oxygen on the metabolism and survival of herring and plaice larvae. Neth. J Sea Res 7:345–362

De Sola R (1931) Sex determination in a species of the Kinosternidae, with notes on sound production in reptiles. Copeia 1931:124–125

De Vorsey L (2007) Americas: exploration voyages, 1500–1620 (Eastern Coast). In: Hattendorff JB (ed) The Oxford encyclopedia of maritime history. Oxford University Press, Oxford, pp 28–36

Deevey GB (1956) Oceanography of Long Island Sound, 1952-1954: Zooplankton. Bull Bingham Oceangr Collect 15:133–155

Dennison WC (1987) Effects of light on seagrass photosynthesis, growth and depth distribution. Aquatic Bot 27:15–26

Dennison WC, Alberte RS (1985) Role of daily light period in the depth distribution of *Zostera marina* (eelgrass). Mar Ecol Prog Ser 25:51–61

Dennison WC, Orth RJ, Moore KA, Stevenson JC, Carter V, Kollar S, Bergstrom P, Batiuk RA (1993) Assessing water quality with submersed aquatic vegetation. Bioscience 43:86–94

Deubler EE Jr, Posner GS (1963) Responses of postlarval flounders, *Paralichthys lethostigma*, to water of low oxygen concentrations. Copeia 2:312–317

Diaz RJ, Rosenberg R (1995) Marine benthic hypoxia: a review of its ecological effects and the behavioral responses of benthic macrofauna. Oceanogr Mar Biol Ann Rev 33:245–303

Diaz RJ, Rosenberg R (2008) Spreading dead zones and consequences for marine ecosystems. Science 321:926–929

DiGiovanni RA, Durham K, Spangler-Martin D (2000) Marine mammal and sea turtle stranding program 1996 through 2000 report to NYS DEC

DiGiovanni RA, Durham KF, Wocial JN, Chaillet AM (2009) An increase in gray seal (*Halichoerous grypus*) sightings in New York waters. Abstract for the 18th biennial conference on the Biology of Marine Mammals. Quebec, Canada

DiMichele L, Powers D (1984) The relationship between oxygen consumption and hatching in *Fundulus heteroclitus*. Physiol Zool 57:46–51

Doeller JE, Gaschen BK, Parrino V, Kraus DW (1999) Chemolithoheterotrophy in a metazoan tissue: sulfide supports cellular work in ciliated mussel gills. J Exp Biol 202:1953–1961

Dominion Nuclear Connecticut (DNC) (2010) Unpublished temperature data, Don Landers, Millstone Environmental Laboratory (personal communication)

Doney SC, Fabry VJ, Feely RA, Kleypas JA (2009) Ocean acidification: the other CO_2 problem. Ann Rev Mar Sci 1:169–192

Donnelly JP, Bertness MD (2001) Rapid shoreward encroachment of salt marsh cordgrass in response to accelerated sea-level rise. Proc Nat Acad Sci U S A 98:14218–14223

Dove A, LoBue C, Bowser P, Powell M (2004) Excretory calcinosis: a new fatal disease of wild American lobster *Homarus americanus*. Dis Aquat Org 58:215–221

Dove A, Allam B, Powers J, Sokolowski M (2005) A prolonged thermal stress experiment on the American lobster *Homarus americanus*. J Shellfish Res 24:761–765

Draxler AFJ, Robohm RA, Wieczorek D, Kapereiko D, Pitchford S (2005) Effect of habitat bio-geochemicals on survival of lobsters (*Homarus americanus*). J Shellfish Res 24(3):821–824

Dreyer GD, Niering WA (eds) (1995) Tidal marshes of Long Island Sound: ecology, history and restoration. Connecticut College Arboretum Bull 34:77 pp

Dring MJ (1992) Biology of marine plants. Cambridge University Press, Cambridge

Dring M, Wagner A, Lüning K (2001) Contribution of the UV component of natural sunlight to photoinhibition of photosynthesis in six species of subtidal brown and red seaweeds. Plant, Cell Environ 24:1153–1164

Duarte CM (1991) Seagrass depth limits. Aquat Bot 40:363–377

Duarte CM (1995) Submerged aquatic vegetation in relation to different nutrient regimes. Ophelia 41:87–112

Duarte CM, Middelburg JJ, Caraco N (2005) Major role of marine vegetation on the oceanic car-bon cycle. Biogeosci 2:1–8

Dubilier N (1988) H_2S-A settlement cue or a toxic substance for Capitella sp. I larvae. Biol Bull 174:30–38

Dunton KJ, Jordaan A et al (2010) Abundance and distribution of Atlantic sturgeon (Acipenser oxyrinchus) within the Northwest Atlantic Ocean, determined from five fishery-independent surveys. Fish Bull 108(4):450–465

Ducheney P, Murray Jr, RF, Waldrip JE, Tomichek CA (2006) Fish passage at Hadley Falls: past, present, and future. http://www.kleinschmidtusa.com/pubs/hadleyfalls_fishpassage.htm

Eby LA, Crowder LB (2002) Hypoxia-based habitat compression in the Neuse River Estuary: con-text-dependent shifts in behavioral avoidance thresholds. Can J Fish Aquat Sci 59(6):952–9655

Eby L, Crowder L, McClellan CM, Peterson CH, Powers MJ (2005) Habitat degradation from intermittent hypoxia: impacts on demersal fishes. Mar Ecol Prog Ser 291:49–261

Edwards RJ, Wright AJ, van de Plassche O (2004a) Surface distributions of salt-marsh foraminif-era from Connecticut, USA: modern analogues for high-resolution sea level studies. Mar Micropaleontol 51:1–21

Edwards RJ, van de Plassche O, Gehrels WR, Wright AJ (2004b) Assessing sea-level data from Connecticut, USA using a foraminiferal transfer function for tide level. Mar Micropaleontol 51:239–255

Egan B, Yarish C (1988) The distribution of the genus *Laminaria* (Phaeophyta) at its southern limit in the western Atlantic Ocean. Botanica Mar 31:155–161

Egan B, Yarish C (1988) The distribution of the genus *Laminaria* (Phaeophyta) at its southern limit in the western Atlantic Ocean. Botanica Mar 31: 155–161

Egan B, Yarish C (1990) Productivity and life history of *Laminaria longicruris* de la Pyl. at its southern limit in the Western Atlantic Ocean. Mar Ecol Prog Ser 76:263–273

Egan B, Vlasto A, Yarish C (1989) Seasonal acclimation to temperature and light in *Laminaria longicruris* de la Pyl. (Phaeophyta). J Exp Mar Biol Ecol 129:1–16

Egan B, Garcia-Esquivel Z, Brinkhuis BH, Yarish C (1990) Genetics of morphology and growth in *Laminaria* from the North Atlantic Ocean-Implications for biogeography. In: Garbary DJ, South GR (eds) Evolutionary biogeography of the marine algae of the North Atlantic. NATO ASI 22. Springer, Berlin, pp 147–172

Elmer WH, Marra RE (2011) New species of *Fusarium* associated with dieback of *Spartina alterniflora* in Atlantic salt marshes. Mycologia 103:806–819

ENSR (2007) Baseline bathymetric surveys at the Central and Western Long Island Sound disposal sites, July 2005. DAMOS Contribution 177. US Army Corps of Engineers, New England District, Concord, MA, 85 pp

EOBRT (Eastern Oyster Biological Review Team) (2007) Status review of the eastern oyster (*Crassostrea virginica*). Report to the Natl Marine Fisheries Service, Northeast Regional Office. 16 Feb 2007, 105 pp

EPA/USACE (US Environmental Protection Agency/US Army Corps of Engineers) (1991) Evaluation of dredged material proposed for ocean disposal: testing manual. EPA-503/8-91/001

Falkowski PG, Algeo T, Codispoti L, Deutsch C, Emerson S, Hales B, Huey RB, Jenkins WJ, Kump LR, Levin LA, Lyons TW, Nelson NB, Schofield OS, Summons R, Talley LD, Thomas E, Whitney F, Pilcher CB (2011) Ocean deoxygenation: past, present, and future. EOS 92(46):409–420

Farrell AP, Richards JG (2009) Defining hypoxia: an integrative synthesis of the responses of fish to hypoxia. In: Hypoxia, Richards JG, Farrell AP, Brauner CJ (eds) Fish physiology vol 27. Academic press, New York, pp 487–503

Farrow DRG, Arnold FD, Lombardi ML, Main MB, Eichelberger PD (1986) The national coastal pollution discharge inventory, estimates for Long Island Sound. NOAA Office of Oceanography and Marine Assessment, Rockville 40 pp

Fauchald P, Mauritzen M et al (2006) Density-dependent migratory waves in the marine pelagic ecosystem. Ecology 87(11):2915–2924

Feely RA, Sabine CL, Hernandez-Ayon M, Ianson D, Hales B (2008) Evidence for upwelling of corrosive "acidified" water onto the continental shelf. Science 320:1490–1492

Fisher C (1983) Social organization and change during the early horticultural period in the Hudson River valley. PhD Dissertation, SUNY, Albany

Fisher CR (1990) Chemoautotrophic and methanotrophic symbioses in marine invertebrates. Rev Aquat Sci 2:399–436

Fitzgerald DM, Fenster MS, Argow BA, Buynevich IV (2008) Coastal impacts due to sea level rise. Ann Rev Earth Planet Sci 36:601–647

Foertch J, Swenarton J, Keser M (2009) Multivariate analysis to show trends in the algae. Presentation at the Northeast Algal Soc Annual Meeting 2009, University of Massachusetts, Amherst, MA

Fogarty MJ, Murawski SA (1998) Large-scale disturbance and the structure of marine systems: fisheries impacts on Georges Bank. Ecol App 8:S6–S22

Forbes TL, Lopez GR (1990) The effect of food concentration, body size, and environmental oxygen tension on the growth of the deposit-feeding polychaete, *Capitella* species I. Limnol Oceanogr 35:1535–1544

Forbes TL, Forbes VE, Depledge MH (1994) Individual physiological responses to environmental hypoxia and organic enrichment: implications for early soft-bottom community succession. J Mar Res 52:1081–1100

Francis CA, Roberts KJ, Beman JM, Santoro AE, Oakley BB (2005) Ubiquity and diversity of ammonia-oxidizing Archaea in water columns and sediments of the ocean. Proc Natl Acad Sci U S A 102:14683–14688

Frederiksen M, Edwards M, Richardson AJ, Halliday NC, Wanless S (2006) From plankton to top predators: bottom-up control of a marine food web across four trophic levels. J Anim Ecol 75:1259–1268

Frederiksen M, Furness RW, Wanless S (2007) Regional variation in the role of bottom-up and top-down processes in controlling sandeel abundance in the North Sea. Mar Ecol Prog Ser 337:279–286

Fredette TJ (1998) DAMOS: Twenty years of dredged material disposal site monitoring. Isn't that enough? Chem Ecol 14:231–239

Fredette TJ, French GT (2004) Understanding the physical and environmental consequences of dredged material disposal: history in New England and current perspectives. Mar Poll Bull 49:93–102

Fredette TJ, Germano JD, Carey DA, Murray PM, Kullberg P (1992) Chemical stability of capped dredged material disposal mounds in Long Island Sound, USA. Chem Ecol 7:173–194

Fredette TJ, Kullberg PG, Carey DA, Germano JD, Morton R (1993) Twenty-five years of dredged material disposal site monitoring in Long Island Sound: a long-term perspective, pp 153–161. In: Van Patten MS (ed) Long Island Sound Research Conference Proc, Univ of Connecticut Sea Grant, Pub No CT-SG-93-03. Storrs, CT

Frisk MG, Miller TJ et al (2008) New hypothesis helps explain elasmobranch "outburst" on Georges Bank in the 1980s. Ecol App 18(1):234–245

Frisk MG, Martell SJD et al (2010) Exploring the population dynamics of winter skate (Leucoraja ocellata) in the Georges Bank region using a statistical catch-at-age model incorporating length, migration, and recruitment process errors. Can J Fish Aquat Sci 67(5):774–792

Frisk MG, Duplisea DE et al (2011a) Exploring the abundance-occupancy relationships for the Georges Bank finfish and shellfish community from 1963 to 2006. Ecol App 21(1):227–240

Frisk MG, Miller TJ et al (2011b) Assessing biomass gains from marsh restoration in Delaware Bay using Ecopath with Ecosim. Ecol Model 222(1):190–200

Gainey LF, Shumway SE (1991) The physiological effect of Aureococcus anophagefferens ("brown tide") on the lateral cilia of bivalve mollusks. Biol Bull 181:298–306

Gattuso J-P, Hansson L (eds) (2011) Ocean acidification. Oxford University Press, New York, 326 pp

Gattuso J-P, Frankignoulle M, Wollast R (1998) Carbon and carbonate metabolism in coastal aquatic ecosystems. Ann Rev Ecol Syst 29:405–434

Gazeau F, Quiblier C, Jansen JM, Gattuso JP, Middelburg J, Heip CHR (2007) Impact of elevated CO_2 on shellfish calcification. Geophys Res Lett 34. doi:10.1029/2006GL028554

General Dynamics (1968) Study of means to revitalize the Connecticut fisheries industry. Report prepared by the Marine Sciences Section, Research and Development Department, Electric Boat Division for the CT Research Commission under contract RSA-66-8

Gerard VA (1995) Recent changes in shallow, hardbottom communities in Long Island Sound. In: McElroy A, Zeidner J (eds) Proc Long Island Sound Research Conference: is the Sound getting better or worse? State University of New York. New York Sea Grant Institute, New York, pp 77–79

Gerard VA (1997) The role of nitrogen nutrition in high-temperature tolerance of the kelp Laminaria saccharina (Chromophyta). J Phycol 33:800–810

Gerard VA (1999) Positive interaction between cordgrass, Spartina alterniflora, and the brown alga Ascophyllum nodosum ecad scorpiodes, in a mid-Atlantic coast salt marsh. J Exp Mar Biol Ecol 239:157–164

Gerard VA, Du Bois KR (1988) Temperature ecotypes near the southern boundary of the kelp *Laminaria saccharina*. Mar Biol 97:575–580

Gerard VA, DuBois K, Greene R (1987) Growth responses of two *Laminaria saccharina* populations to environmental variation. Hydrobiologia 151(152):229–232

Gerard VA, Cerrato RM, Larson A (1999) Potential impacts of a western Pacific grapsid crab on intertidal communities of the northwestern Atlantic. Biol Invasions 1:353–361

Gerino M, Aller RC, Cochran JK, Aller JY, Green MA, Hirschberg D (1998) Comparison of different tracers and methods used to quantify bioturbation during a spring bloom: 234-Thorium, luminophores and chlorophyll a. Est Coast Shelf Sci 46:531–547

Germano JD, Rhoads DC (1984) REMOTS® sediment profiling at the field verification program (FVP) disposal site. In: Montgomery RL, Leach JW (eds) Dredging and dredged material disposal proceedings, vol 1. American Society of Civil Engineers, New York, pp 536–544

Germano JD, Rhoads DC, Lunz JD (1994) An integrated, tiered approach to monitoring and management of dredged material disposal sites in the New England region. DAMOS Contribution 87. US Army Corps of Engineers, New England Division, Waltham, MA

Germano JD, Rhoads DC, Valente RM, Carey D, Solan M (2011) The use of sediment profile imaging (SPI) for environmental impact assessments and monitoring studies: lessons learned from the past four decades. Oceanogr Mar Biol Ann Rev 49:247–310

Getchis TS (2005) An assessment of the needs of Connecticut's shellfish aquaculture industry. CTSG-05-02. Connecticut Sea Grant College Program. Groton, CT, 12 pp

Giannini C, Howell P (2010) Connecticut lobster (*Homarus americanus*) population studies. Five-year performance report for National Marine Fisheries Service Interjurisdictional Grant 3IJ-168-NA05NMF4071033, Connecticut Department of Environmental Protection

Gibson M, Wahle R (2005) Reduced recruitment of inshore lobster in Rhode Island in association with an outbreak of shell disease and management implications. In: Tlusky M, Halvorson H, Smolowitz R, Sharma U (eds) Lobster shell disease workshop. Aquat Forum Series 05-1, New England Aquar, Boston, MA, pp 115–130

Gifford DJ, Dagg MJ (1991) The microzooplankton-mesozooplankton link: consumption of planktonic protozoa by the calanoid copepods *Acartia tonsa* Dana and *Neocalanus plumchrus* Murukawa. Mar Microb Food Webs 5:161–171

Giorgi AE (1981) The environmental biology of the embryo, egg mass, and nesting sites of the lingcod, *Ophiodon elongates*. NWAFC processed report 81-6, US Department of Commerce, Northwest and Alaska Fisheries Center, NMFS

Glenn R, Pugh T (2006) Epizootic shell disease in American lobster (*Homarus americanus*) in Massachusetts coastal waters: interactions of temperature, maturity, and intermolt duration. J Crustacean Biol 26(4):639–645

Glibert PM, Anderson DM, Gentien P, Graneli E, Sellner K (2005) The global, complex phenomena of harmful algal blooms. Oceanography 18(2):132–141

Gobler CJ, Lonsdale DJ, Boyer GL (2005) A synthesis and review of causes and impact of harmful brown tide blooms caused by the alga, *Aureococcus anophagefferens*. Estuaries 28:726–749

Gobler CJ, Sañudo-Wilhelmy SA, Buck NJ, Sieracki ME (2006) Nitrogen and silicon limitation of phytoplankton communities across an urban estuary: The East River-Long Island Sound system. Est Coast Shelf Sci 68:127–138

Gobler CJ, Berry DL, Anderson OR, Burson A, Koch F, Rodgers BS, Moore LK, Goleski JA, Allam B, Bowser P, Tang YZ, Nuzzi R (2008) Characterization, dynamics, and ecological impacts of harmful *Cochlodinium polykrikoides* blooms on eastern Long Island, NY, USA. Harmful Algae 7:293–307

Goebel NL, Kremer JN (2007) Temporal and spatial variability of photosynthetic parameters and community respiration in Long Island Sound. Mar Ecol Prog Ser 329:23–42

Goebel NL, Kremer JN, Edwards CA (2006) Primary production in Long Island Sound. Estuar Coasts 29:232–245

Goedkoop W, Gullberg KR, Johnson RK, Ahlgren I (1997) Microbial response of a freshwater benthic community to a simulated diatom sedimentation event: interactive effects of benthic fauna. Microbial Ecol 34:131–143

Goldberg R, Pereira J, Clark P (2000) Strategies for enhancement of natural bay scallop, *Argopecten irradians irradians*, populations: a case study in the Niantic River estuary, Connecticut, USA. Aquac Int 8:139–158 (Kluwer Academic Publishers)

Gooday AJ (1993) Benthic foraminifera (protista) as tools in deep-water paleooceanography: environmental influences on faunal characteristics. Adv Mar Biol 46:1–90

Gooday AJ, Jorissen F, Levin LA, Middelburg JJ, Naqvi SWA, Rabalais NN, Scranton M, Zhang J (2009) Historical records of coastal-eutrophication induced hypoxia. Biogeosciences 6:1707–1745

Gordon BL (1974) The marine fishes of Rhode Island, 2nd edn. Book & Tackle Shop, Watch Hill 136 pp

Gosselin LA, Qian PY (1997) Juvenile mortality in benthic marine invertebrates. Mar Ecol Prog Ser 146:265–282

Gottschall K, Pacileo D (2010) Long Island Sound trawl survey. In: A study of marine recreational fisheries in Connecticut, Job 2.1. Federal aid in sport fish restoration grant F-54-R-29, Connecticut Department of Environmental Protection

Graf G, Bengtsson W, Diesner U, Schulz R, Theede H (1982) Benthic response to sedimentation of a spring phytoplankton bloom: process and budget. Mar Biol 67:201–208

Grant J (1996) The relationship of bioenergetics and the environment to the field growth of cultured bivalves. J Exp Mar Biol Ecol 200:239–256

Grant WS, Root WS (1952) Fundamental stimulus for erythropoiesis. Physiol Rev 32:449–498

Gray IE (1954) Comparative study of the gill area of marine fishes. Biol Bull 107:219–225

Gray JS, Wu RS-S, Or YY (2002) Effects of hypoxia and organic enrichment on the coastal marine environment. Mar Ecol Prog Ser 238:249–279

Greaney GS, Pace AR, Cashon RE, Smith G, Powers DA (1980) Time course of changes in enzyme activities and blood respiratory properties of killifish during long-term acclimation to hypoxia. Physiol Zool 53:136–144

Green AA, Root RW (1933) The equilibrium between hemoglobin and oxygen in the blood of certain fishes. Biol Bull Mar Biol Lab Woods Hole 64:383–404

Green MA, Jones ME, Boudreau CL, Moore RL, Westman BA (2004) Dissolution mortality of juvenile bivalves in coastal marine deposits. Limnol Oceanogr 49:727–734

Green MA, Waldbusser GG, Reilly SL, Emerson K, O'Donnell S (2009) Death by dissolution: sediment saturation state as a mortality factor for juvenile bivalves. Limnol Oceanogr 54:1037–1047

Greenfield DI, Lonsdale DJ (2002) Mortality and growth of juvenile hard clams *Mercenaria mercenaria* during brown tide. Mar Biol 141:1045–1050

Greenfield DI, Lonsdale DJ, Cerrato RM, Lopez GR (2004) Effects of background concentrations of *Aureococcus anophagefferens* (brown tide) on growth and feeding in the bivalve *Mercenaria mercenaria*. Mar Ecol Prog Ser 274:171–181

Greenfield DI, Lonsdale DJ, Cerrato RM (2005) Linking phytoplankton community composition with juvenile-phase growth in the northern quahog *Mercenaria mercenaria* (L.). Estuaries 28:241–251

Greening H, Janicki A (2006) Toward reversal of eutrophic conditions in a subtropical estuary: water quality and seagrass response to nitrogen loading reductions in Tampa Bay, Florida, USA. Environ Manag 38:163–178

Grieshaber MK, Völkel S (1998) Animal adaptations for tolerance and exploitation of poisonous sulfide. Ann Rev Physiol 60:33–53

Grieshaber MK, Hardewig I, Kreutzer U, Pörtner H-O (1994) Physiological and metabolic responses to hypoxia in invertebrates. Rev Physiol Biochem Pharmacol 125:44–147

Gross AC (1966) Vegetation of the Brucker Marsh and the Barn Island Natural Area, Stonington, Connecticut. Master's dissertation, Connecticut College, 103 pp

Grove M, Breitburg DL (2005) Growth and reproduction of gelatinous zooplankton exposed to low dissolved oxygen. Mar Ecol Prog Ser 301:185–198

Guderley H, Pörtner H-O (2010) Metabolic power budgeting and adaptive strategies in zoology: examples from scallops and fish. Can J Zool 88(8):753–763

Gutierrez JL, Jones CG, Strayer DL, Iribarne OO (2003) Mollusks as ecosystem engineers: the role of shell production in aquatic habitats. Oikos 101:79–90

Hales LS, Able KW (1995) Effects of oxygen concentration on somatic and otolith growth rates of juvenile black sea bass (*Cetropristis striata*). In: Sector DH, Dean JM, Campana SE (eds) Recent developments of fish otolith research. University of SC Press, Columbia, pp 135–153

Hall GF, Gray IE, Lepkovskys S (1926) The influence of asphyxiation on the blood constituents of marine fishes. J Biol Chem 67:549–554

Hall CJ, Jordaan A, Frisk MG (2011) The historic influence of dams on diadromous fish habitat with a focus on river herring and hydrologic longitudinal connectivity. Landsc Ecol 26(1):95–107

Hallegraeff GM (1993) A review of harmful algal blooms and their apparent global increase*. Phycologia 32:79–99

Hallegraeff GM, Lucas IAN (1988) The marine dinoflagellate genus *Dinophysis* (Dinophyceae): photosynthetic, neritic and non-photosynthetic, oceanic species. Phycologia 27:25–42

Hall-Spencer JM, Rodolfo-Metalpa R, Martin S, Ransome E, Fine M, Turner SM, Rowley SJ, Tedesco D, Buia M-C (2008) Volcanic carbon dioxide vents reveal ecosystem effects of ocean acidification. Nature 454:96–99. doi:10.1038/nature07051

Hammerson GA (2004) Connecticut wildlife: biodiversity, natural history, and conservation. University pf Press of New England, One Court Street, Lebanon

Hansen PJ, Calado AJ (1999) Phagotrophic mechanisms and prey selection in free-living dino-flagellates. J Eukaryotic Microbiol 46:382–389

Hare JA, Alexander MA, Fogarty MJ, Williams EH, Scott JD (2010) Forecasting the dynamics of a coastal fishery species using a coupled climate-population model. Ecol Appl 20(2):452–464

Hargraves PE, Maranda L (2002) Potentially toxic or harmful microalgae from the northeast coast. Northeast Nat 9:81–120

Harris E, Riley GA (1956) Oceanography of Long Island Sound, 1952-1954. VIII. Chemical composition of the plankton. Bull Bingham Oceanogr Collect 15:315–323

Haska CL, Yarish C, Kraemer G, Blaschick N, Whitlatch R, Zhang H, Lin S (2012) Bait worm packaging as potential vector of invasive species. Biol Invasions 14:481–493. doi:10.1007/s10530-011-0091-y

Hathaway WE (1910) Effects of menhaden fishing upon the supply of menhaden and of the fishes that prey upon them. Proc 4th Internatl Fishery Congress: Washington, 1908. Government Printing Office, Washington, p 8

Hattenrath TK, Anderson DM, Gobler CJ (2010) The influence of anthropogenic nitrogen loading and meteorological conditions on the dynamics and toxicity of *Alexandrium fundyense* blooms in a New York (USA) estuary. Harmful Algae 9:402–412

Hauxwell J, Cebrián J, Valiela I (2003) Eelgrass *Zostera marina* loss in temperate estuaries: relationship to land-derived nitrogen loads and effect of light limitation imposed by algae. Mar Ecol Prog Ser 247:59–73

Hayward BW, Holzmann M, Grenfell HA, Pawlowski J, Triggs CM (2004) Morphological distinction of molecular types in *Ammonia*-towards a taxonomic revision of the world's most commonly misidentified foraminifera. Mar Micropaleontol 50:237–271

He P, Yarish C (2006) The developmental regulation of mass cultures of free living conchocelis for commercial net seeding of *Porphyra leucosticta* from Northeast America. Aquaculture 257:373–381

Heath A (1987) Water Pollution and Fish Physiology. CRC Press, Inc. Boca Raton, Florida

Heck KL Jr, Hays G, Orth RJ (2003) Critical evaluation of the nursery role hypothesis for sea-grass meadows. Mar Ecol Prog Ser 253:123–136

Hégaret H, Shumway SE, Wikfors GH, Pate S, Burkholder JAM (2008) Potential transport of harmful algae though relocation of bivalve mollusks. Mar Ecol Prog Ser 361:169–179

Heil CA, Glibert PM, Fan CL (2005) *Prorocentrum minimum* (Pavillard) Schiller—a review of a harmful algal bloom species of growing worldwide importance. Harmful Algae 4:449–470

Heisler J, Glibert P, Burkholder J, Anderson D, Cochlan W, Dennison W, Dortch Q, Gobler CJ, Heil C, Humphries E, Lewitus A, Magnien R, Marshall H, Sellner Stockwell K, Stoecker D, Suddleson M (2008) Eutrophication and harmful algal blooms: a scientific consensus. Harmful Algae 8:3–13

Hemminga M, Duarte CM (2000) Seagrass ecology. Cambridge University Press, Cambridge

Hendy IL, Pedersen TF (2006) Oxygen minimum zone expansion in the eastern tropical North Pacific during deglaciation. Geophys Res Lett 33:L20602, 5 pp

Herbst JFW (1782–1804) Versuch einer Naturgeschichte der Krabben und Krebse nebst einer systematischen Beschriebung ihrer verschiedenen Arten, vol 1 (1782–1790), 274 pp, vol 2 (1791–1796), 225 pp, vol 3 (1799–1804), 66 pp

Hewitt JE, Thrush SF, Halliday J, Duffy C (2005) The importance of small-scale habitat structure for maintaining beta diversity. Ecology 86:1619–1626

Hines ME, Orem WH, Lyons WB, Jones GE (1982) Microbial activity and bioturbation-induced oscillations in pore water chemistry of estuarine sediments in spring. Nature 299:433–435

Holzmann M, Pawlowski J (1997) Molecular, morphological and ecological evidence for species recognition in *Ammonia* (Foraminifera). J Foram Res 27:311–318

Hoover MD (2009) Connecticut's changing salt marshes: a remote sensing approach to sea level rise and possible salt marsh migration. Master's dissertation, University of Connecticut, Storrs, CT

Horton BP, Edwards RJ (2006) Quantifying Holocene sea-level change using intertidal foraminifera: lessons from the British Isles. Cushman Foundation for Foraminiferal Research, Special Pub No 40, 97 pp

Houghton RA (1985) The effect of mortality on estimates of net above-ground production of *Spartina alterniflora*. Aq Botany 22:121–132

Howarth RW (2008) Coastal nitrogen pollution: a review of sources and trends globally and regionally. Harmful Algae 8:14–20

Howell P, Simpson D (1994) Abundance of marine resources in relation to dissolved oxygen in Long Island Sound. Estuaries 17(2):394–402

Hu Z, Guiry MD, Critchley AT, Duan D (2010) Phylogeographic patterns indicate transatlantic migration from Europe to North America in the red seaweed *Chondrus crispus* (Gigartinales, Rhodophyta). J Phycol 46:889–900

Hudson IR, Pond DW, Billett DSM, Tyler PA, Lampitt RS, Wolff GA (2004) Temporal variations in fatty acid composition of deep-sea holothurians: evidence of bentho-pelagic coupling. Mar Ecol Prog Ser 281:109–120

Hughes GM (1984) Scaling of respiratory areas in relationship to oxygen consumption of vertebrates. Experientia 40:519–524

HydroQual, Inc (1995) Analysis of factors affecting historical dissolved oxygen trends in western Long Island Sound. Job # NENG0040. Management Committee of the Long Island Sound Estuary Study, Stamford, CT, and the New England Interstate Water Pollution Control Commission. HydroQual, Inc, Mahwah, NJ

Iriarte A, Purdie DA (2004) Controlling the timing of major spring bloom events in an UK south coast estuary. Est Coast Shelf Sci 61:679–690

Isaksen MF, Jorgensen BB (1996) Adaptation of psychrophilic and psychrotrophic sulfate-reducing bacteria to permanently cold marine environments. Appl Env Microbiol 62:408–414

Jackson MJ, James R (1979) The influence of bait digging on cockle, *Cerastoderma edule*, population in North Norfolk. J Appl Ecol 16:671–679

Jackson JBC, Kirby MX, Berger WH, Bjorndal KA, Botsford LW, Bourque BJ, Bradbury RH, Cooke R, Erlandson J, Estes JA, Hughes TP, Kidwell S, Lange CB, Lenihan HS, Pandolfi JM, Peterson CH, Steneck RS, Tegner MJ, Warner RR (2001) Historical overfishing and the recent collapse of coastal ecosystems. Science 293:629–637

Janiak D (2010) Comparisons of epifaunal communities associated with the invasive Grateloupia turuturu and native Chondrus crispus in Long Island Sound. CT. Master's dissertation, University of Connecticut

Javaux E, Scott DB (2003) Illustration of modern benthic foraminifera from Bermuda and remarks on distribution in other subtropical/tropical areas. Palaeontologia Electronica 6(4):29, 2.1 MB. http://palaeo-electronica.org/paleo/2003_1/benthic/issue1_03.htm

Jensen FB, Nikinmaa M, Weber RE (1993) Environmental perturbations of oxygen transport in teleost fishes: causes, consequences and compensations. In: Jensen FB, Rankin JC (eds) Fish ecophysiology. Chapman and Hall, London, pp 161–179

St. John JP, Leo WM, Dodge LJ, Gaffoglio R (1991) Modeling analysis of CSO impacts in a tidal basin. Proceedings of the 1991 special conference on environmental engineering, New York (USA), ASCE, pp 457–462

Johnson M, Beckwith EJ, Carey D, Parker E, Smith E, Volk J, Aarrestad P, Huang M, Mariani E, Rozsa R, Simpson D, Yamalis H (2007) An assessment of the impacts of commercial and recreational fishing and other activities to eelgrass in Connecticut's waters and recommendations for management. Connecticut Department of Environmental Protection and Connecticut Department of Agriculture, Hartford 119

Jordaan A, Chen Y, Townsend DW, Sherman S (2010) Identification of ecological structure and species relationships along an oceanographic gradient in the Gulf of Maine using multivariate analysis with bootstrapping. Can J Fish Aq Sci 67(4):701–719

Jordaan A, Crocker J, Chen Y (2011) Linkages among physical and biological properties in tidepools on the Maine coast. Env Biol Fishes 92(1):13–23

Jørgensen BB (1977) The sulfur cycle of a coastal marine sediment (Limfjorden, Denmark). Limnol Oceanogr 22(5):814–832

Jorissen FJ, Fontanier C, Thomas E (2007) Paleoceanographical proxies based on deep-sea benthic foraminiferal assemblage characteristics. In: Hillaire-Marcel C, de Vernal A (eds) Proxies in Late Cenozoic paleoceanography: Pt 2: Biological tracers and biomarkers. Elsevier, Amsterdam, pp 63–326

Jury S, Watson W (2000) Thermosensitivity of the lobster, *Homarus americanus*, as determined by cardiac assay. Biol Bull 199(3):257–264

Kaputa NP, Olson CB (2000) Long Island Sound ambient water quality monitoring program: summer hypoxia monitoring survey 1991-1998 data review. CT Department of Environmental Protection, 45 pp + appendices

Karlsen AW, Cronin TM, Ishman SE, Willard DA, Kerhin R, Holmes CW, Marot M (2000) Historical trends in Chesapeake Bay dissolved oxygen based on benthic Foraminifera from sediment cores. Estuaries 23:488–508

Keeling RF, Garcia H (2002) The change in oceanic O_2 inventory associated with recent global warming. Proc Natl Acad Sci U S A 99:7848–7853

Keeling RF, Kortzinger A, Gruber N (2010) Ocean deoxygenation in a warming world. Ann Rev Mar Sci 2:199–229

Keister JE, Houde ED, Breitburg DL (2000) Effects of bottom-layer hypoxia on abundances and depth distributions of organisms in Patuxent River, Chesapeake Bay. Mar Ecol Prog Ser 205:43–59

Keller AA, Oviatt CA, Walker HA, Hawk JD (1999) Predicted impacts of elevated temperature on the magnitude of the winter-spring phytoplankton bloom in temperate coastal waters: a mesocosm study. Limnol Oceanogr 44:344–356

Keller AA, Taylor C, Oviatt C, Dorrington T, Holcombe G, Reed L (2001) Phytoplankton production patterns in Massachusetts Bay and the absence of the 1998 winter-spring bloom. Mar Biol 138:1051–1062

Kersten M, Britton RH, Dugan PJ, Hafner H (1991) Flock feeding and food intake in little egrets: the effects of prey distribution and behavior. J Anim Ecol 60:241–252

Keser M, Swenarton JT, Vozarik JM, Foertch JF (2003) Decline in eelgrass (*Zostera marina* L.) in Long Island Sound near Millstone Point, Connecticut (USA) unrelated to thermal input. J Sea Res 49:11–26

Keser M, Foertch J, Swenarton J (2010) Rocky shore community changes noted over 30 + years of monitoring at Millstone Environmental Lab, CT, USA. Oral presentation, Session VI, Northeast Algal Symposium, 2010

Kim JK, Kraemer GP, Yarish C (2008) Physiological activity of *Porphyra* in relation to zonation. J Exp Mar Biol Ecol 365:75–85

Kim JK, Kraemer GP, Yarish C (2009) A comparison of growth and nitrate uptake by New England *Porphyra* species from different tidal elevations in relation to desiccation. Phycol Res 57:152–157

Kimble ME, Able KW (2007) Nekton utilization of intertidal salt marsh creeks: Tidal influences in natural *Spartina,* invasive *Phragmites*, and marshes treated for *Phragmites* removal. J Exp Mar Biol Ecol 347:87–101

Kiørboe T, Hansen JLS, Alldredge AL, Jackson GA, Passow U, Dam HG, Drapeau DT, Waite A, Garcia CM (1996) Sedimentation of phytoplankton during a diatom bloom: rates and mechanisms. J Mar Res 54:1123–1148

Kjeilen-Eilertsen G, Trannum H, Jak RG, Smit MGD, Neff J, Durell G (2004) Literature report on burial: derivation of PNEC as component in the MEMW model tool. Report AM 2004/024, ERMS report 9B

Klein AS, Mathieson AC, Neefus CD, Cain DF, Taylor HA, West AL, Hehre EJ, Brodie J, Yarish C, Teasdale B, Wallace AL (2003) Identifications of northwestern Atlantic *Porphyra* (Bangiaceae, Bangiales) based on sequence variation in nuclear SSU and rbcL genes. Phycologia 42(2):109–122

Knebel H, Poppe LJ (2000) Sea-floor environments within Long Island Sound: a regional overview. Thematic Section, J Coast Res 16:533–550

Kneib RT (1986) The role of Fundulus heteroclitus in salt marsh tophic dynamics. Am Zool 26:259–269

Koch EW (2001) Beyond light: physical, geological, and geochemical parameters as possible submersed aquatic vegetation habitat requirements. Estuaries 24:1–17

Koch EW, Beer S (1996) Tides, light and the distribution of *Zostera marina* in Long Island Sound, USA. Aquat Bot 53:97–107

Koch F, Gobler CJ (2009) The effects of tidal export from salt marsh ditches on estuarine water quality and plankton communities. Estuaries Coasts 32:261–275

Komar PD (1998) Beach processes and sedimentation. Prentice Hall, Upper Saddle River 544 pp

Kott P (2004) A new species of *Didemnum* (Ascidiacea, Tunicata) from the Atlantic coast of North America. Zootaxa 732:1–10

Kraeuter JN (2001) Predators and predation, Chapter 11. In: Kraeuter JN, Castagna M (eds) Biology of the hard clam. Elsevier, New York, pp 441–589

Kraeuter JN, Klinck JM, Powell EN, Hofmann EE, Buckner SC, Grizzle RE, Bricelj VM (2008) Effects of the fishery on the northern quahog (=hard clam, *Mercenaria mercenaria* L.) population in Great South Bay, New York: a modeling study. J Shellfish Res 27:653–666

Kramer DL (1987) Dissolved oxygen and fish behavior. Env Biol Fishes 18:81–92

Kramer DL, Manley D, Bourgeois R (1983) The effect of respiratory mode and oxygen concentration on the risk of aerial predation in fishes. Can J Zool 61:653–665

Kranz PM (1974) The anastrophic burial of bivalves and its paleoecological significance. J Geol 82:237–265

Kremer JN, Vaudrey JNP, Branco A (2008) Habitat characterization data set for 10 New England estuaries. "Hammonasett River 2002" http://www.lisrc.uconn.edu/

Kristensen E, Andersen FO, Blackburn TH (1992) Effects of benthic macrofauna and temperature on degradation of macroalgal detritus: the fate of organic carbon. Limnol Oceanogr 37:1404–1419

Kroeger KD (1997) Temperature and dissolved oxygen concentration as factors influencing the secondary production rate of *Nucula annulata* in Long Island Sound. Univ of Connecticut, USA. Master's dissertation

Kudela RM, Gobler CJ (2012) Harmful dinoflagellate blooms caused by *Cochlodinium* sp.: global expansion and ecological strategies facilitating bloom formation. Harmful Algae 14:71–86

Kurihara H, Kato S, Ishimatsu A (2007) Effects of increased seawater pCO_2 on early development of the oyster *Crassostrea gigas*. Aquat Biol 1:91–98

Kurihara H, Asai T, Kato S, Ishimatsu A (2008) Effects of elevated pCO_2 on early development in the mussel *Mytilus galloprovincialis*. Aquat Biol 4:225–233

Kurlansky M (2006) The Big Oyster: History on the half shell. Ballantine Books, pp 307

Lambert G (2009) Adventures of a sea squirt sleuth: unraveling the identity of *Didemnum vexillum*, a global ascidian invader. Aquat Invasions 4:5–28

Landeck Miller RE, Wands JR, Chytalo KN, D'Amico RA (2005) Application of water quality modeling technology to investigate the mortality of lobsters (*Homarus americanus*) in western Long Island Sound. J Shellfish Res 24(3):859–864

Landry MR, Hassett RP (1982) Estimating the grazing impact of marine microzooplankton. Mar Bio 67:283–288

Landry CA, Steele SL, Manning S, Cheek AO (2007) Long term hypoxia suppresses reproductive capacity in the estuarine fish, *Fundulus grandis*. Comp Biochem Physiol Part A 148:317–323

Landsberg JH (2002) The effects of harmful algal blooms on aquatic organisms. Rev Fisheries Sci 10:113–390

Larsson A, Johansson-Sjöbeck M, Fänge R (1976) Comparative study of some of the haematological and biochemical blood parameters in fishes from the Skagerrak. J Fish Biol 9:425–440

Latham R (1917) Migration notes of fishes, 1916, from Orient, Long Island. Copeia 41:17–23

Latham R (1969) Sea turtles recorded in the Southold township region of Long Island. Engelhardtia 1:7

Latimer JS, Rego SA (2010) Empirical relationship between eelgrass extent and predicted watershed-derived nitrogen loading for shallow New England estuaries. Est Coast Shelf Sci 90:231–240

Lee RE (1989) Phycology, 2nd edn. Cambridge University Press, Cambridge, 645 pp

Lee JJ, Anderson OR (1991) Symbiosis in foraminifera. In: Lee JJ, Anderson OR (eds) Biology of foraminifera. Academic Press, London, pp 157–220

Lee JA, Brinkhuis BH (1988) Seasonal light and temperature interaction effects on development of *Laminaria saccharina* (Phaeophyta) gametophytes and juvenile sporophytes. J Phycology 24(2):181–191

Lee JJ, Lee RE (1989) Chloroplast retention in Elphidids (Foraminifera). In: Nardon P, Gianinazzi-Pearson V, Greiner AM, Margulis L, Smith DC (eds) 4th Internatl colloquium on endocytobiology and symbiosis. Institut National de la Recherche Agronomique, Paris, pp 215–220

Lee YJ, Lwiza KMM (2008) Characteristics of bottom dissolved oxygen in Long Island Sound, New York. Est Coast Shelf Sci 76:187–200

Lee JJ, Lanners E, TerKuile B (1988) The retention of chloroplasts by the foraminifer *Elphidium crispum*. Symbiosis 5:45–60

Lee KS, Park SR, Kim YK (2007) Effects of irradiance, temperature, and nutrients on growth dynamics of seagrasses: a review. J Exp Mar Biol Ecol 350:144–175

Lembi CA, Waaland JR (eds) (2007) Algae and human affairs. Cambridge University Press, Cambridge, 600 pp

Levinton JS (1970) The paleoecological significance of opportunistic species. Lethaia 3:69–78

Lewis RR III, Clark PA, Fehring WK, Greening HS, Johansson RO, Paul RT (1998) The rehabilitation of the Tampa Bay Estuary, Florida, USA, as an example of successful integrated coastal management. Mar Poll Bull 37:468–473

Lightfoot KG, Cerrato RM (1988) Prehistoric shellfish exploitation in coastal New York. J Field Archaeol 15(2):141–149

Limburg KE, Waldman JR (2009) Dramatic declines in North Atlantic diadromous fishes. Bioscience 59(11):955–965

Lin S et al (in prep) Spore abundance of the invasive rhodophyte *Grateloupia turuturu* in Long Island Sound using real-time PCR: an assessment of potential dispersal pathway

Liu S, Lin S (2008) Temporal and spatial variation of phytoplankton community in Long Island Sound. Proceedings of the 9th Biennial Long Island Sound Conference, pp 28–34

Liu H, Bidigare RR, Laws E, Landry MR, Campbell L (1999) Cell cycle and physiological characteristics of Synechococcus (WH7803) in chemostat culture. Mar Ecol Prog Ser 189:17–25

Llanso RJ (1991) Tolerance of low dissolved oxygen and hydrogen sulfide by the polychaete *Streblospio benedicti* (Webster). J Exp Mar Biol Ecol 153:165–178

Lloyd R (1961) Effect of low dissolved oxygen concentration on the toxicity of several poisons to rainbow trout (Salmo gairdneri). J Exp Biol 38:447–455

Lobban CS, Harrison PJ (1994) Seaweed ecology and physiology. Cambridge University Press, Cambridge, 365 pp

Loglisci CA (2007) A mass balance approach to understanding copepod fecal pellet cycling in Long Island Sound. Master's dissertation, University of Connecticut

Lohrer AM, Whitlatch RB (1997) Ecological studies on the recently introduced Japanese shore crab (*Hemigrapsus sanguineus*), in Eastern Long Island Sound. In: Balcom N (ed) Proc 2nd Northeast conference on nonindigenous aquatic nuisance species, Connecticut Sea Grant College Program CTSG-97-02. University of Connecticut, Avery Point, pp 49–60

Lomas ML, Moran SB (2011) Evidence for aggregation and export of cyanobacteria and nano-eukaryotes from the Sargasso Sea euphotic zone. Biogeosciences 8:203–216

Long Island Sound Study (1994) Summary of the comprehensive conservation and management plan July 1994. EPA 842-S-94-001

Long Island Sound Study (2002) Habitat restoration initiative. Annual summary for the year 2002: Technical support for coastal habitat restoration

Long Island Sound Study (2003) Long Island Sound habitat restoration initiative, technical support for coastal habitat restoration, Section 3: Submerged aquatic vegetation

Long WC, Seitz RD (2008) Trophic interactions under stress: hypoxia enhances foraging in an estuarine food web. Mar Ecol Prog Ser 362:59–68

Longstaff BJ, Dennison WC (1999) Seagrass survival during pulsed turbidity events: the effects of light deprivation on the seagrasses *Halodule pinifolia* and *Halophila ovalis*. Aquat Bot 65:105–121

Lonsdale DJ, Cosper EM, Kim W-S, Doall M, Divadeenam A, Jonasdottir SH (1996) Food web interactions in the plankton of Long Island bays, with preliminary observations on brown tide effects. Mar Ecol Prog Ser 134:247–263

Lonsdale DJ, Cerrato RM, Caron DA, Schaffner RA (2007) Zooplankton changes associated with grazing pressure of northern quahogs (*Mercenaria mercenaria* L.) in experimental mesocosms. Est Coast Shelf Sci 73:101–110

Loosanoff VL (1958) Some aspects of behavior of oysters at different temperatures. Biol Bull 114:57–70

Lopez E (1979) Algal chloroplasts in the protoplasm of three species of benthic foraminifera: taxonomic affinity, viability, and persistence. Mar Biol 53:201–211

Lopez GR, Levinton JS (2011) Particulate organic detritus and detritus feeders in coastal food webs. In: Wolanski E, McLusky DS (eds) Treatise on estuarine and coastal sciences, vol 6. Academic Press, Waltham, pp 5–21

Loucks ED, Johnson TZ (1991) Hydraulic impacts of I/I on system capacity. In: Krenkel PA (ed) Proceedings of the 1991 special conference on environmental engineering. American Society for Chemical Engineers, New York, pp 114–119

Lucey SM, Nye JA (2010) Shifting species assemblages in the northeast US continental shelf large marine ecosystem. Mar Ecol Prog Ser 415:23–33

Ludsin S, Zhang X, Brandt SB, Roman RM, Boicourt WC, Mason DM, Costantini M (2009) Hypoxia-avoidance by planktivorous fish in Chesapeake Bay: Implications for food web interactions and fish recruitment. J Exp Mar Biol Ecol 381:S121–S131

Lüning K (1990) Seaweeds—their environment, biogeography, and ecophysiology. In: Yarish C, Kirkman H (eds) Edited translation of the German language edition Meeresbotanik: Verbreitung, Okophysiologie und Nutzung der marinen Makroalgen, by Klaus Lüning. Wiley, New York, p 527

Lutins A (1992) Prehistoric fishweirs in eastern North America. Master's dissertation, State University of New York, Binghamton

Lwiza KMM (2008) Climatic aspects of extreme temperatures in coastal waters. Geophys Res Lett 35:L19604

Lynch J, Cottam C (1937) Status of eelgrass (*Zostera marina*) on the North Atlantic coast, January 1937. USDA Wildlife Research and Management leaflet BS-94

MacKenzie CL Jr (1997) The molluscan fisheries of Chesapeake Bay. In: MacKenzie CL Jr et al (eds) The history, present condition, and future of the molluscan fisheries of North and Central America and Europe, vol 1., Atlantic and Gulf CoastsUS Dept of Commerce, NOAA Tech Report NMFS, pp 141–170

Mackin J, Aller RC, Vigil H, Rude P (1991) Nutrient and dissolved oxygen fluxes across the sediment–water interface, in Long Island Sound Study: Final report, Sediment geochemistry and biology. US EPA Contract CE 002870026. Section IV, 1–252

MacLeod R (2009) Marine angler survey. A study of marine recreational fisheries in Connecticut. Federal aid in sport fish restoration, Grant F-54-R-28, Annual performance report, Job 1

Maes J, Stevens M, Breine J et al (2007) Modelling the migration opportunities of diadromous fish species along a gradient of dissolved oxygen concentration in a European tidal watershed. Est Coast Shelf Sci 75:151–162

Mai K, Mercer JP, Donlon J (1992) Comparative studies on the nutrition of two species of abalone, *Haliotis tuberculata* and *H. discus hannai*. In: Van Patten MS (ed) Irish-American technical exchange on the aquaculture of abalone, sea urchins, lobsters and kelp. Connecticut Sea Grant College Program and Martin Ryan Institute, 83 pp

Mangum CP (1997) Adaptation of the oxygen transport system to hypoxia in the blue crab, *Callinectes sapidus*. Amer Zool 37:604–611

Maranda L, Shimizu Y (1987) Diarrhetic shellfish poisoning in Narragansett Bay. Estuaries 10:298–302

Marston M, Villalard-Bohnsack M (2002) Genetic variability and potential sources of *Grateloupia doryphora* (Halymeniaceae, Rhodophyta), an invasive species in Rhode Island waters (USA). J Phycology 38:649–658

Martin S, Rodolfo-Metalpa R, Ransome E, Rowley S, Buia M, Gattuso J, Hall-Spencer J (2008) Effects of naturally acidified seawater on seagrass calcareous epibionts. Biol Lett 4(6):689–692

Mathieson AC, Pederson J, Dawes CJ (2008a) Rapid assessment surveys of fouling and introduced seaweeds in the Northwest Atlantic. Rhodora 110:406–478

Mathieson AC, Pederson JR, Neefus CD, Dawes CJ, Bray TL (2008b) Multiple assessments of introduced seaweeds in the Northwest Atlantic. ICES J Mar Sci 65:730–741

Mathieson AC, Dawes CJ, Pederson J, Gladych RA, Carlton JT (2008c) The Asian red seaweed *Grateloupia doryphora* (Rhodophyta) invades the Gulf of Maine. Biol Invasions 10:985–988

Matthews JH (1927) Fisheries of the North Atlantic. Econ Geogr 3(1):1–22

Maung ES (2010) Hypoxia and benthic invertebrate communities in Western Long Island Sound. Masters dissertation, University of Connecticut, 196 pp

Maurer D, Keck RT, Tinsman JC, Leathem WA, Wethe C, Lord C, Church TM (1986) Vertical migration and mortality of marine benthos in dredged material: a synthesis. Int Revue Gesam Hydrobiol 71:49–63

McCall PL (1975) The influence of disturbance on community patterns and adaptive strategies of the infaunal benthos of central Long Island Sound. PhD dissertation, Yale University, New Haven, CT, 198 pp

McCall PL (1977) Community patterns and adaptive strategies of infaunal benthos of Long Island Sound. J Mar Res 35:221–266

McDermott J (1998) The western Pacific brachyuran (*Hemigrapsus sanguineus*: Grapsidae) in its new habitat along the Atlantic coast of the United States: geographic distribution and ecology. ICES J Mar Sci 55:289–298

McEnroe M (1991) Review of physiological effects of hypoxia on the forage-base organisms of the sound. Final report to Long Island Sound Study, US Environmental Protection Agency, Long Island Sound Study Office, Stamford, CT, 94 pgs

McEnroe M, Allen L (1995) Behavioral and respiratory responses of mummichogs, *Fundulus heteroclitus*, to hypoxia. In: Proceedings of the Long Island Sound Research conference, NY Sea Grant Institute, SUNY Stony Brook, Publ No NYSGI-W-94-001, pp 85–87

McEnroe M, Kroslowitz D (1997) Activity of juvenile winter flounder, *Pleuronectes americanus*, in response to progressive environmental hypoxia. In: Proceedings of the 3rd Long Island Sound Research conference, CT Sea Grant Publ No CTSG-97-08, Groton, CT, pp 53–59

McEnroe M, Dubay S, Boccia T, Hersh M, Gordon P (1995) The fish community of a stressed environment: Milton Harbor, Rye, NY. In: Proceedings of the Long Island Sound Research conference, NY Sea Grant Institute, SUNY Stony Brook, Publ No NYSGI-W-94-001, pp 83–85

McGlathery KJ, Sundbäck K, Anderson IC (2007) Eutrophication in shallow coastal bays and lagoons: the role of plants in the coastal filter. Mar Ecol Prog Ser 348:1–18

McGovern TM, Hellberg ME (2003) Cryptic species, cryptic endosymbionts, and geographical variation in chemical defenses in the bryozoan *Bugula neritina*. Mol Ecol 12:1207–1215

McIvor LM, Maggs CA, Provan J, Stanhope M (2001) RbcL sequences reveal multiple cryptic introductions of the Japanese red alga *Polysiphonia harveyi*. Mol Ecol 10:911–919

McLeese D (1956) Effects of temperature, salinity and oxygen on the survival of the American lobster. J Fish Res Bd Canada 13(2):247–372

McMahon KW, Ambrose WG Jr, Johnson BJ, Sun M-Y, Lopez GL, Clough LM, Carroll ML (2006) Benthic community response to ice algae and phytoplankton in Ny Ålesund, Svalbard. Mar Ecol Prog Ser 310:1–14

McManus GB (1986) Ecology of heterotrophic nanoflagellates in temperate coastal waters. PhD dissertation, SUNY Stony Brook, 174 pp

McNatt RA, Rice JA (2004) Hypoxia-induced growth rate reduction in two juvenile estuary-dependent fishes. J Exp Mar Biol Ecol 311(1):147–156

McRoy CP, McMillan C (1977) Production ecology and physiology of seagrasses. In: McRoy CP, Helfferich C (eds) Seagrass ecosystems: a scientific perspective. Marcel Dekker, Inc., New York, pp 53–87

McVey JP, Stickney R, Yarish C, Chopin T (2002) Aquatic polyculture and balanced ecosystem management: new paradigms for seafood production. In: Stickney RR, McVey JP (eds) Responsible aquaculture. CAB International, Oxon, pp 91–104

Mellergaard S, Nielsen E (1990) Fish disease investigations in Danish coastal waters with special reference to the impact of oxygen deficiency ICES, CM. 1190/E, Mar Env Qual Comm

Meng L, Powell JC, Taplin B (2001) Winter flounder growth rates to assess habitat quality across anthropogenic gradients in Narragansett Bay RI. Estuaries 24:576–584

Meng L, Taylor DL, Serbst J, Powell JC (2008) Assessing habitat quality of Mount Hope Bay and Narragansett Bay using growth, RNA:DNA, and feeding habits of caged juvenile winter flounder (Pseudopleuronectes americanus Waldbaum). Northeast Naturalist 15(1):35–56

Merwin DE (1993) Maritime history of southern New England: the view from Long Island, New York. Bull Archaeol Soc of Connecticut 65:3–18

Metzler K, Rozsa R (1987) Additional notes on the tidal wetlands of the Connecticut River. Newslett Conn Bot Soc 15:1–4

Michaelidis B, Ouzounis C, Paleras A, Portner HO (2005) Effects of long-term moderate hypercapnia on acid-base balance and growth rate in marine mussels *Mytilus galloprovincialis*. Mar Ecol Prog Ser 293:109–118

Michaels AF, Silver MW (1988) Primary production, sinking fluxes and the microbial food web. Deep-Sea Res II 35:473–490

Middaugh DP, Scott GI, Dean JM (1981) Reproductive behavior of the Atlantic silverside, *Menidia menidia* (Pisces, Atherinidae). Env Biol Fish 6:269–2276

Miller W, Egler F (1950) Vegetation of the Wequetequock-Pawcatuck tidal marshes, Stonington, Connecticut. Ecol Monogr 20:143–172

Miller AAL, Scott DB, Medioli FS (1982) *Elphidium excavatum* (Terquem): Ecophenotypic versus subspecific variation. J Foram Res 12:116–144

Miller DC, Poucher SL, Coiro L, Rego S, Munns W (1995) Effects of hypoxia on growth and survival of crustaceans and fishes of Long Island Sound. In: Proceedings of the Long Island Sound research conference: is the sound getting better or worse? New York Sea Grant Inst NYSGI-W_94-001

Miller DC, Poucher SL, Coiro L (2002) Determination of lethal dissolved oxygen levels for selected marine and estuarine fishes, crustaceans, and a bivalve. Mar Biol 140:287–296

Miller AW, Reynolds AC, Sobrino C, Riedel GF (2009) Shellfish face uncertain future in high CO2 world: Influence of acidification on oyster larvae and growth in estuaries. Plos ONE 4(5): e5661. doi:10.1371/journal.pone.0005661

Mills EL, Scheuerell MD, Carlton JT, Strayer DL (1997) Biological invasions in the Hudson River Basin: an inventory and historical analysis. NY State Museum Circ No 57, 51 pp

Millstone Environmental Laboratory (2009) Monitoring the marine environment of Long Island Sound at Millstone Power Station, Waterford. Dominion Nuclear Inc, Connecticut 284 pp

Minchinton TE, Bertness MD (2003) Disturbance-mediated competition and the spread of *Phragmites australis* in a coastal marsh. Ecol Appl 13:1400–1416

Moodley L, Hess C (1992) Tolerance of infaunal benthic foraminifera for low and high oxygen concentrations. Biol Bull 183:94–98

Moodley L, Middelburg JJ, Soetaert K, Boschker HTS, Herman PMJ, Heip CHR (2005) Similar rapid response to phytodetritus deposition in shallow and deep-sea sediments. J Mar Res 63:457–469

Moore KA (1991) Field studies on the effects of variable water quality on temperate seagrass growth and survival. In: Kenworthy WJ, Haunert DE (eds) The light requirements of seagrass. NOAA Tech Mem NMFS-SEFC-287, pp 42–57

Moore KA, Wetzel RL (2000) Seasonal variations in eelgrass (*Zostera marina* L.) response to nutrient enrichment and reduced light availability in experimental ecosystems. J Exp Mar Biol Ecol 244:1–28

Moore KA, Neckles HA, Orth RJ (1996) Zostera marina (eelgrass) growth and survival along a gradient of nutrients and turbidity in the lower Chesapeake Bay. Mar Ecol Prog Ser 142:247–259

Morreale SJ, Standora EA (1992) Habitat use and feeding activity of juvenile Kemp's ridleys in inshore waters of the Northeastern US. In: Proceedings of the 11th annual workshop on sea turtle conservation and biology. NOAA Tech Mem NMFS-SEFSC-302, pp 75–77

Morreale SJ, Standora EA (2005) Western North Atlantic waters: crucial developmental habitat for Kemp's ridley and loggerhead sea turtles. Chelonian Conserv Biol 4:4872–4882

Morreale SJ, Meylan A, Sadove SS, Standora EA (1992) Annual occurrence and winter mortality of marine turtles in New York waters. J Herpetol 26(3):301–308

Moss DD (1965) A history of the Connecticut River and its fisheries. Connecticut Board of Fisheries and Game, 15 pp

Mullen T, Nevis K, O'Kelly C, Gast R, Frasca S Jr (2005) Nuclear small-subunit ribosomal RNA gene-based characterization, molecular phylogeny and PCR detection of the Neoparamoeba from western Long Island Sound lobster. J Shellfish Res 24(3):719–732

Murawski SA (1993) Climate-change and marine fish distributions-forecasting from historical analogy. Trans Am Fish Soc 122(5):647–658

Murdy E, Birdsong R, Musick J (eds) (1997) Fishes of Chesapeake Bay. Smithsonian Institution Press, Washington, DC

Murphy RC (1916) Long Island turtles. Copeia 33:56–60

Murray JW (1976) Comparative studies of living and dead benthic foraminiferal distributions. In: Hedley RH, Adams CG (eds) Foraminifera, vol 2. Academic Press, New York, pp 45–109

Murray JW (1991) Ecology and paleoecology of benthic foraminifera. Longman Scientific and Technical Publishers, Harlow 451 pp

Murray JW (2006) Ecology and applications of benthic foraminifera. Cambridge University Press, Cambridge

Murray JW (2007) Biodiversity of living benthic foraminifera: how many species are there? Mar Micropaleont 64:163–176

Murray PM, Carey DA, Fredette TJ (1994) Chemical flux of pore water through sediment caps. Dredging'94, Proceedings of the 2nd international conference, sponsored by Waterways Committee of the Waterway, Port, Coastal and Ocean Division, ASCE. Buena Vista, FL, pp 1008–1016

Myers AA (1969) A revision of the amphipod genus *Microdeutopus* Costa (Gammaridea: Aoridae). Bull Brit Museum Nat Hist 17:93–148

Myers RA, Baum JK, Shepherd TD, Powers SP, Peterson CH (2007) Cascading effects of the loss of apex predatory sharks from a coastal ocean. Science 315:1846–1850

Myre PL, Germano JD (2007) Field verification program (FVP) disposal mound monitoring survey 2005. DAMOS Contribution No 175. US Army Corps of Engineers, New England District, Concord, MA, 71 pp

Neckles HA, Short FT, Barker S, Kopp BS (2005) Disturbance of eelgrass *Zostera marina* by commercial mussel *Mytilus edulis* harvesting in Maine: dragging impacts and habitat recovery. Mar Ecol Prog Ser 285:57–73

Neefus CD, Allen BP, Baldwin HP, Mathieson AC, Eckert RT, Yarish C, Miller MA (1993) An examination of the population genetics of *Laminaria* and other brown algae in the laminariales using starch gel electrophoresis. Hydrobiologia 260(261):67–79

Neefus C, Mathieson AC, Yarish C, Klein A, West A, Teasdale B, Hehre EJ (2000) Five cryptic species of *Porphyra* from the Northwest Atlantic. J Phycol 36(3):73

Neefus C, Mathieson AC, Bray TL, Yarish C (2008a) The occurrence of three introduced Asiatic species of *Porphyra* (Bangiales, Rhodophyta) in the northwestern Atlantic. J Phycol 44:1399–1414

Neefus C, Mathieson AC, Bray TL, Yarish C (2008b) The distribution, morphology, and ecology of three introduced Asiatic species of *Porphyra* (Bangiales, Rhodophyta) in the Northwestern Atlantic. J Phycol 44(6):1399–1414

Neely A, Zajac R (2008) Applying marine protected area design models in large estuarine systems. Mar Ecol Prog Ser 373:11–23

Nettleton JC, Mathieson AC, Thornber C, Neefus CD, Yarish C (2012) (in preparation) Introduction and distribution of *Gracilaria vermiculophylla* (Ohmi) Papenfuss (Rhodophyta, Gracilariales) in New England, USA

Neuenfeldt S (2002) The influence of oxygen saturation on the distributional overlap of predator (cod, *Gadus morhua*) and prey (herring, *Clupea harengus*) in the Bornholm basin of the Baltic Sea. Fish Oceanogr 11:11–17

New York State Seagrass Task Force (2009) Final report of the New York State seagrass task force: recommendations to the New York State Governor and Legislature. http://www.dec. ny.gov/docs/fish_marine_pdf/finalseagrassreport.pdf

Newell RIE, Koch EW (2004) Modeling seagrass density and distribution in response to changes in turbidity stemming from bivalve filtration and seagrass sediment stabilization. Estuaries 27:793–806

Nichols GE (1920) The vegetation of Connecticut, VII. The plant associations of depositing areas along the Seacoast. Bull Torrey Botan Club 47:511–548

Niering WA, Warren RS (1980) Vegetation patterns and processes in New England salt marshes. Bioscience 30:301–307

Niklitschek EJ (2001) Bioenergetics modeling and assessment of suitable habitat for juvenile Atlantic and shortnose sturgeons in Chesapeake Bay. PhD dissertation, University of Maryland, College Park, MD

Niklitschek EJ, Secor DH (2005) Modeling spatial and temporal variation of suitable nursery habitat for Atlantic sturgeon in the Chesapeake Bay. Est Coast Shelf Sci 64:135–148

Niklitschek EJ, Secor DH (2009) Dissolved oxygen, temperature and salinity effects on the ecophysiology and survival of juvenile Atlantic sturgeon in estuarine waters. I. Laboratory results. J Exp Mar Biol Ecol 383:S150–160

Nikulina A, Polovodova I, Schoenfeld J (2008) Foraminiferal response to environmental changes in Kiel Fjord, SW Baltic Sea. eEarth 3:37–49. www.electronic-earth.net/3/37/2008/

Nixon SW, Fulweiler RW, Buckley BA, Granger SL, Nowicki BL, Henry KM (2009) The impact of changing climate on phenology, productivity, and benthic-pelagic coupling in Narragansett Bay. Est Coast Shelf Sci 82:1–18

NRC (National Research Council) (1993) Managing wastewater in coastal urban areas. National Academy Press, Washington, DC

Nuttall MA, Jordaan A, Cerrato RM, Frisk MG (2011) Identifying 120 years of decline in eco-system structure and maturity of Great South Bay, New York using the Ecopath modelling approach. Ecol Model 222:3335–3345

Nutting CC (1901) The hydroids of the Woods Hole region. Bull U S Fish Comm 1899:325–386

Nybakken JW (2001) Marine biology: an ecological approach. Benjamin Cummings, San Francisco 516 pp

Nydick K, Bidwell A, Thomas E, Varekamp JC (1995) A sea-level rise curve for Guilford, Connecticut, USA. Mar Geol 124:137–159

Nye JA, Link JS, Hare JA, Overholtz WJ (2009) Changing spatial distribution of fish stocks in relation to climate and population size on the northeast United States continental shelf. Mar Ecol Prog Ser 393:111–129

NYS Seagrass Task Force (2009) Final Report of the New York State Seagrass Task Force: rec-ommendations to the New York State Governor and Legislature. http://www.dec.ny.gov/docs/fish_marine_pdf/finalseagrassreport.pdf

O'Shea ML, Brosnan TM (2000) Trends in indicators of eutrophication in western Long Island Sound and the Hudson-Raritan Estuary. Estuaries 23:877–901

Ocean Surveys Inc (2010) Final report–Eighteen-month benthic biology survey. A component of the benthic monitoring study for the Long Island replacement cable project December 2009, Sheffield Harbor and Long Island Sound, Norwalk, CT, Report No 07ES077.6B, 23 pp

O'Donnell J, Dam HG, Bohlen WF, Fitzgerald W, Gay PS, Houk AE, Cohen DC, Howard-Strobel MM (2008) Intermittent ventilation in the hypoxic zone of western Long Island Sound during the summer of 2004. J Geophys Res 113:C09025. doi:10.1029/2007JC004716

Officer CB, Smayda TJ, Mann R (1982) Benthic filter feeding—a natural eutrophication control. Mar Ecol Prog Ser 9:203–210

Olesen B, Sand-Jensen K (1993) Seasonal acclimatization of eelgrass *Zostera marina* growth to light. Mar Ecol Prog Ser 94:91–99

Oliver JD, Van Den Avyle MJ, Bozeman EL (1989) Species profiles: Life – histories and environ-mental requirements of coastal fishes and invertebrates- bluefish. US Fish Wildlife Div Biol Serv Biol Rept 82/11.96, pp 4–15

Olsson IC, Greenberg LA et al (2006) Environmentally induced migration: the importance of food. Ecol Lett 9(6):645–651

Orth RJ, Moore KA (1988) Distribution of *Zostera marina* L. and *Ruppia maritima* L. sensu lato along depth gradients in the lower Chesapeake Bay, USA. Aquat Bot 32:291–305

Orth RJ, Batiuk RA, Bergstrom PW, Moore KA (2002) A perspective on two decades of policies and regulations influencing the protection and restoration of submerged aquatic vegetation in Chesapeake Bay, USA. Bull Mar Sci 71:1391–1403

Orth RJ, Carruthers TJB, Dennison WC, Duarte CM, Fourqurean JW, Heck KL, Hughes AR, Kendrick GA, Kenworthy WJ, Olyarnik S, Short FT, Waycott M, Williams SL (2006) A global crisis for seagrass ecosystems. Bioscience 56:987–996

Osman RW (1977) The establishment and development of a marine epifaunal community. Ecol Monogr 47:37–63

Oviatt CA (2004) The changing ecology of temperate coastal waters during a warming trend. Estuaries 27:895–904

Oviatt C, Keller A, Reed L (2002) Annual primary production in Narragansett Bay with no bay-wide winter-spring phytoplankton bloom. Est Coast Shelf Sci 54:1013–1026

Paerl HW, Pinckney JL, Fear JM, Peierls BL (1998) Ecosystem responses to internal and water-shed organic matter loading: consequences for hypoxia in the eutrophying Neuse river estu-ary, North Carolina, USA. Mar Ecol Prog Ser 166:17–25

Palermo MR (1991) Design requirement for capping. Tech Note DRP-5-03, US Army Engineer Waterways Experiment Station, Vicksburg, MS. http://el.erdc.usace.army.mil/elpubs/pdf/drp5-03.pdf

Parker FL 1(1948) Foraminifera of the continental shelf from the Gulf of Maine to Maryland. Bull Museum Comp Zool, Harvard College, 100(2):214–240

Parker FL (1952) Foraminiferal distribution in the Long Island Sound-Buzzards Bay area. Bull Museum Comp Zool, Harvard College 106(10):428–473

Parker CA, O'Reilly JE (1991) Oxygen depletion in Long Island Sound: a historical perspective. Estuaries 14:248–264

Parrino V, Kraus DW, Doeller JE (2000) ATP production from the oxidation of sulfide in gill mitochondria of the ribbed mussel *Geukensia demissa*. J Exp Biol 203:2209–2218

Paterson AW, Whitfield AK (2000) Do shallow water habitats function as refugia for juvenile fishes? Est Coast Shelf Sci 51:359–364

Pavela JS, Ross JL, Chittenden ME Jr (1983) Sharp reductions in abundances of fishes and benthic macroinvertebrates in the Gulf of Mexico off Texas associated with hypoxia. Northeast Gulf Sci 6:167–173

Pawlowski J, Holzmann M (2008) Diversity and geographic distribution of benthic foraminifera: a molecular perspective. Biodivers Conserv 17:317–328

Pearce J, Balcom N (2005) The 1999 Long Island Sound lobster mortality event: findings of the comprehensive research initiative. J Shellfish Res 24(3):691–698

Pearson TH, Rosenberg R (1978) Macrobenthic succession in relation to organic enrichment and pollution of the marine environment. Oceanogr Mar Biol Ann Rev 16:229–311

Pedersen A, Kraemer G, Yarish C (2008) Seaweed of the littoral zone at Cove Island in Long Island Sound: annual variation and impact of environmental factors. J Appl Phycol 20(5):869–882

Pederson J, Bullock R, Carlton J, Dijkstra J et al (2005) Marine invaders in the northeast. Rapid assessment survey of non-native and native marine species of float dock communities, Aug 2003. MIT Sea Grant College Program Publ No 05-3, Cambridge, MA, 40 pp

Pellegrino P, Hubbard W (1983) Baseline shellfish data for the assessment of potential environmental impacts associated with energy activities in Connecticut's coastal zone, vols I and II. Report to the State of Connecticut, Dept of Agriculture, Aquaculture Div, Hartford, CT, 177 pp

Pereira R, Yarish C (2010) The role of *Porphyra* in sustainable culture systems: physiology and applications. In: Israel A, Einav R (eds) Role of seaweeds in a globally changing environment. Springer, Dordrecht, pp 339–354

Perez-Dominguez R, Holt SA, Holt GJ (2006) Environmental variability in seagrass meadows: effects of nursery environment cycles on growth and survival in larval red drum *Sciaenops ocellatus*. Mar Ecol Prog Ser 321:41–53

Perry SF, Jonz MG, Gilmour KM (2009) Oxygen sensing and the hypoxic ventilatory response. Fish Physiol 27:193–253

Petersen JK, Petersen GI (1990) Tolerance, behavior, and oxygen consumption in the sand goby, *Pomatochistus minutis* (Pallas), exposed to hypoxia. J Fish Biol 37:921–933

Peterson WT (1986) The effect of seasonal variations in stratification on plankton dynamics in Long Island Sound. In: Bowman MJ, Yentsch CM, Peterson WT (eds) Tidal mixing and plankton dynamics. Lecture notes on coastal and estuarine studies, vol 17. Springer, New York, pp 297–320

Peterson CH (1991) Intertidal zonation of marine invertebrates in sand and mud. Am Sci 79:237–249

Peterson WT, Bellantoni DC (1987) Relationships between water column stratification, phytoplankton cell size and copepod fecundity in Long Island Sound and off central Chile. S Afr J Mar Sci 5:411–421

Peterson WT, Dam HG (1996) Pigment ingestion and egg production rates of the calanoid copepod *Temora longicornis*: implications for gut pigment loss and omnivorous feeding. J Plankton Res 18:855–861

Pickerell C, Vaudrey J (2010) The NY/CT Long Island Sound Eelgrass Restoration Project. Proposal funded by The National Fish and Wildlife Foundation, Sound Futures Fund. Project no. 24369, 17 pp

Pickerell CH, Schott S, Petersen K (2007) Eelgrass establishment at a high energy site: a case study along Long Island's north shore. Estuarine Research Fed, 4–8 Nov 2007, Providence, RI

Pickerell C, Brousseau L, Vaudrey J, Yarish C, Fonseca M (2011) Development and application of a Long Island Sound GIS-based eelgrass habitat suitability index model. Proposal funded by New England Interstate Water Pollution Control Commission and the Long Island Sound Study

Pihl L, Baden S, Diaz RJ (1991) Effects of periodic hypoxia on distribution of demersal fish and crustaceans. Mar Biol 108:349–360

Pihl L, Baden S, Diaz RJ, Schaffner LC (1992) Hypoxia induces structural changes in the diet of bottom-feeding fish and crustacean. Mar Biol 112:349–361

Pillet L, de Vargas C, Pawlowski J (2011) Molecular identification of sequestered diatom chloroplasts and kleptoplastidy in foraminifera. Protist 162:394–404

Platon E, Sen Gupta BK, Rabalais NN, Turner RE (2005) Effect of seasonal hypoxia on the benthic foraminiferal community of the Louisiana inner continental shelf: the 20th century record. Mar Micropaleont 54:263–283

Plumb JA, Grizzle JM, Defigueriedo J (1976) Necrosis and bacterial infection in channel catfish (*Ictalurus punctatus*) following hypoxia. J Wild Dis 12:247

Poindexter-Rollings ME (1990) Methodology for analysis of subaqueous sediment mounds. US Army Corps of Engineers, Waterways Experiment Station, Vicksburg, MS. Tech Report D-90-2, 110 pp

Pomeroy LR, D'Elia CF, Schaffner LC (2006) Limits to top-down control of phytoplankton by oysters in Chesapeake Bay. Mar Ecol Prog Ser 325:301–309

Poppe LJ, Lewis RS, Knebel HJ, Haase EA, Parolski KF, DiGiacomo-Cohen ML (2001) Sidescan sonar images, surficial geologic interpretations, and bathymetry of the Long Island Sound sea floor in New Haven Harbor and New Haven dumping ground, Connecticut. US Geological Survey Geological Investigations Series Map I-2736

Poucher SL, Coirio L (1997) Test reports: effects of low dissolved oxygen on saltwater animals. Memorandum to DC Miller, US Environmental Protection Agency, Atlantic Ecology Division, Narragansett, RI

Powers J, Lopez G, Cerrato R, Dove A (2004) Effects of thermal stress on Long Island Sound lobster, H. americanus. Proceedings of the Long Island Sound Lobster Research initiative working meeting, 3–4 May 2004. University of Connecticut, Avery Point, Groton, CT

Provan J, Booth D, Todd NP, Beatty GE, Maggs CA (2008) Tracking biological invasions in space and time: elucidating the invasive history of the green alga *Codium fragile* using old DNA. Divers Distrib 14(2):343–354

Rabalais NN, Turner RE (eds) (2001) Coastal hypoxia: consequences for living resources and ecosystems, vol 58. American Geophysics Union, Washington, DC

Rabalais NN, Diaz RJ, Levin LA, Turner RE, Gilbert D, Zhang J (2010) Dynamics and distribution of natural and human-caused hypoxia. Biogeosciences 7:585–619

Rafinesque CS (1817) Synopsis of four new genera and ten new species of Crustacea, found in the United States. Am Monthly Mag 2:40–43

Rafinesque CS (1819) Descriptions of species of sponges observed on the shores of Long Island. Am J Sci 1(11):149–151

Raillard O, Menesguen A (1994) An ecosystem box model for estimating the carrying-capacity of a macrotidal shellfish system. Mar Ecol Prog Ser 115:117–130

Ram ASP, Nair S, Chandramohan D (2003) Seasonal shift in net ecosystem production in a tropical estuary. Limnol Oceanogr 48:1601–1607

Rasmussen E (1977) The wasting disease of eelgrass (*Zostera marina*) and its effect on environmental factors and fauna. In: McRoy CP, Helfferich C (eds) Seagrass ecosystems. Marcel Dekker, New York, pp 1–51

Rasmussen H, JØrgensen BB (1992) Microelectrode studies of seasonal oxygen-uptake in a coastal sediment-role of molecular diffusion. Mar Ecol Prog Ser 81:289–303

Rawson PD, Lindell S, Guo X, Sunila I (2010) Cross-breeding for improved growth and disease resistance in the eastern oyster. Northeastern Regional Aquaculture Center, Publication No 206-2010, 6 pp

Reid RN, Frame AB, Draxler AF (1979) Environmental baselines in Long Island Sound, 1972-1973. National Oceanic and Atmospheric Administration, Technical Report SSRF-738, 31 pp

Rhoads DC (1994) Analysis of the contribution of dredged material to sediment and contaminant fluxes in Long Island Sound. DAMOS Contribution 88, US Army Corps of Engineers, New England District, Concord, MA, 45 pp

Rhoads DC, Germano JD (1982) Characterization of organism-sediment relations using sediment profile imaging: an efficient method of remote ecological monitoring of the seafloor (REMOTS) System. Mar Ecol Prog Ser 8:115–128

Rhoads DC, Germano JD (1986) Interpreting long-term changes in benthic community structure: a new protocol. Hydrobiologia 142:291–308

Rhoads DC, McCall PL, Yingst JY (1978) Disturbance and production on the estuarine seafloor. Am Sci 66:577–586

Rhoads DC, Boyer LF, Welsh BL, Hampson GR (1984) Seasonal dynamics of detritus in the benthic turbidity zone (BTZ): implications for bottom-rack molluscan mariculture. Bull Mar Sci 36:36–549

Richards JG (2011) Physiological, behavioral and biochemical adaptations of intertidal fishes to hypoxia. J Exp Biol 214:191–199

Riedel B, Zuschin M, Haselmair A, Stachowitsch M (2008) Oxygen depletion under glass: behavioural responses of benthic macrofauna to induced anoxia in the Northern Adriatic. J Exp Mar Biol Ecol 367:17–27

Richards JG, Farrell AP, Brauner CJ (2009) In: Hypoxia, Richards JG, Farrell AP, Brauner CJ (eds) Fish physiology, vol 27. Academic Press, New York

Riley GA (1941) Plankton studies, III. Long Island Sound. Bull Bingham Oceanogr Coll 7:1–93

Riley GA (1956a) Oceanography of Long Island Sound, 1952–1954. IX. Production and utilization of organic matter. Bull Bingham Oceanogr Coll 15:324–344

Riley GA (1956b) Production and utilization of organic matter. In: Oceanography of long island sound vol 15. Bull. Bingham Oceanographic Collection. Peabody Museum of Natural History, Yale Univ. pp 324–343

Riley GA, Conover SAM (1956) Oceanography of Long Island Sound, 1952-1954. III. Chemical oceanography. Bull Bingham Oceanogr Coll 15:47–61

Riley GA, Conover SM (1967) Phytoplankton of Long Island Sound 1954-1955. Bull Bingham Oceanog Coll 19:5–33

Ripley JL, Foran CM (2007) Influence of estuarine hypoxia on feeding and sound production by two sympatric pipefish species. Mar Environ Res 63:350–367

Ritter C, Montagna PA (1999) Seasonal hypoxia and models of benthic response in a Texas bay. Estuaries 22:7–20

Robinette HR (1983) Species profiles: life histories and environmental requirements of coastal fishes and invertebrates (Gulf of Mexico)—bay anchovy and striped anchovy. US Fish and Wildlife Div Biol Serv FWS/OBS-82/11.14. pp 15

Robohm R, Draxler A, Sherrell R, Wieczorek D, Kapareiko D, Pitchford S (2005) Effects of environmental stressors on disease susceptibility in American lobsters: a controlled laboratory study. J Shellfish Res 24(3):773–779

Rombough PJ (1988) Respiratory gas exchange, aerobic metabolism, and effects of hypoxia during early life. In: Hoar WS, Randall DJ (eds) Fish physiology, vol XI: The Physiology of Developing Fish. Part A. Eggs and Larvae. Academic Press, New York, pp 59–161

Romero J, Lee K-S, Pérez M, Mateo MA, Alcoverro T (2006) Nutrient dynamics in seagrass ecosystems. In: Larkum AWD, Orth RJ, Duarte CM (eds) Seagrasses: biology, ecology and conservation. Springer, New York, pp 227–254

Ross P (1902) A history of Long Island from earlier settlement to the present time, vol 1. The Lewis Publishing Company, New York 1080 pp

Roussel JM (2007) Carry-over effects in brown trout (Salmo truttta): hypoxia on embryos impairs predator avoidance by alevins in experimental channels. Can J Fish Aquat Sci 64:79–786

Rozsa R (1994) Long term decline of Zostera marina in Long Island Sound and Fishers Island Sound. Office of Long Island Sound Programs, CT Dept of Environmental Protection, p 10 (Little Narr Bay page). http://www.lisrc.uconn.edu/eelgrass/index.html

Rozsa R, Metzler KJ, Fell P (2001) Ecology of the lower Connecticut River: plants, animals and their habitats. In: Dreyer G, Caplis M (eds) The living resources and habitats of the Lower Connecticut River. The Connecticut College Arboretum, pp 29–47

Rudnick DT, Oviatt CA (1986) Seasonal lags between organic carbon deposition and mineralization in marine sediments. J Mar Res 44:815–837

Ryland JS, Hayward PJ (1991) Marine flora and fauna of the northeastern United States. Erect Bryozoa. NOAA Tech Report NMFS 99, 48 pp

Ryland JS, Bishop JDD, De Blauwe H, El Nagar A, Minchin D, Wood CA, Yunnie ALE (2011) Alien species of *Bugula* along the Atlantic coasts of Europe. Aquat Invasions 6:17–31

Saffert H, Thomas E (1997) Living foraminifera and total populations in salt marsh peat cores: Kelsey Marsh (Clinton, CT) and the Great Marshes (Barnstable, MA). Mar Micropaleontol 33:175–202

Salisbury J, Green M, Hunt C, Campbell J (2008) Coastal acidification by rivers: a new threat to shellfish? EOS, Trans Am Geophys Union 89:513

Saltonstall K (2002) Cryptic invasion by a non-native genotype of the common reed, *Phragmites australis*, into North America. Proc Nat Acad Sci U S A 99:2445–2449

Sanders HL (1956) Oceanography of Long Island Sound, 1952-1954. X. Biology of marine bottom communities. Bull Bingham Oceanogr Coll 15:345–414

Sanders HL (1969) Marine benthic diversity and the stability-time hypothesis. In: Woodwell GM, Smith HH (eds) Diversity and stability in ecological systems, Brookhaven symposium in biology, No 22. Brookhaven National Laboratory, Upton, pp 71–81

Sanger D, Arendt M, Chen Y, Wenner E, Holland A, Edwards D, Caffrey J (2002) A synthesis of water quality data: National Estuarine Research Reserve System-wide monitoring program (1995–2000). South Carolina Department of Nat Resources, Marine Resources Div, p 135

Say T (1817) An account of the crustacea of the United States. J Acad Nat Sci Phil 1:57–63

Schimmel S, Benyi S, Strobel C (1999) An assessment of the ecological condition of Long Island Sound, 1990–1993. Env Monitor Assess 56:27–49

Schneider CW (2010) Report of a new invasive alga in the Atlantic United States: "*Heterosiphonia*" *japonica* in Rhode Island. J Phycol 46:653–657

Schneider CW, Suyemoto M, Yarish C (1979) An annotated checklist of Connecticut seaweeds. Connecticut geological and natural history survey, Connecticut Dept of Environmental Protection 24 pp

Schoenfeld J, Alve E, Geslin E, Jorissen F, Korsun S, Spezzaferri S (2012) The FOBIMO (FOraminiferal Bio-Monitoring) initiatve: towards a standardized protocol for soft-bottom benthic foraminiferal monitoring studies. Mar Micropaleonology 94–95:1–13

Schoepf JD (1788) Beschriebung einiger nordamerikanishen fische vorzuglick aus den neu yorkischen seewasser. Schriften der Berliner Gesellschaft Naturforschender Freunde, Berlin 8:138–194

Schwartz FJ, Hogarth WT, Weinstein MP (1981) Freshwater fishes of the Cape fear estuary, North Carolina, and their distribution in relation to environmental factors. Brimleyana 7:17–37

Scott DB, Medioli FS (1978) Vertical zonations of marsh foraminifera as acute indicators for sea level studies. Nature 272:528–531

Scott DB, Medioli FS (1980) Quantitative studies of marsh foraminiferal distributions in Nova Scotia: implications for sea level studies. Cushman Foundation for Foraminiferal Research, Special Pub No 17

Scott DB, Medioli FS, Schafer CT (2001) Monitoring in coastal environments using foraminifera and thecamoebian indicators. Cambridge University Press, New York 177 pp

Schurmann H, Steffensen JF (1992) Lethal oxygen levels at different temperatures and the preferred temperature during hypoxia of the Atlantic cod, Gadus morhua L. J Fish Biol 41(6):927–34

Schurmann H, Steffensen JF (1997) Effects of temperature, hypoxia and activity on the metabolism of juvenile Atlantic cod, Gadus morhua L. J Fish Biol 50:1166–1180

Secor DH, Gunderson TE (1998) Effects of hypoxia and temperature on survival, growth, and respiration of juvenile Atlantic sturgeon, *Acipenser oxyrinchus*. Fish Bull 96:603–613

Secor DH, Niklitschek EJ (2001) Hypoxia and sturgeon: report to the Chesapeake Bay program dissolved oxygen criteria team. University of Maryland Center for Environmental Studies, Chesapeake Biol Lab. Technical Report Series No TS-314-01-CBL

Sen Gupta BK, Platon E (2006) Tracking past sedimentary records of oxygen depletion in coastal waters: use of the *Ammonia-Elphidium* foraminiferal index. J Coast Res Special Issue 39:1351–1355

Sen Gupta BK, Turner RE, Rabalais NN (1996) Seasonal oxygen depletion in continental-shelf waters of Louisiana: historical record of benthic foraminifers. Geology 24:227–230

Shang E, Yu R, Wu R (2006) Hypoxia affects sex differentiation and development, leading to a male-dominated population in zebrafish (*Danio rerio*). Environ Sci Tech 40:3118–3122

Sherr EB, Sherr BF, Fallon RD, Newell SY (1986) Small aloricate ciliates as a major component of the marine heterotrophic nanoplankton. Limnol Oceanogr 31:177–183

Shiah F-K, Ducklow HW (1994) Temperature regulation of heterotrophic bacterioplankton abundance, production, and specific growth rate in Chesapeake Bay. Limnol Oceanogr 39:1243–1258

Shiah F-K, Gong G-C, Chen T-Y, Chen C-C (2000) Temperature dependence of bacterial specific growth rates on the continental shelf of the East China Sea and its potential application in estimating bacterial production. Aquat Microb Ecol 22:55–162

Shimps EL, Rice JA, Osbourne JA (2005) Hypoxia tolerance in two estuarydependent fish. J Exp Mar Biol Ecol 325:146–162

Shimshock N, Sennefelder G, Dueker M, Thurberg F (1992) Patterns of metal accumulation in *Laminaria longicruris* from Long Island Sound. Arch Envir Contam Toxicol 22:305–312

Short FT, Burdick DM (1996) Quantifying eelgrass habitat loss in relation to housing development and nitrogen loading in Waquoit Bay, Massachusetts. Estuaries 19:730–739

Short FT, Neckles HA (1999) The effects of global climate change on seagrasses. Aquat Bot 63:169–196

Short FT, Muehlstein LK, Porter D (1987) Eelgrass wasting disease: causes and recurrence of a marine epidemic. Biol Bull 173:557–565

Short FT, Davis RC, Kopp BS, Short CA, Burdick DM (2002) Site selection model for optimal transplantation of eelgrass *Zostera marina* in the northeastern US. Mar Ecol Prog Ser 227:253–267

Shumway DL, Warren CW, Douderoff P (1964) Influence of oxygen concentration and water movement on the growth of steelhead trout and coho salmon embryos. Trans. Am Fish Soc 93:342–356

Shumway SE (1983) Factors affecting oxygen consumption in the coot clam *Mulinia lateralis* (Say). Ophelia 22(2):143–171

Shumway SE, Scott TM, Shick JM (1993) The effects of anoxia and hydrogen sulphide on survival, activity and metabolic rate in the coot clam, *Mulinia lateralis* (Say). J Exp Mar Biol Ecol 71(2):135–146

Shupack B (1934) Some foraminifera from western Long Island and New York Harbor. Amer Museum Novitates 737:1–12

Silva AJ, Brandes HG, Uchytil CJ, Fredette TJ, Carey DA (1994) Geotechnical analysis of capped dredged material mounds. Dredging'94, Proceedings of the 2nd international conference, sponsored by Waterways Committee of the Waterway, Port, Coastal and Ocean Div, ASCE. Buena Vista, FL, pp 410–419

Smetacek V (1984) The food supply to the benthos. In: Fashem MRJ (ed) Flows of energy and material in marine ecosystems. NATO conference series, vol IV. Plenum Press, New York, pp 517–547

Smith K, Able K (2003) Dissolved oxygen dynamics in salt marsh pools and its potential impacts on fish assemblages. Mar Ecol Prog Ser 258:223–232

Smith E, Howell P (1987) The effects of trawling on American lobsters, *Homarus americanus*, in Long Island Sound. Fish Bull 85(4):737–744

Smith LM, Whitehouse S, Oviatt CA (2010) Impacts of climate change on Narragansett Bay. Northeast Nat 17:77–90

Sniesko SF (1973) The effect of environmental stress on outbreaks of infectious disease in fish. Fish Biol 6:197–208

Sogard SM (1994) Use of suboptimal foraging habitats by fishes: consequences to growth and survival. In: Stouder DJ, Fresh KL, Feller RJ (eds) Theory and application of fish feeding ecology. University of South Carolina press, Columbia, pp 103–132

Sollid J, De Angekis P, Gunderson K, Nilsson GE (2003) Hypoxia induces adaptive and reversible morphological changes in crucian carp gills. J Exp Biol 206:3667–3673

Somero GN, Childress JJ, Anderson AE (1989) Transport, metabolism and detoxification of hydrogen sulfide in animals from sulfide-rich marine environments. Critical Rev Aquat Sci 1:591–614

Sorte CJB, Williams SL, Carlton JT (2010) Marine range shifts and species introductions: comparative spread rates and community impacts. Global Ecol Biogeogr 19:303–316

Sousa WP (1979) Disturbance in marine intertidal boulder fields: the nonequilibrium maintenance of species diversity. Ecology 60:1225–1239

Spendelow JA, Nichols JD, Nisbet ICT, Hayes H, Cormons GD (1995) Estimating annual survival and movement rates of adults within a metapopulation of Roseate Terns. Ecology 76:2415–2428

Stachowicz JJ, Whitlatch RB (2005) Multiple mutualists provide complementary benefits to their seaweed hosts. Ecology 86:2418–2427

Stachowicz JJ, Terwin JR, Whitlatch RB, Osman RW (2002) Linking climate change and biological invasions: ocean warming facilitates nonindigenous species invasions. Proc Natl Acad Sci U S A 99:15497–15500

Staehr PA, Borum J (2011) Seasonal acclimation in metabolism reduces light requirements of eelgrass (Zostera marina). J Exp Mar Biol Ecol 407:139–146

Stead RA, Thompson RJ (2003) The effect of the sinking spring diatom bloom on digestive processes of the cold-water protobranch Yoldia hyperborea. Limnol Oceanogr 48:157–167

Stead RA, Thompson RJ (2006) The influence of an intermittent food supply on the feeding behaviour of Yoldia hyperborea (Bivalvia: Nuculanidae). J Exp Mar Biol Ecol 332:37–48

Stead RA, Thompson RJ, Jaramillo JR (2003) Absorption efficiency, ingestion rate, gut passage time and scope for growth in suspension- and deposit-feeding Yoldia hyperborea. Mar Ecol Prog Ser 252:150–172

Steenbergen J, Steenbergen S, Shapiro H (1978) Effects of temperature on phagocytosis in Homarus americanus. Aquaculture 14:23–30

Steinback JMK (1999) The ocean beach of Fire Island National Seashore, New York: spatial and temporal trends and the effects of vehicular disturbance. Master's dissertation, Marine Sciences Research Center, Stony Brook Univ, Stony Brook, NY, 252 pp

Steneck RS, Carlton JT (2001) Human alterations of marine communities. In: Bertness MD, Gaines SD, Hay ME (eds) Marine community ecology. Sinauer Associates, Sunderland, p 550

Steward JS, Green WC (2007) Setting load limits for nutrients and suspended solids based upon seagrass depth-limit targets. Est Coasts 30:657–670

Stierhoff KL, Targett TE, Grecay PA (2003) Hypoxia tolerance of the mummichog: the role of access to the water surface. J Fish Biol 63:580–592

Stierhoff KL, Targett TF, Miller KL (2006) Ecological responses of juvenile summer and winter flounder to hypoxia: experimental and modeling analyses of effects on estuarine nursery quality. Mar Ecol Prog Ser 315:255–266

Stierhoff KL, Targett TE, Power JH (2009a) Hypoxia-induced growth limitation of juvenile fishes in an estuarine nursery: assessment of small-scale temporal dynamics using RNA:DNA. Can J Fish Aquatic Sci 66:1033–1047

Stierhoff KL, Targett TE, Miller K (2009b) Ecophysiological responses of juvenile summer and winter flounder to hypoxia: experimental and modeling analyses of effects on estuarine nursery quality. Mar Ecol Prog Ser 325:255–266

Stierhoff KL, Tyler RM, Targett TE (2009c) Hypoxia tolerance of juvenile weakfish (Cynoscion regalis): laboratory assessment of growth and behavioral avoidance responses. J Exp Mar Biol Ecol 381:S173–S179

Stramma L, Johnson GC, Sprintall J, Mohrholz V (2008) Expanding oxygen-minimum zones in the tropic oceans. Science 320:655–658

Sun MY, Aller RC, Lee C (1994) Spatial and temporal distributions of sedimentary chloropigments as indicators of benthic processes in Long Island Sound. J Mar Res 52:149–176

Talbot CW, Able KW (1984) Composition and distribution of larval fishes in New Jersey high marshes. Estuaries 7(4A):434–433

Talmage SC, Gobler CJ (2009) The effects of elevated carbon dioxide concentrations on the metamorphosis, size, and survival of larval hard clams (*Mercenaria mercenaria*), bay scallops (*Argopecten irradians*), and Eastern oysters (*Crassostrea virginica*). Limnol Oceanogr 54:2072–2080

Talmage SC, Gobler CJ (2010) Effects of past, present, and future ocean carbon dioxide concentrations on the growth and survival of larval shellfish. Proc Natl Acad Sci U S A 107:17246–17251

Tango P, Butler W, Lacouture R, Eskin R, Goshorn D, Michael B, Magnien R, Beatty W, Brohawn K, Wittman R, Hall S (2002) An unprecedented bloom of Dinophysis acuminata in Chesapeake Bay. In: Steidinger KA, Landsberg JH, Tomas CR, Vargo GA (eds) Xth international conference on harmful algae. St Pete Beach, FL, p 358

Taylor GT (1982) The role of pelagic protozoa in nutrient cycling: a review. Ann Inst Oceanogr (Suppl), Paris 58:227–241

Taylor JC, Miller JM (2001) Physiological performance of juvenile southern flounder, *Paralichthys lethostigma* (Jordan and Gilbert, 1884), in chronic and episodic hypoxia. J Exp Mar Biol Ecol 258:195–214

Taylor GT, Scranton MI, Iabichella M, Ho T-Y, Thunell RC, Muller-Karger F, Varela R (2001) Chemoautotrophy in the redox transition zone of the Cariaco Basin: a significant midwater source of organic carbon production. Limnol Oceanogr 46:148–163

Taylor GT, Way J, Scranton MI (2003a) Transport and planktonic cycling of organic carbon in the highly urbanized Hudson River estuary. Limnol Oceanogr 48:1779–1795

Taylor GT, Way J, Yu Y, Scranton MI (2003b) Patterns of hydrolytic ectoenzyme activity among bacterioplankton communities in the lower Hudson River and Western Long Island Sound estuaries. Mar Ecol Prog Ser 263:1–15

Taylor GT, Thunell RC, Varela R, Benitez-Nelson C, Scranton MI (2009) Hydrolytic ectoenzyme activity associated with suspended and sinking organic particles above and within the anoxic Cariaco Basin. Deep-Sea Res I 56:1266–1283

Thomas E (2007) Cenozoic mass extinctions in the deep sea; what disturbs the largest habitat on Earth? In: Monechi S, Coccioni R, Rampino M (eds) Large ecosystem perturbations: causes and consequences. Geol Soc Am Special Paper 424:1–24

Thomas P, Rahman MS (2012) Extensive reproductive disruption, ovarian masculinization and aromatase suppression in Atlantic croaker in the northern Gulf of Mexico hypoxic zone. Proc R Soc B 279:28–38

Thomas E, Varekamp JC (1991) Paleo-environmental analyses of marsh sequences (Clinton, CT): evidence for punctuated rise in relative sea level during the latest Holocene. J Coast Res 11:125–158

Thomas, E., Lugolobi, F., Abramson, I., and Varekamp, J. C., 2001. Reconstructing Long Island Sound Environmental Changes and Their Influence on the Biota. EOS Suppl., Trans. AGU , F 773

Thomas E, Varekamp JC (2011) Benthic foraminifera in Long Island Sound. NEERS spring meeting, 5–7 May 2011, Port Jefferson, Long Island, NY

Thomas E, Gapotchenko T, Varekamp JC, Mecray EL, Buchholtz ten Brink MR (2000) Benthic foraminifera and environmental changes in Long Island Sound. J Coast Res 16:641–655

Thomas E, Abramson I, Varekamp JC, Buchholtz ten Brink MR (2004) Eutrophication of Long Island Sound as traced by benthic foraminifera. Proceedings of the 6th Biennial LIS research conference, pp 87–91

Thomas H, Bozec Y, Elkalay K, de Baar HJW (2004b) Enhanced open ocean storage of CO_2 from shelf sea pumping. Science 304:1005–1008

Thomas P, Rahman S, Kummer J, Lawson S (2006) Reproductive endocrine dysfunction in Atlantic croaker exposed to hypoxia. Soc Environ Mar Env Res 62:S249–S252

Thomas P, Rahman S, Khan I, Kummer J (2007) Widespread endocrine disruption and reproductive impairment in an estuarine fish population exposed to seasonal hypoxia. Proc Biol Sci 274:2693–2701

Thomas E, Varekamp JC, Cooper S, Sangiorgi F, Donders T (2009) Proxies for eutrophication in Long Island Sound. 'Coasts and estuaries in a changing world,' CERF 2009, Portland, OR (1–5 Nov 2009)

Thomas E, Varekamp JC, Cooper S, Sangiorgi F, Donders T (2010) Microfossil proxies for anthropogenic environmental changes in Long Island Sound. Abstract 76-6, GSA Programs and Abstracts 42(1):175–176

Thompson RJ (1984) The reproductive cycle and physiological ecology of the mussel *Mytilus edulis* in a subarctic, non-estuarine environment. Mar Biol 79:277–288

Thompson WG, Thomas E, Varekamp JC (2000) 1500 years of sea level rise in Long Island Sound. Proceedings of the 4th Biennial LIS Research Conf, pp 139–148

Tielens AGM, Rotte C, van Hellemond JJ, Martin W (2002) Mitochondria as we don't know them. Trends Biochem Sci 27(11):564–572

Timmerman C, Chapman L (2004) Patterns of hypoxia in a coastal salt marsh: implications for ecophysiology of resident fishes. Fla Sci 67:80–91

Tiner R, Bergquist H, Halavik T, MacLachlan A (2007) 2006 eelgrass survey for eastern Long Island Sound, Connecticut and New York. US Fish and Wildlife Service, National Wetlands Inventory Program, Northeast Region, Hadley MA, 24 pp + appendices

Tiner R, McGuckin K, Fields N, Fuhrman N, Halavik T, MacLachlan A (2010) 2009 Eelgrass survey for Eastern Long Island Sound, Connecticut and New York. US Fish and Wildlife Service, National Wetlands Inventory Program, Northeast Region, Hadley, MA, 15 pp + appendix

Torgersen T, DeAngelo E, O'Donnell J (1997) Calculations of horizontal mixing rates using [222]Rn and the controls on hypoxia in western Long Island Sound. Estuaries 20:328–345

Tsutsumi H (1990) Population persistence of *Capitella* sp. (Polychaeta; Capitellidae) on a mud flat subject to environmental disturbance by organic enrichment. Mar Ecol Prog Ser 63:147–156

Tsutsumi H, Wainright S, Montani S, Saga M, Ichihara S, Kogure K (2001) Exploitation of a chemosynthetic food resource by the polychaete *Capitella* sp. I. Mar Ecol Prog Ser 216:119–127

Tyler RM, Targett TE (2007) Juvenile weakfish (*Cynoscion regalis*) distribution in relation to diel-cycling dissolved oxygen in an estuarine tributary. Mar Ecol Prog Ser 333:257–269

Ukeles R, Sweeney BM (1969) Influence of dinoflagellate trichocysts and other factors on the feeding of *Crassostrea virginica* larvae on *Monochrysis lutheri*. Limnol Oceanogr 14:403–410

USASAC (US Atlantic Salmon Assessment Committee) (2011) Annual report of the US Atlantic salmon assessment committee-2010 activities. Report no 23, Portland, ME. http://www.nefsc.noaa.gov/USASAC/Reports/USASAC2011-Report%2323-2010-Activities.pdf

USEPA (2000) Ambient aquatic life water quality criteria for dissolved oxygen (saltwater): Cape Cod to Hatteras. EPA-822-R-00012. US Environmental Protection Agency, Washington, DC

USEAP (2003) Biological evaluation for the issuance of ambient water quality criteria for dissolved oxygen, water clarity and chlorophyll a for Chesapeake Bay and its tidal tributaries. US Environmental Protection Agency, Region III, Chesapeake Bay Program Office

USFWS (2011) US Fish and Wildlife Service Connecticut River coordinator's office report. www.fws.gov/r5crc

Vadas RL (1977) Preferential feeding: an optimization strategy in sea urchins. Ecol Monogr 47:337–371

Vadas RL (1985) Herbivory. In: Littler MN, Littler DS (eds) Handbook of phycological methods: ecological field methods: Macroalgae. Cambridge University Press, Cambridge, pp 531–572

Valente RM (2004) The role of seafloor characterization and benthic habitat mapping in dredged material management: a review. J Mar Env Eng 7:185–215

Valente RM, Carey DA, Read LB, Esten ME (2012) Monitoring survey at the Central Long Island Sound disposal site, October 2009. DAMOS Contribution No 184, US Army Corps of Engineers, New England District, Concord, MA, 90 pp

Valente RM, Rhoads DC, Germano JD, Cabelli VJ (1992) Mapping of benthic enrichment patterns in Narragansett Bay, Rhode Island. Estuaries 15(1):1–17

Valiela I, Cole ML (2002) Comparative evidence that salt marshes and mangroves may protect seagrass meadows from land-derived nitrogen loads. Ecosystems 5:92–102

Van Name WG (1912) Simple ascidians of the coasts of New England and neighboring British provinces. Proc Boston Soc Nat Hist 34:439–619

Van Name WG (1945) The North and South America ascidians. Bull Am Museum Nat Hist 84:1–476

Van Patten MS (2006) Seaweeds of Long Island Sound. Foreword by Charles Yarish, Connecticut Sea Grant 104 pp

Van Patten MS, Yarish C, O'Muirchairtaigh I (1993) Effects of temperature on reproduction in the Atlantic kelp, *Laminaria longicruris*, in the North Atlantic Ocean. In: Van Patten M (ed) Irish-American technical exchange on the aquaculture of abalone, sea urchins, lobsters and kelp. Connecticut Sea Grant College Program and Martin Ryan Institute, pp 50–51

Vaquer-Sunyer R, Duarte CM (2008) Thresholds of hypoxia for marine biodiversity. Proc Natl Acad Sci U S A 105:15452–15457

Vaquer-Sunyer R, Duarte CM (2010) Sulfide exposure accelerates hypoxia-driven mortality. Limnol Oceanogr 55(3):1074–1082

Vaquer-Sunyer R, Duarte CM (2011) Temperature effects on oxygen thresholds for hypoxia in marine benthic organisms. Global Change Biol 17:1788–1797

Varekamp JC, Thomas E (1998) Sea level rise and climate change over the last 1000 years. EOS 79:69–75

Varekamp JC, Thomas E, Van de Plassche O (1992) Relative sea-level rise and climate change over the last 1500 years (Clinton, CT, USA). Terra Nova 4:293–304

Varekamp, J. C., Thomas, E., and Groner, M., 2005. The late Pleistocene – Holocene History of Long island Sound, Seventh Biennial LIS *Research Conference Proceedings*, 2004, p. 27–32

Varekamp JC, Thomas E, Lugolobi F, Buchholtz ten Brink MR (2004) The paleo-environmental history of Long Island Sound as traced by organic carbon, biogenic silica and stable isotope/trace element studies in sediment cores. Proceedings of the 6th Biennial LIS Research Conference, pp 109–113

Varekamp JC, Altabet M, Thomas E, Ten Brink M, Andersen N, Mecray E (2009) Hypoxia in Long Island Sound-Since when and why. 'Coasts and Estuaries in a Changing World,' CERF 2009, Portland, OR (1–5 Nov 2009)

Varekamp JC, Thomas E, Altabet M, Cooper S, Brinkhuis H, Sangiorgi F, Donders T, Buchholtz ten Brink M (2010) Environmental change in Long Island Sound in the recent past: eutrophication and climate change. Final report, LISRF grant No CWF 334-R (FRS #525156), 54 pp. http://www.wesleyan.edu/ees/JCV/LobstersReport final.pdf

Vaudrey JMP (2008a) Establishing restoration objectives for eelgrass in Long Island Sound, Part I: Review of the seagrass literature relevant to Long Island Sound. Final grant report to the Connecticut Dept of Environmental Protection, Bureau of Water Protection and Land Reuse and the US Environmental Protection Agency, Groton, CT, 58 pp

Vaudrey JMP (2008b) Establishing restoration objectives for eelgrass in Long Island Sound, Part II: Case studies. Final grant report to the Connecticut Department of Environmental Protection, Bureau of Water Protection and Land Reuse and the US Environmental Protection Agency, Groton, CT

Vaudrey JMP, Kremer JN, Branco BF, Short FT (2010) Eelgrass recovery after nutrient enrichment reversal. Aquat Bot 93:237–243

Verity PG, Wassman P, Frischer ME, Howard-Jones MH, Allen AE (2002) Grazing of phytoplankton by microzooplankton in the Barents Sea during early summer. J Mar Syst 38:109–123

Verrill AE (1898) Descriptions of new American actinians, with critical notes on other species, I. Am J Sci (4)6:493–498

Verrill AE, Smith SI, Harger O (1873) Catalogue of the marine invertebrate animals of the southern coast of New England, and adjacent waters. Report of the US Fish Commission 1871–1872:537–747

Vigil HL (1991) The fate of dissolved oxygen in Long Island Sound bottom waters. Master's dissertation, State University of New York, Stony Brook

Völkel S, Grieshaber MK (1994) Oxygen-dependent sulfide detoxification in the lugworm *Arenicola marina*. Mar Biol 118:137–147

Voyer RA, Hennecky RJ (1972) Effects of dissolved oxygen on the two life stages of the mummichog. Prog Fish Culturist 34:222–225

Wada M, Wu SS, Tsutsumi H, Kita-Tsukamoto K, Hyung-Ki D, Nomura H, Ohwada, K, Kogure K (2006) Effects of sodium sulfide on burrowing activity of *Capitella* sp. I and bacterial respiratory activity in seawater soft-agar microcosms. Plankton Benthos Res 1(2):117–122

Waldbusser GG, Voigt EP, Bergschneider H, Green MA, Newell RIE (2011) Biocalcification in the Eastern Oyster (*Crassostrea virginica*) in relation to long-term trends in Chesapeake Bay pH. Estuaries Coasts 34:221–231

Walker DI, Lukatelich RJ, Bastyan G, McComb AJ (1989) Effect of boat moorings on seagrass beds near Perth, Western Australia. Aquat Bot 36:69–77

Walker A, Bush P, Wilson T, Chang E, Miller T, Horst M (2005) Metabolic effects of acute exposure to methoprene in the American lobster, *Homarus americanus*. J Shellfish Res 24(3):787–794

Wall CC, Peterson BJ, Gobler CJ (2008) Facilitation of seagrass *Zostera marina* productivity by suspension-feeding bivalves. Mar Ecol Prog Ser 357:165–174

Walters GR, Plumb JA (1980) Environmental stress and bacterial infection in channel catfish, Ictalurus punctatus. J Fish Biol 17:177

Wannamaker CM, Rice JZ (2000) Effects of hypoxia on movements and behavior of selected estuarine organisms from the southeastern United States. J Exp Mar Biol Ecol 249:145–163

Warkentine BE, Rachlin JW (2010) The first record of *Palaemon macrodactylus* (Oriental shrimp) from the eastern coast of North America. Northeast Nat 17:91–102

Warnock N, Elphick CS, Rubega MA (2001) Biology of marine birds: Shorebirds. In: Burger J, Schreiber BA (eds) Biology of marine birds. CRC Press, Boca Raton, pp 581–615

Warren RS, Niering WA (1993) Vegetation change on a northeast tidal marsh-interaction of sea-level rise and marsh accretion. Ecology 74:96–103

Warren RS, Fell PE, Rozsa R, Brawley AH, Orsted AC, Olson ET, Swamy V, Niering WA (2002) Salt marsh restoration in Connecticut: 20 years of science and management. Restor Ecol 10:497–513

Waycott M, Duarte CM, Carruthers TJB, Orth RJ, Dennison WC, Olyarnik S, Calladine A, Fourqurean JW, Heck KL Jr, Hughes AR, Kendrick GA, Kenworthy WJ, Short FT, Williams SL (2009) Accelerating loss of seagrasses across the globe threatens coastal ecosystems. Proc Natl Acad Sci U S A 106:12377–12381

Wazniak CE, Hall MR, Carruthers TJB, Sturgis B, Dennison WC, Orth RJ (2007) Linking water quality to living resources in a mid-Atlantic lagoon system, USA. Ecol Appl 17:S64–S79

Weber RE (1988) Intraspecific adaptation of hemoglobin function in fish to oxygen availability. In: Addink ADF, Spronk N (eds) Exogenous and endogenous influences on metabolic and neural control. Pergamon Press, Oxford, pp 87–102

Weber RE, Jensen FB (1988) Functional adaptations in hemoglobins from ectothermic vertebrates. Ann Rev Physiol 50:161–179

Weber JM, Kramer DL (1983) Effects of hypoxia on surface access on growth, mortality, and behavior of juvenile guppies, *Poecilia reticulate*. Can J Fish Aquat Sci 40:1583–1588

Weinstein MP (1979) Shallow marsh habitats as primary nurseries for fishes and shellfish, Cape Fear River, North Carolina. Fish Bull 77:339–357

Weisberg SB, Wilson HT, Himchak P, Baum T, Allen R (1996) Temporal trends in abundance of fish in the tidal Delaware River. Estuaries 19:723–729

Weiss HM (1995) Marine animals of southern New England and New York. State Geological and Natural History Survey of Connecticut

Welschmeyer NA, Lorenzen CT (1985) Chlorophyll budgets: Zooplankton grazing and phytoplankton growth in a temperate fjord and the central Pacific gyres. Limnol Oceanogr 30:1–21

Welsh BL (1995) Hypoxia in Long Island Sound: one researcher's perspective. Long Island Sound Research Conference: Is the Sound Getting Better or Worse? New York Sea Grant Institute, Stony Brook, NY

Welsh BL, Eller FC (1991) Mechanisms controlling summertime oxygen depletion in Western Long Island Sound. Estuaries 14:265–278

Weltzien F, Doeving KB, Carr WES (1999) Avoidance reaction of yolk-sac larvae of the inland silverside *Menidia beryllina* (Atherinidae) to hypoxia. J Exp Biol 202(20):2869–2876

Whitlatch RB (1982) The ecology of New England tidal flats: a community profile. US Fish and Wildlife Service, Biol Services Program, Washington, DC, FWS/OBS-81/01, 125 pp

Whitlatch RB, Osman RW (2000) Geographical distributions and organism-habitat associations of shallow-water introduced marine fauna in New England. In: Pederson J (ed) Marine bioinvasions. Proceedings of the 1st national conference, 4–27 Jan 1999, MIT Sea Grant College Program, Massachusetts Institute of Technology, Cambridge, MA, pp 61–65

Wikfors GH (2005) A review and new analysis of trophic interactions between *Prorocentrum minimum* and clams, scallops, and oysters. Harmful Algae 4:585–592

Wikfors GH, Smolowitz RM (1993) Detrimental effects of a *Prorocentrum* isolate upon hard clams and bay scallops in laboratory feeding studies. In: Smayda TJ, Shimizu Y (eds) Toxic phytoplankton blooms in the sea. Elsevier, New York, pp 447–452

Wilber DH, Clarke DG, Rees SI (2007) Responses of benthic macroinvertebrates to thin-layer disposal of dredged material in Mississippi Sound, USA. Mar Poll Bull 54:42–52

Williams PJ leB (1981) Incorporation of microheterotrophic processes into the classical paradigm of the planktonic food web. Kieler Meeresforsch 5:1–28

Wilson RE, Crowley H, Brownawell B, Swanson RL (2005) Simulation of transient pesticide concentrations in Long Island Sound for late summer 1999 with a high resolution coastal circulation model. J Shellfish Res 24(3):865–875

Wilson RE, Swanson RL, Crowley HA (2008) Perspectives on long-term variations in hypoxic conditions in western Long Island Sound. J Geophys Res 113:C12011. doi:10.1029/200 7JC004693

Witek JC (1990) An outline of the aboriginal archaeology of Shelter Island, New York. Bull Archaeol Soc Connecticut 53:39–58

Witte U, Aberle N, Sand M, Wenzhöfer F (2003) Rapid response of a deep-sea benthic community to POM enrichment: an in situ experimental study. Mar Ecol Prog Ser 251:27–36

Wolfe DA, Monahan R, Stacey PE, Farrow DRG, Robertson A (1991) Environmental quality of Long Island Sound: assessment and management issues. Estuaries 14:224–236

Wu RSS (2002) Hypoxia: from molecular responses to ecosystem responses. Mar Poll Bull 45:35–45

Wu RSS (2009) Effects of hypoxia on fish reproduction and development, Chapter 3. In: Richards J, Farell A, Brauner C (eds) Fish physiology: Hypoxia. Elsevier, San Diego, 517 pp

Wu RSS, Zhou BS, Randall DJ, Woo NYS, Lam PKS (2003) Aquatic hypoxia is an endocrine disruptor and impairs fish reproduction. Environ Sci Technol 37:1137–1141

Yarish C, Baillie PW (1989) Ecological study of an impounded estuary, Holly Pond, Stamford, CT. Submitted to the Stamford Environmental Protection Board, Stamford, CT and the Coastal Area Management Program, Hartford, CT, 117 pp + appendices

Yarish C, Edwards P (1982) Field and cultural studies on the seasonal and horizontal distribution of estuarine red algae of New Jersey. Phycologia 21:112–124

Yarish C, Edwards P, Casey S (1980) The effects of salinity, and calcium and potassium variations on the growth of two estuarine red algae. J Exp Mar Biol Ecol 47(3):235–249

Yarish C, Breeman AM, van den Hoek C (1984) Temperature, light and photoperiod responses of some Northeast American and west European endemic rhodophytes in relation to their geographic distribution. Helgoländer Meeresunters 38:273–304

Yarish C, Breeman AM, van den Hoek C (1986) Survival strategies and temperature responses of seaweeds belonging to different biogeographic distribution groups. Botanica Mar 24:215–230

Yarish C, Kirkman H, Lüning K (1987) Lethal exposure times and preconditioning to upper temperature limits of some temperate North Atlantic red algae. Helgoländer Meeresunters 41:323–327

Yarish C, Brinkhuis BH, Egan B, Garcia-Esquivel Z (1990) Morphological and physiological bases for *Laminaria* selection protocols in Long Island Sound. In: Yarish C, Penniman CA, van Patten M (eds) Economically important marine plants of the Atlantic: their biology and cultivation. Connecticut Sea Grant College, Groton, pp 53–94

Yarish C, Wilkes R, Chopin T, Fei XG, Mathieson AC, Klein AS, Friel D, Neefus CD, Mitman GG, Levine I (1998) Domesticating indigenous *Porphyra* (nori) species for commercial cultivation in Northeast America. World Aquacul 29(4):26–29

Yarish C, Chopin T, Wilkes R, Mathieson AC, Fei XG, Lu S (1999) Domestication of nori for Northeast America: The Asian experience. Bull Aquacul Assoc Can 99(1):11–17

Yarish C, Linden RE, Capriulo G, Koch EW, Beer S, Rehnberg J, Troy R, Morales EA, Trainor FR, DiGiacomo-Cohen M, Lewis R (2006) Environmental monitoring, seagrass mapping and biotechnology as means of fisheries habitat enhancement along the Connecticut coast. Univ of Connecticut, Stamford, p 105

Yarish C, Whitlatch RB, Kraemer G, Lin S (2010) Multi-component evaluation to minimize the spread of aquatic invasive seaweeds, harmful algal bloom microalgae, and invertebrates via the live bait vector in Long Island Sound. Report to US Environmental Protection Agency. http://digitalcommons.uconn.edu/ecostam_pubs/2/

Yarish C, Whitlatch RB, Kraemer GP, Lin S (2010) Final report to Connecticut Sea Grant on Project R/LR-17, Spread and impacts of the non-indigenous rhodophycean alga, *Grateloupia turuturu*, on Long Island Sound, 29 pp

Yasumoto T, Oshima Y, Sugawara W (1980) Identification of *Dinophysis-Fortii* as the causative organism of diarrhetic shellfish poisoning. Bull Japanese Soc Sci Fish 46:1405–1411

York JK, Costas BA, McManus GB (2011) Microzooplankton grazing in green water-Results from two contrasting estuaries. Estuaries Coasts 34:373–385. doi:10.1007/s12237-010-9336-8

Zajac RN (1996) Ecologic mapping and management-based analyses of benthic habitats and communities in Long Island Sound. Chapters I, II and III. Final report, Long Island Sound Research Fund Grant CWF-221-R, Connecticut Dept of Environmental Protection, Office of Long Island Sound Programs, Hartford, CT

Zajac RN (1998a) Spatial and temporal characteristics of selected benthic communities in Long Island Sound and management implications. Final report, Long Island Sound Research Fund Grant CWF317-R, Connecticut Dept of Environmental Protection, Office of Long Island Sound Programs, Hartford, CT

Zajac RN (1998b) A review of research on benthic communities conducted in Long Island Sound and assessment of structure and dynamics. In: Poppe LJ, Polloni C (eds) Long Island Sound Environmental Studies, US Geological Survey, Open-File Report 98-502. http://128.128.240.33/cdroms/ofr98-502/chapt4/rz1cont.htm

Zajac RN (1999) Understanding the seafloor landscape in relation to assessing and managing impacts on coastal environments. In: Gray JS, Ambrose W Jr, Szaniawska A (eds) Biogeochemical cycling and sediment ecology. Kluwer Publishing, Dordrecht, pp 211–227

Zajac RN (2001) Organism-sediment relations at multiple spatial scales: implications for community structure and successional dynamics. In: Aller JY, Woodin SA, Aller RC (eds) Organism-sediment interactions. University of South Carolina Press, Columbia, pp 119–139

Zajac RN, Whitlatch RB (1982) Responses of estuarine infauna to disturbance. II. Spatial and temporal variation in succession. Mar Ecol Prog Ser 10:15–27

Zajac RN, Lewis RS, Poppe LJ, Twichell DC, Vozarik J, DiGiacomo-Cohen ML (2000a) Relationships among sea-floor structure and benthic communities in Long Island Sound at regional and benthoscape scales. J Coast Res 16:627–640

Zajac RN, Lewis RS, Poppe LJ, Twichell DC, Vozarik J, DiGiacomo-Cohen ML, (2000b) Benthic community geographic information system (GIS) data layers for Long Island Sound, Chapter 10. In: Paskevich VF, Poppe LJ (eds) Georeferenced mapping and bottom photography in Long Island Sound. US Geological Survey Open File-Report 00-304, CD-ROM. http://pubs.usgs.gov/openfile/of00-304/

Zajac RN, Lewis RS, Poppe LJ, Twichell DC, Vozarik J, DiGiacomo-Cohen ML (2003) Responses of infaunal populations to benthoscape structure and the potential importance of transition zones. Limnol Oceanogr 48:829–842

Zajac RN, Seals B, Simpson D (2008) Food webs in Long Island Sound: review, synthesis and potential applications. Final report-EPA Grant No. LI-97101401 submitted to the US Environmental Protection Agency, Long Island Sound Study, 87 pp (Available at the EPA LISS website)

Zimmerman RC, Smith RD, Alberte RS (1987) Is growth of eelgrass nitrogen limited? A numerical simulation of the effects of light and nitrogen on the growth dynamics of Zostera marina. Mar Ecol Prog Ser 41:167–176

Zulkosky A, Ruggieri J, Terracciano S, Brownawell B, McElroy A (2005) Acute toxicity of resmethrin, malathion and methoprene to larval and juvenile American lobsters (Homarus americanus) and analysis of pesticide levels in surface waters after Scourge™, Anvil™, and Altosid™ application. J Shellfish Res 24(3):795–804

Chapter 7
Synthesis for Management

Mark A. Tedesco, R. Lawrence Swanson, Paul E. Stacey, James S. Latimer, Charles Yarish and Corey Garza

7.1 Introduction

In the foreword to Tom Andersen's "environmental history" of Long Island Sound (LIS), *This Fine Piece of Water* (Andersen 2002), Robert F. Kennedy, Jr., describes "a region of mythical productivity" observed by the first explorers.

M. A. Tedesco (✉)
Long Island Sound Office, US Environmental Protection Agency, 888 Washington Blvd, Stamford, CT 06904-2152, USA
e-mail: tedesco.mark@epa.gov

R. L. Swanson
School of Marine and Atmospheric Sciences, Stony Brook University, Stony Brook, NY 11794, USA

P. E. Stacey
Great Bay National Estuarine Research Reserve,
New Hampshire Fish and Game Department, Durham, NH 03824, USA

J. S. Latimer
Office of Research and Development, US Environmental Protection Agency, Narragansett, RI 02882, USA

C. Yarish
Department of Ecology and Evolutionary Biology, University of Connecticut, Stamford, CT 06901-2315, USA

C. Garza
Division of Science and Environmental Policy, California State University at Monterey Bay, Seaside, CA 93955, USA

J. S. Latimer et al. (eds.), *Long Island Sound*, Springer Series on
Environmental Management, DOI: 10.1007/978-1-4614-6126-5_7,
© Springer Science+Business Media New York 2014

They smelled aromas from Long Island's flowers before sighting land and found four hundred bird species, many of which are gone today. Henry Hudson's lieutenant Robert Juett described rivers choked with salmon (probably striped bass) and mullet. Giant dolphin pods schooled in the East River and New York Harbor. F. Scott Fitzgerald, one of Long Island's most faithful chroniclers in recalling its legendary abundance, suggested that the Sound appeared to the first Dutch sailor as the "fresh green breast of the new world," compelling him to hold his breath in "an aesthetic contemplation he neither understood nor desired, face to face for the last time in history with something commensurate with his capacity for wonder."

This bounty of Nature provided an ample supply of resources long before the complications of human use and consequent competition over ecosystem services became mainstream social and economic concerns. Ironically, the geographic construct of LIS and its watershed provided both abundant resources and ideal conditions for their exploitation from human habitation. Lewis (Chap. 2, in this volume) describes the geologic processes that molded the landscape and influenced the early patterns of human development along the Sound's shoreline and major river valleys. Weigold and Pillsbury's (Chap. 1, in this volume) history of land use surrounding LIS is also a history of the pollution and habitat loss that followed as a consequence. They describe how the "taming" of the land by settlers altered the geologic blueprint described by Lewis (Chap. 2, in this volume), reshaping the landscape to fuel the needs and desires of society. The use of LIS as a "nautical highway" (Weigold and Pillsbury, Chap. 1, in this volume) provided the means for trade and ultimately drew commerce from Boston to New York City and satellite harbors along the Connecticut coast.

The settlement and development of the coastline and watershed was a societal success story of human opportunity and adaptation; it also gave rise to pollution from agriculture, industry, and the effluvia of human populations. The urbanization of the Sound's watershed (Fig. 7.1), initiated in the nineteenth century, expanded quickly in the twentieth century. This was made possible by improved transportation systems and a growing middle-class population that was supplanted in urban centers, particularly New York City, by mass immigrations from Europe. At the western border of LIS evolved the world center of commerce, New York City, whose associated economic and social developments ultimately sprawled eastward along the Westchester County, Long Island, and Connecticut shorelines.

The sobriquets applied to LIS follow this theme, emphasizing human habitation as inherent to its character: The American Mediterranean according to Daniel Webster, *The Urban Sea* (Koppelman et al. 1976), and more recently, in the words of Prof. Glenn Lopez of Stony Brook University, "the first 21st century estuary." This last nickname suggests a broader narrative—a trajectory of change being duplicated in estuaries throughout the country: from natural paradise of plenty to natural resource extraction, the felling of forests for agriculture to industrialization, suburbanization, and, with the environmental movement, regulation and restoration. But the conclusion is not foregone. Today, new threats beckon—climate change, invasive species, emerging contaminants—while the old adversaries of habitat loss and nutrient pollution stubbornly remain.

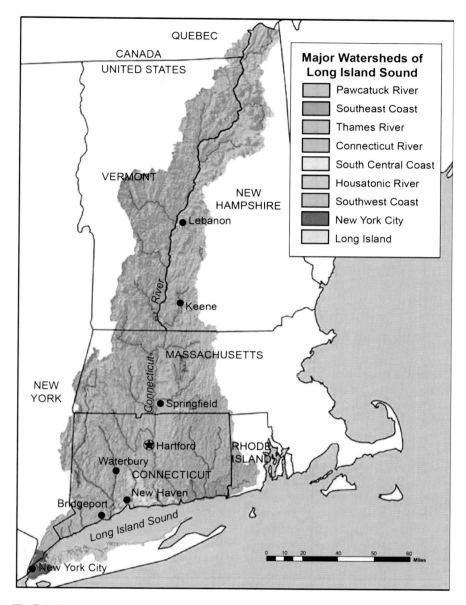

Fig. 7.1 The major watersheds draining to Long Island Sound

How this narrative concludes should matter to anyone concerned with our coastal environment. Long Island Sound is a harbinger of the fate of coastal waters and economies everywhere. Through all its changes, LIS, ever resilient, remains productive, providing leisure and livelihood. Lessons learned from efforts to understand and restore the LIS ecosystem are transferable to other coastal ecosystems and economies. For while the historical transitions observed in other estuaries may have started after or lagged behind LIS, the twenty-first century beckons for all.

This chapter begins with a brief review of the development of scientific knowledge in and around LIS. We then highlight some of the insights into the function of the LIS ecosystem that are described in detail in the preceding chapters, examine their linkages, and propose a set of actions to improve management of LIS. To accomplish this, it is essential to recognize the link between the health of the watershed and the health of LIS. This relationship is complex, especially in the densely-populated watersheds of LIS. Natural ecosystems are in a continual state of fluctuation, and our activities have accelerated the pace of change—physically, chemically, and biologically. Particular emphasis is placed on changes that may be driven by broader climate patterns and on the implications of such insights to the ecosystem and to regulatory and management approaches. Our hope is not to provide a definitive synthesis for management, but to foster continued dialogue among scientists, environmental managers, economists, and the LIS community on actions needed to provide for a resilient and healthy ecosystem for generations to come.

7.2 Science Foundations

Perhaps the first use of "Long Island Sound" on a map was by Robert Ryder in 1670 (on file in the New York Historical Society, February 2012). This may also be the first map of an "extensive area of colonial America" created using surveying instruments and techniques (Allen 1997). Only the western end of the map remains, but two-thirds of LIS is depicted and "Long Island Sound" is clearly labeled at the map's right edge. The scale of the Sound shoreline appears to be reasonably accurate, and many of the place names familiar today are identified including Eatons Neck, Huntington, Norwalk, Stamford, and Hell Gate.

Safe navigation to promote commerce was the impetus for the first systematic studies of the Sound. Long Island Sound was a relatively protected route to New York City and became increasingly important in the early nineteenth century despite the requirement to navigate the dangerous waters of Hell Gate. Steam power had allowed vessels to increase speed as well as tonnage, and thus vessel drafts became deeper. But the controlling depth for vessels entering New York Harbor from the Atlantic Ocean was about 7.3 m in the vicinity of the Sandy Hook, NJ to Rockaway, NY transect (Swanson et al. 1982; Klawonn 1977). Since ocean dredging was not yet technically feasible, and dredging a channel at Sandy Hook did not commence until 1885 (Klawonn 1977), entering New York Harbor through Hell Gate became a more desirable route. Still, as late as the 1840s, Charles Dickens, on a trip to the US observed, "The Sound, which has to be crossed on this passage, is not always a very safe or pleasant navigation, and has been the scene of some unfortunate accidents" (Dickens 1985).

Edmund Marsh Blunt and his sons published several maps of Long Island and its waters in the early 1800s, culminating in an 1830 chart of its north shore (Sound-side) that Allen (1997) judges as "relatively large scale" and "accurately done." President Thomas Jefferson created what was to become the Coast Survey (later the Coast and Geodetic Survey {C&GS} and now part of NOAA) in 1807. However, it was not able to undertake its mission until much later because of the War of 1812 and lack of Congressional support (Allen 1997). According to Allen (1997), Long Island and the surrounding waters were chosen as the location of the survey's earliest work because of Long Island's proximity to New York Harbor and the tortuous waters throughout the area.

These earliest surveys provide our first glimpses into the geomorphology and tidal phenomena of the Sound. They even afford some insight concerning the sediment characteristics of the waterways, as that was important information for secure anchoring. The surveys are designated as topographic and hydrographic. Topographic surveys are of the terrestrial land masses; hydrographic surveys are of the sea floor. The Coast Survey's topographic survey from Norwalk to Fairfield, CT was completed in 1835 at a scale of 1:10,000 (US Coast and Geodetic Survey Undated-c). Topographic surveys in the vicinity of Throgs Neck were completed in 1837, also at a scale of 1:10,000 (US Coast and Geodetic Survey Undated-c). Hydrographic surveys, completed at a scale of 1:10,000, covered the Sound floor from Flushing to New Rochelle, NY by 1837. The Race was charted by 1839 at 1:20,000. Both types of surveys were then published as maps or nautical charts at a smaller scale (US Coast and Geodetic Survey Undated-a, b). Initial hydrographic surveys of most of the Sound were completed by the 1840s. Sounding poles and lead lines were the depth measuring devices until the World War I era when the use of the velocity of sound in sea water became the preferred technique to measure depth (Shalowitz 1964).

The Battery, at the southern tip of Manhattan Island in New York City, is the site of the second longest time series of tidal observations (1856 to present) in the US. These data are commonly used to compare recent sea level changes on the East Coast. The earliest tide observations in LIS were made at Bridgeport and Black Rock Harbors, CT in 1835, followed by City Island and Throgs Neck, NY in 1837. These early observations contributed to the Coast Survey's first published government tide tables in 1853 (Marmer 1926).

Surface tidal current observations commenced in the 1840s (Stratford Shoals, Execution Rocks, Hart Island, and Stepping Stones) using floats and log lines. To the east, early measurements using floats were made near Plum Island, Cornfield Point, and the Thames River mouth. The first mechanical metered measurements were taken in the 1890s. The C&GS in 1934 released the first edition of *Tidal Current Charts, Long Island Sound and Block Island Sound*. From 1965 to 1967, the C&GS R/V MARMER resurveyed the tidal currents of LIS and Block Island Sound (BIS) at about 160 locations. These data, collected by the instruments of the day, were used to modernize and amplify the published tidal current tables and charts (Conover 1966).

Temperature (T) and salinity (S) measurements were recorded at the locations of C&GS current measuring stations dating back to the summers of 1929 and 1930. At that time, observations were obtained at some 70 locations from The Race to Hell Gate, including some in the Thames and Connecticut Rivers. Le Lacheur and Sammons (1932) summarized these data going back nearly a century in *Tides and Currents in Long Island and Block Island Sounds*. The authors also discussed tidal theory as it related to LIS, including that the tidal wave form was a combination of a standing and a progressive wave, and that Coriolis affected the tidal ranges on the north side of the Sound relative to the south side.

In the mid-nineteenth century, the geologic conditions that created Long Island were not yet understood. It was in the 1840s that Dr. Louis Agassiz's theories of glaciation were beginning to be published and accepted (Gilluly et al. 1959). Benjamin Thompson (1849), perhaps the preeminent natural scientist of his time in the area and a Setauket, NY resident, believed that the Island was "reclaimed from the sea," that is, it essentially washed up from the ocean floor. As explained by Lewis (Chap. 2, in this volume), it wasn't until the studies of Dana (1870, 1890) that Veatch (1906) and Fuller (1914) were able to describe the fluvial processes that created the Sound and establish that Long Island was, in fact, the residue of the Pleistocene ice advances and retreats. Fuller's (1914) *The Geology of Long Island* is a particularly important early study.

Weigold and Pillsbury (Chap. 1, in this volume) elucidate the importance of LIS's fishery resources dating back to the colonial period. It is evident from this analysis that scientific studies did not drive the early understanding of harvest effects as much as observations of reduced resources and implementation of trial and error policy actions to remediate stock declines. In fact, most of the problems that confront fisheries today were experienced more than a century ago. Overharvesting, unintended by-catch mortality, pollution, and public health were issues then, as were hindered migration routes for such species as salmon, shad, and alewives by construction of dams and spillways. And, as is true today, there were conflicts between commercial and recreational fishers on access and harvest rights.

A significant boost to better understanding the plight of living marine resources was provided by the establishment of the US Fish Commission by the Congress in 1871. For example, the 1895 *Bulletin of the Commission* reports on the status of the Atlantic menhaden (*Brevoortia tyrannus*) fishery as well as food fishes in much of the northeast, including LIS, based upon surveys of the previous year. Today, the National Marine Fisheries Service (NMFS) is a descendant of the original commission, specifically focused on marine waters and their fisheries. Currently, the only federal marine laboratory on the Sound is the NMFS lab at Milford, CT, established there in 1931, with a focus on aquaculture, specifically bivalve shellfish. From ongoing measurements in Milford Harbor, the laboratory has T data since 1948 and S data since 1953. More recently, regional fish commissions were established in 1942 (Atlantic States Marine Fisheries Commission) and in 1976 under the Magnuson-Stevens Fishery Conservation and Management Act (the New England Fishery Management Council and the Mid-Atlantic Marine Fishery Council) in order to understand and address regional fisheries that crossed jurisdictional boundaries.

Just as the federal government dedicated resources to investigate the condition of the states' fisheries resources, so too did state governments. New York and

Connecticut both established fish commissions in the 1860s to examine the conditions of their fresh- and saltwater fisheries. Of particular concern to both was improving the deteriorating eastern oyster (*Crassostrea virginica*) beds in the western part of LIS. State commission reports document the changing conditions of the fisheries and the Sound in the late nineteenth century.

The oldest continuous water quality monitoring program in the US, which commenced in 1909, is New York City's Harbor Survey Program. Originally established in response to public outcry about fouled waterways, the program has provided a rich database of unparalleled value to researchers and managers in the New York City region. The length of the data sets provides a unique view of how the water quality of New York Harbor and the westernmost parts of the Sound has changed over the years. Even during the program's earliest days, low dissolved oxygen concentrations were a problem in harbor waters. It is perhaps the first recorded instance for the marine environment of the US that the issues we now identify as hypoxia and anoxia were identified, significant problems that persist today (NYCDEP 2010).

The Bingham Oceanographic Laboratory at Yale University (1930–1966) conducted, documented, and synthesized some of the most important research concerning the physics, chemistry, and biology of LIS undertaken until the advent of the USEPA Long Island Sound Study (LISS). The work of that institution continues to provide a foundation for our understanding of many of the processes that define and impact the Sound today. In fact, the important reviews by Parker and O'Reilly (1991) and Capriulo et al. (2002) use the laboratory's studies as the basis for analysis. Gordon Riley and colleagues conducted a number of cruises in LIS and BIS in 1946, and his article about the hydrography of LIS (Riley 1952) was the most comprehensive synthesis of the physical oceanography of the Sound in its time, reviewing non-tidal currents, conservation of mass and salt, and lateral diffusion.

Predating the hydrographic work, Riley (1941) and the Bingham Oceanographic Laboratory initiated some of the earliest, if not the earliest, plankton studies of LIS. These studies were expanded upon in the 1950s and summarized by Riley and Conover (1967) in a volume of the *Bulletin of the Bingham Oceanographic Collection*. This volume contained a number of important papers. Richards and Riley (1967) introduced their studies of epifauna. Riley (1967a, b) mathematically modeled nitrate and phosphate in coastal waters, and, building on the work of Harris (1959), provided an early review of nutrients and nutrient cycling in the LIS ecosystem that has important theoretical and quantitative value today. Riley (1967a, b) also returned to physical oceanography with a paper on transport and mixing. Richards (1963) devoted an entire volume to the demersal fishes of LIS and continued to be actively involved in LIS research into the first years of the LISS.

These efforts served as a backbone of one of the first syntheses of the oceanography and environmental condition of LIS, published as *The Urban Sea* (Koppelman et al. 1976). More recently, with the rapidly escalating national and global interest in nutrient management, the early works of Riley and others provide one of the very few windows into the past through which resource and management goals can be viewed. Despite this trove of research, and spatially limited work by some programs such as the New York City Harbor Survey Program, it was not until 1985 with the advent of the LISS that concerted and intensive water quality research and monitoring efforts began in earnest.

7.3 Physical and Biogeochemical Setting

Long Island Sound is functionally an estuary, generally defined as "…a semi-enclosed coastal body of water which has a free connection with the open sea and within which sea water is measurably diluted with fresh water derived from land drainage" (Cameron and Pritchard 1963). While the Sound is an estuary, it is an unusual estuary. It has entrances connecting to the ocean near its head (the East River) and its mouth (The Race). Physical oceanographers consider Hell Gate as the head, as it is there that stratification breaks down and the water column becomes well mixed due to turbulence (Fig. 7.2). The Connecticut River, which contributes 75 % of the gauged freshwater flow into LIS (O'Donnell et al. Chap. 3, in this volume), discharges along the Sound's northern flank but only a short distance from its mouth (about 15 % up estuary from its mouth to its head). The East River is not a river but rather a hydraulic strait in which the flow is controlled by a mismatch in the tidal phase and the tidal range at either end (Bowman 1976; Swanson et al. 1982).

The unusual character of the Sound is further defined by the long-term flux of salt, which is from the Sound toward New York Harbor (i.e., upstream and out of the estuary, contrary to typical estuarine patterns), as is the net flux of water. More consistent with what one normally thinks about an estuary, the flux of fresh water is from New York Harbor into the Sound (Blumberg and Pritchard 1997). As

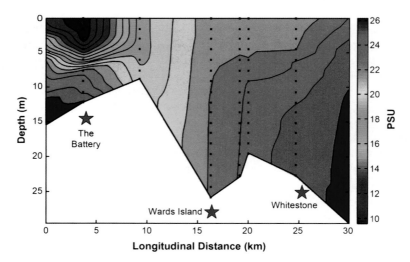

Fig. 7.2 Salinity section from Upper Bay of New York Harbor through the East River to Willets Point collected during the US C&GS MARMER survey of New York Harbor in 1959 (provided by R. Wilson). The Battery is located at the southern tip of Manhattan and Wards Island is near Hell Gate

pointed out by O'Donnell et al. (Chap. 3, in this volume), historical changes in the basin, such as engineering projects to improve navigation safety in the East River and the increase in the discharge of treated sewage into the East River, now on the order of 3.8 billion liters per day, have modified water properties, decreasing S in western Long Island Sound (WLIS) and increasing exchange between the Sound and the East River, as confirmed in core studies by Varekamp et al. (2010).

As discussed in O'Donnell et al. (Chap. 3, in this volume), the unique aspects of salt transport within LIS make it difficult to classify according to an estuarine scheme proposed by Hansen and Rattray (1966) that considers only the relative contributions of advective and diffusive mechanisms. Crowley (2005) found that the westward salt flux in LIS associated with the Stokes transport (mass movement in direction of wave propagation from oscillatory flows) was a dominant contributor to the total westward salt flux, and that the contribution from estuarine gravitational circulation was significant but not dominant. Nonetheless, it is enlightening to use contemporary results to place LIS in the Hansen and Rattray classification. Based on the analysis of S and non-tidal velocity sections by Crowley (2005), and her analysis of the relative contributions of advective and dispersive mechanisms for salt transport, LIS can be classified as a Type 2 (partially mixed) estuary. It is influenced by Coriolis in which gravitational circulation makes a significant contribution to the upstream salt flux. As a comparison, both Chesapeake Bay and Narragansett Bay are partially mixed (Knauss 1997), while the entrance to Delaware Bay is considered to be vertically homogeneous (Williams 1962).

Long Island Sound also does not fit well into the standard geological classifications of an estuary. It is neither a coastal plain estuary (drowned river valley) nor a fjord. Like its sister sounds along the southern New England coast (Fishers Island, Block Island, Rhode Island, Martha's Vineyard and Nantucket), it was shaped by streams and glaciers and became a marine system as sea level rose. Long Island Sound also has a feature that oceanographers would consider to be acting like a "sill" close to its mouth. However, according to Lewis (Chap. 2, in this volume), the "sill" developed after marine waters entered the Sound and does not fit the geologic definition of a structural (typically bedrock) ridge separating basins. It developed as glacial deposits in the easternmost Sound were eroded more than their western counterparts. The sill sits high above the underlying bedrock, has glacial Lake Connecticut deposits under it, and it is slowly migrating westward (Fenster et al. 2006) as marine erosion continues to reshape the basin.

Lewis (Chap. 2, in this volume) describes the recent advances in our understanding of the forces shaping LIS, laying out the geologic framework of the ecosystem, while noting that there remains some ambiguity concerning the nature and timing of marine transgressions into the Sound. The flanks of the Sound are geologically dissimilar. The northern shore consists largely of crystalline rock, while the southern shore consists of more permeable glacial deposits from which there is minimal surface runoff. Groundwater seepage is most likely the greatest source of fresh water to LIS from Long Island. Thus, dissimilar morphological and topographical features of the lands bordering the Sound play an important role in its physical oceanographic processes. The thalweg (the deep channel of a watercourse) is parallel to the mainland (coastal Connecticut), atypical of estuaries other than lagoonal systems. Most US estuaries incise a coast; LIS does not.

Lewis (Chap. 2, in this volume) states that the average depth of the Sound is 20 m, which is greater than the other major estuaries on the US East Coast. Because of this depth, LIS has not been dredged for navigation along its length, also unique for major US Atlantic Coast estuaries. Long Island Sound demarks a change from the sediment-dominated Atlantic coastline that extends southward to Florida from the south shore of Long Island to the largely bedrock-dominated shorelines northward of Rhode Island (with the exception of Cape Cod and some parts of Rhode Island dominated by glacial deposits). Interestingly, based on data from Ranheim and Bokuniewicz (1991), LIS appears to have only a weak turbidity maximum (around Cable and Anchor Reef), unlike the strong maximums experienced in Chesapeake Bay and the Hudson-Raritan Estuary. This may be partly a consequence of the relatively small flux of fresh water at the head of the LIS estuary.

The size and shape of LIS are such that it is a co-oscillating tidal basin with a frequency close to that of the semidiurnal tide. There is amplification of the tidal range from its mouth to its head. The minimum mean tidal range occurs at The Race (0.7 m) and the maximum in the vicinity of Throgs Neck (2.2 m). The effect of Coriolis produces a slightly greater tidal range on the north side of the Sound than on the south side. High water occurs only about 2 h later at Throgs Neck than at The Race (a distance of 106 km). The corresponding tidal currents reach their maximum near The Race (approximately 2.6 m/s) and minimum near Throgs Neck, just to the east of the East River tidal strait where the hydraulic current, reversing at tidal frequency, is again swift (maximum approximately 2.6 m/s).

Long Island Sound experiences one of the largest intra-annual T ranges for estuaries throughout the world—28 °C (roughly between −1° and 27 °C). Albion (1939) asserted that during the early days of steamboats "service was generally suspended" in the western end of the Sound "from mid-November to mid-March" because of ice. The Sound has frozen over on occasion in the recent past. The last such occurrence was January into February 1977 (US Coast Guard 2006), during a period of winters so cold some wondered if we were entering the next "little ice age" (Gwynne 1975). The US Coast Guard (2006) states that in 1977 "most of LIS was frozen over," and the "waters at Execution Rocks on the western end of the Sound were solid ice," measuring "2–3 feet thick in certain portions of the Sound."

The large T range, in conjunction with varying S (temporally and spatially), creates the seasons in the water column of the Sound. Typically, the water column is well mixed in winter. With spring and the freshening of the surface waters with lower density riverine water, S-driven density stratification commences. Shortly thereafter, solar heating drives the water column stratification process as warmer surface waters are less dense than cooler, deeper waters. Wind mixing to depth all but stops, and dissolved oxygen (DO) drawdown begins. By fall, surface waters cool, autumnal storms enhance mixing, and the process begins anew.

Cuomo et al. (Chap. 4, in this volume) highlight how sediment geochemistry patterns reflect the specific geological and physical characteristics of the major basins, as well as embayments and harbors. These geochemical processes influence the sources, fate, and effects of chemical species, which in turn affect benthic organisms. Overall, the sub-tidal, fine-grained sediments found in WLIS and in the majority of harbors and embayments are characterized by reducing conditions. A large amount of the organic carbon present in the sediments of WLIS undergoes anaerobic, microbially-mediated decomposition rather than aerobic decomposition, or direct consumption by benthos, which dominates in eastern Long Island Sound (ELIS). While aerobic decomposition or direct consumption by benthos also characterizes the sediments of central Long Island Sound (CLIS), this basin is transitional. Cuomo et al. (Chap. 4, in this volume), therefore, recommend that the sediment geochemistry of CLIS be closely monitored to detect trends in carbon deposition and shifts in the balance between aerobic and anaerobic decomposition. Trends in nutrient control and climate change may tug CLIS in opposing directions; sentinel monitoring may provide insight into how LIS responds more generally to these changes.

Cuomo et al. (Chap. 4, in this volume) also emphasize that remineralization of organic matter in the sediments of WLIS and, possibly, CLIS provides additional N to the bottom waters of the Sound. This source of N, however, is poorly constrained in present models and is not measured by ongoing monitoring programs. While Lopez et al. (Chap. 6, in this volume) demonstrate that most of the oxygen demand in LIS occurs in the water column, additional sediment nutrient rate measurements, both spatially and temporally, would help elucidate this balance.

The Sound's numerous embayments add to its complexity. Infrastructure (e.g., wastewater treatment facilities (WWTF), industries, transportation corridors, stormwater sewers), concentrated along bay and harbor shores, is a source for pollutants that can impact the Sound. Each embayment is a micro-system controlled by its own geomorphology and anthropogenic modifications. Some have narrow, tortuous entrances through baymouth bars; in other cases, embayment mouths have been hardened with breakwaters or jetties. Some have been deepened by dredging, altering the natural flushing. Placement of the dredged material in some instances has further changed flow characteristics.

Yet, despite their importance for recreation and commerce, the Sound's embayments are relatively understudied, and little is known about their interaction with the main basins. Do embayments transport pollutants into LIS or do geology and modifications restrict transport of water and sediment? Perhaps embayments are sinks for many sediment-bound pollutants, which get into the Sound only after they are dredged or during large storm events. This is particularly relevant to the issue of dredging because in WLIS and CLIS, tidal resuspension of sediment dominates net long-term influx of sediment to LIS from the watershed. Understanding the sources and deposition rates of sediments that need to be dredged to keep open recreational marinas or federal shipping channels would help target sediment management programs.

Another issue of management interest is the degree to which embayments cycle nutrients internally, mitigating impacts upon the greater Sound. Sea level rise may alter exchange between the Sound and the bays. For example, sea level rise may change flood-dominated embayments to ebb-dominated or vice versa—conditions that were mechanically achieved, for better or worse ecologically, over the past century. Stony Brook Harbor, one of Long Island's most pristine bays, is strongly flood-dominated (Swanson and Wilson 2005). This harbor fills semi-daily with oxygen-rich surface waters from Smithtown Bay. Dredging of the entrance channel, which has never occurred, to 5 m would allow hypoxic bottom waters from Smithtown Bay to flow into the harbor during summers (Swanson and Wilson 2005; Bauer 2012). Clarifying the functional relationships between these water bodies and the Sound would enhance the overall effectiveness of management.

7.4 Pollutant Sources, Conditions, and Management

7.4.1 Metals

Varekamp et al. (Chap. 5, in this volume) describe the changing landscape and human and industrial legacy using metal profiles captured in LIS sediments. Sediment depth profiles chronicle pollutant loads over time as they increased rapidly during the Industrial Revolution and decreased with the advent of stronger state regulations, changing economies and consumer demands, and eventually the federal Clean Water Act (CWA). In recent decades, pollutant sources shifted from industry to sewage and urban runoff, as heavy industry moved overseas and human populations around LIS grew.

As described in Chap. 5 (Varekamp et al., in this volume), metals distributions in sediments reflect the proximity of sources and parallel grain size and organic carbon distribution in LIS. In spatially comprehensive surveys of sediments (e.g., Mecray and Buchholtz ten Brink 2000), Ag, Cu, Cd, and Hg were found to be enriched in LIS sediments, and increased in concentration with proximity to New York City (Mitch and Anisfeld 2010). Sediments are finer grained and higher in organic carbon content toward New York City (Knebel et al. 2000), and metal concentrations generally exceed natural concentrations, sometimes by many times. Silver is highly enriched in some locations, which Sañudo-Wilhelmy and Flegal (1992) link to WWTF effluent. Varekamp et al. (2005) report Housatonic River sediments are invariably enriched in Cu, Zn, and Cr, and estimate that the level of enrichment from metal sources from that basin could explain up to 20 % of the Cr and Cu in many LIS sediment samples.

The coastal embayments of LIS that are characterized by low energy, depositional dynamics, and proximity to sources are enriched in metals relative to those sources (Breslin and Sanudo-Wilhelmy 1999). Because harvestable shellfish resources historically are located near many of the most urban harbors of LIS where their food sources are stimulated by nutrient inputs from major tributaries, O'Connor (1996) cautioned that metals in shellfish may be of public health and environmental concern.

Sediment core data, as reported in Chap. 5 (Varekamp et al. in this volume), provide a distinct historical record of change in metals concentrations in LIS sediments. Radioisotope-dated sediment cores show metal concentrations slowly increased as development and industry grew in the early to mid-1800s; accumulation rates increased with the growth of industry through 1980. Between 1980 and 2010, stricter water pollution laws helped abate metals discharges into LIS and its major tributaries, and heavy industry moved overseas; the sediment core record shows concomitant declines in metals.

Mercury (Hg) has a rich history as both a pollutant and as an intensively-researched case study. Mercury discharges were related to industries, especially the hat-making industry, and sewage. Interestingly, the sedimentary record shows the signal from wet periods when contaminated sediments washed from the watersheds to LIS created "peaks" in the record, particularly in the late 1800s and early 1900s, but also in the 1960s–1970s. Varekamp et al. (Chap. 5, in this volume) attribute 20–30 % of the Hg to hat making and up to 25 % to sewage, with the balance coming from sediments contributed by the Connecticut River and other tributaries from watershed sources, including atmospheric deposition.

Changing consumer demands and stronger environmental regulation were probably most responsible for lowering sediment burdens of metals including Hg. However, Hg remains a concern because of potential redistribution of legacy contaminants in the sediments and continuing inputs of atmospheric sources from burning of fossil fuels. Persistent fish tissue contamination with Hg throughout the Northeast is symptomatic of this problem. All the New England states and New York have consumption advisories for Hg, primarily in freshwater fishes (CTDEP et al. 2007). A 2007 Northeast Regional Mercury TMDL summarizes the many pollution prevention steps taken by all the participating states to control sources of mercury in recent years. Those actions virtually shut down active, waterborne sources associated with sewage, especially from dental amalgam, other medical and scientific uses, and general products with practicable alternatives (King et al. 2008; NEIWPCC et al. 2007). Despite the regional efforts to control Hg, windborne sources from outside the watershed are linked clearly by modeling and source studies (NESCAUM 2008) as the major contemporary contributors to Hg pollution in the Northeast (CTDEP et al. 2007).

7.4.2 Organic Contaminants

Varekamp et al. (Chap. 5, in this volume) identify the primary sources of organic contaminants to LIS: WWTFs, urban runoff, combined sewer overflows (CSOs), and atmospheric deposition. Unregulated sources originating from everyday products used in homes and yards and from transportation infrastructure and vehicle emissions, however, contribute contaminants to surface runoff and ground water. Clean Water Act requirements, especially permitting of industrial and increasingly stormwater sources, have reduced the inflow of toxic organic compounds (USEPA 2011a). Product bans, most notably persistent chlorinated pesticides and polychlorinated biphenyls (PCBs), have eliminated new sources, but legacy sources are still active. Many organic compounds are volatile, allowing global distribution to levels of concern in some cases. Riverine inputs, particularly in industrial and urban settings, extend the range of contaminant sources. Major waterways, especially navigable waters, have served as ports and sources of power that were ideal hosts for industrial activity and sites for waste disposal. Broad use and combustion of petroleum in various fractions for fuels, lubricants, and a multitude of carbon-based products have led to wide distribution of polycyclic aromatic hydrocarbons (PAHs) that often reside in fine sediments of harbors and waterways.

The sediment organic contaminant database reported by Varekamp et al. (Chap. 5, in this volume) is not as rich as the database for metals. From a comprehensive evaluation of organic contaminants begun in 1995 by the USGS, Varekamp et al. report generally higher levels of organic contaminants in WLIS, particularly for PAHs and PCBs, which share a similar distribution pattern (See Figs. 5.18 and 5.19 in Chap. 5, in this volume). The Mitch and Anisfeld (2010) data review also revealed a general increase in organic contaminants toward the western end of LIS, especially for PAHs and PCBs, but also for DDT.

High sediment concentrations of organic compounds were reported in some coastal embayments, especially New Haven Harbor (Varekamp et al., Chap. 5, in this volume). NOAA National Status & Trends Program data identify the Housatonic River, Mamaroneck Harbor, and Throgs Neck as areas of high (>85 % percentile of national levels) PCBs and PAHs. Mitch and Anisfeld (2010) found those same harbors to exceed the 85 % percentile for almost all pesticides in sediments, and Yang et al. (2007) further reported that pesticides had not consistently diminished in concentration in the past two decades. Despite these observations, the data were insufficient to detect significant differences between nearshore and open water sites for LIS. This may be due to sample distribution, especially because the preponderance of the USEPA National Coastal Assessment (NCA) sites, and thus analytical data, are from open waters of WLIS; no sediment core data similar to studies of Varekamp et al. (2005) for metals are available for organic contaminants.

Polychlorinated biphenyls are a continuing concern because they bioaccumulate in fish tissues, particularly larger, predatory species such as bluefish and striped bass (Varekamp et al., Chap. 5, in this volume). These migratory species can accumulate PCBs from sources throughout their range, particularly the Hudson River and LIS. As a result, Connecticut and New York have issued consumption advisories for those species taken from LIS, although their respective advisories vary as a result of different interpretations of health risk. Both states continue to remediate active sources of PCBs and act to minimize redistribution of contaminated sediments dredged from harbors. Nevertheless, the pool of PCBs in LIS and regional locations that contributes to tissue contamination is still large enough to be of concern for human health. A recent survey of fish tissues from target species in LIS, including bluefish and striped bass, however, shows reductions in the levels of PCBs in both species (Skinner et al. 2009). As PCBs no longer are manufactured in the United States, burial and decay of PCBs are expected to result in a continuation of the downward trend.

7.4.3 Toxicity of Contaminants in Sediments

The expansive list of trace metals and organic compounds has overwhelmed regulators' ability to test for toxic effects of individual pollutants. The USEPA has maintained a list of fewer than 130 "priority pollutants" that provide insight into classes of pollutants that are expected to be toxic to humans, fish, and wildlife if threshold concentrations and exposures are exceeded (Copeland 1993). However, hundreds of additional chemical compounds, both inorganic and organic, are regulated under the Resource Conservation and Recovery Act (RCRA) and the Comprehensive Environmental Response, Compensation, and Liability Act (CERCLA, commonly known as Superfund) as "toxic pollutants" in a risk-assessment framework (USEPA 2008; 2011b). The LISS has focused assessment and management primarily on legacy contaminants and contaminants that exceed levels of concern in sediments and/or tissues of living organisms, especially those that might be consumed by humans and present a public health risk, as described in Varekamp et al. (Chap. 5).

A few national surveys have been conducted that test sediment toxicity in LIS and several of its embayments, summarized by Varekamp et al. (Chap. 5, in this volume). Primary sources of data include the NOAA National Status & Trends (NS&T) Program (O'Connor et al. 2006; Kimbrough et al. 2009) and the USEPA NCA (USEPA 2004; 2007; 2012). While "effects response levels" for bulk chemical analyses have been used to determine potential toxic impact levels on ecosystems from sediment burdens (see Figs. 5.18 and 5.19 in Chap. 5, this volume), whole sediment bioassays are used more commonly to test general toxicity. These tests more realistically assess the toxic impacts on living organisms compared to effects response levels derived from bulk chemical analyses. Tests are conducted in a controlled laboratory setting and assess the effects of the "mix" of chemicals that may reside in complex depositional areas of LIS under ambient physical and chemical conditions directly on sensitive marine species (See Sect. 5.6.2, in this volume).

The earliest toxicity surveys of consequence were NOAA's NS&T studies, which employed a variety of organism sediment assays in the main body of LIS and its embayments, reported by Wolfe et al. (1991). The use of an amphipod, *Ampelisca abdita*, under a 10-day exposure regime provided a relative indication of toxicity from surveyed sites. In general, embayments yielded the highest level of toxicity, with 80 % of those sites exhibiting sediment toxicity (see Fig. 5.26 in Chap. 5, in this volume); open water LIS sediments showed relatively little toxicity. More recently (2000–2006), work using *A. abdita* tests in the USEPA NCA showed that sediment toxicity still persists in many embayments, though there are indications that levels of toxicity are declining compared to earlier surveys (see Fig. 5.27 in Chap. 5, in this volume).

7.4.4 Nutrients

In many ways, nutrient enrichment has redefined estuarine pollution biology and management for much of the nation in the last 30 years (Bricker et al. 1999, 2007). Long Island Sound is no exception, with the impacts of N enrichment garnering the majority of LISS research, monitoring, and management dollars. Seasonal hypoxia, a symptom of cultural eutrophication, has been the focus of research since surveys in the mid-1980s showed it to be a predominant problem (Welsh and Eller 1991; Parker and O'Reilly 1991) affecting large portions of WLIS and into CLIS each year (see Fig. 5.2a, Chap. 5, in this volume). Subsequent research and analysis has shown more expansive effects of nutrient enrichment, often interacting with other chemical, physical, and biological changes.

Nutrient fluxes have been the subject of intensive monitoring and study to support the most effective mix of management options that would meet state water quality standards for DO. Nutrient inputs are variable over time and space and subject to differences in delivery efficiency to LIS because of attenuation processes during riverine transport and in the Sound itself. Therefore, long-term monitoring data analyses conducted by the USGS and others reported in Chap. 5 (Varekamp et al. this volume) have been extremely valuable. Mullaney et al. (2002, 2009) reviewed N loads from major tributaries discharging into LIS from 1974 to 2008 to characterize nutrient loads and identify trends. Significant downward trends in total N loading were observed at seven of 11 sites, with total loads from the seven sites falling from 21 to 15 Mkg/yr, more than a 25 % decline (see Table 5.9, Chap. 5, in this volume). Control of N from WWTFs in both Connecticut and New York has yielded substantial N reductions, with the discharge from both states combined falling from 31 to 22 Mkg/yr. The Chap. 5 analysis also shows the relationship between land cover and total N yields, which ranged from 3.0 kg/ha-yr in the least developed watersheds to 21.4 kg/ha-yr in watersheds with dense development and WWTFs contributing N.

Groundwater contributions of N can be a perplexing problem especially in the sandy, porous soils of Long Island, which lacks the major riverine delivery routes that exist to the north of LIS, and in some CT embayments (e.g., Niantic Bay). The concentrations of N in the groundwater aquifers of Long Island increased between 1987 and 2005 (Suffolk County 2010), mainly due to increased development and contributions from on-site wastewater treatment systems (e.g., cesspools and septic systems). This points out that, despite management efforts, continuing population increases contributing to nonpoint sources of N have increased loading. This concern is diminished somewhat by the relatively small area of the LIS watershed on Long Island, which contributes a total N load of less than 1 Mkg/yr (Scorca and Monti 2001)—less than 2 % of the total N load delivered to LIS by combined sources in Connecticut and New York (CTDEP and NYSDEC 2000). Although a relatively small load, it is still of management interest and subject to a load allocation to meet the TMDL requirements. It is also indicative of the challenge of N contributions from on-site wastewater treatment systems throughout the watershed. Varekamp et al. (Chap. 5, in this volume) point out that further research is needed to understand the relative importance of groundwater sources of N, their origin (e.g., contributions of subsurface disposal compared to fertilizers), and the transport mechanisms in ground water, including travel times and attenuation from source to delivery to LIS and its embayments.

Chapter 5 also provides a brief look at the distribution of nutrients in the water column (Varekamp et al. in this volume) and its relationship to features of primary productivity in the Sound. A strong east–west gradient is apparent, with highest N concentrations in the far WLIS. Seasonal cycles also are readily apparent, as inorganic N builds during the fall and early winter, and then declines rapidly in late winter as the "spring" bloom of phytoplankton occurs, also apparent in chlorophyll a peaks at that time (Lopez et al., Chap. 6, in this volume). Generally, N becomes limiting after the spring bloom, and remains so during much of the summer. Research by Capriulo et al. (2002) and Gobler et al. (2006) supports N limitation observations and further demonstrate that NO_3/NH_3 ratios can determine dominant phytoplankton groups by preference for one N species over another.

In addition to the short-term trends, Varekamp et al. (Chap. 5, in this volume) evaluated the longer historic record of primary productivity in the Sound using sediment core data. Using organic carbon and N concentrations, organic carbon accumulation rates were estimated as a proxy for nutrient loading. The analysis was able to discern changes over time that could be related to landscape changes, such as deforestation as early as the 17th century, and primary productivity increases in LIS through about 1850. The picture becomes more complicated at that point (See Fig. 5.45, Chap. 5, in this volume). Productivity apparently decreased between 1850 and 1900, followed by an increase until 1950, then a decrease to the relatively low values of today. Recent sediment conditions reflect changes in labile carbon from terrestrial sources versus marine carbon sources and changes in how nutrients from land runoff and WWTFs affect the levels and type of organic carbon generation.

7.4.5 Pollutant Management

With this complexity of cause and effect related to pollutant release into the environment, it is often difficult to identify problem pollutants and their interactions, and institute remedial actions before the damage is done. While nutrients are a natural and necessary ingredient in a healthy LIS ecosystem, enrichment of nutrient levels can alter community structure, favoring species adapted to those conditions. Legacy pollutants such as PCBs are a good example of the damage a single type of pollutant can cause. More difficult are some of the emerging contaminants discussed briefly by Varekamp et al. (Chap. 5, in this volume). Polybrominated diethyl ethers (PBDEs), used as fire retardants in clothing and other everyday products, are persistent, accumulate in living tissues, and are known to have effects on living organisms, although the risk to humans is somewhat unclear. High levels were reported in mussel tissues in the LIS estuary (Varekamp et al. Chap. 5, in this volume). Because of structural similarities to PCBs, a known carcinogen, bans—sometimes for various forms—have been instituted in many nations.

Endocrine disrupting compounds (EDCs) have come under increased scrutiny for likely effects on estuarine fishes and invertebrates, particularly related to reproductive success of some species. A wide range of substances has potential to disrupt endocrine systems of organisms, with consequences including changes in sex ratios and general decline in population health and viability (Duffy et al.

2009; Laufer and Baclaski 2012). Broad classes of substances can act as EDCs and mimic hormones in estuarine organisms. These include pharmaceuticals, personal care products, detergents, pesticides, and even some metals.

Varekamp et al. (Chap. 5, in this volume) conclude their analysis with the summary statement, "Long Island Sound represents an estuarine system significantly impacted by anthropogenic activities." Although acknowledging that water quality has improved in the past 30 years following better regulation and management of pollutant sources, they note there still is cause for concern, especially with legacy pollutants, emerging pollutants of concern, nutrients, and climate change. They highlight nutrients as a "pervasive and disrupting" problem—one clearly associated with human presence, lifestyle, and economy, with cultural eutrophication showing few signs of improvement and at greater risk from a changing climate. Continued research and monitoring is needed to better understand and more effectively manage these issues. Nutrient and hypoxia monitoring should be continued with added attention to changes in pH, T, and system processes caused by climate change. Metals and organic compounds in sediments and the biota need to be monitored to track their fate and ensure that public health is protected while consuming seafood. Like nutrients, the dynamics of toxic pollutants may change, including release from sediments as T and pH vary. Finally, Varekamp et al. (Chap. 5, in this volume) call for monitoring plans and research to address emerging contaminants.

An important issue not treated within Chap. 5 is contamination of LIS by human pathogens, primarily from untreated or inadequately treated human sewage or from wild and domestic animal wastes. This is attributable to the fact that the obstacles to reducing pathogen impairments are not primarily related to gaps in estuarine science, and, therefore, generally have not been the focus of coastal research institutions. As described in the LISS's Comprehensive Conservation and Management Plan (1994), because of the public health implications to humans, management programs target source reductions to improve water quality, while minimizing human exposure through monitoring, bathing beach closures, and shellfish harvesting restrictions. Achieving source reductions requires continuing investments to abate CSOs, address failing on-site wastewater treatment systems, and control storm water. Recent successes include the completion in 2011 of a Sound-wide no-discharge zone for boater waste with the addition of portions of New York's waters to the Connecticut ban. Further recommendations for managing pathogens to LIS are included in the broader recommendations in Sect. 7.7.

7.5 Biology and Ecology

7.5.1 Historical and Current Conditions

Historic habitat loss and overfishing exert structural and functional changes in coastal ecosystems that reverberate into the present (Jackson et al. 2001). But understanding the extent of those changes and setting targets for protection and restoration of biological communities are often hindered by a lack of data on historical abundances and understanding of ecosystem processes. This certainly is true for LIS. Lopez et al. (Chap. 6, in this volume) note that the current extent of many habitats can be quantified, but what has been lost and how much of that loss can be recovered is largely unknown or undocumented. To some extent, ecosystems and environmental conditions can be reconstructed through core studies, but this is not an option for littoral communities and seaweeds, for which there has been little monitoring of species type, distribution, or abundance. It applies even to comparatively well-studied habitats such as tidal wetlands and seagrasses, which have been subject to regulation through legislation and have been targeted for restoration, or to well-studied sites, such as around power plants, that do not encompass the full suite of ecosystem components or stressors.

Tidal wetlands loss along the Connecticut coast estimated from 1880s Coast and Geodetic Surveys was 30 % prior to legislated protections (Dreyer and Niering 1995). Losses vary along the coast, with more than 60 % lost along the more densely developed western part of the Connecticut shoreline. Although no estimates exist for the loss of tidal wetlands along the entire Sound shoreline of New York, it is likely that similarly high losses occurred along the western portion. Tidal marshes that weren't filled or dredged were ditched for mosquito control, which altered hydrology and modified the marsh plant and animal communities. The loss of seagrass communities has not been quantified, although it is clear from historic accounts that eelgrass (*Zostera marina*) was common throughout LIS prior to the 1931 die-off from a fungal infection referred to as wasting disease. Eelgrass recovery was limited to the eastern portions of LIS (see Fig. 6.7 in Chap. 6, in this volume). Comparing the habitat requirements of eelgrass with current water quality confirms that WLIS will not support eelgrass and that conditions in CLIS are marginal at best. Recent surveys supported by the LISS indicate that the majority of eelgrass is found east of Rocky Neck State Park in Connecticut and around Orient Point and Fishers Island in New York. Conducted in 2002, 2006, and 2009, these surveys need to be continued to document the spatial and temporal variability of eelgrass in LIS and to better understand responses to water quality conditions.

Our understanding of the plankton in LIS is greatly aided by comparisons with the pioneering work of Riley (1941, 1956) and others. Importantly, Lopez et al. (Chap. 6, in this volume) point out that nutrient and chlorophyll *a* levels have not changed in the open Sound over the past 60 years. The overall pattern is of a decreasing gradient from west to east in concentrations of N, chlorophyll *a*, and zooplankton. This trend, however, is dominated by comparisons between WLIS and ELIS, with concentrations variable in much of CLIS.

Pioneering research in LIS also has improved understanding of sediment biogeochemistry in deep-water estuarine habitat and the successional dynamics of benthic communities in response to disturbance (Rhoads et al. 1978; Rhoads and Germano 1982). Lopez et al. (Chap. 6, in this volume) provide an overview of the characteristics of the soft-sediment communities in LIS, noting general west to east gradients, with variations at smaller scales. Species richness is relatively low in the WLIS and CLIS basins, starts to increase in the eastern portion of the CLIS basin, and increases sharply east of the Connecticut River into ELIS (See Fig. 6.25 in Chap. 6, in this volume). Despite a number of Sound-wide surveys that provide a general understanding of the composition and structure of the seafloor communities of LIS (Pellegrino and Hubbard 1983; Reid et al. 1979; Schimmel et al. 1999), the temporal and spatial dynamics of benthic communities, both seasonally and inter-annually, and their response to environmental stressors (particularly in poorly characterized embayments) are not well understood. A more thorough understanding of the benthic communities in LIS will facilitate interpretation of temporal changes and assignment of causation to human and natural stressors, including climate change and ocean acidification.

Dredged sediment was historically disposed at many sites in LIS, commonly located just outside harbors adjacent to the dredged channel. Today disposal is confined to four open water sites, which have been evaluated systematically, confirming that sediment disposal has immediate effects upon sessile epifauna and infauna. There do not appear to be long-term ecological effects, however, resulting from the disposal activities (Germano et al. 2011). Although dredged sediment disposal can modify habitat by altering grain-size composition and sediment-transport conditions by changing seabed elevation, there do not appear to be general changes in benthic processes or habitat at the disposal sites.

Foraminifera shells preserved in sediments have been used as proxies to reconstruct environmental changes over long time scales. *Ammonia* spp. and *Elphidium* spp. are the most common foraminifera genera in LIS, and higher ratios in *Ammonia* to *Elphidium* abundance have been interpreted as indicators of eutrophication (see Fig. 6.39 in Chap. 6, in this volume). Varekamp et al. (2010) have documented an increase in the ratio of *Ammonia* to *Elphidium* abundance, but they cite a number of possible causes including decreases in diatom productivity, an increase in an invasive cryptogenic species of *Ammonia*, or different hypoxic tolerances.

Of all the historic changes in the LIS ecosystem, perhaps those associated with declines in fishery and wildlife resources resonate the most with the public. Both Weigold and Pillsbury (Chap. 1, in this volume) and Lopez et al. (Chap. 6, in the volume) recount the historic changes driven in large part by overharvesting, habitat loss, and pollution. Diadromous fish populations suffered first from habitat loss and pollution, with current Atlantic salmon (*Salmo salar*) and American shad (*Alosa sapidissima*) runs a fraction of historic numbers along with declines in other herring species. Menhaden were harvested intensively all along the Atlantic coast. Oyster reefs in LIS were exploited with little thought of sustainability. The decline in these species has altered the energy flow in food webs, with likely cascading effects upon coastal ecosystem function (Nuttall et al. 2011).

The eastern oyster is of particular importance, providing multiple services as a habitat forming species in addition to its value as a fishery. Aggregates of shell topped by living oysters provide habitat for associated species, can lessen shoreline erosion by reducing wave energy, and biofilter the water column (Peterson et al. 2003; Piazza et al. 2005; Dame et al. 1984). Beck et al. (2011) estimate oyster reefs have declined worldwide by 85 %, with a lesser loss in the United States. In a review of changes in oyster habitat in 24 estuaries across the United States, Zu Emergassen et al. (2012) estimate a loss of 64 %, with greater declines in biomass. Data on the current extent of oyster habitat in LIS were not available, preventing a comparison with historic estimates. While the extensive areas leased for oyster aquaculture in LIS increase the area and biomass of oysters beyond the limits of the historical natural beds, it is still likely that both the number and function of oysters in LIS are diminished compared to historic levels.

The geographic location and orientation of LIS is such that it hosts an interesting and unique assemblage of both cold- and warm-T tolerant species (Lopez et al., Chap. 6, in this volume). Some of the Arctic-cold-to-temperate-Atlantic assemblage includes species that originated in the Pacific and traveled to the Atlantic during paleo-migration. During the summer, many warm-temperate Atlantic species and some subtropical species appear, some of which may be endemic while others travel north via the Gulf Stream. Historic commercial and recreational catch records provide insight into the changes in populations over the centuries, and contemporary trends in finfish populations over more than the past two decades are aided by trawl surveys conducted by the Connecticut Department of Energy and Environmental Protection (CTDEEP) (Gottschall and Pacileo 2010). Since 1984, the overall biomass of finfishes in LIS has been relatively stable, but changes have occurred in individual types of fishes and likely in multi-species groups representing different guilds (Howell and Auster 2012). This is seen in the decrease in the overall abundance of fishes caught in the spring survey (driven by declines in epibenthic species such as winter flounder (*Pseudopleuronectes americanus*) and windowpane flounder (*Scophthalmus aquosus*), despite an increase in demersal species such as butterfish (*Peprilus triancanthus*) and scup (*Stenotomus chrysops*)) and an increase in the fall (particularly scup, butterfish, and weakfish (*Cynoscion regalis*)). Cold-adapted species have declined in abundance, particularly in spring; while warm-adapted have increased (see Fig. 6.49 in Chap. 6, in this volume). Section 7.6 of this chapter discusses the possible implications of climate on these trends.

Invasive species have also exerted a strong influence on the diversity and distribution of flora and fauna in LIS, dating back to the European exploration of the Americas. Some common species considered native are in fact introduced, including the common periwinkle (*Littorina littorea*) and a common red alga (*Neosiphonia harveyi*). Despite the descriptions of the modern biota of LIS by Weiss (1995) and Weiss et al. (1995), and by Schneider et al. (1979) for marine algae, defining the extent of past invasions is hampered by the lack of a scholarly atlas of the historical biota of LIS upon which to make comparisons. As a result, the number of non-native species in LIS, particularly for microscopic, cryptic, or invertebrate taxa, is speculative. Increased globalization, however, has accelerated species introductions, some of which become invasive, thus altering the ecology of LIS. Temperature increases enhance the survival of some introduced species and extend the range into LIS of other species.

7.5.2 Cross-Cutting Issues

Lopez et al. (Chap. 6, in this volume) review the interactions among physical, geochemical, and biological processes contributing to hypoxia in LIS. Clearly, inorganic nutrient loadings, primarily from wastewater discharges and land runoff, and photosynthetic production fuel the biological respiration that drives the system to hypoxia. Planktonic community respiration dominates benthic oxygen demand and respiration rates can deplete DO in days if not for physical ventilation. Despite a general understanding, work remains to more fully elucidate the mechanisms linking nutrient loading, physical forces, and the biology of the system.

The fate of primary production and linkages to hypoxia need to be better understood; there are uncertainties about the sinking and horizontal transport of primary production biomass, and imbalances between the sources and sinks of estimated carbon budgets in WLIS belie an incomplete understanding of ecosystem function. The Systemwide Eutrophication Model (SWEM) developed to support hypoxia management currently underestimates rates of both production and respiration, and improving its formulation is a priority for future model refinement and application. Remineralization of organic carbon and nitrification of ammonium from sewage treatment plant discharges can contribute to hypoxia, but direct measurements using ^{15}N tracer and other methods are needed to assess their importance. Future modeling should also consider the consequences of changes in climate affecting the timing and fate of primary production upon oxygen dynamics.

In general, food web dynamics of LIS are relatively poorly known. Ecosystem modeling (e.g., Hakanson and Boulion 2003) would help relate trends in harvest and survey abundances to the biology of natural resources. The mechanisms involving important variables such as loss of keystone species, fishing pressure, T, and habitat alteration need to be better understood. Likewise, there is a need to better understand trophic linkages between primary production and apex predators. And perhaps surprisingly, considering much of the foundational work conducted in LIS, measurements of biomass and productivity of subtidal macrophytes and deep-water benthos for incorporation into food web models are sparse.

Expanded monitoring to assess critical habitats and biological communities is needed to develop a more holistic understanding of how the ecosystem will respond to continued nutrient reductions, invasive species, and climate change. Recent efforts by the LISS to increase monitoring of variables that provide sentinel insight to changes in biological communities, particularly from climate-driven changes, may be helpful in this regard. Continued and strengthened monitoring of key habitats and biological communities combined with carefully designed observations and experiments are needed to support models that will integrate our understanding of LIS.

7.6 Climate Changes Affecting LIS

Changes in the annual march of the water column seasons impact the biological processes of the Sound and hence potentially the entire ecosystem. Some organisms may adapt to these changes, but others may not. Growth and reproduction may be altered even if organisms can survive seasonal changes in T and S. Mobile organisms may alter their migration patterns and may be excluded from important food sources or spawning areas. Thus, simply altering the extremes of T and S or the duration of their seasonality can have significant impacts on many organisms.

Organisms living at the extremes of their geographical range may be particularly susceptible to stress from disease, hypoxia, or changes in physical conditions such as acidification. For example, seaweed populations may shift from the cool-temperate kelp (e.g., *Saccharina* spp.) and rockweed (*Fucus* spp.) species to more warm-temperate red algae (e.g., the red weed *Gracilaria tikvahiae*) or brown seaweed (e.g., *Sargassum* spp.). The impact of climate change, ocean acidification, and introduced species upon the Sound's native flora is often anecdotal and points to the need for long-term monitoring in LIS.

As discussed by Lopez et al. (Chap. 6, in this volume), the American lobster population in LIS, living at the limit of its southern range, was decimated in 1999 (Pearce and Balcom 2005). Exceptionally high bottom water temperatures in the deep regions of the Sound seem to have played a significant role in the mortality, perhaps even triggering it. In that year, the bottom water T abruptly increased about 2 °C at station D3 (off shore from Norwalk, CT) from August 4 to September 1 (Wilson and Swanson 2005). Much of this increase was associated with two water column destratifying events that mixed warm surface water to the bottom. The wind event of August 29, when a cold front passed over the area and the water column became isothermal, was the more important of the two. This type of environmental event could become more common as the ecosystem adjusts to a rapidly changing physical oceanographic setting.

Long Island Sound has been experiencing changes continuously in climate that affect its physical functioning as well as its ecosystem. If the rate of change increases rapidly, what were subtle changes in ecosystem reactions could now stress the ability of natural and human systems to adapt. For example, the Union of Concerned Scientists (2007) report that Northeast US air temperatures have been rising about 0.3 °C every ten years since 1970 and 0.7 °C in winter over the same period. They project that in the next several decades temperatures are likely to increase an additional 0.8 °C in summer and 1.9 °C in winter.

Perhaps one of the most important observations of the physical oceanography of LIS by O'Donnell et al. (Chap. 3, in this volume) is that the forces driving its physical functioning are quite variable; consequently the responses to these forces fluctuate considerably as well. Despite the variability, there are some generalities that can be made about the behavior of the Sound and what might occur in the context of climate change. For example, in recent research examining recurring hypoxia in LIS, Wilson et al. (2008) showed that in WLIS there was roughly a 1.5 °C change in the difference between summer-averaged surface and bottom water temperatures (ΔT) over the period 1946–2006 (Fig. 7.3). Most of that difference occurred because bottom waters are cooling. Over the same period, the summer-averaged bottom DO concentrations declined about 2 mg O_2/L, making hypoxic conditions more prevalent. Cooler bottom waters are a consequence of increased periods of water column stratification, partly in response to a long-term change in the directionality of regional winds. Over the period of investigation, the

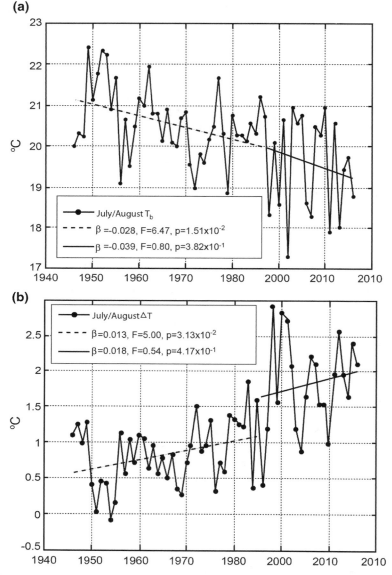

Fig. 7.3 *Top* Time series for bottom T at NYC DEP E10 averaged over July and August. *Bottom* Time series for surface-bottom T at NYC DEP E10 averaged over July and August

direction of the summertime wind field shifted about 30° from 245 °T (degrees True) to 215 °T—the latter being more conducive to maximizing water column stratification. It is not clear whether this change in wind direction is related to climate change or the increasing heat island effect caused by continuing development throughout the area. The observed colder bottom water temperatures occur because stratification is established earlier in the year and water column mixing is reduced. Thus, vernal warming of bottom waters does not take place as far into the spring as it did a few decades ago.

It is instructive to look at bottom water temperatures in the Sound—particularly during winter. As part of an analysis related to the 1999 lobster mortalities, the evolution of Sound-wide temperatures from 1991 to 2003 was examined for depths <20 m and those >20 m. The variability in winter minimum temperatures is noteworthy: the range over the 12 years was about 3 °C for depths <20 m and 4 °C for depths >20 m. The variability in summer maxima was somewhat less, being more on the order of 2–2.3 °C. As a result, changes in winter water column temperatures attributable to climate change may be as great as or greater than the often-discussed increases in summer maximum temperatures. Any significant long-term changes could not only alter the seasonal development of water column structure but also have a pronounced impact on survivability of organisms in the Sound.

Besides noting a change in the directionality of the winds, it is useful to examine wind speed. Robert Chant has plotted the north–south and east–west components of the daily average wind for the years 1947–1977 and 1977–2007 (Fig. 7.4). For both periods, the minimum in the average daily wind speed in the east–west direction (roughly aligning with the major axis of the Sound) occurs around day 130 in the calendar. But during the period 1977–2007, that minimum speed approached 0 m/s, about 1 m/s less than the minimum for the period 1947–1977. The minimum wind speed of 1 m/s for the period 1947–1977 is reached about 30 days earlier in the year for the period 1977–2007. Thus, the speed of the east–west component of the daily average wind speed is now reduced relative to the past and the historic minimum value is nearly a month earlier. The north–south component of mean daily wind speed is also about 0.5 m/s less between days 100–135 from 1977–2007 than it was over the period of 1947–1977. The vigor with which LIS surface waters are mixed to depth in spring is reduced, and this minimum wind value occurs earlier in the year than in the past.

Our local climate, as annually summarized for Central Park by *The New York Times*, has experienced about a 13 % increase in average annual precipitation over the last 20 years and perhaps as much as 20 % over 40 years. That increase is rather evenly distributed throughout the year. However, the big impact may be the form in which that precipitation is occurring (e.g., rain vs. snow). The Hudson River discharge has annually peaked in April or later about 79 % of the time over the 33 year period from 1946 through 1978. But for the last half of the record through 2010, this occurred only 61 % of the time; 39 % of the peak flows were in March or earlier (US Geological Survey 2011).

The pattern for the Hudson River is consistent with broader regional patterns. Hodgkins et al. (2003) analyzed the 66 yearly records of stream flow in 27 unregulated rivers in New England. They found that for rivers dominated by snowmelt the timing of the winter-spring center of volume advanced 1–2 weeks earlier in the year. O'Donnell et al. (2010b) have found that winter-spring center of volume of the Connecticut River is also occurring earlier in the calendar year at a rate of 9 ± 2 days

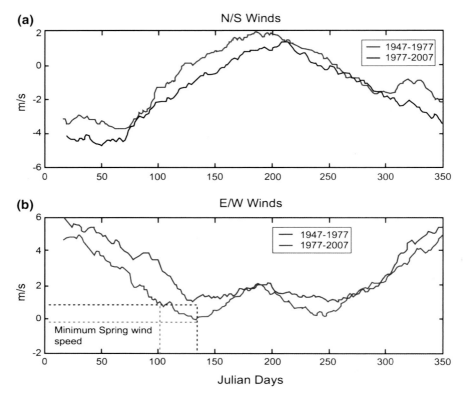

Fig. 7.4 The north–south and east–west components of wind speed (m/s over the calendar year (0–350 days) for the periods of 1947–1977 and 1977–2007 at LaGuardia Airport. *Source* Robert Chant, Institute of Marine and Coastal Sciences, Rutgers University

per 100 years, though the number of dams makes interpretations of causes more difficult. The Union of Concerned Scientists (2007), in a review of predicted climate changes in the State of New York, observes that snow is now wetter than in the past and opines that the snow season could be "cut roughly in half this century."

Water column stratification in the Sound typically commences when the spring riverine freshet contributes fresh water to the system followed by vernal warming. Thus, haline stratification may be initiated earlier now compared to the past due to the earlier freshet. All signals point to climate change contributing measurably to earlier water column stratification and therefore a longer period over which bottom waters are isolated from mixing with surface waters. The period of mixing of DO to depth from wind stress in LIS is reduced. As a consequence, there is now a longer period during the year that DO drawdown can take place and hypoxic conditions can be established. In the future, one might expect that the spring freshet will come still earlier or maybe disappear altogether if precipitation stored as snow declines. Either way, this will have a pronounced impact on water column structure in the Sound.

Changes in the timing of the spring freshet will also impact the delivery of allochthonous carbon and suspended sediment to LIS. For example, the organic carbon and sediment load of the Hudson River originates primarily from the nontidal portion of the Hudson (above the dam at Troy, NY) (Limburg et al. 1986). According to Wall et al. (2008), most of the suspended sediment load is transported through the lower Hudson from November through April—much of it in November and December and then again with the freshet in March and April. If the distribution of flow from rivers discharging to the Sound is altered as a consequence of climate change, one can expect that the discharge of organic/suspended sediment will likewise be modified. Water column transparency will change and conceivably primary production in the Sound as well.

It is revealing that the Hudson River, with its high anthropogenic N loading, is only "moderately susceptible" to eutrophication (Howarth et al. 2000) while the river flow is critical to that condition in LIS. Howarth et al. (2000) point out that the Hudson has a short residence time (days) for water and dissolved material, and that the water column is light-limited, thus inhibiting primary production. The Hudson River not only contributes to haline stratification in WLIS, but also controls gravitational circulation through variations in the river's discharge. Hao and Wilson (2007) ascertained that the residence time of WLIS was on the order of months, and that period depended upon the degree of water column stratification and the strength of gravitational circulation. Large freshwater discharge from the Hudson increases the latter, reducing residence time, but increasing haline stratification and decreasing mixing.

The Connecticut River is the largest source of fresh water to LIS and is important in freshening its mouth at the eastern end (O'Donnell et al., Chap. 3, in this volume). The thin plume of the river is greatly affected by the local tidal currents and during high river discharge an ebbing current can be observed some 20 km to the east, near The Race (O'Donnell et al., Chap. 3, in this volume). Apparently, not as much is known about the transport of carbon and suspended sediment in the Connecticut as compared with the Hudson. Knebel and Poppe (2000) contend that LIS has a "nearly 100 % …trapping efficiency" for fine-grained bottom sediments discharged by the Connecticut River, which are generally advected to the west, settling there. However, the gross transport processes of the Connecticut and how they might change in a warming climate are likely to be similar to those of the Hudson, since the climatic characteristics of their drainage basins are similar. Reductions in the spring freshet will likely cause the salt wedge to move upstream and change the characteristics and distribution of the brackish- and fresh-tidal marshes, which are designated as a wetland of international importance under the Ramsar Convention (an intergovernmental treaty established in Ramsar, Iran in 1971). The reduction in the spring freshet and earlier arrival will likely cause the loss of prolonged spring flood in the upstream fresh-tidal marshes, reducing or eliminating their value as important stopover sites for spring migrating waterfowl.

Climate change also could affect delivery of fresh water to the Sound through changes between precipitation (P) and evaporation (E). Koppelman et al. (1976), using data from 1953–1954, estimated that P minus E for LIS to be about 31 cm where $P:E$ is 1:0.75. Krug et al. (1990) show that the annual precipitation rate in Connecticut is roughly twice that of evapotranspiration. Long Island is much the same (Central Pine Barrens Joint Planning and Policy Commission 1995). However, Kowalsick (2012) estimated monthly evapotranspiration rates from 2006 to 2009 during the growing season for Long Island and found spring and fall rates were one-third of summer rates. Since precipitation rates are fairly uniform throughout the year, the difference between precipitation and evapotranspiration is quite variable. Changes in either with climate change could significantly alter the delivery of fresh water to the Sound.

The Union of Concerned Scientists (2007) postulates that our region will experience stronger and more frequent precipitation events because of global warming. This will undoubtedly translate into an increase in temporary shellfish bed and beach closures. Depending on the frequency of such events, permanent closures might also increase. Changes in cloud cover could have a pronounced effect on eutrophication and ecological functioning in LIS by altering primary productivity and thus influencing eutrophication. In Narragansett Bay, Nixon et al. (2009) associate decreases in annual and summer mean phytoplankton abundance with warming of the water, especially during winter, and to increased cloudiness. The subsequent decline in organic matter deposition to the bay has reduced benthic metabolism.

Sallenger et al. (2012) identified the northeast coast of the United States as experiencing accelerated sea level rise relative to the global mean. The average rate of relative sea level rise as determined from four NOAA tide gauges (New London, Bridgeport, Kings Point/Willets Point, Montauk) around the Sound over the period of 1986–2010 is 4.6 ± 1.4 mm/y (Fig. 7.5). This rate, determined using three-year running averages of annual sea level values at the respective stations, is significant ($p = 0.01$). Thus, over the period of the LISS, relative sea level has risen about 11.5 cm. Climate prediction models suggest that this rate may increase substantially. There is concern that low-lying areas will be flooded and that wetlands will be lost. It is also likely that erosional processes along the Long Island shore will be altered as a consequence of this projected sea level rise.

While it is evident from O'Donnell et al. (Chap. 3, in this volume) that LIS's wave field will not be altered due to sea level rise, its directionality could change as the wind field shifts. Some areas may be more exposed to the influence of waves, others less. The Long Island shoreline is perhaps the most vulnerable as the steep slopes of sand, gravel, and glacial till slump due to undercutting of the toe of these steep bluffs by wind waves generated at the higher sea level. Shepard and Wanless (1971) reported that in the eighteenth and nineteenth centuries prominent headlands were eroded some 150 m in the vicinity of Oak Neck on Long Island. Davies et al. (1973) found a bluff recession rate in the 20th century of some 0.5 m/y with a range of 0–1.6 m/y at 19 locations from Oak Neck Point near Oyster Bay to Orient Point, a distance of some 97 km. This bluff erosion feeds the littoral drift and the beaches of the north shore (Bokuniewicz and Tanski 1983). For example, the bluffs of Nissequogue currently feed Long Beach, causing the spit to prograde to the east-northeast about 1.8 m/y (Swanson and Bowman, in preparation). Bokuniewicz and Tanski (1983), in their sediment budget resulting from bluff erosion, estimate that about 85 % of the eroded

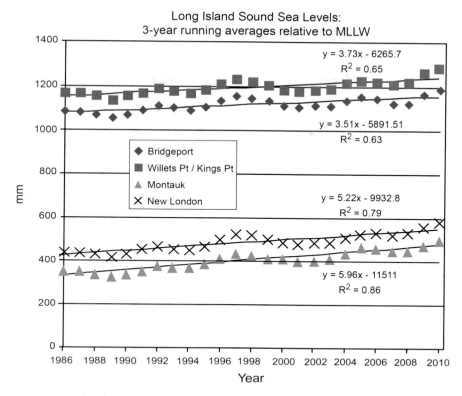

Fig. 7.5 Three-year running averages of mean sea level relative to mean low low water (MLLW) at Bridgeport, Willets Point/Kings Point, Montauk and New London and their respective relative sea level rise rates, 1986–2010

material ends up in the deep waters of the Sound with about 3 % going to wetlands; very little permanently remains on the beaches, hence they are generally receding.

There are additional, less obvious, possible impacts of alterations to the physics of the Sound associated with climate change. For example, sea level rise will cause the water table to rise in coastal areas as a consequence of saltwater intrusion, causing low-lying areas along the shoreline to experience a greater likelihood of flooding than that due to sea level alone. Or as sea level rises, the tombolo that forms Truman Beach on the north fork of Long Island will become a mere shoal, making the physics of the mouth of the Sound even more complex than at present.

The changes in climate that influence the physical oceanography of LIS will reverberate throughout the ecosystem. As a result, the magnitude and time frames over which physical processes operate need to be better understood and accounted for to support management. This is particularly true in the design aspects of coastal development and related infrastructure initiatives. It is also true for diagnosing and responding to LIS's lobster mortality events and the chronic annual hypoxia, since both appear to be closely tied to altered physical drivers that may further change in the coming decades.

7.7 Integrating Science and Management

7.7.1 Management Jurisdiction

Although functionally an estuary, according to a 1985 US Supreme Court decision, LIS is a juridical bay. The court judged Long Island an extension of the mainland, considering physical connections (bridges, tunnels, etc.) and historical and political ties rather than strict oceanographic or geological interpretations of whether or not it is an island (Swanson 1989). The closing line (demarcates inland waters from the 3-nautical mile marginal seas of the states) at the mouth of the Sound, as determined by the court, runs from Montauk Point, NY to Watch Hill Point, RI. This ruling is important because it established the Sound as "internal waters" in a legal and navigational sense, which means that it falls, for the most part, under the authority of the applicable coastal states (Westerman 1987)—Connecticut, New York, and to a limited degree Rhode Island. Thus, the states have primary responsibility for many aspects pertaining to the management of the Sound and will need to take leaderships roles in implementing such initiatives as ecosystem-based management (EBM) and coastal and marine spatial planning (CMSP).

7.7.2 Ecosystem-Based Management

7.7.2.1 Ecosystem-Based Management Concepts

President Obama's Executive Order 13547 (2010) adopted the recommendations of the Interagency Ocean Policy Task Force (Council on Environmental Quality 2010), making ecosystem-based management (EBM) a foundational principle of federal management of US coastal waters and resources. The executive order was preceded nationally by the recommendations of the Pew Oceans Commission (2003) and the US Commission on Ocean Policy (2004), and in New York State by the passage of the New York Ocean and Great Lakes Ecosystem Conservation Act (2006), which adopted EBM policies for the management of New York's marine and Great Lakes ecosystems and established a conservation council to implement them. Although the specific approaches for EBM programs vary, the general tenets for marine systems can be summarized as follows (McLeod et al. 2005):

- Plan on an ecosystem level, considering cumulative impacts and long-term changes.
- Involve multiple stakeholders and integrate the full spectrum of ecosystem services supporting human wants and needs.
- Develop cross-jurisdictional goals through formal agreements among multiple levels of government and the private sector.

- Implement programs through coordinated strategies that recognize shared responsibility across levels of government and the private sector, but delineate clear responsibilities for accountability.
- Incorporate adaptive management that acknowledges uncertainty in our understanding, fosters testing of alternate management approaches, and allows for adjustments based on improved information.
- Establish long-term observation, modeling, and research programs that support adaptive management by improving ecosystem understanding and evaluating the effectiveness of management action.

These concepts provide a framework for both analysis and management that accounts for the complex interrelationships of human society, economics, and the environment. Societal needs and the economic consequences of activities to ecosystem services that society relies upon are vital elements to be integrated into management. Otherwise, EBM is just another form of environmental management, to be treated as a lower priority that can be addressed after tackling more pressing social and economic needs, instead of an integral part of meeting those needs.

How can these principles be best put into practice at the estuary scale? How can science be integrated across disciplines to interpret and forecast changes in coastal systems? What obstacles exist in facilitating cross-jurisdictional planning and permitting (e.g., for energy production or transmission proposals)? How do existing statutory and policy requirements within those jurisdictions support or conflict with these principles? For example, do policies for the development and implementation of pollution total maximum daily loads (TMDLs) under the CWA provide mechanisms for adaptive management? How can estuary-specific restoration efforts complement and benefit from regional fishery management?

We review these questions from the perspective of LIS. For context, we present essential elements of the CWA, providing specific examples of current program policies in synchrony or in conflict with the principles of EBM, along with recommendations for integration within the constraints of current technology limitations, statutes, regulations, and policies. We then recommend a general framework and specific actions to support ongoing management. Our intent is to demonstrate the utility of this analysis for LIS and help inform the debate on how to strengthen coastal ecosystem protection nationally.

7.7.2.2 The Context for Ecosystem-Based Management on the Estuary Scale

Estuarine and coastal systems have been impaired primarily from overharvesting, habitat loss and degradation, and pollution (Lotze et al. 2006). Invasive species and arguably climate change have had a lesser impact to date but may become more influential in the future (Lotze et al. 2006). To address these stressors successfully, EBM must foster integrated and coordinated responses among fisheries, coastal zone, and water quality management programs (Rosenberg and McLeod

2005). Each of these programs, however, operates at different scales, none specifically at the level of the LIS ecosystem. For example, fisheries management will be regional, reflecting the migratory nature of many fish stocks. Coastal zone management for LIS is implemented on a state scale by Connecticut and New York under authority of the Coastal Zone Management Act through NOAA. Water quality management is both state and regional, reflecting a LIS watershed that extends beyond the coastal states of Connecticut, New York, and Rhode Island through portions of Massachusetts, New Hampshire, and Vermont. Each of these programs is conducted under separate federal authorizing statutes and administered by different agencies, primarily NOAA for fisheries and coastal management, and USEPA for water quality. Other federal agencies have significant roles: the US Department of Agriculture for nonpoint source runoff, the USGS for water resource research and monitoring, and the US Fish and Wildlife Service for habitat and species management are just three examples. These federal programs often are delegated or administered jointly with state agencies.

Although created prior to the common use of EBM as a term of reference, the National Estuary Program (NEP) was established in 1987 under Section 320 of the CWA to support comprehensive and inclusive planning with an ecosystem focus. Additional legislation specific to LIS that passed in 1990 further strengthened USEPA's role in coordinating implementation strategies through cross-jurisdictional partnerships. Reflecting its origin within the CWA, however, the NEP legislation directs development of a CCMP that "…recommends priority corrective actions and compliance schedules addressing point and nonpoint sources of pollution to restore and maintain the chemical, physical, and biological integrity of the estuary…," a narrower construct than current EBM approaches would take in considering threats to ecosystem services. In practice, most estuary programs have targeted additional stressors of ecosystem function, including degradation and loss of habitat, and invasive species.

Despite the integrating intentions of the above-mentioned legislation, even a comprehensive program cannot be all-inclusive; consideration of the ecosystem's scale and impairments in relation to the scale and function of existing management programs is needed to set priorities and effectively direct management responses. For example, the LISS, one of 28 NEPs, decided early not to involve itself organizationally with fishery management, recognizing that there are existing multijurisdictional planning programs (the New England and Mid-Atlantic fishery management councils, established under the Magnuson-Stevens Fishery Conservation and Management Act of 1976, and the Atlantic States Marine Fishery Commission, established in 1942) that operate at the appropriate regional scale. Instead, the LISS has focused on water and habitat quality within LIS, which can affect species abundance and diversity. While avoiding duplication of effort, this approach has led to criticism by some that the LISS has not been engaged directly in the restoration of one of the most public of estuarine ecosystem services—abundant and diverse populations of commercially- and recreationally-important species. The lesson is that estuary management programs do not need to tackle all aspects of the ecosystem. An assessment of gaps or limitations to effective EBM of the specific ecosystem services in question is needed before designing an effective program.

Another example of priority-setting for LIS is in the area of water pollution. Although contamination from pathogens and toxic substances initially was considered to cause the major impairments to human use and ecosystem health of LIS, water quality surveys supported by LISS in 1986 and 1987 documented hypoxia as more severe and covering a larger area than previously measured (Welsh and Eller 1991). The need to address nutrient levels on an ecosystem level, considering assimilative capacity, vaulted hypoxia management as a priority over pathogen and toxic substance effects, both of which had already been markedly reduced through national management programs (e.g., technology-based limits for industry, combined sewer overflow remediation, product bans, and wastewater pretreatment requirements). For example, air and water releases of industrial chemicals from sources in the Connecticut and New York portions of the LIS watershed declined by more than 90 % between 1988 and 2008 (LISS 2010). Releases watershed-wide declined by 74 % from 2001 to 2010 (USEPA 2010). Product bans for PCBs in 1979 and for lead in gasoline as part of the 1990 Clean Air Act amendments have led to reductions of these chemicals in the sediments and biota of LIS (Varekamp et al., Chap. 5, in this volume). Controls of phosphorus (P) in response to localized instances of eutrophication of freshwater resources had been in place for decades, but watershed sources of N to the Sound remained without assessment or control. Here, too, an assessment of current environmental conditions and gaps in existing management programs was essential to setting priorities for LIS.

7.7.2.3 Total Maximum Daily Loads and Ecosystem-Based Management

At the heart of EBM is the recognition of human society as an integral component of ecosystems and their functioning. The goal of management is to maximize the services to human society that the ecosystem supports, recognizing that there are conflicts and tradeoffs in the provision of those services. But for managing water quality, federal water quality standards regulation established under the CWA requires states to specify the appropriate water uses to be achieved and protected for recreation and the protection and propagation of wildlife. States then adopt criteria to protect these designated water uses as part of the state water quality standards package. The concepts of ecosystem services as defined under EBM and of designated uses as applied under the CWA bear some resemblance, but they are not interchangeable, and for water quality management of LIS, the latter has primacy. States can designate new uses, or create subcategories of existing uses, that require less-stringent criteria, but only through a formal assessment called a use attainability analysis. A detailed review of water quality regulations under the CWA is beyond the scope of this discussion; however, the key point is that these regulations are not antiquated notions of how to manage water quality. These statutes and resulting regulations are the legal, enforceable framework in the United States. For water quality management, the concepts of EBM must be adapted to the CWA (either as it currently stands or as modified through legislation).

The application of adaptive management, another key tenet of EBM, to water quality protection also is instructive. By explicitly acknowledging uncertainty in our understanding of both the natural and social sciences relating to estuarine management, adaptive management ideally fosters integration of science into management through the testing, measurement, and assessment of alternate management approaches. Adaptive management is not organic to many of the implementing statutes under which water quality management is conducted. Perhaps most relevant to LIS are the policies regarding numeric criteria and the development and implementation of TMDLs for pollutants that impair water quality. These policies and regulatory guidelines can constrain the flexibility necessary for effective adaptive management (National Research Council 2011). Before undertaking the effort needed for successful application of adaptive management, consideration of its benefits is warranted. Adaptive management within a TMDL framework may not be justified when the degree of uncertainty is low or when the costs of errors resulting from that uncertainty are acceptable (Shabman et al. 2007).

Total maximum daily loads require precision in identifying allocations to attain water quality standards, but the science upon which such calculations are based, let alone how comprehensively the science is applied, is often filled with uncertainty. The TMDL program considers uncertainty in specifically allocating loading to a Margin of Safety (MOS), which is meant to set allocations conservatively to attain water quality standards. As a result, the TMDL framework is at odds immediately with EBM's emphasis on acknowledging uncertainty, as well as adaptive management protocols that address that uncertainty. The MOS concept was developed primarily with toxic contaminants in mind to account for the uncertainty of synergistic effects and bioaccumulation or biomagnification in living organisms. It can be argued that the MOS for nutrients extends both above and below a target concentration or load based upon weight-of-evidence determinations set to attain a trophic condition, or secondary indicator (e.g., DO), that has a range of outcomes to nutrient control that can be quite broad. "The lower the better" paradigm for xenobiotics does not necessarily apply to nutrients. For example, excessive nutrient control could change nutrient balances, and thus the limiting nutrient in LIS, which could have major consequences for ecosystem structure and function.

To reconcile precision and uncertainty, USEPA TMDL guidance wisely allows for "phased TMDLs" in situations wherein limited existing data are used to develop a TMDL and wherein the use of additional data would likely increase the accuracy of the TMDL load calculation. In such cases the TMDL, in addition to allocating and implementing controls to attain water quality standards, can include schedules for additional data collection and analysis to refine the TMDL in a phased approach. A phased TMDL may include activities to test and evaluate alternate management approaches, but it still must set allocations among pollution sources, with a reasonable assurance of being implemented and with a MOS, necessary to achieve water quality standards. Hence, there remains some tension between the requirement for specificity in setting firm allocations among sources in TMDLs, a departure from which would require USEPA approval, and an adaptive implementation approach that might over time identify alternate approaches to meeting water quality standards, or bring new information to bear on their attainability. There remains a need for clearer guidance on provisions for flexible allocations, particularly adjustments between wasteload (permitted point sources) and load allocations (unpermitted nonpoint sources). Ideally, the uncertainty in predicted outcomes would be acknowledged in TMDLs and in the scope and pace of investments (e.g., advanced sewage treatment) to achieve those outcomes.

Increased data collection and analysis (including research into ecosystem structure and function) would likely improve the accuracy of calculations of most TMDLs, but they are particularly relevant to the management of eutrophication in coastal waters. As the science of coastal eutrophication still is evolving (Cloern 2001), with the relationship between nutrient loading and adverse impacts in estuaries being complex and often nonlinear (Kemp et al. 2009), often it is difficult to predict how the water body will react to load reductions. Regime shifts—changes in food webs and nutrient processing—can alter the assimilative capacity of estuaries to nutrient loading, and the trajectory and rate of response to controls (Duarte et al. 2009). The drastic decline in bivalve populations or beds of submerged aquatic vegetation in many coastal systems, resulting in a loss of ecosystem services related to nutrient uptake, processing, and denitrification, may in effect move the goal posts for restoration efforts (Newell 2004; Kemp et al. 2005; Testa and Kemp 2012).

Predicting the response of coastal systems to local forcing (e.g., nutrient reductions) is complicated further by larger-scale influences driven by climate change. Fulweiler et al. (2010) observed declines in sediment oxygen demand and dissolved inorganic N fluxes in the upper portion of Narragansett Bay related to decreases in primary production and shifts in bloom phenology caused not by local nutrient management but by climate change. Management decisions that do not account for these influences will likely not universally achieve the predicted outcomes. In LIS, managing outcomes will need to take into account the changes in physical forcing that are lengthening the stratification season and increasing susceptibility to hypoxia.

7.7.2.4 The Long Island Sound Total Maximum Daily Load and Ecosystem-Based Management

Hypoxia (low dissolved oxygen), fueled by nutrient enrichment, is considered Long Island Sound's most pressing environmental problem (LISS 1994, Lopez et al., Chap. 6, in this volume). Both Connecticut and New York have DO water quality standards and numeric criteria applicable to LIS. Historical and current water quality monitoring and modeling show that during portions of the summer months substantial areas of the deep waters of WLIS do not meet states' DO criteria. Because of these identified CWA "impairments," the two states were in a position of having to develop a TMDL for DO. With research and monitoring fostered by the LISS since the mid-1980s, planning focused on N as the most important nutrient to mitigate the development of excess phytoplankton blooms and consequent hypoxia (Howarth and Marino 2006), an approach supported by subsequent monitoring and modeling of LIS. As there are no adopted numeric nutrient criteria in Connecticut or New York applicable to LIS, the LISS supported the states' efforts to translate numeric DO criteria endpoints into allowable N loads within a TMDL.

In 2000, after more than a decade of effort, Connecticut and New York submitted to the USEPA a TMDL (CTDEP and NYSDEC 2000) that set allocations for N to improve DO conditions. The allocations for N were set at a 58.5 % reduction from an estimated 1990 baseline of controllable point and nonpoint sources originating in Connecticut and New York (10 % from urban and agricultural runoff and the balance from point sources, primarily WWTFs). The TMDL also set planning targets for N control in the tributary states of Massachusetts, New Hampshire, and Vermont (a 25 % reduction from point sources and a 10 % reduction in nonpoint sources), and considered reductions in atmospheric deposition of N from air emission control programs (an 18 % reduction based upon model forecasts of national and regional Clean Air Act initiatives). Despite this unprecedented proposal for N control, small portions of the Sound would still not meet state water quality standards, a requirement of a TMDL. To fill the gap, the TMDL proposed assessment of alternative technologies including mechanical aeration of the residual, non-complying areas or bioextraction of nutrients by aquaculture of shellfish and seaweed.

At the time of its approval, and for a decade after until the completion of the Chesapeake Bay nutrient TMDL in 2010, the LIS TMDL was the most complex and comprehensive in the nation, setting N allocations for more than 100 WWTFs in two states and making recommendations for allocations in three other watershed states. Its development spurred policy innovations in three areas of management: equivalency factors for pollutant trading, compliance schedules within pollutant discharge permits beyond their 5-year term, and incorporation of alternative technologies (e.g., bioextraction of nutrients by shellfish and seaweed aquaculture).

The LIS TMDL established a flexible framework for the states to modify wasteload allocations among wastewater treatment plants without USEPA approval through pollutant trading approaches. By including equivalency factors in the TMDL that accounted for the relative impact of different sources on water quality based on geographic location, states could reallocate N source reductions among individual sources as long as the same or better predicted water quality improvement would result. This allowed flexibility to dischargers in meeting permit limits and allowed them to plan for N removal at the time when economics were most favorable to their facility, without a negotiated permit compliance schedule. This flexibility does not extend to reallocations between load and wasteload allocations, which require USEPA approval in a revised TMDL.

The 2000 LIS TMDL was constructed as a phased TMDL to allow for ongoing work to address uncertainty and support adjustments over time. Marine DO criteria (USEPA 2000) that had been concurrently under development would provide a firmer scientific basis for state water quality standards. Additional modeling was in the works, and plans for more inclusive management throughout the basin were proposed and discussed with a Connecticut River Nitrogen Work Group, coordinated by the New England Water Pollution Control Commission (NEIWPCC).

Although a TMDL is not a strict enforcement tool, underlying enforceable mechanisms mandate that CWA National Pollutant Discharge Elimination System (NPDES) permit limits meet the requirements of the TMDL for pollutant-load reductions. For non-regulated sources, primarily nonpoint sources, "reasonable assurances" must be provided by the states that identify how the load allocation will be met. The TMDL also should include an implementation schedule, although USEPA approval of a TMDL does not include a review and approval of implementation plans.

Connecticut and New York have chosen different approaches and uses of regulatory authorities to meet the wasteload allocation from WWTFs in their respective states. Both include innovations in permitting that take advantage of economics to minimize costs to municipalities and their rate payers. The New York State Department of Environmental Conservation (NYSDEC) has issued traditional NPDES permits that limit N loads for individual New York WWTFs, but has used or allowed a "bubble" approach by grouped, regulated administrative entity—New York City, Westchester County, Nassau County, and Suffolk County. This has provided a distinct advantage for N control for New York City in particular, where WWTFs, including related CSO loads, provide a high percentage of the N load to the "edge of Sound" along the East River. Because a portion of the load is swept into New York Harbor by prevailing tidal currents in the East River (Blumberg and Pritchard 1997), four of the six facilities along the East River, i.e., those most proximate to WLIS, were targeted for more aggressive management, providing relief for the two facilities closer to New York Harbor. Due to the cost and complexity of the upgrades, NYSDEC has established separate consent agreements with both New York City and Westchester County that extend the time for full compliance to 2017.

Connecticut chose a different alternative to managing 79 municipal facilities distributed throughout the state by implementing a relatively new concept, pollutant "trading," using a watershed-based N general permit and an oversight Nitrogen Credit Advisory Board to run a Nitrogen Credit Exchange (NCE). Although not a true free-market enterprise, the NCE provides an incentive for those facilities with low marginal costs for N removal to sell credits to those facilities with higher marginal treatment costs. The NCE has, as of 2012, been in operation for 10 years, and currently is on schedule to meet the 2014 target for N control from those facilities.

Through implementation of the TMDL, nutrient inputs to LIS for all major input categories (WWTFs, atmospheric deposition, tributaries) have decreased significantly in several watersheds (See Sect. 7.4.4 and Varekamp et.al., Chap. 5, in this volume). This is the result of reductions from WWTF inputs as well as regional reductions in atmospheric emissions. Water quality programs to address polluted runoff also are evolving, with green infrastructure and low impact development the most notable examples. The downward trend in N loading is occurring even as the human population in the watershed has increased by about 3.5 % from 2000–2010 and development in Connecticut and New York portions of the watershed has increased by about 15 % in the 1985–2010 period. However, some caution is warranted. Tributary concentrations of N generally show no significant trend in recent years, with evidence of an uptick in delivered loads with higher flow. Trends in concentrations or fluxes of N in groundwater also are much less definitive, but there is clear evidence of increase groundwater concentrations of N in Suffolk County, NY. Delivery of groundwater to LIS is subject to time lags attributable to transport, which will delay the response to changes in management or land use (Mullaney, 2007).

As discussed in earlier chapters, the science of the mechanisms of eutrophication and hypoxia in the Sound has advanced in the past decade. High-frequency, time-series measurements, in particular, have helped elucidate the physical factors that influence the seasonal and annual variability in hypoxia. Core data present a reconstruction of hypoxic conditions dating back centuries. Current evaluations of the TMDL incorporate a number of technical enhancements (updated computer modeling, refined load estimates, etc.) as well as revised water quality standards for DO. Any future revisions of the TMDL, however, still will need to deal with considerable uncertainty and would benefit from phasing and adaptive management. To be successful, adaptive management must be embraced fully, particularly to incorporate management of waterbody processing of nutrients, wherein much of the uncertainty lies, in addition to watershed loading. Adaptive management also requires incorporation of flexible allocation schemes that retain specificity and accountability. And, perhaps most significantly, adaptive management must incorporate schedules reflective of the extensive time lag between implementation of a practice and the realization of the environmental benefit. For example, reduced nutrient delivery to receiving water through groundwater sources can be delayed years after effective control practices are put into place (Bachman et al. 1998; Focazio et al. 1998).

Modeling and monitoring supported the view that P can limit primary production in some areas of the Sound during portions of the year; nevertheless, the 2000 TMDL focused on N as the primary limiting nutrient. With the national initiative to develop numeric nutrient criteria for P to protect freshwater systems from eutrophication, however, it is likely that future P loading from the watershed also will decline from current levels. States increasingly are required to incorporate P permit limits for discharges to wastewater-dominated fresh waters. For example, in 2011, Connecticut came to agreement with the USEPA to incorporate P limits on all municipal discharges to non-tidal fresh waters upon permit renewal, which will lower the total load of P from that source substantially (http://www.ct.gov/dep/cwp/view.asp?a=2719&q=474130&depNav_GID=1654). Decreases in P loading from the LIS watershed from implementation of freshwater P criteria should be considered in evaluating water quality responses. Likewise, watershed reductions in N to achieve downstream targets for LIS should be evaluated for benefits to inland waters, since there is evidence of benefits to dual nutrient controls (Lewis et al. 2011).

Despite overall progress in reducing the inputs of nutrients to the Sound, analysis to date has not shown a significant improvement in hypoxia. This is attributable to a number of factors that are articulated within this volume, including pressures on the Sound related to large-scale climatic shifts (wind direction and magnitude, increased water temperatures and associated stratification). The response in hypoxic condition to regional management efforts focused on N may not be immediate, and intermediate components such as nutrient levels, phytoplankton production, water clarity, and eelgrass must continue to be monitored and evaluated, with management likely adapted. For example, in 1987, Mumford Cove, an embayment in eastern Connecticut, contained a near monoculture of Sea lettuce (*Ulva lactuca*), a green macroalgae, with no seagrass found throughout the cove (Curtis and Dunbar 1985). By 1988, less than one year after the diversion of a wastewater treatment plant discharge from the cove to the Thames River, the biomass of sea lettuce was reduced by 99 % (French et al. 1989). Within 10 years after diversion, eelgrass was once again the dominant primary producer in the cove (see case study at http://www.lisrc.uconn.edu/eelgrass/Mumford.html).

Watershed loading targets allocated among sources, with schedules for implementation, are needed in combination with policies and programs to enhance populations of shellfish and other filter feeders, restore submerged aquatic vegetation, and restore wetlands. Tradeoffs among components of the ecosystem and the services provided will need to be acknowledged. For example, reducing phytoplankton production through nutrient loading controls may increase the distribution of submerged aquatic vegetation while simultaneously reducing growth of some species of bivalves (Wall et al. 2011). Similarly, the indirect benefits of environmental quality management aimed at LIS support the local economy and lifestyle throughout the watershed. Concepts of healthy watersheds can address a multitude of ecosystem services and garner more local support if the community benefits are realized, as well as the downstream benefits that are shared regionally (USEPA 2012). Flexibility should continue to be given to a phased TMDL for LIS that addresses uncertainty and crafts an adaptive systems approach that integrates watershed load reduction programs with enhanced nutrient processing to attain water quality standards, restore designated uses, and more broadly restore ecosystem services basin-wide.

7.7.2.5 Management Application

Boesch et al. (2000) assessed the application of scientific knowledge about estuaries to site-specific management, reviewing case studies, and highlighting the factors supporting or impeding success. Among successful traits are the availability of long-term scientific data to document problems and an institutional structure that includes a multiple range of stakeholders—agencies, scientists, user groups, conservation organizations, and the public—to develop response strategies. With the maturation over the past quarter century of collaborative programs managing coastal and estuarine waters, both nationally (e.g., Chesapeake Bay, Gulf of Mexico, South Florida, 28 National Estuary Programs) and internationally, best practices have emerged for science, collaboration, communication, and implementation. One such practice is to synthesize the underlying science as a foundation for management (see Abal et al. 2005, Dennison et al. 2009; Desbonnet and Costa-Pierce 2008; Levinton and Waldman 2006).

While the science of LIS has advanced greatly over the past quarter century, better informing planning and management efforts, there remain gaps that hinder effective management. These science gaps and the priority needs, identified in prior chapters, are summarized in Figs. 7.6 and 7.7. While the gaps and research needs are drawn from the estuarine, freshwater, and terrestrial scientific communities, this book reflects a bias toward the estuary. Further dialogue and collaboration among these communities, as well as social scientists, will be needed to integrate watershed-waterbody scale processes. No attempt is made to prioritize among these needs, which will vary depending on the mission and resources of individual programs.

The synthesis of LIS attempted here is particularly timely, with planning underway to update and revise both the LIS CCMP (LISS 1994) and the N TMDL (CTDEP and NYSDEC 2000). Combined with a community vision for LIS developed in 2011 by the LISS Citizens Advisory Committee, it fulfills key elements of enhancing application of EBM to LIS.

Boesch et al. (2006) emphasize that true synthesis requires that the science be integrated and applied to questions facing management, supporting formulation of response strategies. Based on our interpretation of both the underlying state of the science of LIS, the general tenets for EBM, and our interpretation of the public's vision and hopes for the waterbody, we offer ten recommendations for enhancing management of LIS (Fig. 7.8).

Geology
- Reduce ambiguity regarding the thickness of in-place Coastal Plain vs. glacial deposits along the north shore of Long Island leads, which to a potential misunderstanding of groundwater flow and discharges to LIS.
- Better understand the nature and timing of marine transgressions to the Sound.
- Better understand potential seismicity in LIS.

Physical Oceanography
- Better characterize lateral structure with particular emphasis on the smaller bays and inlets. Characterizing the exchange between LIS and its bays, harbors, and inlets will require higher resolution measurements.
- Undertake additional seafloor observations to better understand sediment-water interactions.
- Reduce uncertainty in understanding local effects of global climate change by increasing measurements to better describe changes in all habitats.
- Undertake high resolution bathymetry and topography measurement to support predictions of coastal hazards and flooding from storms/sea level rise coupled with higher density sea level measurements.

Geochemistry
- Better characterize the role of sediment processes in the sources, transport, and fate of nutrients and other compounds.
- Clarify the relationship between embayments and LIS proper, including linkages between phytoplankton production in the main stem and organic matter accumulation in embayment sediments.

Pollution
- Better understand the dynamics of pH and dissolved inorganic carbon.
- Evaluate risk from emerging contaminants.
- Characterize the sediment load and sources as part of a dredged material management program.

Fig. 7.6 Gaps and key research needs to support management

Ecology/Biology

- Increase quantitative data on spatial and temporal features in the littoral zone to quantify what has been lost and better evaluate how to restore.

- Tidal Marshes:

  a Better understand the potential link between pathogens and other possible causes of sudden vegetation dieback and to evaluate its effect on important marsh fauna and flora.
  b Determine the importance of tidal pools to important marsh fauna.
  c Better evaluate the contributions of restored tidal marshes to marsh birds, plants, and animal communities.
  d Evaluate the ecological implications of the spread of *Phragmites australis*, an important non-native invader to LIS salt marshes.

- Better evaluate the response of seagrass to stress.

- Fate of primary production and linkages to hypoxia need to be better understood. There are uncertainties to sinking and horizontal export of primary production and imbalances between sources and sinks of carbon in WLIS.

- Deep water benthos:

  a Seasonal and year-to-year temporal dynamics need to be better studied and understood (particularly for coarse grain sediment successional dynamics, response to infrastructure disturbance, and succession in hypoxic areas).
  b Better understand spatial dynamics and relationships to different stressor/pressures (including characterization in embayments).
  c Additional measurements of biomass, productivity, and community composition are needed for incorporation into food web models.

- Fish, Shellfish, Wildlife: The mechanisms involving important variables such as loss of keystone species, fishing pressure, temperature, habitat alternation need to be better understood.

- Cross-cutting issues:

  a Remineralization of organic carbon and nitrification of ammonium from sewage treatment plant discharges can contribute to hypoxia, but direct measurements using 15N tracer and other methods will be needed to assess their importance.
  b Food web dynamics are relatively poorly known; need to better understand trophic linkages between primary production and apex predators.
  c Need to better understand the consequences of changes in climate affecting timing and fate of primary production.

Fig. 7.6 (continued)

Embrace sustainability. The modern environmental movement has been successful in decoupling the rise in population and economic output from similar rates of increase in the generation of air and water pollution. Increases in population and the economy over the past 30–40 years have been accompanied by slower rates of increase or an overall decrease in many air and water pollutants from regulated sources, mainly through application of pollution control technologies and product bans. But environmental problems remain, caused by more diffuse, unregulated sources of pollution and from landscape changes.

Sustainable development, defined as "meeting the needs of the present without compromising the ability of future generations to meet their own needs,"

Physical Oceanography
- Additional attention should be paid to developing assimilative models that incorporate hydrography and circulation. Better wind, river and ocean boundary data are necessary components.

Geochemistry
- CLIS and embayments will be sensitive to changes in the overall system (including climate change) and should be emphasized for monitoring/studies of carbon organic content and inputs, redox oscillations, benthos, sediment accumulation and bacterial biomass.

Pollution
- Need to characterize regulatory effectiveness by continuing periodic sampling of sediments for xenobiotics, toxicity, and benthic condition (e.g., NCA).
- Monitor sediment loads as part of a sediment management program to alleviate dredging demands.
- Continue to evaluate management effectiveness by monitoring nutrient inputs and evaluating flow-normalized trends.

Ecology/Biology
- Continue to monitor the spatial and temporal variability of seagrass.
- Monitor components of the biological community, e.g. seaweeds, benthic animals.
- Improved ecosystem modeling is needed to help relate trends in harvest and survey abundances to the biology of natural resources.

Fig. 7.7 Monitoring and modeling needs to support management

1. Embrace sustainability
2. Prioritize management of existing pollution sources and impairments
3. Establish baselines of historical condition and magnitudes of change
4. Integrate climate change across programs
5. Enhance positive feedback loops
6. Improve eutrophication and ecological modeling and monitoring
7. Design adaptive management framework
8. Conduct marine spatial planning
9. Improve data management and interpretation
10. Reconnect people to the Sound.

Fig. 7.8 Recommendations to enhance EBM of Long Island Sound

(Bruntland 1987) can maintain ecosystem services while providing for other societal needs. Sustainability might strike some as out of place leading the list of science and management actions for LIS. But managing aquatic ecosystems requires a combined watershed and waterbody perspective. As the costs of mandating remediation of past environmental sins often are steep and the timeframes long, it is appropriate to start with actions that can have the most benefit relative to cost. Development of technological infrastructure and implementation of pollutant-control regulation are usually reactions to environmental problems that exist already. Their implementation typically is three to four decades late.

Practices to develop in a sustainable manner can be accomplished with more modest upfront costs compared to remediating impaired watersheds, thus reducing the long-term costs to society. For example, low impact development can maintain predevelopment hydrology, greatly reducing stormwater, and flooding concerns and at much lower costs than post-development treatment practices (Clausen 2007; Dietz and Clausen 2008). The cost of conservation can be small compared to the cost of restoration.

Hardin (1993) advises that we must learn to "live within limits." Restoration of historical services supported by the LIS ecosystem will not be achieved if new development further adds to stressors that are a consequence of past development. This is not to suggest that the obstacles to sustainability—technological, social, political, and economic—are trivial, or that sustainability can eliminate tradeoffs in the provisioning of ecosystem services. Developed landscapes will not function as natural landscapes. Nevertheless, development for housing and commerce will remain a desired human use of landscapes; doing so sustainably will lessen its impact on water resources and reduce the scope of post-development remediation. Although place-based ecosystem restoration programs usually will not lead or coordinate broader sustainability efforts, they can and should acknowledge, promote, and support sustainable practices and can become catalysts for change with far-reaching benefits.

Interestingly, a side effect of improving water, sediment, and habitat quality in LIS is increased pressure to redevelop areas that previously were not considered desirable locations because of hazardous waste contamination, industrial activities, odors, or debris. Redevelopment projects must be seen as opportunities to enhance sustainability, with foresight to climate change adaptation, stormwater infiltration, public access, and habitat protection.

Prioritize management of existing pollution sources and impairments. Second in priority to preventing or minimizing new impairments is systematically addressing the drivers of existing impairments. Core CWA programs to attain water quality standards through pollutant source identification and remediation, along with habitat protection and restoration, will continue to be the backbone of efforts to restore LIS. Current federal and state programs, both national and LIS-specific, must continue to advance environmental protection. Consider that national emission control programs to attain air quality standards are vital to achieving the N TMDL for LIS. And many of the notable successes in reducing contamination from toxic substances are the consequence of national product bans and discharge control programs. Planning efforts must continue to integrate, watershed-wide, multiple sources of stress on the ecosystem—wastewater treatment plants, on-site wastewater treatment systems, stormwater runoff, combined sewer overflows, agriculture, atmospheric deposition, habitat loss and modification, and home and yard care practices—in prioritizing limited resources to remediate impairments.

Establish baselines of historical condition and magnitudes of change. As one of their "top 40 priorities for science to inform" conservation and management policy, Fleishman et al. (2011) point out that scientific, political, and policy communities have not done well in developing and committing to the creation of "reliable scientific metrics for detecting chronic long-term changes in ecosystems."

Understanding the historical (and pre-historical) condition of ecosystem functions and services helps support the development of appropriate restoration targets. These reference points, combined with assessments of current conditions and the drivers of degradation, provide the context for restoration strategies (Lotze et al. 2006). In an urban estuary like LIS, restoration to historical conditions, even if known, is most likely no longer feasible. Instead, restoration efforts will need to consider what can be accomplished practically within existing conditions and the expanded human presence (Bain et al. 2007).

For example, in 1996 the LISS established a Habitat Restoration Initiative to restore and protect 12 priority LIS habitat types that have been degraded or are under threat from development and pollution. The initiative set 10-year acre and river-mile restoration targets and has tracked progress toward their attainment. This approach has been successful where broad-brush habitat requirements can be surmised from incomplete historical accounts. The extent of historical access for diadromous fish to spawn in LIS tributaries was estimated and the difference between historic and current access estimates was then partitioned into river miles that potentially are restorable versus those that cannot be restored because of current tributary and watershed conditions. These estimates provide a stronger basis to establish targets for restoring river-mile access for shad, alewife (*Alosa pseudoharengus*), and blueback herring (*A. aestivalis*). Where quantitative or even qualitative estimates of the (pre)historical extent and condition for specific habitats are lacking, restoration targets cannot be put in the context of actual ecosystem restoration.

Integrate climate change across programs. Section 7.6 of this chapter discusses some of the changes affecting LIS attributable to climate change, including potential changes or losses in commercially-valuable species, increased susceptibility to hypoxia, and changes in the timing and intensity of riverine discharges. These climate and environmental change signals warrant integration with existing science programs and adaptation measures. Monitoring will be needed to provide insight into the direction and rates of changes. Habitat protection and restoration programs also will need to consider storm surge and sea level rise in project planning.

Adaptation strategies for climate change often rely upon management practices that protect, simulate, or restore natural ecosystem features and functions. As a result, these strategies yield multidimensional management benefits that build ecosystem resilience against a range of environmental stressors. Multiple climate adaptation initiatives are underway in Connecticut and New York. The Connecticut Governor's Steering Committee on Climate Change created an Adaptation Subcommittee in accordance with the requirements of Public Act 08-98. This subcommittee is assessing the impacts of climate change on Connecticut's infrastructure, natural resources and ecological habitats, public health, and agriculture and developing recommendations for changes to programs and laws that would enable state and local government to adapt to such impacts. The New York Governor's Climate Action Council, formed through Executive Order No. 24 in 2009, developed a Climate Action Plan that includes recommendation for adaptation.

One challenge will be to develop coastal management policies and guidelines relating to where, when, and how to armor shorelines or retreat in the face of higher sea levels or more intense storm surges. Long Island Sound is developed intensively along its shoreline, and redevelopment of previously degraded industrial sites is becoming attractive for commercial or residential use. Demands for protection of property and public safety will exert enormous pressures to armor and protect low-lying coastal areas against damage from the combined effects of storm surges and sea level rise (Strauss et al. 2012; Tebaldi et al. 2012).

The question of whether to construct seawalls or other shore-protection devices to stabilize the existing shoreline is scientifically complex; implementation will be costly and contentious, socially and politically. Even now erosion is essential to feed the existing beaches and littoral drift. But, as the rates of erosion increase, do we "stand or retreat?" Who has the burden of paying for shoreline protection? Is there political will to enforce setbacks to allow for shoreline accretion and avulsion as sea level rises?

One fact is absolutely clear: coastlines and their associated features naturally adjust to rising sea levels by migrating landward. It is possible that more extensive bluff erosion will occur as a consequence of waves undercutting the bluff toe at a higher elevation. But efforts to prevent erosion often have unintended consequences, and ultimately yield to the sea in the long run. Resource managers should be conservative and keep development away from vulnerable coastal features, rather than have to expend resources to protect threatened areas in desperation.

Enhance positive feedback loops. Attempts to reverse coastal eutrophication primarily focus on reducing land-based sources of nutrients, such as fertilizer applications, WWTF dischargers, and air emissions. But as noted, historical alterations in habitat quality, food webs, and community structure in coastal systems also alter nutrient processing, thus modifying the ecosystem response to reduced nutrient loads (Duarte et al. 2009). One example of a systems approach integrates watershed load reduction programs with enhanced nutrient processing in coastal systems. This may prove to be more effective at restoring ecosystem services at less cost than load reduction programs alone, helping to arrive at cost-effective, affordable, and equitable solutions. Six areas merit particular attention.

- Enhance shellfish and macroalgae aquaculture as a means to bioextract nutrients.
- Protect and restore submerged aquatic vegetation through enhanced water quality and targeted plantings and management.
- Protect and restore wetlands, considering possible marine transgression from sea level rise.
- Protect and restore vegetated buffers along streams and coastlines.
- Restore molluscan beds, particularly for species such as oysters that create subhabitats.
- Account for integrated societal and economic benefits derived from this approach.

In addition to improving the nutrient-assimilative capacity of LIS, enhancing habitat and food web functions will benefit the provisioning of other ecosystem services, including flood prevention, species abundance and diversity, and harvestable resources.

Improve eutrophication and ecological monitoring and modeling. Water quality monitoring programs should be maintained to support improved understanding
of hypoxia, identification of status and trends, and diagnostic analyses for management, including modeling. Mechanistic models of eutrophication in LIS have
been relied upon to apply scientific knowledge to a range of specific management
questions relating to nutrient control (HydroQual, Inc. 1996). The Systemwide
Eutrophication Model, or SWEM (HydroQual, Inc. 1999), originally developed by
the New York City Department of Environmental Protection in the late 1990s, currently is being used by the LISS to evaluate alternatives for hypoxia management.
A LISS-funded enhancement of the model is underway and will implement a number of recommendations for improving it (O'Donnell et al. 2010a), including:

- Transition to a community modeling framework that provides open source access
  to SWEM to facilitate external assessment and enhancement of the model.
- Foster model collaborations, including nesting of nearshore areas and embayments within SWEM to better resolve lateral circulation and exchange.
- Link SWEM to watershed and groundwater models to better refine nutrient and
  water budgets.
- Modify and use SWEM to evaluate nutrient bioextraction by aquaculture of filter feeding shellfish and seaweeds (especially the warm-temperate rhodophytes
  such as graceful red weed and the cool-temperate kelp species).
- Evaluate strategies to improve modeled hydrographic fields and circulation using
  existing monitoring data and improve eutrophication assessments of the Sound.

Models can provide insight into ecosystem processes, but they must be used conservatively; they are only as good as the assumptions and the data that feed them.
Given the uncertainties of any particular model (boundary conditions, mixing coefficients, etc.), employing a suite of models to analyze situations and make management decisions would enhance management skill over relying solely on one.

Effective modeling requires a high level of sophistication that links atmospheric
deposition models with watershed and estuarine models. Successful integration
must be supported by intensive monitoring of all media and research into ecosystem structure and function that verify model accuracy and predictive capability
through biological endpoints. Future development of ecological models to relate
physical and biogeochemical processes and energy transfer from source through
higher trophic levels, including fish production, also is desirable.

Monitoring of LIS could be strengthened, including: water column carbon
biomass, rates of production/respiration, and remineralization of organic matter
within the water column and from the sediments of WLIS that resupplies N and
P. Adding redundant buoy instrumentation would minimize data gaps in temporal coverage. Increased monitoring and study of CLIS and LIS's bays can provide
sentinel insight into how LIS responds to complex interactions between nutrient
control programs and changing drivers mediated through climate change. The
CLIS is a transitional basin between the more divergent conditions present in

WLIS and ELIS, and responses there may be more easily detected and diagnostic of forcing variables. Likewise, the embayments of LIS are the most intensively used portions of the waterbody, and often have strong constituencies advocating for their protection. They provide enhanced opportunities to involve a community directly in resource protection. Both CLIS and LIS's embayments should be priorities for monitoring and study, and citizen involvement in structured and quality assured data collection can stretch limited financial resources. Research funded by the LISS (Vaudrey and Yarish 2010) comparing the status of eight embayments will provide a cross-system comparison and a base line for future study. This work can be supported by local community organizations, often with the assistance of volunteers.

Design an adaptive management framework. The uncertainties associated with eutrophication in LIS warrant an adaptive management approach. Successful application, however, will require a commitment from LISS partners, particularly federal and state regulatory agencies, to understand and incorporate its principals. Research and monitoring priorities must explicitly address uncertainties relevant to decision making and be designed to test cause-and-effect relationships necessary to instruct the adaptive process. Flexibility within a phased TMDL is essential to support innovation, testing, and evaluation. So is incorporation of actions within a systems approach to meeting ecosystem objectives while supporting related socioeconomic goals and outcomes as well.

Conduct coastal marine spatial planning (CMSP). Long Island Sound could derive particular benefit from CMSP because of the characteristics that differentiate it, both from other planning sub-regions and within the Northeast and Mid-Atlantic regions. CMSP provides an opportunity to enhance EBM benefits because of its nexus to a socioeconomic context. Although a single estuary and ecosystem, LIS is administratively divided not just between two states, but between two ocean planning regions. As an estuarine sub-region surrounded by a large, urban population, and hosting intense human uses, it presents a different set of resource, use, and governance issues than regional-scale ocean management. For example, planning would be driven less by offshore renewable energy needs than by the need to manage a crossroads of energy and telecommunications infrastructure such as cables and pipelines overlaid on a historic maritime-transportation network and placed within a defined estuarine ecosystem.

Coastal and marine spatial planning for LIS will need to address conflicts arising from proposals for energy-related infrastructure proactively, including but not limited to liquid natural gas platforms, cable crossings, tidal energy turbines, etc. These efforts also will be relevant to EBM and stewardship of marine resources, including issues such as adaptation to climate change (sea level rise and coastal hazards, alterations in food webs, shifts in high-value living resources, acidification), allocation of areas for aquaculture, as well as planning for dredged material management.

An opportunity exists with the establishment of a LIS research and restoration fund (Cable Fund) established by a settlement agreement regarding two electrical cable crossings of LIS. The Cable Fund is targeting research that improves scientific understanding of the seafloor environment, emphasizing seafloor mapping, in order to prevent or mitigate the effects of current or potential energy-related infra-structure or other uses. To accomplish this goal, cooperative partnerships must be formed to marry management needs with the proper data collection, management, and interpretation, with an emphasis on mapping the bathymetry and surficial geology of the seafloor in LIS to help increase the understanding of seafloor habi-tats and improve resource management.

Improve data management and interpretation. Data sets that can support spa-tially-driven environmental assessments are proliferating. Available data, however, often are housed in multiple locations and in differing formats, making integra-tive assessments and gap identification time consuming and expensive. Specific expertise, software, or equipment is often required to support analyses. A useful step would be to complete a GIS needs assessment to identify existing and needed data, user-groups creating, housing, and using GIS data, and data products and resources needed to meet data-assessment objectives. Further, the assessment should outline the key data sets that are available and make recommendations on how to increase public access and best support future environmental assessments.

Management and interpretation of time-series water quality data must be given a high priority. All agency and university observing system data should be readily available to facilitate analyses and distribution to other users. Local embayment monitoring programs should be integrated with the other data sets to assist in plan-ning, quality assurance, and analysis.

Improved spatial data sharing could be achieved through regional efforts. For example, an EcoSpatial Information Database (ESID) currently is being devel-oped for the North- Mid- and South-Atlantic by the Bureau of Ocean Emergency Management, Regulation, and Enforcement (within the US Department of Interior). The system is designed to accept ecological information for additional marine and coastal areas such as LIS. Likewise, the software system could be adapted into a database specific to LIS. It would also provide an excellent frame-work for incorporating social and economic indicators that interact with environ-mental goals and objectives.

Reconnect people to the Sound. Protection and restoration of LIS compete with many other public needs and quality-of-life objectives: safety and security, eco-nomic opportunity, education, and entertainment, among many. Yet, the ecosystem services that LIS and its surrounding environs provide are integral to our economy and lifestyle. There are intrinsic and highly recognizable personal values of LIS. Public uses—fishing, boating, swimming, passive recreation among others—that contribute to a sense of place and personal connection to the resource are vital to sustaining public support for its protection. The "Urban Sea," however, cannot be set aside as a reserve or returned to pristine condition; success will be defined by a LIS and watershed where human leisure and livelihoods are sustained and inter-twined with a vibrant, resilient ecosystem.

Continuing to support and develop public access and use of LIS will help build and maintain a supportive LIS constituency. One obstacle to public use is a shore-line predominantly in private ownership. This is particularly applicable to urban areas. Improved environmental conditions in urban areas will increase demand for access and provide opportunities to broaden and strengthen the constituency for LIS. Maintaining safe navigational access for the boating community, while positing some environmental challenges, also is important for maintaining a maritime tradition and economy long characteristic of the "Urban Sea." Efforts must continue to develop acceptable options for managing and disposing of dredged material.

Without strong public and political support, investments to maintain and improve existing, let alone new, water infrastructure will be a challenge to sustain. Some elements of our water infrastructure, such as the pipes conveying storm water and sewage, date back hundreds of years. Although impressive for the craftsmanship with which they were constructed, time has taken a toll, resulting in leaky systems that compromise the effectiveness of nutrient and pathogen treatment. The design life of WWTFs constructed in the 1970s is approaching or has past senescence. And many on-site wastewater treatment systems in older homes do not meet current standards for public health protection. Addressing these issues are fundamental components of federal and state water quality protection programs; they pose financial and administrative challenges more than technical.

Examples of regulation and management having protected ecosystem services with benefits to human health and the environment need to be documented and communicated. Among the examples given in this chapter, the chemical bans and use restrictions on Pb, Hg, DDT, and PCBs stand out as having markedly reduced exposures in the environment, leading to lower body burdens in LIS organisms. The resurgence of ospreys nesting along LIS is a direct result, as are lower risks to humans from cancer and other health concerns from consuming fish caught from LIS. National pollution control programs have improved water quality in many urban areas to such a degree that surrounding land values have increased and public recreational uses are on the rise. One such example is the reopening in 2011 of 2,500 acres in Hempstead Harbor for shellfishing for the first time in more than 40 years, a result of numerous water quality improvement efforts, elimination of many industrial uses around the harbor, and water quality monitoring and shellfish tissue testing. Recognition of the successes achieved to date will help affirm the benefits to continued investments in environmental protection. Outreach efforts must expand the conversation to sectors of the public that have been less involved, considering social media and other technologies, and be science-based.

Finally, there are opportunities to redefine normal, accepted practices in a sustainability framework and instill them in our culture so that what is customary also contributes to our economy and lifestyle while protecting our ecosystem. Residential landscapes, for example, that are more compatible with our climate and water resources can encourage yard practices that cost less money and require less time than traditional yards. Sustainability practices can be adopted without added direct costs or burdens when supported with science-based information and outreach, and the potential future management savings are real and consequential. One only needs to consider the construction and remediation costs of cleaning up the singular issue of nutrient enrichment in LIS to understand that current practices on the land create an economic deficit of essential ecosystem services that is as real as the national debt, and will eventually have to be repaid with interest.

7.8 The Urban Sea Revisited

Long Island Sound is a creation of climate change since the last ice age, and subsequently evolved to its present geomorphology as a result of fluctuations of climate and human actions. Measured physical and chemical variables, as well as the biological community, have been altered by climate changes over the last century. And so we can project with considerable certainty that the Sound, its processes, and its ecological functioning will respond to future climate change regardless of the cause. This ecosystem will be quite different a half-century into the future. Species migrations and seasons of availability may shift. Sizes of individuals and stocks may change. Some species may disappear, supplanted by others. We have been slow to recognize the changes in LIS driven by climate over the past century. These changes have been subtle relative to strong inter-annual and inter-decadal signals, and to the significant consequences of anthropogenic activity—port and industrial development, dredging, hardening of the shoreline, destruction of wetlands, diversion of water courses, industrial and sewage pollution, and fishing pressure. But the impact of storm surge, exacerbated by sea level rise, will be anything but subtle, as illustrated by the devastation of Hurricane Sandy, which made landfall on October 29, 2012 in southern New Jersey. The storm surge in parts of WLIS and the New York-New Jersey Harbor rose four meters or more above mean low water, resulting in billions of dollars in damages to the region's infrastructure. We must be prepared to deal with climate change-driven shifts and manage them so that new resources may become robust and the ecosystem services we rely upon are sustained.

Sewage discharges, whether from septic systems or WWTFs, remain a threat to the Sound, and solutions warrant innovative and forward thinking. Hypoxia, harmful algal blooms, shellfish bed closures, fish consumption warnings, and swimming restrictions all are linked to sewage. Long Island's groundwater aquifer, in particular, is threatened severely by sewage and land use. Although the region has made great strides in sewage treatment and regulatory controls, there remain challenges in designing, financing, and administering acceptable advances in wastewater treatment. Water use, reclamation, and disposal can be designed in ways that will reduce impacts (e.g., composting toilets, gray water irrigation). We can ill afford to waste potable water to relocate sewage wastes, and the environmental consequences of doing so are onerous. Getting sewage out of water would be a positive step for improving the long-term integrity of LIS, and thus should be a top priority. Society needs a clean, renewable water supply.

Eliminating discharge of polluted storm water into the Sound is also a necessary long-term goal. The challenge is that of preventing offending materials from getting into urban runoff and overland flow, and preserving pre-development hydrology, with a large portion of precipitation infiltrating into the ground to recharge aquifers and provide base flow for streams (Arnold and Gibbons 1996). To this end, stormwater management technologies should be designed to mimic natural systems, diverting the first polluted flush of storm water but retaining the natural flow and timing of pulses of fresh water, which are vital for the functioning of the Sound's oceanographic processes. Clean storm water recharges our coastal systems; it is this flow that drives estuarine circulation and establishes salinity regimes important for ecological functioning.

The outlook for the future of the quality of the Sound, its waters, ecological functioning, and aesthetic pleasures is actually quite positive, particularly if we eliminate sewage pollution. Over the last three to four decades, considerable investment has been made to understand the science of the Sound and to monitor changes in numerous, quantitative metrics of Sound health. These have formed the basis for enlightened management of an extremely complex ecosystem that spans multiple political boundaries. It should remain an imperative to continue to support estuarine research and monitoring to stay abreast of critical changes in the Sound ecosystem and to understand why they are happening.

One can take pride in the many efforts being undertaken to improve the condition of the Sound. Development of public engagement has been critical in this regard. Concerted efforts to reduce discharges of polluting materials from point and nonpoint sources, create vessel no-discharge zones, protect wetlands, buffer watercourses with vegetation, and improve public access are notable achievements. The Sound has benefitted from the national effort to ban lead from gasoline and stop production of PCBs. And who would have ever imagined that osprey and eagles would repopulate the area as a result of banning DDT—a nation-wide action that was initiated in Suffolk County, NY.

We can reduce and ameliorate consequences of society's insults to the Sound more effectively. Green products and green development increasingly are available to lessen our ecological footprint. We can, with strategic investments, adapt to and accommodate accelerating climate change. Conservation measures, marine spatial planning, and soft, environmentally acceptable coastal engineering technologies are some of the techniques that recently have been developed or improved. Smartly applied, they can help to assure that Long Island Sound, the "Urban Sea," retains its eminence as a beautiful, productive, enjoyable place to live and work in the 21st century.

References

Abal EG, Bunn SE, Dennison WC (eds) (2005) Healthy watersheds, healthy catchments: making the connection in South East Queensland, Australia. Moreton Bay waterways and catchment Partnership, Brisbane, p 240

Albion RG (1939) The rise of New York port, 1815-1860. Charles Scribner's Sons, New York, p 485

Allen DY (1997) Long Island maps & their makers. Amereon House, Mattituck, p 153

Andersen T (2002) This fine piece of water: an environmental history of Long Island Sound. Yale University Press, New Haven

Arnold CL Jr, Gibbons CJ (1996) Impervious surface coverage: the emergence of a key environmental indicator. J Am Plan Assoc 62(2)

Bachman LJ, Lindsey BD, Brakebill JW, Powars DS (1998) Groundwater discharge and baseflow nitrate loads of nontidal streams, and their relation to a hydrogeomorphic classification of the Chesapeake Bay Watershed, Middle Atlantic Coast: US Geological Survey Water-Resources Investigations Report 98-4059, p 71

Bain M, Lodge J, Suszkowski DJ, Botkin D, Brash A, Craft A, Diaz R, Farley K, Gelb Y, Levinton JS, Matuszeski W, Steimle F, Wilber P (2007) Target ecosystem characteristics for the Hudson Raritan Estuary: technical guidance for developing a comprehensive ecosystem restoration plan. A report to the Port Authority of NY/NJ. Hudson River Foundation, New York, NY, p 106

Bauer C (2012) Physical processes contributing to localized, seasonal hypoxic conditions in the bottom waters of Smithtown Bay, Long Island Sound, New York. Dissertation, School of Marine and Atmospheric Sciences, Stony Brook University, Stony Brook, NY

Beck MW, Brumbaugh RD, Airoldi L, Coen LD, Crawford C, Defeo O, Edgar GJ, Hancock B, Kay M, Lenihan H, Luckenbach MW, Toropova CL, Zhang G, Guo X (2011) Oyster reefs at risk and recommendations for conservation, restoration and management. Bioscience 61(2):107–116

Blumberg AF, Pritchard DW (1997) Estimates of transport through the East River, New York. J Geophys Res 120(C3):5685–5703

Boesch DF, Burger J, D'Elia CF, Reed DJ, Scavia D (2000) Scientific synthesis in estuarine management. In: Hobbie JE (ed) Estuarine science: a synthetic approach to research and practice. Island Press, Washington DC, pp 507–526

Bokuniewicz H, Tanski JJ (1983) Sediment partitioning at an eroding coastal bluff. Northeast Geol 5(2):73–81

Bowman MJ (1976) The tides of the East River, New York. J Geophys Res 81(9):1609–1616

Breslin VT, Sañudo-Wilhelmy SA (1999) High spatial resolution sampling of metals in the sediment and water column in Port Jefferson Harbor, NY. Estuaries 22(3A):669–680

Bricker SB, Clement CG, Pirhalla DE, Orlando SP, Farrow DRG (1999) National Estuarine Eutrophication Assessment: effects of nutrient enrichment in the nation's estuaries. National Oceanic and Atmospheric Administration, Silver Springs, p 71

Bricker SB, Longstaff B, Dennison W, Jones A, Boicourt K, Wicks C, Woerner J (2007) Effects of nutrient enrichment in the nation's estuaries: a decade of change. National Oceanic and Atmospheric Administration, Silver Springs 328 p

Bruntland G (ed) (1987) Our common future: the world commission on environment and development. Oxford University Press, Oxford

Cameron WM, Pritchard DW (1963) Estuaries. In: Hill MN (ed) The sea. Wiley, New York, pp 306–324

Capriulo GM, Smith G, Troy R, Wikfors G, Pellet J, Yarish C (2002) The planktonic food web structure of a temperate zone estuary, and its alternation due to due to eutrophication. Hydrobiologia 475(476):263–333

Central Pine Barrens Joint Planning and Policy Commission (1995) Central Pine Barrens comprehensive land use plan, vol 2: existing conditions. Chapter 4, Hydrology and water quality overview. Central Pine Barrens Joint Planning and Policy Commission, Great River, NY, pp 47–60. http://pb.state.ny.us/cpb_plan_Vol2/Vol2.pdf. Accessed on 10 Jan 2012

Clausen JC (2007) Jordan Cove watershed project (2007) Section 319 project final report. http://jordancove.uconn.edu/jordan_cove/publications/final_report.pdf. Accessed 30 Aug 2011

Cloern JE (2001) Our evolving conceptual model of the coastal eutrophication problem. Mar Ecol Prog Ser 210:223–253

Connecticut Department of Environmental Protection, Maine Department of Environmental Protection, Massachusetts Department of Environmental Protection, New Hampshire Department of Environmental Services, New York State Department of Environmental Conservation, Rhode Island Department of Environmental Management, Vermont Department of Environmental Conservation and New England Interstate Water Pollution Control Commission (2007) Northeast regional mercury total maximum daily load. NEIWPCC, Lowell, p 97

Connecticut Department of Environmental Protection, New York State Department of Environmental Conservation (2000) A total maximum daily load analysis to achieve water quality standards for dissolved oxygen in Long Island Sound. http://longislandsoundstudy.net/wp-content/uploads/2010/03/Tmdl.pdf

Conover B (1966) USC&GS MARMER, ASU-89, Charming workhorse of the Coast and Geodetic Survey. Tidings 1(6):14–17, 38–39 (Tidings Publishing Company, Norwalk, CT)

Copeland C (1993) Toxic pollutants and the Clean Water Act: current issues. Washington, DC, UNT Digital Library. http://digital.library.unt.edu/ark:/67531/metacrs89/. Accessed 24 May 2012

Council on Environmental Quality (2010) Final recommendations of the interagency ocean policy taskforce, July 19, 2010. 77 p. http://www.whitehouse.gov/files/documents/OPTF_FinalRecs.pdf. Accessed 28 Sept 2012

Crowley H (2005) The seasonal evolution of thermohaline circulation in Long Island Sound. PhD Dissertation, Marine Sciences Research Center, Stony Brook University, Stony Brook, NY, p 142

Curtis MD, Dunbar LE (1985) Water quality analysis of Mumford Cove final report: model development and waste load allocation. University of Connecticut, Connecticut Department of Environmental Protection, Water Compliance Unit, Storrs, p 55

Dame RF, Zingmark RG, Haskin E (1984) Oyster reefs as processors of estuarine materials. J Exp Mar Biol Ecol 83:239–247

Dana JD (1870) Origin of some of the topographic features of the New Haven region. 671 Trans Conn Acad Sci II:42–112

Dana JD (1890) Long Island Sound in the Quaternary Era, with observations on the submarine Hudson River channel. Am J Sci 40(Third Series):425–437

Davies DS, Axelrod EW, O'Connor JS (1973) Erosion of the north shore of Long Island. Tech Report Series 18. Marine Sciences Research Center, SUNY at Stony Brook, Stony Brook, p 97

Dennison WC, Thomas JE, Cain CJ, Carruthers TJB, Hall MR, Jesien RV, Wazniak, CE, Wilson DE (2009) Shifting sands: environmental and cultural change in Maryland's coastal bays. University of Maryland Center for Environmental Science. Integration and Application Network Press, Cambridge, p 396

Desbonnet A, Costa-Pierce BA (eds) (2008) Science for ecosystem-based management: Narragansett Bay in the 21st century. Springer series on environmental management. Springer, New York, p 570

Dickens C (1985) American notes. St Martins Press, New York, p 232

Dietz ME, Clausen J (2008) Stormwater runoff and export changes with development in a traditional and low impact subdivision. J Environ Manage 87:560–566

Dreyer G, Niering W (eds) (1995) Tidal marshes of Long Island Sound. Ecology, history and restoration. Conn Coll Arboretum Bull 35: p 73

Duarte CM, Conley DJ, Carstensen J, Sanchez-Camacho J (2009) Return to "neverland": shifting baselines affect eutrophication restoration targets. Estuar Coasts 32:29–36

Duffy TA, McElroy AE, Conover DO (2009) Variable susceptibility and response to estrogenic chemicals in *Menidia menidia*. Mar Ecol Prog Ser 380:245–254

Executive Order 13547. Stewardship of the ocean, our coasts, and the Great Lakes (2010). http://www.whitehouse.gov/files/documents/2010stewardship-eo.pdf Accessed 18 Feb 2011

Fenster MS, Fitzgerald DM, Moore MS (2006) Assessing decadal-scale changes to a giant sand wave field in eastern Long Island Sound. Geology 34(2):89–92

Fleishman E, Blockstein DE, Hall JA et al (2011) Top 40 priorities for science to inform US conservation and management policy. Bioscience 61(4):290–300

Focazio MJ, Plummer LN, Böhlke JK, Busenburg E, Bachman LJ, Powars DS (1998) Preliminary estimates of residence times and apparent ages of groundwater in the Chesapeake Bay watershed, and water quality data from a survey of springs: US Geological Survey Water-Resources Investigations Report 97-4225, p 75

French D, Harlin MM, Gundlach E, Pratt S, Rines H, Jayko K, Turner C, Puckett S (1989) Mumford Cove water quality: 1988 monitoring study and assessment of historical trends. Applied Science Associates, Narragansett, p 126

Fuller ML (1914) The geology of Long Island, New York. US Geol Surv Prof Pap 82: p 231

Fulweiler RW, Nixon SW, Buckley BA (2010) Spatial and temporal variability of benthic oxygen demand and nutrient regeneration in an anthropogenically impacted New England estuary. Estuar Coasts 33(6):1377–1390. doi:10.1007/s12237-009-9260-y

Germano JD, Rhoads DC, Valente RM, Carey D, Solan M (2011) The use of Sediment Profile Imaging (SPI) for environmental impact assessments and monitoring studies: lessons learned from the past four decades. Oceanog Mar Biol Ann Rev 5110(49):247–310

Gilluly J, Water AC, Woodford AO (1959) Principles of geology, 2nd edn. WH Freeman and Company, San Francisco p 534

Gobler CJ, Sañudo-Wilhelmy SA, Buck NJ, Sieracki ME (2006) Nitrogen and silicon limitation of phytoplankton communities across an urban estuary: the East River-Long Island Sound system. Estuar Coast Shelf Sci 68:127–138

Gottschall K, Pacileo D (2010) Long Island Sound trawl survey. In: A study of marine recreational fisheries in Connecticut, Job 2.1. Federal aid in sport fish restoration grant F-54-R-29, Connecticut Department of Environmental Protection

Gwynne P (1975) The cooling world. Newsweek, 28 April, p 64

Hakanson L, Boulion VV (2003) A general dynamic model to predict biomass and productivity of phytoplankton in lakes. Ecol Model 165:285–301

Hansen DV, Rattray M Jr (1966) New dimensions in estuary classification. Limnol Oceanogr XI(3):319–326

Hao Y, Wilson RE (2007) Modeling the spatial patterns of residence time in Long Island Sound. Final report to the LIS STAC Graduate Fellowship Program. Marine Sciences Research Center, Stony Brook University, Stony Brook, p 16

Hardin G (1993) Living within limits. Oxford University Press, New York p 339

Harris E (1959) The nitrogen cycle in Long Island Sound. Bull Bingham Oceanogr Coll 17(1):31–65

Hodgkins GA, Dudley RW, Huntington TG (2003) Changes in the timing of high river flows in New England over the 20th century. J Hydrol 278:242–250

Howarth RW, Marino R (2006) Nitrogen as the limiting nutrient for eutrophication in coastal marine ecosystems: evolving views over 3 decades. Limnol Oceanogr 51:364–376

Howarth RW, Swaney DP, Butler TJ, Marino R (2000) Climatic control on eutrophication of the Hudson River Estuary. Rapid Communication. Ecosystems 3:210–215

Howell P, Auster PS (2012) Phase shift in an estuarine finfish community associated with warming temperatures. Mar Coast Fish : Dyn Manage 4(1):484–485

HydroQual, Inc (1996) Water quality modeling analysis of hypoxia in Long Island Sound using LIS3.0. Report prepared for the New England Interstate Water Pollution Control Commission and the Management Committee of the Long Island Sound Estuary Study

HydroQual, Inc (1999) Newtown creek WPCP Project East River Water Quality Plan, Task 10.0—Systemwide Eutrophication Model (SWEM), subtasks 10.1–10.7. Reports prepared under contract to Greeley and Hansen for the City of New York Department of Environmental Protection

Jackson JBC, Kirby MX, Berger WH, Bjorndal KA, Botsford LW, Bourque BJ, Bradbury RH, Cooke R, Erlandson J, Estes JA, Hughes TP, Kidwell S, Lange CB, Lenihan HS, Pandolfi JM, Peterson CH, Steneck RS, Tegner MJ, Warner RR (2001) Historical overfishing and the recent collapse of coastal ecosystems. Science 293:629–638

Kemp WM, Boynton WR, Adolf JE, Boesch DF, Boicourt WC, Brusch G, Cornwell JC, Fisher TR, Glibert PM, Hagy JD, Harding LW, Houde ED, Kimmel DG, Miller WD, Newell REE, Roman MR, Smith EM, Stevenson JC (2005) Eutrophication of Chesapeake Bay: historical trends and ecological interactions. Mar Ecol Prog Ser 303:1–29

Kemp WM, Testa JM, Conley DJ, Gilbert D, Hagy JD (2009) Temporal responses of coastal hypoxia to nutrient loading and physical controls. Biogeosciences 6:2985–3008

Kimbrough KL, Johnson WE, Lauenstein GG, Christensen JD, Apeti DA (2009) An assessment of polybrominated diphenyl ethers (PBDEs) in sediments and bivalves of the US coastal zone. Silver Spring, MD, NOAA Technical Memorandum NOS NCCOS 94, p 87

King S, Miller P, Goldberg T, Graham J, Hochbunn S, Weinert A, Wilcox M (2008) Reducing mercury in the northeast United States. EM Mag AirWaste Manage Assoc May 2008:9–13

Klawonn MJ (1977) Cradle of the corps. US Army Corps of Engineers, New York, p 310

Knauss JA (1997) Introduction to physical oceanography, 2nd edn. Waveland Press, Long Grove p 309

Knebel, HJ, Lewis RS, Varekamp JC (eds) (2000) Regional processes, conditions and characteristics of the Long Island Sound sea floor. J Coast Res 16(3):519–662

Knebel HJ, Poppe LJ (2000) Sea-floor environments within Long Island Sound: a regional overview. J Coast Res 16(3):553–550

Koppelman LE, Weyl PK, Gross MG, Davies DS (1976) The Urban Sea: Long Island Sound. Praeger Spec Stud p 223

Kowalsick T (2012) Growing degree days, soil temperature, precipitation, and evapotranspiration. Cornell Cooperative Extension of Suffolk County. http://ccesuffolk.org/growing_degree_days_soil_temperature_precipitation_andevapotranspiration_rates_reports. Accessed 10 Jan 10 2012

Krug WR, Gebert WA, Graczyk DJ, Stevens DL, Rochelle BP, Church MR (1990) Map of mean annual runoff for northeastern, southeastern, and mid-Atlantic water years 1951–1980. US Geological Survey Water-Resources Investigations WRI Report 88-4094, p 11

Laufer H, Baclaski B (2012) Alkylphenols affect lobster (*Homarus americanus*) larval survival, molting and metamorphosis. Inv Reprod Dev 56:66–71

Le Lacheur EA, Sammons JC (1932) Tides and currents in Long Island and Block Island Sounds. Special publication 174. Coast and Geodetic Survey. US Government Printing Office, Washington DC, p 184

Levinton JS, Waldman JR (eds) (2006) The Hudson River estuary. Cambridge University Press, Cambridge, p 471

Lewis MW Jr, Wurtsbaugh WA, Paerl HW (2011) Rationale for control of anthropogenic nitrogen and phosphorus to reduce eutrophication of inland waters. Environ Sci Technol 45:10300–10305

Limburg KE, Moran MA, McDowell WH (1986) The Hudson River ecosystem. Springer, New York, p 331

Long Island Sound Study (1994) The comprehensive conservation and management plan for Long Island Sound. 168 pp. http://longislandsoundstudy.net/wp-content/uploads/2011/10/management_plan.pdf

Long Island Sound Study (2010) Sound Health: status and trends in the health of Long Island Sound. 16 pp. http://longislandsoundstudy.net/2010/12/sound-health-2010/

Lotze JK, Lenihan Bourque HS, Bradbury BJ, Cooke RH, Cooke RG, Kay MC, Kidwell SM, Kirby MX, Peterson CH, Jackson JBC (2006) Depletion, degradation, and recovery potential of estuaries and coastal seas. Science 312:1806–1809

Marmer HA (1926) The Tide. D. Appleton and Company, New York, p 282

McLeod KL, Lubchenko J, Palumbi SR, Rosenberg AA (2005) Scientific consensus statement on marine ecosystem-based management. http://www.compassonline.org/sites/all/files/document_files/EBM_Consensus_Statement_v12.pdf. Accessed 18 Feb 2011

Mecray EL, Buchholtz ten Brink MR (2000) Contaminant distribution and accumulation in the surface sediments of Long Island Sound. J Coast Res 16(3):575–590

Mitch AA, Anisfeld SC (2010) Contaminants in Long Island Sound: data synthesis and analysis. Estuar Coast 33:609–628

Mullaney JR (2007) Nutrient loads and ground-water residence times in an agricultural basin in north-central Connecticut: U.S. Geological Survey Scientific Investigations Report 2006–5278, p 45

Mullaney JR, Schwarz GE, Trench ECT (2002) Estimation of nitrogen yields and loads from basins draining to Long Island Sound, 1988–1998: US Geological Survey Water-Resources Investigations Report 02–4044, p 84

Mullaney JR, Lorenz DL, Arntson AD (2009) Chloride in groundwater and surface water in areas underlain by the glacial aquifer system, northern United States. US Geological Survey Scientific Investigations Report 2009–5086, p 41

National Research Council (2011) Achieving nutrient and sediment reduction goals in the Chesapeake Bay: an evaluation of program strategies and implementation. National Academy Press, Washington, DC. ISBN 978-0-309-21079-9

NEIWPCC, NESCAUM and NEWMOA (2007) Northeast states succeed in reducing mercury in the environment. New England Interstate Water Pollution Control Commission, Northeast States for Coordinated Air Use Management and Northeast Waste Management Officials' Association. Fact Sheet, 2 p

NESCAUM (2008) Sources of mercury deposition in the northeastern United States. Northeast states for coordinated air use management, Boston, p 75

New York Ocean and Great Lakes Ecosystem Conservation Act (2006) Environmental Conservation Law. Article 14. http://www.oglecc.ny.gov/media/ECL_Article%2014.pdf. Accessed 22 March 2012

Newell RIE (2004) Ecosystem influences of natural and cultivated populations of suspension-feeding bivalve mollusks: a review. J Shellfish Res 23:51–61

Nixon SW, Fulweiler RW, Buckley BA, Granger SL, Nowicki BL, Henry KM (2009) The impact of changing climate on phenology, productivity, and benthic-pelagic coupling in Narragansett Bay. Estuar, Coast, Shelf Sci 82:1–18. doi:10.1016/j.ecss.2008.12.016

Nuttall MA, Jordaan A, Cerrato RM, Frisk MG (2011) Identifying 120 years of decline in ecosystem structure and maturity of Great South Bay, New York using the Ecopath modelling approach. Ecol Model 222:3335–3345

NYCDEC (2010) New York Harbor survey program. Celebrating 100 years. 1909–2009. New York City Department of Environmental Protection, New York, p 32

O'Connor TP (1996) Trends in chemical concentrations in mussels and oysters collected from the US coast from 1986 to 1993. Mar Environ Res 41(2):183–200

O'Connor TP, Lauenstein GG (2006) Trends in chemical concentrations in mussels and oysters collected along the US coast: update to 2003. Mar Environ Res 62:261–285

O'Donnell J, Dam HG, McCardle GM, Fake T (2010a) Final report: simulation of Long Island Sounds with the Systemwide Eutrophication Model (SWEM)—inter-annual variability and sensitivity. http://longislandsoundstudy.net/wpcontent/uploads/2010/02/LI97127101Final-ReportV2.pdf

O'Donnell J, Morrison J, Mullaney J (2010b) The expansion of the Long Island Sound Integrated Coastal Observing System (LISICOS) to the Connecticut River in support of understanding climate change. Final Report to the CTDEP, LIS License Plate Fund

Parker CA, O'Reilly JE (1991) Oxygen depletion in Long Island Sound: a historical perspective. Estuaries 14(3):248–264

Pearce J, Balcom N (2005) The 1999 Long Island Sound lobster mortality event: findings of the comprehensive research initiative. J Shellfish Res 24(3):691–697

Pellegrino P, Hubbard W (1983) Baseline shellfish data for the assessment of potential environmental impacts associated with energy activities in Connecticut's coastal zone, vols I and II. Report to the State of Connecticut, Department of Agriculture, Aquaculture Division, Hartford, CT, p 177

Peterson CH, Grabowski JH, Powers SP (2003) Estimated enhancement of fish production resulting from restoring oyster reef habitat: quantitative valuation. Mar Ecol Prog Ser 264:249–264

Pew Oceans Commission (2003) America's living oceans: charting a course for sea change. A report to the Nation. Pew Oceans Commission, Arlington, Virginia, www.pewoceans.org

Piazza BP, Banks PD, La Peyre MK (2005) The potential for created oyster shell reefs as a sustainable shoreline protection strategy in Louisiana. Restor Ecol 13:499–506

Ranheim R, Bokuniewicz H (1991) Observations and temperature, conductivity and suspended sediment concentrations in Long Island Sound, 1990. Special data report #7, Reference #91-03, Marine Sciences Research Center, State University of New York, Stony Brook, NY

Reid RN, Frame AB, Draxler AF (1979) Environmental baselines in Long Island Sound, 1972–1973. National Oceanic and Atmospheric Administration, Technical Report SSRF-738, p 31

Rhoads DC, Germano JD (1982) Characterization of organism-sediment relations using sediment profile imaging: an efficient method of remote ecological monitoring of the seafloor (REMOTS) system. Mar Ecol Prog Ser 8:115–128

Rhoads DC, McCall PL, Yingst JY (1978) Disturbance and production on the estuarine seafloor. Am Sci 66:577–586

Richards SW (1963) The demersal fish population of Long Island Sound. Bull Bingham Oceanogr Coll 19(2):5–101

Richards SW, Riley GA (1967) The benthic epifauna of Long Island Sound. Bull Bingham Oceanogr Coll 19(2):89–135

Riley GA (1941) Plankton studies, III. Long Island Sound. Bull Bingham Oceanogr Coll 7:1–93

Riley GA (1952) Hydrography of the Long Island and Block Island Sounds. Bull Bingham Oceanogr Coll 13(3):5–39

Riley GA (1956) Oceanography of Long Island Sound, 1952-1954. IX. Production and utilization of organic matter. Bull Bingham Oceanogr Coll 15:324–344

Riley GA (1967a) Transport and mixing processes in Long Island Sound. Bull Bingham Oceanogr Coll 19(2):35–61

Riley GA (1967b) Mathematical model of nutrient conditions in coastal waters. Bull Bingham Oceanogr Coll 19(2):72–88

Riley GA, Conover SM (1967) Phytoplankton of Long Island Sound, 1954-1955. Bull Bingham Oceanogr Coll 19(2):5–34

Rosenberg AA, Mcleod KL (2005) Implementing ecosystem-based management approaches to management for the conservation of ecosystem services. Mar Ecol Prog Ser 300:241–296

Sallenger AH, Doran KS, Howd PA (2012) Hotspot of accelerated seal-level rise on the Atlantic Coast of North America. Nature Climate Change, 5 pp www.nature.com/natureclimatechange.

Sanudo-Wilhelmy SA, Flegal AR (1992) Anthropogenic silver in the Southern California Bight: a new tracer of sewage in coastal waters. Environ Sci Technol 26:2147–2151

Schimmel S, Benyi S, Strobel C (1999) An assessment of the ecological condition of Long Island Sound, 1990–1993. Environ Monitor Assess 56:27–49

Schneider CW, Suyemoto M, Yarish C (1979) An annotated checklist of Connecticut seaweeds. Connecticut Geological and Natural History Survey. Connecticut Department of Environmental Protection, p 24

Scorca MP, Monti J (2001) Estimates of nitrogen loads entering Long Island Sound from ground water and streams on Long Island, New York, 1985–1996. US Geological Survey Water Resources Investigations Report 00-4196, p 29

Shabman L, Reckhow K, Beck MB, Benaman J, Chapra S, Freedman P, Nellor M, Rudek J, Schwer D, Stiles T, Stow C (2007) Adaptive implementation of water quality improvement plans: opportunities and challenges. Nicholas Institute for Environmental Policy Solutions Report #NI-R-07-03. Duke University, Durham

Shalowitz AL (1964) Shore and sea boundaries, vol 2. US Coast and Geodetic Survey. US Government Printing Office, Washington, DC, 749 pp

Shepard FP, Wanless HR (1971) Our changing coastlines. McGraw-Hill Book Company, New York, p 579

Skinner LC, Kane MW, Gottschall K, Simpson DA (2009) Chemical residue concentrations in four species of fish and the American lobsters from Long Island Sound, Connecticut and New York 2006 and 2007. Report to the Environmental Protection Agency

Strauss BH, Ziemlinski R, Weiss JL, Overpeck JT (2012) Tidally adjusted estimates of topographic vulnerability to sea level rise and flooding for the contiguous United States. Environ Res Lett 7(2012):014033, p 12

Suffolk County (2010) Draft Suffolk County comprehensive water resources management plan. Submitted to Suffolk County by Camp Dresser & McKee. http://www.suffolkcountyny.gov/Departments/HealthServices/EnvironmentalQuality/WaterResources/ComprehensiveWaterResourcesManagementPlan.aspx. Accessed 31 July 2012

Swanson RL (1989) Is Long Island an island? Long Island Hist J 2(1):118–128

Swanson RL, Bowman M (in preparation) Between Stony Brook Harbor tides. State University of New York Press, Stony Brook

Swanson RL, Wilson RE (2005) Stony Brook Harbor hydrographic study. MSRC Special Report #128. Marine Sciences Research Center, Stony Brook University, Stony Brook, NY, p 17

Swanson RL, Parker CA, Meyer MC, Champ MA (1982) Is the East River a river or Long Island an island? NOAA Technical Report NOS 93. National Oceanic and Atmospheric Administration, Rockville, p 23

Tebaldi C, Strauss BH, Zervas CE (2012) Modelling sea level rise impacts on storm surges along US coasts. Environ Res Letts

Testa JM, Kemp WM (2012) Hypoxia-induced shifts in nitrogen and P cycling in Chesapeake Bay. Limnol Oceanogr 57(3):835–850. doi:10.4319/lo.2012.57.3.0835

Thompson BF (1849 edition) History of Long Island from its discovery and settlement to the present time, vol 1. Ira J Friendman, Inc, Port Washington (reprinted 1862), p 538

Union of Concerned Scientists (2007) New York confronting climate change in the US Northeast. www.climatechoices.org. Accessed 19 July 2011

US Coast and Geodetic Survey (Undated-a) Hydrographic index No 63 A. Long Island Sound and Vicinity, pp 1834–1836

US Coast and Geodetic Survey (Undated-b) Hydrographic index No 63 B. Long Island Sound and Vicinity, p 1837

US Coast and Geodetic Survey. (Undated-c) Topographic index No 4A. Long Island Sound and Vicinity, pp 1834–1845

US Commission on Ocean Policy (2004) An ocean blueprint for the 21st century. Final report to the President and Congress. Washington, DC, ISBN:#0–9759462–0–X

US Geological Survey (2011) USGS surface-water monthly statistics for the nation, USGS 01358000 Hudson River at Green Island NY. http://waterdata.usgs.gov/nwis/monthly/?Referred-module=sweamp;site_...amp;rdb-compression=fileeamp;submitted_form=parameter_selection_list. Accessed 5 March 2012

USEPA (2000) Ambient aquatic life water quality criteria for dissolved oxygen (saltwater): Cape Cod to Cape Hatteras. Environmental Protection Agency, p 49

USEPA (2004) National coastal condition report II. EPA-620/R-03/002. US Environmental Protection Agency, Washington, DC, p 286

USEPA (2007) National Estuary Program Coastal Condition Report. EPA-842/B-06/001. US Environmental Protection Agency, Washington, DC, p 445

USEPA (2008) Superfund environmental indicators guidance. Human exposure revisions. US Environmental Protection Agency, Washington DC, p 80

USEPA (2010) Toxics release inventory. http://www.epa.gov/tri/tridata/tri10/nationalanalysis/tri-lae-long-island.html

USEPA (2011a) 2010 toxic release inventory national analysis overview. US Environmental Protection Agency, Washington DC p 34

USEPA (2011b) RCRA orientation manual 2011. Resource Conservation and Recovery Act. EPA530-F-11.003. US Environmental Protection Agency, Washington DC, p 241

USEPA (2012) Identifying and protecting healthy watersheds. Concepts, assessments and management approaches. EPA 841-B-11-002. US Environmental Protection Agency, Washington DC, p 296

Varekamp, JC, Mecray EL, Zierzow T (2005) Once spilled, still found: Metal contamination in Connecticut wetlands and Long Island Sound sediment from historic industries. In: Whitelaw DM, Visiglione GR (eds) Our changing coasts. E. Elgar Publishers, Chapter 9, pp 122–147

Varekamp JC, Thomas E, Altabet M, Cooper S, Brinkhuis H, Sangiorgi F, Donders T, Buchholtz ten Brink M (2010) Environmental change in Long Island Sound in the recent past: eutrophication and climate change. Final Report, LISRF grant #CWF 334-R 6535 (FRS #525156), 54 pp. http://www.wesleyan.edu/ees/JCV/LobstersReportfinal.pdf6536

Vaudrey JMP, Yarish C (2010) Comparative analysis of eutrophic condition and habitat status in Connecticut and New York embayments of Long Island Sound. CT Sea Grant, and NY Sea Grant, Project number R/CE-32-CTNY

Veatch AC (1906) Underground water resources of Long Island, New York. US Geol 1122 Surv Prof Pap 44:19–32

Wall GR, Nystrom EA, Litton S (2008) Suspended sediment transport in the freshwater reach of the Hudson River estuary in eastern New York. Estuar Coasts 31:542–553. doi:10.1007/s12237-008-9050-y

Wall CC, Peterson BJ, Gobler CJ (2011) The growth of estuarine resources (*Zostera marina, Mercenaria mercenaria, Crassostrea verginica, Argoecten irradians, Cyprinodon variegates*) in response to nutrient loading and enhanced suspension feeding by adult shellfish. Estuar Coasts. doi:10.1007/s12237-011-9377-7

Weiss HM (1995) Marine animals of Southern New England and New York: identification guide to common nearshore and shallow macrofauna. State Geological and Natural History Survey of Connecticut. Connecticut Department of Environmental Protection. Bulletin 115: ISBN 0-942081-06-4

Weiss HM, Glemboske D, Philips K, Roper P, Rosso A, Sweeney T, Vittarellis A, Wahle L, Weiss J (1995) Plants and animals of Long Island Sound: a documented checklist, bibliography, and computer data base. Project Oceanology, Groton

Welsh BL, Eller FC (1991) Mechanisms controlling summertime oxygen depletion in Western Long Island Sound. Estuaries 14:265–278

Westerman GS (1987) The juridical bay: its designation and delimitation in international law. Oxford University Press, New York 304 pp

Williams J (1962) Oceanography. Little, Brown and Company, Toronto, p 242

Wilson RE, Swanson RL (2005) A perspective on bottom water temperature anomalies in Long Island Sound during the 1999 lobster mortality event. J Shellfish Res 24(3):825–830

Wilson RE, Swanson RL, Crowley HA (2008) Perspectives on long-term variations in hypoxic conditions in western Long Island Sound. J Geophys Res 113:C12011. doi:10.1029/2007JC004693

Wolfe DA, Monahan R, Stacey PE, Farrow DRG, Robertson A (1991) Environmental quality of Long Island Sound: assessment and management issues. Estuaries 14:224–236

Yang L, Li X, Crusius J, Jans U, Melcer ME, Zhang P (2007) Persistent chlordane concentrations in Long Island South sediment; implications from chlordane, 210Pb, and 137Cs profiles. Environ Sci Technol 41:7723–7729

Zu Ermgassen PSE, Spalding MD, Blake B, Coen LD, Dumbauld B, Geiger S, Grabowski JH, Grizzle R, Luckenbach M, McGraw K, Rodney W, Ruesink JL, Powers SP, Brumbaugh R (2012) Historical ecology with real numbers: past and present extent and biomass of an imperiled estuarine habitat. Proceedings of the Royal Society B 2012, vol 279, pp 3393–3400. Accessed 13 June 2012. doi:10.1098/rspb.2012.0313

Glossary

Adaptive management Management Framework whereby on-going knowledge acquisition, monitoring, and evaluation lead to continuous improvement in the identification of priority management actions and their implementation

Adsorbed Bound to **sediment** particles in an exchangeable form. Often referring to **nutrients**

Advection Horizontal movement of a fluid such as that driven by currents. Also refers to the horizontal movement of anything carried by that fluid (e.g., salt, heat, pollutants) by the same mechanism

Agglutination Formation of a mass of particles. In biology frequently refers to a clumping of red blood cells or bacteria in response to an antibody

Aerobic In the presence of oxygen, e.g., aerobic decomposition of **organic** matter leads to the production of carbon dioxide (CO_2) (see Anaerobic)

Allochthonous Inputs of energy or **nutrients** that are derived from outside of the system (e.g., leaf litter from **riparian** zones entering streams)

Ambient Background environmental condition

Ammonium Reduced form of **dissolved inorganic nitrogen** (NH_4^+ and an important nutrient for plants (see: Nitrogen. Nitrate, Nitrite)

Amphiatlantic A species that is found on both sides (east and west) of the Atlantic Ocean basin

Anadromous see diadromous

Anaerobic In the absence of oxygen, e.g., anaerobic decomposition of organic matter leads to the production of methane (CH_4) (see Aerobic)

Anammox Acronym for **anaerobic ammonium** oxidation, a bacterially mediated process by which **nitrite** and **ammonium** are converted directly into nitrogen gas (bypassing the formation of **nitrate**). This process is not well studied,

J. S. Latimer et al. (eds.), *Long Island Sound*, Springer Series on
Environmental Management, DOI: 10.1007/978-1-4614-6126-5,
© Springer Science+Business Media New York 2014

but may be responsible for as much as half of the **nitrogen** removal from the coastal ocean

Anoxic In the absence of oxygen. In some cases, management may functionally define water below a certain threshold (e.g., 1 mg/l) as anoxic, since it supports very little life, but to be truly anoxic, the concentration must drop to zero

Anthropogenic Caused by or resulting from human activities

Authigenic a mineral or sedimentary deposit that was formed in the place where it is presently located (see: **cosmogenic**, **cryptogenic**)

Autochthonous Inputs of matter or energy to a system that are sourced internally (e.g **primary productivity** from **phytoplankton**)

Autotrophic Any organism (animal or plant) that is capable of synthesizing its own food from inorganic substances using light (**photosynthesis**) or chemical (**chemosynthesis**) energy. (see **Heterotrophic**, **Chemosynthesis**, **Photosynthesis**)

Baroclinic a region with clear density gradients (often caused by changes in temperature or salinity) separating distinct air or water masses (see **barotropic**)

Barotropic a region where air or water masses are relatively uniform, and changes in density are due mostly to changes in pressure

Baseflow Flow of water entering stream channels from groundwater sources

Basin morphology The size, shape, and structure of a water body or **watershed**

Benthic Pertaining to or living on the seafloor or river bottom (see **Pelagic**, **Demersal**)

Benthic microalgae microscopic plants which inhabit the sediment surface or interstitial water, mostly **diatoms** and **dinoflagellates**

Benthoscape the landscape of the seafloor, or a type of study which takes into account the complexity of the seafloor environment

Bioassays A controlled experimental treatment used to test the response of a biological indicator to additions of **nutrients** or pollutants

Biodiversity The number and variety of organisms found within a specified geographic region

Bioextraction see Bioremediation

Biomass The total amount of plant and/or animal material within a given area or region. Typically expressed in terms of wet or dry weight per unit area

Bioremediation The use and engineering of biological processes to solve environmental problems (e.g., bacterial nitrogen removal from wastewater (BNR) or creation of shellfish habitat to improve water quality)

Biota All living organisms within an area or region includes both plants and animals

Bioturbation The stirring or mixing of sediment or soil by organisms

Body burden The total amount of a chemical, mineral, or radioactive substance that has been absorbed into the tissues of an animal

Brackish Of intermediate salinity lower than that of full strength seawater

Catadromous see diadromous

Chlorophyll Primary pigment group that captures light for **photosynthesis**, found in cells of plants and photosynthetic bacteria. Measurement of the specific pigment Chlorophyll *a* (Chl *a*) in a water sample is often used as a surrogate for **primary productivity**

Conceptual diagram High-level diagrams that convey generic and fundamental information about the major ecosystem functions and processes (biological, physical, and chemical components) of a particular location. Typically uses boxes to represent storage terms and arrows to represent flows between the storage terms

Combined sewers Sewer systems that are designed carry both rainwater run-off and municipal **sewage** in a single pipe to a **wastewater treatment facility** (WWTF). During heavy storms or snow melts, this type of system can overwhelm the capacity of the WWTF, resulting in discharge of some of the water (including some raw **sewage**) directly into nearby waterbodies through combined sewer overflow pipes or CSO's

Compensatory (fisheries mortality) An increase in population growth rate corresponding to a reduction in population, often associated with reduction of competition

Coriolis effect The observed deflection (to the right in the northern hemisphere, left in the southern) of an object (or air or water mass) moving along the surface of the earth which is caused by the rotation of the frame of reference (earth in this case)

Cosmogenic An object that originated outside earth

Cryptogenic An object whose source location is unknown

Cultural eutrophication see eutrophication

Cuesta A ridge with a gentle slope on one side and a steep slope or cliff on the other side

Cyanobacteria Primitive, photosynthetic bacteria occurring as a single cell or in filaments, some of which are often capable of **nitrogen fixation** (often referred to as blue-green algae)

Chemosynthesis The creation of organic compounds by energy derived from chemical reactions (e.g., reduction of methane or sulfate), typically in the absence of sunlight. (see: **photosynthesis**, **primary production**)

Demersal Living close to the bottom of a body of water. Often used to refer to fishes such as cod, scup, or bass

Denitrification Conversion, carried out by **anaerobic** bacteria, of biologically available, oxidized form of inorganic **nitrogen** (NO_3^-, nitrate) to nitrogen gas (N_2) (see **Nitrification**)

Deposit feeder An animal that feeds on organic material contained in the sediment. (see **detritivore**)

Detritivore A heterotrophic organism that obtains nutrition by feeding on decaying organic material (**detritus**)

Detritus Fragments of dead and decomposing organic mater. NOTE: Can also be used to refer more generally to waste or debris of any kind

Diatoms A group of unicellular, pelagic and benthic microalgae, which are characterized by the presence of an intricate exoskeleton made predominantly from silicate. Often form long chains of individual cells. One of the predominant types of **phytoplankton** found in marine environments

Diadromous A species that spends part of its life cycle in fresh water, and part in salt water. Diadromous species can be **anadromous,** living in the ocean and migrating to fresh water to breed (e.g., salmon, alewife, herring) or, less commonly **catadromous**, living in fresh water but migrating to the ocean to breed (e.g., American eel)

Diagenesis The physical and chemical changes that take place in deposited sediment during its conversion into (sedimentary) rock

Diel Denoting or relating to a period of 24 h. Distinct from **diurnal**, which refers to a pattern of light and darkness

Diffuse loads See **nonpoint source**

Dinoflagellate A group of unicellular algae, characterized by two flagellae, whip-like appendages used for propulsion. Some are autotrophic, while others are heterotrophic. Some species of dinoflagellates are responsible for "red tides"—a type of **harmful algal bloom** (HAB)

Dissolved inorganic nitrogen (DIN) The sum of Nitrate (NO_3^-) Nitrite (NO_2^-) and Ammonium (NH_4^+). These species are the most readily biologically available forms of nitrogen for primary production, and therefore are of key management concern

Diurnal Denoting or relating to the day/night cycle

Drainage Basin See **watershed**

Ecosystem A cohesive system formed by the interactions between a community of living organisms in a particular area with each other and the nonliving environment around them

Ecotypic A group of organisms within a species which are adapted to their particular environmental conditions such that they respond differently physically or behaviorally than other members of their species

Ectotherm An animal whose body temperature varies with the environment around it. Commonly referred to as cold blooded

Ekman transport The net transport of surface water perpendicular (to the right in the northern hemisphere, left in the southern) to the direction of the wind. caused by **the coriolis effect**

Epifauna Animals that live on top of the bottom substrate rather than in it (e.g., crabs, starfish, mussels) (see **infauna**)

Epiphyte A plant that grows on top of another plant, animal, or inanimate object (e.g., rock) but derives its nutrition from the air or water around it

Estuary Ecosystem occurring in the region of mixing of fresh and salt water in the lower reaches of a river

Eulerian (framework) A method of discussing or measuring fluid dynamics where the observer is stationary and measures the velocity and direction of the fluid moving past a fixed position (e.g.an anchored buoy). (see **Lagrangian**)

Eutrophication The addition of excess organic matter to an ecosystem. **Cultural eutrophication** refers to the process by which humans cause eutrophication through the contribution of excess nitrogen and phosphorus from sewage, run-off, and fertilizer use which can cause increases in algae or phytoplankton production

Euryhaline An organism capable of tolerating a wide range of salinities

Eurythermic An organism capable of tolerating a wide range of temperatures (see **Stenothermic**)

Facies The rock or stratified body distinguished from others by its appearance, composition, and conditions of formation

Fetch The distance over which wind or waves travel unobstructed by land

Fines Very small particles in a mixture of particles of varying sizes

Flushing Exchange of water from one location to another (see **Residence lime**)

Fluvial Relating to or produced by a river or stream

Flu A continuous flow of a substance from a point source. Often used to refer to the amount of a dissolved substance contributed by flow from a source, represented mathematically as flow rate times concentration

Gametophyte The gamete producing stage in a plant that exhibits alternating generations of sexual and asexual reproduction (e.g., kelp) (see: **sporophyte**)

Gamont See gametophyte

Geostrophic Caused by or related to the force created by the rotation of the earth. Often used in relation to currents. Geostrophic balance occurs when the **coriolis** force is balanced by gravitational forces

Glacial mantle A deposition of sand and silt over barren glacial sediments caused by increased winds created by temperature differences between the glacier and the barren land

Guano The excrement of sea birds, often collected from rocks and used as fertilizer. Also used colloquially to refer to manmade fertilizer made from fish

Habitat The physical and chemical environment in which a plant or animal lives

Harmful algal bloom (HAB) A bloom of algae (often phytoplankton) that causes negative impacts to other species often through use of toxins, but also through mechanical or other means

Heavy Metals A loosely defined term often used to refer to the group of metals and metalloids that are associated with contamination or ecotoxicity. Typically includes transition metals, lathanoids, actinoids, and some metalloids

Heterotrophic An organism that obtains its nutrition by ingestion of organic molecules. Inclusive of all consumers (predators and **detritivores**) (see Autotrophic)

Hydrodynamics The study of movement of water and the interactions of the body of water with its boundaries

Hydrography The science of surveying and charting bodies of water

Hydrophobic A substance that repels or fails to mix with water

Hydrophillic A substance that attracts, absorbs, or mixes easily with water

Hypoxic Low in dissolved oxygen. While no universal threshold exists for what is considered hypoxia most organizations use an operational definition of less than approximately 3 mg/l of oxygen. (see: **anoxia**)

Hysteresis The phenomenon in which changes in a physical property lag behind changes in the effect causing it

Indicator Species A species whose presence, abundance, or condition in a given environment is indicative of the condition of the ecosystem as a whole

Infauna Animals that spend most of their time buried or partially buried in the sediment

Interfluve A ridge of land dividing two river valleys

Interstitial The spaces between particles of sand or gravel in the bottom of a water body. **Porewater** is water contained in this region

Intertidal The area along the coast that is periodically inundated by the tides during high tide, and exposed to air during low tides. Typically defined as the region between **mean high water** and **mean low water**

Invertebrates Animals without backbones or spinal columns

Isobaths Areas or contours on a map indicating constant depth

Isohalines Areas or contours on a map indicating constant salinity

Isopycnals Areas or contours on a map indicating constant density

Isotherms Areas or contours on a map indicating constant temperature

Labile Unstable, readily changed by heat, oxidation or other process, highly mobile, non refractory

Lagrangian (framework) A representation of fluid dynamics that follows an individual parcel of fluid, observing the trajectory of that parcel (e.g., a floating drifter buoy that moves with the current) (see Eulerian)

Legacy contaminants Pollutants or chemicals, often produced by industry, which remain in the system long after they are discharged, such that their ecological impact continues even after discharge has been curtailed

Littoral zone The range of nearshore environment that is subject to periodic inundation by tides and periodic exposure to air. Commonly referred to as the intertidal zone

Macroalgae Large multi-cellular marine plants including anchored (e.g., kelp) and floating species (e.g., sargassum) green algae, red algae and brown algae. Algae belong to one of the two subkingdoms of plants characterized by the absence of specialized tissues or organs

Mean high water The maximum water level or area of inundation achieved by the average high tide

Mean low water The minimum water level or area of inundation achieved by the average low tide

Mesohaline Of moderate salinity. Often defined as the range of salinities between approximately 5–18 parts per thousand.

Meiospore Haploid spores found in flowers of angiosperms. Spores are often able to survive long periods of dispersal in adverse conditions

Metric tons A unit of weight equaling 1000 kg, or approximately 1.1 imperial tons (2200 lb)

Mixotrophic An organism capable of switching between or using multiple sources of energy, e.g., both **photosynthesis** and **heterotrophy** or **photosynthesis** and **chemosynthesis**

Monitoring A series of continuous measurements of water quality or other parameters made with the goal of detecting changes in the environment

Morphology See basin morphology

Morphotypes A group or strain within a single species that are distinguishable from other such groups or strains because of morphological characteristics

Morraine An accumulation of boulders, stones and debris carried and deposited by a glacier

Nauplius The free swimming first larval stage of many crustaceans

Nekton A collective term used to refer to all marine and freshwater organisms that can swim relatively independently of currents

Nitrate The most abundant oxidized form of **dissolved inorganic nitrogen** (NO_3^-) in the coastal marine environment, an important **nutrient** required for photosynthesis {see Ammonium, Nitrogen, Nitrite)

Nitrification The conversion, carried out by aerobic bacteria, of ammonium (NH_4^+), to the oxidized forms of inorganic nitrogen, nitrite (NO_2^-) and nitrate (NO_3^-)

Nitrite A less-abundant intermediate oxidized form of nitrogen (NO_2^-). Can be toxic to fish at high concentrations (see Ammonium, Nitrogen, Nitrate)

Nitrogen The most abundant element in Earth's atmosphere. Constitutes approximately 78 % of the air we breathe. Nitrogen is an essential nutrient for all organisms, forming a component of many proteins and amino acids, but virtually all of the nitrogen on earth is in the form of dinitrogen gas (N_2) which cannot be used by most organisms. These organisms are instead dependent on the much rarer **dissolved inorganic nitrogen,** which is frequently the nutrient that limits **primary production** in **marine ecosystems. Anthropogenic** activities contribute a large amount of nitrogen to coastal marine ecosystems, primarily through sewage discharge, agricultural fertilizer, and industrial emissions. (see Ammonium, Nitrate, Nitrite)

Nitrogen fixation The conversion of nitrogen gas (N_2), which is biologically unavailable to most organisms, to ammonia (NH_3) a process carried out by a select group of bacteria called nitrogen fixers (see Denitrification)

Nonpoint source A source of, for example nutrients or sediment, which is not restricted to a clearly identifiable discharge location like a river, pipe, or culvert (see also Diffuse Source, Point Source) **Nonpoint source** of pollution such as sediment or nutrients such as runoff, groundwater inputs or atmospheric fallout. (see Point Source, Nonpoint Source)

Northeaster (Nor'easter) A storm or gale with winds blowing from the northeast. Term is commonly used in the northeastern United States

Normoxic Referring to water with concentration of dissolved oxygen roughly in equilibrium with the surface air. While no universal threshold exists, water with above 5mg/l of dissolved oxygen is generally considered normoxic

Nutrients Essential elements required by an organism for growth. In a marine context, this term is typically used to refer to nitrogen and phosphorus, but can also include silica (required by diatoms) and micronutrients such as iron, zinc, magnesium, etc

Oblique tow A tow made by pulling the net at a slow tow speed from the sea floor to the surface. Under this configuration, the angle between the net and sea floor is maintained at 45 degrees

Oligohaline Water with salinity between 0.5 and 5 parts per thousand

Oligotrophic Characterized by low primary productivity and organic matter. Frequently with low concentrations of dissolved inorganic nutrients, and high dissolved oxygen

Onsite wastewater treatment system see septic system

Orogenies The process of mountain formation by folding or faulting of Earth's crust

Oxic see **normoxic**

Pelagic Pertaining to or living in the water column (see **Benthic**)

Perrenate Surviving from one growing season to the next, often by use of a dormant phase. Frequently, but not exclusively applied to plants

Phosphorus An essential nutrient for all organisms naturally contributed to marine systems primarily from the weathering of rocks. Phosphorus readily binds up into forms that are not biologically available and is typically the nutrient limiting **primary productivity** in freshwater and **oligohaline** environments. Humans contribute phosphorus to marine systems primarily from detergents and industrial surfactants, but also from sewage and fertilizer

Photosynthesis The process carried out by plants and some bacteria, in which light energy is harvested by pigments (mostly chlorophyll) and utilized to convert carbon dioxide and water into organic molecules (sugars) and oxygen. This process requires **nutrients** such as **nitrogen** and **phosphorus** as well as several other trace nutrients (iron, manganese, zinc, etc.)

Phytoplankton Microscopic, **photosynthetic** plants that are usually single celled, but can be multicellular or form long chains of single celled organisms. Phytoplankton form the base of most marine food webs

Plankton Organisms that are not able to significantly locomote in relation to the currents around them

Planetary vorticity See vorticity

Poikilotherm See **ectotherm**

Point source A specific localized and stationary source of a pollutant (e.g., nutrients, sediment, toxic metals) such as a pipe, culvert, or outfall (see Nonpoint Source, Diffuse Source)

Polyhaline Referring to estuarine or brackish waters between 18 and 30 parts per thousands salinity

Porewater Interstitial water found between the sediment grains (see **interstitial**)

Primary productivity Production of organic compounds from atmospheric or oceanic carbon dioxide. Principally through the process of **photosynthesis**, but also by **chemosynthesis**

Productivity The rate at which biomass (of plants or animals) is produced. Includes both primary (**autotrophic**) and secondary (**heterotrophic**) productivity

Pycnocline A layer in a body of water in which the density changes rapidly with respect to depth

Ravinement The formation of ravines. An irregular junction that marks a break in sedimentation

Residence Time Average length of time that water, or compounds dissolved or suspended in the water, remains in a certain location (see **Flushing**)

Residual circulation The circulation remaining after the tidal circulation is removed

Runoff Nonpoint source flows of water into a stream, lake or estuary: typically from a rainfall event where rate of accumulation exceeds losses from infiltration and evapo-transpiration

Scarp A steep bank or slope. Can also be used to refer to the process which creates this type of formation

Sea grass Marine flowering plants, which are generally rooted in the sediments. *Zostera marina* (eelgrass) is the most abundant sea grass in LIS, and is an important habitat for many species of fish and invertebrates

Secondary treatment See Wastewater treatment

Semidiurnal Occurring approximately once every 12 h. The predominant tidal pattern in LIS is semidiurnal

Septic system A tank based system, typically below ground, where waste is decomposed by anaerobic bacteria

Sewage Liquid or solid waste products from domestic or industrial processes carried away by sewers. Can also be used to refer exclusively to human feces and urine

Shoaling Becoming shallower. Can also refer to a school of fish moving into shallow water

Significant wave height The average height (crest to trough) of the largest third of waves passing through a given area in a given period of time

Significant wave period The average period (time from crest to crest) of the largest third of waves passing through a given area in a given period of time

Somatic Of or relating to the body

Sporophyte The spore producing (usually diploid) stage of an organism exhibiting alternate generations

Strata A horizontal layer of material, especially one of several parallel layers arranged one on top of another

Stokes transport The motion resulting from flow of a parcel in a current and/ or waves. The difference between the **Eulerian** and **Lagrangian** velocity of a parcel

Stable isotopes Naturally occurring and non- radioactive isotopes of common elements, such as carbon (^{12}C:^{13}C) and nitrogen (^{14}N:^{15}N). Useful tracers in the study of ecosystem processes and in the detection of sewage nitrogen

Stipe A stalk or stem, particularly of a seaweed

Stratification (vertical) Physical layering of the water column resulting from density differences primarily due to temperature or salinity differences

Successional The gradual process of ecosystem development brought about by changes in community composition towards a climax community of a particular geographic region

Sulfidic Of, relating to, or containing sulfide

Suspension feeder Animals that feed by filtering or straining particles of suspended organic matter and prey from the water column. Commonly referred to as filter feeders

Taxa General taxonomic term for a sub-group of organisms (e.g., species, genus. family, etc.)

Terrane An area or region bounded by faults that exhibits distinct **stratigraphy,** structure, or geologic history

Terriginous Made of material originating from and or/eroding from the land (see also: **Authigenic**)

Thalweg A line connecting the lowest points of successive cross-sections along the course of a valley or river. The line of fastest flow for a river

Till Unstratified sediments deposited by melting glaciers or ice sheets

Tributary A river or stream that flows into a larger river or lake

Tubiculous Organisms that form or live in burrows

Turbidity Measure of the amount of suspended particulate matter in water that is inversely related to water clarity

Varves A layer or series of layers in the sediment of a body of water, with each varve representing the sediment laid down in one year. A varve often consists of a dark and a light band deposited during different seasons of the year

Vorticity Vorticity is a measure of the local 'spin' or rotation of a fluid

Watershed The region draining into a river, river system, or other body of water

Wastewater treatment A process designed to clean and treat raw sewage to remove contaminants and pathogens. Generally a three part process, consisting of primary treatment involving screening and settlement of large particles, secondary treatment, involving anaerobic digestion of organic sludge. Water is then chlorinated and/or treated with UV sterilization to remove bacterial contaminants and discharged into the receiving water body. In some cases tertiary or advanced wastewater treatment is added to remove inorganic **nutrients** (nitrogen and/or phosphorus) from effluent prior to discharge

Zooplankton Heterotrophic microscopic animals that live in the water column and feed on phytoplankton, bacteria, and other zooplankton (includes protists, animals and larvae of animals)

Index

J. S. Latimer et al. (eds.), *Long Island Sound*, Springer Series on
Environmental Management, DOI: 10.1007/978-1-4614-6126-5,
© Springer Science+Business Media New York 2014

Printed by Publishers' Graphics LLC
KSO131213.15.16.5